POWER VACUUM TUBES

HANDBOOK

Second Edition

ELECTRONICS HANDBOOK SERIES

Series Editor:
Jerry C. Whitaker
Technical Press
Morgan Hill, California

PUBLISHED TITLES

AC POWER SYSTEMS HANDBOOK, SECOND EDITION
Jerry C. Whitaker

POWER VACUUM TUBES HANDBOOK, SECOND EDITION
Jerry C. Whitaker

FORTHCOMING TITLES

THE ELECTRONIC PACKAGING HANDBOOK
Glenn R. Blackwell

THE RESOURCE HANDBOOK OF ELECTRONICS
Jerry C. Whitaker

ELECTRONIC SYSTEMS MAINTENANCE HANDBOOK
Jerry C. Whitaker

POWER VACUUM TUBES HANDBOOK

Second Edition

Jerry C. Whitaker

CRC Press
Boca Raton   London   New York   Washington, D.C.

Acquiring Editor:	Nora Konopka
Project Editor:	Carol Whitehead
Marketing Manager:	Barbara Glunn, Jane Lewis, Arline Massey, Jane Stark
Cover design:	Jonathan Pennell
PrePress:	Gary Bennett
Manufacturing:	Carol Slatter

Library of Congress Cataloging-in-Publication Data

Whitaker, Jerry C.
　　Power vacuum tubes handbook / Jerry C. Whitaker. -- 2nd ed.
　　　　p.　cm. -- (Electronics handbook series)
　　Includes bibliographical references and index.
　　ISBN 0-8493-1345-7 (alk. paper)
　　1. Vacuum-tubes. 2. Power electronics. I. Title. II. Series.
TK7871.72.W47 1999
621.3815'12--dc21

99-21062
CIP

Preface

The phrase "high technology" is perhaps one of the more overused descriptions in our technical vocabulary. It is a phrase generally reserved for discussion of integrated circuits, fiber optics, satellite systems, and computers. Few people would associate high technology with vacuum tubes. The notion that vacuum tube construction is more art than science may have been true 10 or 20 years ago, but today it's a different story.

The demand on the part of industry for tubes capable of higher operating power and frequency, and the economic necessity for tubes that provide greater efficiency and reliability, have moved power tube manufacturers into the high-tech arena. Advancements in tube design and construction have given end users new transmitters and RF generators that allow industry to grow and prosper.

If you bring up the subject of vacuum tubes to someone who has never worked on a transmitter or high-power RF generator, you are likely to get a blank stare and a question: "Do they make those anymore?" Although receiving tubes have more-or-less disappeared from the scene, power tubes are alive and well and are performing vital functions in thousands of divergent applications. Solid-state and tube technologies each have their place, each with its strengths and weaknesses. It should be noted that even receiving tubes are staging somewhat of a comeback in high-end audio applications.

Tube design and development, although accompanied by less fanfare, is advancing as are developments in solid-state technology. Power tubes today are designed with an eye toward high operating efficiency and high gain/bandwidth properties. Above all, a tube must be reliable and provide long operating life. The design of a new power tube is a lengthy process that involves computer-aided calculations and advanced modeling.

Despite the inroads made by solid-state technology, the power vacuum tube occupies—and will continue to occupy—an important role in the generation of high-power radio frequency energy in the high-frequency regions and above. No other device can do the job as well. Certainly, solid-state cannot, especially if cost, size, and weight are important considerations.

The field of science encompassed by power vacuum tubes is broad and exciting. It is an area of growing importance to military and industrial customers, and a discipline in which significant research is now being conducted.

Power vacuum tubes include a wide range of devices, each for specific applications. Devices include power grid tubes (triodes, tetrodes, and pentodes) and microwave power tubes (klystrons, traveling wave tubes, gyrotrons, and numerous other high-frequency devices). Research is being conducted for both tube classes to extend output power and maximum frequency, and to improve operating efficiency.

This book examines the underlying technology of each type of power vacuum tube device in common use today and provides examples of typical applications. New development efforts also are reported, and the benefits of the work explained.

This Second Edition of *Power Vacuum Tubes* is directed toward engineering personnel involved in the design, specification, installation, and maintenance of high-power equipment utilizing vacuum tubes. Basic principles are discussed, with emphasis on how the underlying technology dictates the applications to which each device is dedicated. Supporting mathematics are included where appropriate to explain the material being discussed. Extensive use of technical illustrations and schematic diagrams aid the reader in understanding the fundamental principles of the subject.

Today's modern power tube is unlike the power tubes in use a decade ago. And with the trend in industry toward operation at higher power levels and higher frequencies, the vacuum tube is certain to remain on the scene for a long time to come.

Jerry C. Whitaker

About the Author

Jerry Whitaker is a technical writer based in Morgan Hill, California, where he operates the consulting firm *Technical Press*. Mr. Whitaker has been involved in various aspects of the communications industry for more than 20 years. He is a Fellow of the Society of Broadcast Engineers and an SBE-certified Professional Broadcast Engineer. He is also a member and Fellow of the Society of Motion Picture and Television Engineers, and a member of the Institute of Electrical and Electronics Engineers. Mr. Whitaker has written and lectured extensively on the topic of electronic systems installation and maintenance.

Mr. Whitaker is the former editorial director and associate publisher of *Broadcast Engineering* and *Video Systems* magazines. He is also a former radio station chief engineer and TV news producer.

Mr. Whitaker is the author of a number of books, including:

- *AC Power Systems*, 2nd Edition, CRC Press, 1998.

- *DTV: The Revolution in Electronic Imaging*, McGraw-Hill, 1998.

- Editor-in-Chief, *NAB Engineering Handbook*, National Association of Broadcasters, 1998.

- Editor-in-Chief, *The Electronics Handbook*, CRC Press, 1996.

- Coauthor, *Communications Receivers: Principles and Design*, McGraw-Hill, 1996.

- *Electronic Displays: Technology, Design, and Applications*, McGraw-Hill, 1994.

- Coauthor, *Interconnecting Electronic Systems*, CRC Press, 1992.

- Coeditor, *Television Engineering Handbook*, revised edition, McGraw-Hill, 1992.

- Coeditor, *Information Age Dictionary*, Intertec/Bellcore, 1992.

- *Maintaining Electronic Systems*, CRC Press, 1991.

- *Radio Frequency Transmission Systems: Design and Operation*, McGraw-Hill, 1990.

- Coauthor, *Television and Audio Handbook for Technicians and Engineers*, McGraw-Hill, 1990.

Mr. Whitaker has twice received a Jesse H. Neal Award *Certificate of Merit* from the Association of Business Publishers for editorial excellence. He also has been recognized as *Educator of the Year* by the Society of Broadcast Engineers.

Acknowledgement

The author wishes to express appreciation to Varian Associates, whose support was invaluable in the preparation of this book.

Contents

Preface   v

About the Author   vii

Chapter 1: Power Vacuum Tube Applications   1
  1.1   Introduction   1
    1.1.1   Vacuum Tube Development   1
    1.1.2   Standardization   10
    1.1.3   Transmission Systems   11
  1.2   Vacuum Tube Applications   11
    1.2.1   Market Overview   12
    1.2.2   AM Radio Broadcasting   14
    1.2.3   Shortwave Broadcasting   14
    1.2.4   FM Radio Broadcasting   15
    1.2.5   TV Broadcasting   16
    1.2.6   Satellite Transmission   18
    1.2.7   Radar   20
    1.2.8   Electronic Navigation   23
    1.2.9   Microwave Radio   29
    1.2.10 Induction Heating   30
    1.2.11 Electromagnetic Radiation Spectrum   31
  1.3   Bibliography   34

Chapter 2: Modulation Systems and Characteristics   37
  2.1   Introduction   37
    2.1.1   Modulation Systems   37
    2.1.2   Principles of Resonance   38
    2.1.3   Frequency Source   43
    2.1.4   Operating Class   49
    2.1.5   Broadband Amplifier Design   51
    2.1.6   Thermal and Circuit Noise   53
  2.2   Amplitude Modulation   55
    2.2.1   High-Level AM Modulation   58
    2.2.2   Vestigial-Sideband Amplitude Modulation   59
    2.2.3   Single-Sideband Amplitude Modulation   60
    2.2.4   Quadrature Amplitude Modulation (QAM)   62
  2.3   Frequency Modulation   63
    2.3.1   Modulation Index   63
    2.3.2   Phase Modulation   68
    2.3.3   Modifying FM Waves   68
    2.3.4   Preemphasis and Deemphasis   69

2.3.5   Modulation Circuits   69
2.4   Pulse Modulation   74
   2.4.1   Digital Modulation Systems   74
   2.4.2   Pulse Amplitude Modulation   75
   2.4.3   Pulse Time Modulation (PTM)   75
   2.4.4   Pulse Code Modulation   78
   2.4.5   Delta Modulation   79
   2.4.6   Digital Coding Systems   79
   2.4.7   Baseband Digital Pulse Modulation   81
   2.4.8   Spread Spectrum Systems   82
2.5   References   84
2.6   Bibliography   84

Chapter 3: Vacuum Tube Principles   87
3.1   Introduction   87
3.2   Characteristics of Electrons   87
   3.2.1   Electron Optics   88
   3.2.2   Thermal Emission From Metals   89
   3.2.3   Secondary Emission   90
   3.2.4   Diode   92
   3.2.5   Triode   94
   3.2.6   Tetrode   99
   3.2.7   Pentode   102
   3.2.8   High-Frequency Operating Limits   107
3.3   Vacuum Tube Design   110
   3.3.1   Device Cooling   112
   3.3.2   Cathode Assembly   115
   3.3.3   Grid Structures   121
   3.3.4   Plate Assembly   129
   3.3.5   Ceramic Elements   129
   3.3.6   Tube Construction   132
   3.3.7   Connection Points   133
   3.3.8   Tube Sockets   133
3.4   Neutralization   134
   3.4.1   Circuit Analysis   135
   3.4.2   Circuit Design   136
   3.4.3   Grounded-Grid Amplifier Neutralization   141
   3.4.4   Self-Neutralizing Frequency   144
   3.4.5   Neutralization Adjustment   147
3.5   References   148
3.6   Bibliography   148

Chapter 4: Designing Vacuum Tube Circuits   151
4.1   Introduction   151

	4.1.1	Class A Amplifier	152
	4.1.2	Class B and AB Amplifiers	152
	4.1.3	Class C Amplifier	152
4.2		Principles of RF Power Amplification	153
	4.2.1	Drive Power Requirements	156
	4.2.2	Mechanical and Electrical Considerations	158
	4.2.3	Bypassing Tube Elements	160
	4.2.4	Parasitic Oscillations	162
	4.2.5	Shielding	167
	4.2.6	Protection Measures	169
4.3		Cavity Amplifier Systems	171
	4.3.1	Bandwidth and Efficiency	171
	4.3.2	Current Paths	173
	4.3.3	The 1/4-Wavelength Cavity	176
	4.3.4	The 1/2-Wavelength Cavity	181
	4.3.5	Folded 1/2-Wavelength Cavity	186
	4.3.6	Wideband Cavity	187
	4.3.7	Output Coupling	187
	4.3.8	Mechanical Design	194
4.4		High-Voltage Power Supplies	197
	4.4.1	Silicon Rectifiers	197
	4.4.2	Operating Rectifiers in Series	198
	4.4.3	Operating Rectifiers in Parallel	200
	4.4.4	Silicon Avalanche Rectifiers	201
	4.4.5	Thyristor Servo Systems	201
	4.4.6	Polyphase Rectifier Circuits	210
	4.4.7	Power Supply Filter Circuits	213
4.5		Parameter Sampling Circuits	217
4.6		References	221
4.7		Bibliography	221

Chapter 5: Applying Vacuum Tube Devices — 223

5.1		Introduction	223
5.2		AM Power Amplification Systems	223
	5.2.1	Control Grid Modulation	224
	5.2.2	Suppressor Grid Modulation	225
	5.2.3	Cathode Modulation	226
	5.2.4	High-Level AM Amplification	226
	5.2.5	Pulse Width Modulation	230
5.3		Linear Amplification	232
	5.3.1	Device Selection	233
	5.3.2	Grid-Driven Linear Amplifier	233
	5.3.3	Cathode-Driven Linear Amplifier	234
	5.3.4	Intermodulation Distortion	234
5.4		High-Efficiency Linear Amplification	236

5.4.1   Chireix Outphasing Modulated Amplifier   237
5.4.2   Doherty Amplifier   238
5.4.3   Screen-Modulated Doherty-Type Amplifier   240
5.4.4   Terman-Woodyard Modulated Amplifier   242
5.4.5   Dome Modulated Amplifier   242
5.5   Television Power Amplifier Systems   244
5.5.1   System Considerations   245
5.5.2   Power Amplifier   247
5.6   FM Power Amplifier Systems   250
5.6.1   Cathode-Driven Triode Amplifier   251
5.6.2   Grounded-Grid vs. Grid-Driven Tetrode   252
5.6.3   Grid-Driven Tetrode/Pentode Amplifiers   253
5.6.4   Impedance Matching into the Grid   255
5.6.5   Neutralization   257
5.7   Special-Application Amplifiers   258
5.7.1   Distributed Amplification   258
5.7.2   Radar   258
5.8   References   263
5.9   Bibliography   263

Chapter 6: Microwave Power Tubes   265

6.1   Introduction   265
6.1.1   Linear-Beam Tubes   266
6.1.2   Crossed-Field Tubes   266
6.2   Grid Vacuum Tubes   270
6.2.1   Planar Triode   271
6.2.2   High-Power UHF Tetrode   272
6.2.3   Diacrode   274
6.3   Klystron   275
6.3.1   Reflex Klystron   276
6.3.2   The Two-Cavity Klystron   278
6.3.3   The Multicavity Klystron   281
6.3.4   Beam Pulsing   295
6.3.5   Integral vs. External Cavity   296
6.3.6   MSDC Klystron   300
6.4   Klystrode/Inductive Output Tube (IOT)   310
6.4.1   Theory of Operation   312
6.4.2   Electron Gun   312
6.4.3   Grid Structure   314
6.4.4   Input Cavity   315
6.4.5   Output Cavity   316
6.4.6   Application Considerations   317
6.4.7   Continuing Research Efforts   319
6.5   Constant Efficiency Amplifier   320
6.5.1   Theory of Operation   320

6.6   Traveling Wave Tube   322
  6.6.1   Theory of Operation   323
  6.6.2   Operating Efficiency   329
  6.6.3   Operational Considerations   330
6.7   Crossed-Field Tubes   331
  6.7.1   Magnetron   332
  6.7.2   Backward Wave Oscillator   337
  6.7.3   Strap-Fed Devices   338
  6.7.4   Gyrotron   340
6.8   Other Microwave Devices   344
  6.8.1   Quasiquantum Devices   345
  6.8.2   Variations on the Klystron   345
6.9   Microwave Tube Life   347
  6.9.1   Life-Support System   347
  6.9.2   Protection Measures   348
  6.9.3   Filament Voltage Control   350
  6.9.4   Cooling System   350
  6.9.5   Reliability Statistics   351
6.10   References   352
6.11   Bibliography   353

Chapter 7: RF Interconnection and Switching   **335**
7.1   Introduction   355
  7.1.1   Skin Effect   355
7.2   Coaxial Transmission Line   356
  7.2.1   Electrical Parameters   357
  7.2.2   Electrical Considerations   362
  7.2.3   Coaxial Cable Ratings   363
  7.2.4   Mechanical Parameters   371
7.3   Waveguide   372
  7.3.1   Propagation Modes   372
  7.3.2   Ridged Waveguide   375
  7.3.3   Circular Waveguide   375
  7.3.4   Doubly Truncated Waveguide   376
  7.3.5   Impedance Matching   377
  7.3.6   Installation Considerations   380
  7.3.7   Cavity Resonators   382
7.4   RF Combiner and Diplexer Systems   383
  7.4.1   Passive Filters   384
  7.4.2   Four-Port Hybrid Combiner   387
  7.4.3   Non-Constant-Impedance Diplexer   389
  7.4.4   Constant-Impedance Diplexer   391
  7.4.5   Microwave Combiners   399
  7.4.6   Hot Switching Combiners   399
  7.4.7   Phased-Array Antenna Systems   406

7.5   High-Power Isolators   413
   7.5.1   Theory of Operation   413
   7.5.2   Applications   416
7.6   References   418
7.7   Bibliography   419

Chapter 8: Cooling Considerations   421
8.1   Introduction   421
   8.1.1   Thermal Properties   421
   8.1.2   Heal Transfer Mechanisms   422
   8.1.3   The Physics of Boiling Water   424
8.2   Application of Cooling Principles   427
   8.2.1   Forced-Air Cooling Systems   427
   8.2.2   Water Cooling   434
   8.2.3   Vapor-Phase Cooling   440
   8.2.4   Temperature Measurements   453
   8.2.5   Air-Handling System   456
8.3   Operating Environment   456
   8.3.1   Air-Handling System   456
   8.3.2   Air Cooling System Design   459
   8.3.3   Site Design Guidelines   462
   8.3.4   Water/Vapor Cooling System Maintenance 468
8.4   References   470
8.5   Bibliography   471

Chapter 9: Reliability Considerations   473
9.1   Introduction   473
   9.1.1   Terminology   473
9.2   Quality Assurance   475
   9.2.1   Inspection Process   476
   9.2.2   Reliability Evaluation   476
   9.2.3   Failure Analysis   477
   9.2.4   Standardization   478
9.3   Reliability Analysis   478
   9.3.1   Statistical Reliability   480
   9.3.2   Environmental Stress Screening   483
   9.3.3   Latent Defects   487
   9.3.4   Operating Environment   489
   9.3.5   Failure Modes   489
   9.3.6   Maintenance Considerations   490
9.4   Vacuum Tube Reliability   491
   9.4.1   Thermal Cycling   492
   9.4.2   Tube-Changing Procedure   492
   9.4.3   Power Tube Conditioning   494

9.4.4 Filament Voltage   497
9.4.5 Filament Voltage Management   499
9.4.6 PA Stage Tuning   501
9.4.7 Fault Protection   502
9.4.8 Vacuum Tube Life   504
9.4.9 Examining Tube Performance   505
9.4.10 Shipping and Handling Vacuum Tubes   509
9.5   Klystron Reliability   510
9.5.1 Cleaning and Flushing the Cooling System   510
9.5.2 Cleaning Ceramic Elements   514
9.5.3 Reconditioning Klystron Gun Elements   514
9.5.4 Focusing Electromagnet Maintenance   516
9.5.5 Power Control Considerations   516
9.6   References   519
9.7   Bibliography   519

Chapter 10: Device Performance Criteria   523

10.1 Introduction   523
10.2 Measurement Parameters   524
10.2.1 Power Measurements   524
10.2.2 Decibel Measurement   531
10.2.3 Noise Measurement   532
10.2.4 Phase Measurement   533
10.2.5 Nonlinear Distortion   535
10.3 Vacuum Tube Operating Parameters   547
10.3.1 Stage Tuning   548
10.3.2 Amplifier Balance   551
10.3.3 Parallel Tube Amplifiers   552
10.3.4 Harmonic Energy   552
10.3.5 Klystron Tuning Considerations   553
10.3.6 Intermodulation Distortion   555
10.3.7 VSWR   559
10.4 RF System Performance   563
10.4.1 Key System Measurements   563
10.4.2 Synchronous AM in FM Systems   563
10.4.3 Incidental Phase Modulation   567
10.4.4 CarrierAmplitude Regulation   569
10.4.5 Site-Related Intermodulation Products   570
10.5 References   572
10.6 Bibliography   572

Chapter 11: Safe Handling of Vacuum Tube Devices   575

11.1 Introduction   575
11.2 Electric Shock   575

11.2.1 Effects on the Human Body 576
11.2.2 Circuit Protection Hardware 577
11.2.3 Working with High Voltage 577
11.2.4 First Aid Procedures 580
11.3 Operating Hazards 582
11.3.1 OSHA Safety Considerations 582
11.3.2 Beryllium Oxide Ceramics 586
11.3.3 Corrosive and Poisonous Compounds 586
11.3.4 FC-75 Toxic Vapor 587
11.3.5 Nonionizing Radiation 587
11.3.6 X-Ray Radiation Hazard 591
11.3.7 Implosion Hazard 591
11.3.8 Hot Coolant and Surfaces 591
11.3.9 Polychlorinated Biphenyls 591
11.4 References 596
11.5 Bibliography 596

Chapter 12: Reference Data 599

Chapter 13: Glossary 613

Index of Figures 673

Index of Tables 689

Cited References 691

Subject Index 701

This book is dedicated to my daughter
Alexis Ann Whitaker
My greatest joy is watching you grow up

Power Vacuum Tube Applications

1.1   Introduction

The continuing demand for energy control devices capable of higher operating power, higher maximum frequency, greater efficiency, and extended reliability have pushed tube manufacturers to break established performance barriers. Advancements in tube design and construction have given engineers new RF generating systems that allow industry to grow and prosper. Power grid vacuum tubes have been the mainstay of transmitters and other RF generation systems since the beginning of radio. Today, the need for new gridded and microwave power tubes is being met with new processes and materials.

Although low-power vacuum tubes have been largely replaced by solid-state devices, vacuum tubes continue to perform valuable service at high-power levels and, particularly, at high frequencies. The high-power capability of a vacuum device results from the ability of electron/vacuum systems to support high-power densities. Values run typically at several kilowatts per square centimeter, but may exceed 10 MW/cm^2. No known dielectric material can equal these values. For the foreseeable future, if high power is required, electron/vacuum devices will remain the best solution.

It is worthwhile to point out that certain devices within the realm of *receiving tubes* still continue to find application within high-end audio systems. A select group of tubes never went out of style because of their intrinsic benefits, at least as perceived by elements of the audiophile community. These "golden devices" include the 12AT7, 12AU7, 12AX7, 6L6, 6V6, and even the 5U4. These components are used today not just in classic 1960s-era audio amplifiers, but also in new microphone preamps and power amplifiers manufactured for sale to discriminating customers.

1.1.1   Vacuum Tube Development

Receiving tubes have more or less disappeared from the scene (the foregoing notwithstanding) because of the development of transistors and integrated circuits. Power grid and microwave tubes, however, continue to push the limits of technology. Power tubes are an important part of RF technology today.

From 1887—when Heinrich Hertz first sent and received radio waves—to the present, an amazing amount of progress has been made by engineers and scientists. The public takes for granted today what was considered science fiction just a decade or two ago. The route from the primitive spark-gap transmitters to the present state of the art has been charted by the pioneering efforts of many. It is appropriate to review some of the milestones in electron tube development. Much of the fundamental work on power vacuum devices can be traced to early radio broadcasting, which—along with telephone technology—has brought the nations of the world closer together than the early pioneers of the art could have imagined. More than 80 years have passed since Charles D. (Doc) Herrold founded a *voice station* (as it then was known) at San Jose, CA. Developments since have been the result of many inspired breakthroughs and years of plain hard work.

Pioneer Developers

In 1895, 21-year-old Guglielmo Marconi and his brother Alfonso first transmitted radio signals across the hills behind their home in Bologna, Italy. Born in 1874 to an Italian merchant and a Scotch-Irish mother, young Marconi had learned of Hertzian waves from August Righi, a professor at the University of Bologna. Convinced that such waves could be used for wireless communication, Marconi conducted preliminary experiments using a spark-gap source and a coherer detector. Unable to interest the Italian government in his invention, Marconi took his crude transmitter and receiver to England, where he demonstrated his wireless system to officials of the British Post Office. Marconi received a patent for the device in July 1897. With the financial support of his mother's relatives, Marconi organized the Wireless Telegraph and Signal Company that same year to develop the system commercially. Regular transatlantic communications commenced in 1903 when a Marconi station at Cape Cod, MA, sent a short message from President Theodore Roosevelt to King Edward VII in England.

The invention of the vacuum tube diode by J. Ambrose Fleming in 1904 and the triode vacuum tube amplifier by Lee De Forest in 1906 launched the electronics industry as we know it. The De Forest invention was pivotal. It marked the transition of the vacuum tube from a passive to an active device. The new "control" electrode took the form of a perforated metal plate of the same size and shape as the existing anode, positioned between the filament and the anode. Encouraged by the early test results, De Forest worked to perfect his invention, trying various mechanical arrangements for the new grid.

With the invention of the control grid, De Forest had set in motion a chain of events that led the vacuum tube to become the key element in the emerging discipline of electronics. Early experimenters and radio stations took this new technology and began developing their own tubes using in-house capabilities, including glassblowing. As the young electronics industry began to grow, vacuum tubes were produced in great quantity and standardized (to a point), making it possible to share new developments and applications. A major impetus for standardization was the U.S. military, which required vacuum tubes in great quantities during World War I. Pushed by the navy, a standardized design, including base pins and operating parameters, was forged. The economic benefits to both the tube producers and tube consumers were quickly realized.

Most radio stations from 1910 through 1920 built their own gear. For example, at the University of Wisconsin, Madison, special transmitting tubes were built by hand as needed to keep radio station 9XM, which later became WHA, on the air. The tubes were designed, constructed, and tested by Professor E. M. Terry and a group of his students in the university laboratories. Some of the tubes also were used in wireless telephonic experiments carried on with the Great Lakes Naval Training Station during 1918, when a wartime ban was imposed on wireless broadcasts.

It took many hours to make each tube. The air was extracted by means of a mercury vapor vacuum pump while the filaments were lighted and the plate voltage was on. As the vacuum increased, the plate current was raised until the plate became red-hot. This *out-gassing* process was primitive, but it worked. The students frequently worked through the night to get a tube ready for the next day's broadcast. When completed, the device might last only a few hours before burning out.

Plate dissipation on Professor Terry's early tubes, designated #1, #2, and so on, was about 25 W. Tube #5 had a power output of about 50 W. Tubes #6 to #8 were capable of approximately 75 W. Tube #8 was one of the earliest handmade commercial products.

The addition of a "screen" grid marked the next major advancement for the vacuum tube. First patented in 1916 by Dr. Walter Schottky of the Siemens & Halske Company (Germany), the device garnered only limited interest until after World War I when the Dutch firm, Philips, produced a commercial product. The "double-grid" Philips *type Q* was introduced in May 1923. Later variations on the initial design were made by Philips and other manufacturers.

Early in the use of the tetrode it was determined that the tube was unsuited for use as an audio frequency power amplifier. Under certain operating conditions encountered in this class of service, the tetrode exhibited a negative-resistance characteristic caused by secondary emission from the anode being attracted to the positively charged screen. Although this peculiarity did not affect the performance of the tetrode as an RF amplifier, it did prevent use of the device for AF power amplification. The pentode tube, utilizing a third ("suppressor") grid was designed to overcome this problem. The suppressor grid provided a means to prevent the secondary emissions from reaching the screen grid. This allowed the full capabilities of the tube to be realized.

Philips researchers Drs. G. Holst and B. Tellegen are credited for the invention, receiving a patent for the new device in 1926.

Radio Central

The first major project that the young Radio Corporation of America tackled was the construction of a huge radio transmitting station at Rocky Point, NY. The facility, completed in 1921, was hailed by President Harding as a milestone in wireless progress. The president, in fact, put the station into operation by throwing a switch that had been rigged up at the White House. Wireless stations around the globe had been alerted to tune in for a congratulatory statement by the president.

For a decade, this station—known as *Radio Central*—was the only means of direct communications with Europe. It was also the "hopping off" point for messages transmitted by RCA to Central and South America.

The Rocky Point site was famous not only for its role in communications, but also for the pioneers of the radio age who regularly visited there. The guest book lists such pioneers as Guglielmo Marconi, Lee De Forest, Charles Steinmetz, Nikola Tesla, and David Sarnoff. Radio Central was a milestone in transatlantic communications.

Originally, two antenna structures stood at the Rocky Point site, each with six 410 ft towers. The towers stretched over a 3-mile area on the eastern end of Long Island.

The facility long outlived its usefulness. RCA demolished a group of six towers in the 1950s; five more were destroyed in early 1960. The last tower of the once mighty Radio Central was taken down on December 13, 1977.

WLW: The Nation's Station

Radio station WLW has a history as colorful and varied as any in the United States. It is unique in that it was the only station ever granted authority to broadcast with 500 kW. This accomplishment pushed further the limits of vacuum tube technology.

The station actually began with 20 W of power as a hobby of Powell Crosley, Jr. The first license for WLW was granted by the Department of Commerce in 1922. Crosley was authorized to broadcast on a wavelength of 360 meters with a power of 50 W, three evenings a week. Growth of the station was continuous. WLW operated at various frequencies and power levels until, in 1927, it was assigned to 700 kHz at 50 kW and remained there. Operation at 50 kW began on October 4, 1928. The transmitter was located in Mason, OH. The station could be heard as far away as Jacksonville, FL, and Washington, DC.

The superpower era of WLW began in 1934. The contract for construction of the enormous transmitter was awarded to RCA in February 1933. Tests on the unit began on January 15, 1934. The cost of the transmitter and associated equipment was approximately $400,000—not much today, but a staggering sum in the middle of the Great Depression.

At 9:02 p.m. on May 2, 1934, programming was commenced with 500 kW of power. The superpower operation was designed to be experimental, but Crosley managed to renew the license every 6 months until 1939. The call sign W8XO occasionally was used during test periods, but the regular call sign of WLW was used for programming.

"Immense" is the only way to describe the WLW facility. The antenna (including the flagpole at the top) reached a height of 831 ft. The antenna rested on a single ceramic insulator that supported the combined force of 135 tons of steel and 400 tons exerted by the guys. The tower was guyed with eight 1-7/8-in cables anchored 375 ft from the base of the antenna.

The main antenna was augmented by a directional tower designed to protect CFRB, Toronto, when the station was using 500 kW at night. The directional system was unique in that it was the first designed to achieve both horizontal directivity and vertical-angle suppression.

A spray pond in front of the building provided cooling for the system, moving 512 gallons of water per minute. Through a heat exchanger, the water then cooled 200 gallons of distilled water in a closed system that cooled the transmitting tubes.

The transmitter consumed an entire building. Modulation transformers, weighing 37,000 lb each, were installed in the basement. Three plate transformers, a rectifier filter reactor, and a modulation reactor were installed outside the building. The exciter for the transmitter produced 50 kW of RF power. A motor-generator was used to provide 125 V dc for control circuits.

The station had its own power substation. While operating at 500 kW, the transmitter consumed 15,450,000 kWh per year. The facility was equipped with a complete machine shop because station personnel had to build much of the ancillary hardware needed. Equipment included gas, arc and spot welders, a metal lathe, milling machine, engraving machine, sander, drill press, metal brake, and a table saw. A wide variety of electric components were also on hand.

WLW operated at 500 kW until March 1, 1939, when the FCC ordered the station to reduce power to 50 kW. The station returned to superpower operation a few times during World War II for government research. The days when WLW could boast of being "the nation's station," however, were in the past.

UHF: A New Technical Challenge

The early planners of the U.S. television system thought that 13 channels would more than suffice. The original channel 1 was from 44 MHz to 50 MHz, but because of possible interference with other services, it was dropped before any active use. There remained 12 channels for normal broadcasting. Bowing to pressure from various groups, the FCC revised its allocation table in 1952 to permit UHF-TV broadcasting for the first time. The new band was not, however, a bed of roses. Many people went bankrupt, building UHF stations only to find few receivers were available to the public. UHF converters soon became popular. The first converters were so-called matchbox types that were good for one channel only. More expensive models mounted on top of the TV receiver and were tunable. Finally, the commission issued an edict that all TV set manufacturers had to include UHF tuning in their receivers. This move opened the doors for significant market penetration for UHF broadcasters.

The klystron has been the primary means of generating high-power UHF-TV signals since the introduction of UHF broadcasting.

Birth of the Klystron

Quietly developed in 1937, the klystron truly revolutionized the modern world. Indeed, the klystron may have helped save the world as we know it. More than 50 years after it was first operated in a Stanford University laboratory by Russell Varian and his brother Sigurd, the klystron and its offspring remain irreplaceable, even in the age of solid-state microelectronics.

The Varian brothers were unusually bright and extremely active. Mechanically minded, they produced one invention after another. Generally, Sigurd would think up an idea, Russell would devise a method for making it work, then Sigurd would build the device.

Through the influence of William Hansen, a former roommate of Russell and a physics professor at Stanford University, the Varians managed to get nonpaying jobs as research associates in the Stanford physics lab. They had the right to consult with members of the faculty and were given the use of a small room in the physics building.

Hansen's role, apparently, was to shoot down ideas as fast as the Varians could dream them up. As the story goes, the Varians came up with 36 inventions of varying impracticality. Then they came up with idea number 37. This time Hansen's eyes widened. On June 5, 1937, Russell proposed the concept that eventually became the klystron tube. The device was supposed to amplify microwave signals. With $100 for supplies granted by Stanford, Sigurd built it.

The device was simple: A filament heated by an electric current in turn heated a cathode. A special coating on the cathode gave off electrons when it reached a sufficiently high temperature. Negatively charged electrons attracted by a positively charged anode passed through the first cavity of the klystron tube. Microwaves in the cavity interacted with the electrons and passed through a narrow passage called a *drift tube*. In the drift tube, the electrons tended to bunch up; some speeded up, some slowed down. At the place in the drift tube where the bunching was most pronounced, the electrons entered a second cavity, where the stronger microwaves were excited and amplified in the process.

The first klystron device was lit up on the evening of August 19, 1937. Performance was marginal, but confirmed the theory of the device. An improved klystron was completed and tested on August 30. The name for the klystron came from the Greek verb *klyzo*, which refers to the breaking of a wave—a process much like the overtaking of slow electrons by fast ones.

The Varians published the results of their discovery in the *Journal of Applied Physics*. For reasons that have never been clear, their announcement immediately impressed British scientists working in the same field; but was almost entirely ignored by the Germans. The development of the klystron allowed British and American researchers to build smaller, more reliable radar systems. Klystron development paralleled work being done in England on the magnetron.

The successful deployment of microwave radar was accomplished by the invention of the cavity magnetron at Manchester University in the late 1930s [10]. It was one of Britain's "Top Secrets" handed over to the Americans early in the war. The cavity magnetron was delivered to the Radiation Laboratory at MIT, where it was incorporated into later wartime radar systems. During the Battle of Britain in May 1940, British defenses depended upon longer wavelength radar (approximately 5 m), which worked but with insufficient resolution. The magnetron provided high-power microwave energy at 10 cm wavelengths, which improved detection resolution enormously.

Armed with the magnetron and klystron, British and American scientists perfected radar, a key element in winning the Battle of Britain. So valuable were the secret devices that the British decided not to put radar in planes that flew over occupied Europe lest one of them crash, and the details of the components be discovered.

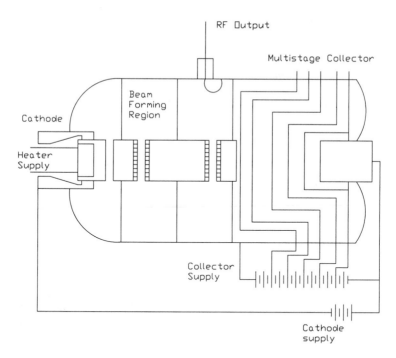

Figure 1.1 Simplified drawing of a multistage depressed-collector klystron using six collector elements.

After the war, the Varians—convinced of the potential for commercial value of the klystron and other devices they had conceived—established their own company. For Stanford University, the klystron represents one of its best investments: $100 in seed money and use of a small laboratory room were turned into $2.56 million in licensing fees before the patents expired in the 1970s, three major campus buildings and hundreds of thousands of dollars in research funding.

At about the same time the Varian brothers were working to perfect the klystron, Andrew Haeff of RCA Laboratories was developing what would come to be known as the *inductive output tube* (IOT). The crucial feature in both devices was that the electron beam, having given up a large part of its energy in a small, low-capacitance gap, tended to spread because of space-charge effects (the mutual repulsion of electrons). The characteristics of electron flow make it possible to use multiple collectors with differing potentials in tubes with large energy spread in the spent electron beam, such as the klystron and IOT. This *multistage depressed-collector* (MSDC) technique results in a considerable increase in operating efficiency. One of the earliest patents on the MSDC was issued in 1943 to Charles Litton, who later formed a company that would become Litton Industries. Figure 1.1 shows the basic MSDC structure, developed from the original Litton patent.

Nuclear Magnetic Resonance (NMR)

NMR, a technique that has revolutionized chemistry throughout the world, had its origins in experiments carried out on opposite sides of the North American continent in the late 1940s. As is often the case with basic discoveries in science, the original intent of the work was far removed from the ultimate practical application. NMR was first demonstrated during the winter of 1945 by professors Felix Bloch and William Hansen and their associates at Stanford University. At the same time another group was working independently at Harvard University, directed by Dr. E. M. Purcell. The Nobel Prize for physics was awarded to Bloch and Purcell in 1952 in recognition of their pioneering work in NMR, which allows chemists to make structural determinations of substances, and thereby see how a molecule is put together.

The Transistor Is Born

In December 1947, Dr. William Shockley of Bell Laboratories changed the course of history by demonstrating to his colleagues a newly discovered device that exhibited what he called the *transistor effect*. From this demonstration, and a later one at the Bell Labs in New York City on June 30, 1948, sprang one of the most important inventions of the 20th century—the working transistor. (The phrase "transistor" came about from a contraction of "transfer resistor.") For their development efforts, Bell Telephone scientists John Bardeen, Walter H. Brattain, and William Shockley received the Nobel Prize for physics in 1956.

The first experimental transistor made its debut as a 3/16-in-diameter, 5/8-in-long metal cylinder. Today, transistors are fabricated in dimensions finer than the wavelength of light. The first integrated circuit, made more than 30 years ago, had two transistors and measured 7/16-in-long and 1/16-in-thick. Today, a VHSIC (very high speed integrated circuit) die with more than 10 million transistors is in production for use in desktop computer systems.

The transistor was an economical and durable alternative to receiving vacuum tubes. The tubes were big, fragile, and had a relatively short operating life. They consumed lots of power and, as a result, got very hot. Transistors offered a way to make a product that was compact, efficient, and reliable.

What made transistors distinctive was that they were fabricated from a single solid material that either insulated or conducted, depending on the purity of the base material. A solitary transistor could be scribed on nothing more than a chip of germanium (subsequently supplanted by silicon). Eager to exploit the riches of solid-state technology, engineers drew up complicated circuits, cramming boards of discrete devices into tight little islands.

Satellite Technology

The phrase "live via satellite" is commonplace today. Communications satellites have meant not only live coverage of world events, but also more service to more people at lower cost. The potential for use of satellites for communications was first demonstrated in 1960 with the launching by the United States of Echo 1 and Echo 2. These passive reflector satellites bounced radio signals across the Atlantic. This type of satellite, however, left a lot to be desired for communications purposes. Echo was superseded by Telstar 1, an active repeater satellite, launched 2 years later. Telstar demonstrated that color video signals could be reliably broadcast across the oceans and, in doing so, captured the interest and imagination of the public.

The first live transatlantic telecast was relayed by Telstar on July 10, 1962. The picture was of the American flag fluttering in front of the sending station at Andover, ME. More panoramic telecasts, showing life in widely distant places, were exchanged between the United States and Europe 13 days later.

With the potential of international communications becoming increasingly apparent, Congress passed the Communications Satellite Act in 1962. The legislation created, among other things, Comsat, which was to establish, in cooperation with organizations in other countries, a global commercial communications satellite system as quickly as possible.

U.S. initiative under the Communications Satellite Act, combined with growing international interest in the new technology, led to the formation of Intelsat in 1964. Acting as a technical manager of Intelsat during its initial growth period, Comsat developed Intelsat's first geosynchronous commercial communications satellite, *Early Bird*. The project brought to reality the concept envisaged some 20 years earlier by Arthur C. Clarke, the noted British science fiction writer.

Early Bird (also known as Intelsat 1) was launched from Cape Kennedy on April 2, 1965 and placed into synchronous orbit 22,300 miles above the coast of Brazil. The launch marked the first step toward a worldwide network of satellites linking the peoples of many nations. Early Bird, the only mode of live transatlantic television, provided in July 1965 the first live telecast (via Intelsat satellite) to the United States, an American-vs.-Soviet track meet. All of these breakthroughs relied on new, compact, efficient, and lightweight microwave power tubes.

Although a dramatic improvement over the transatlantic telecommunications facilities at the time, Early Bird was nonetheless limited in capacity and capability. For example, in order for the only TV channel to be operative, all 240 voice channels had to be shut down. Furthermore, the cost of Early Bird time was high.

Following Early Bird's introduction to service in the Atlantic Ocean region, the challenge remained to develop a global network. The next step toward that goal was taken on July 11, 1967, with the successful launch of Intelsat 2, which established satellite communications between the U.S. mainland and Hawaii. Two years later Intelsat III was launched for Indian Ocean region service, thereby completing the provisions of global coverage.

Fortunately, global coverage capacity was in place just in time for the TV audience—estimated to be the largest in world history—to see man set foot on the moon. The satellite system that had been a vision by Arthur C. Clarke two decades earlier, and a formidable legislative mandate more than a decade earlier, had emerged as a reality.

1.1.2 Standardization

Throughout the history of product development, design standardization has been critically important. To most engineers, the term "standards" connotes a means of promoting an atmosphere of interchangeability of basic hardware. To others, it evokes thoughts of a slowdown of progress, of maintaining a status quo—perhaps for the benefit of a particular group. Both camps can cite examples to support their viewpoints, but no one can seriously contend that we would be better off without standards. The standardization process has been an important element in the advancement of power vacuum tube technology.

In 1836 the U.S. Congress authorized establishment of the Office of Weights and Measures (OWM) for the primary purpose of ensuring uniformity in customs house dealings. The Treasury Department was charged with its operation. As advancements in science and technology fueled the industrial revolution, it was apparent that standardization of hardware and test methods was necessary to promote commercial development and to compete successfully in the world market. The industrial revolution in the 1830s introduced the need for interchangeable parts and hardware. Wide use of steam railways and the cotton gin, for example, were possible only with mechanical standardization.

By the late 1800s, professional organizations of mechanical, electrical, and chemical engineers were founded with this aim in mind. The American Institute of Electrical Engineers developed standards based on the practices of the major electrical manufacturers between 1890 and 1910. Because such activities were not within the purview of the OWM, there was no government involvement during this period. It took the pressures of war production in 1918 to cause the formation of the American Engineering Standards Committee (AESC) to coordinate the activities of various industry and engineering societies. This group became the American Standards Association (ASA) in 1928.

Parallel development occurred worldwide. The International Bureau of Weights and Measures was founded in 1875, the International Electrotechnical Commission (IEC) in 1904, and the International Federation of Standardizing Bodies in 1926. Following World War II, this group was reorganized as the International Standards Organization (ISO). Today, representatives of approximately 54 countries serve on the ISO's 145 technical committees.

The International Telecommunications Union (ITU) was founded in 1865 for the purpose of coordinating and interfacing telegraphic communications worldwide. Today, its 164 member countries develop regulations and voluntary recommendations relating to telecommunications systems.

Many of the early standards relating to vacuum tubes in the United States were developed by equipment manufacturers, first under the banner of the Radio Manufacturers Association (RMA), then the Radio, Electronic, and Television Manufacturers Association (RETMA), and now the Electronic Industries Association (EIA). The Institute of Radio Engineers (the forerunner of the IEEE) was responsible for measurement standards and techniques.

Standards usually are changed only through natural obsolescence. Changes in basic quantities, such as units of length and volume, are extremely difficult for the general public to accept. In 1900 nearly all members of the scientific, commercial, and engineering communities supported the change to the metric system. But the general public finds the idea of changing ingrained yardsticks for weight and measure as unpalatable as learning to speak a new language in a native country. The effort to convert to the metric system is, in fact, still under way, with various degrees of success.

1.1.3 Transmission Systems

Radio communication is the grandfather of vacuum tube development. Amplitude modulation (AM) was the first modulation system that permitted voice communications to take place. This simple scheme was predominant throughout the 1920s and 1930s. Frequency modulation (FM) came into regular broadcast service during the 1940s. TV broadcasting, which uses amplitude modulation for the visual portion of the signal and frequency modulation for the aural portion of the signal, became available to the public during the mid-1940s. These two basic approaches to modulating a carrier have served the communications industry well for many decades. Although the basic schemes still are used today, numerous enhancements have been made.

Technology also has changed the rules by which the communications systems developed. AM radio, as a technical system, offered limited audio fidelity but provided design engineers with a system that allowed uncomplicated transmitters and simple, inexpensive receivers. FM radio, on the other hand, offered excellent audio fidelity but required a complex and unstable transmitter (in the early days) and complex, expensive receivers. It is no wonder that AM radio flourished while FM remained relatively stagnant for at least 20 years after being introduced to the public. It was not until transistors and later integrated circuits became commonly available that FM receivers gained consumer acceptance.

TV broadcasting evolved slowly during the late 1940s and early 1950s. Color transmissions were authorized as early as 1952. Color receivers were not purchased in large numbers by consumers, however, until the mid-1960s. Early color sets were notorious for poor reliability and unstable performance. They also were expensive. As with FM, all that changed with the introduction of transistors and, later, integrated circuits. Most color TV receivers produced today consist of an integrated circuit (IC) chip set numbering 8 to 15 devices.

1.2   Vacuum Tube Applications

The range of uses for power grid and microwave tubes is wide and varied. Some of the more common applications include the following:

- AM radio broadcasting, with power levels up to 50 kW

- Shortwave radio, with power levels of 50 kW to several megawatts

- FM radio broadcasting, with power levels up to 100 kW

- Television, with operating power levels up to 5 MW

- Radar, with widely varying applications ranging from airborne systems to over-the-horizon networks

- Satellite communications, incorporating the ground segment and space segment

- Industrial heating and other commercial processes

Figure 1.2 charts the major device types in relation to power and frequency.

1.2.1 Market Overview

There is a huge installed base of power vacuum tubes. Users cover a wide range of disciplines and facilities. The power tube market can be divided into the following general segments:

- *Medical.* Applications include diagnostics, such as X ray and CAT scanners, and treatment, such as radio therapy. Devices used include X ray tubes and klystrons. This important area of power tube application has seen continuous growth over the past two decades.

- *Scientific.* Applications include various types of research, including particle accelerators for nuclear research. A wide variety of devices are used in this area. Specialized units can be found in "big science" projects, such as the huge Stanford linear accelerator. The worldwide scientific market for microwave power tubes continues to grow as researchers reach for higher power levels and higher frequencies of operation.

- *Electronic Warfare.* Applications include ship- and air-based radar systems, eavesdropping systems, and radio-signal jamming equipment. Electronic warfare applications are a driving force in power microwave tube development. For example, the *Aegis* weapons system ship uses hundreds of microwave power tubes.

- *Communications.* Applications include space- and ground-based satellite systems, microwave relay, and specialized point-to-point communications systems. Power tubes play an important role in all types of high-power communications systems.

- *Radar.* Applications include ground-, ship-, and air-based radar for both commercial and military applications.

- *Industrial Heating.* Applications center on material heating and chemical processing, and environmental uses such as smokestack scrubbing.

- *Broadcasting.* Applications are TV and radio broadcasting, including shortwave. High-efficiency UHF-TV is an area of rapid development.

Despite the inroads made by solid-state technology, power vacuum tube applications continue to span a broad range of disciplines around the world.

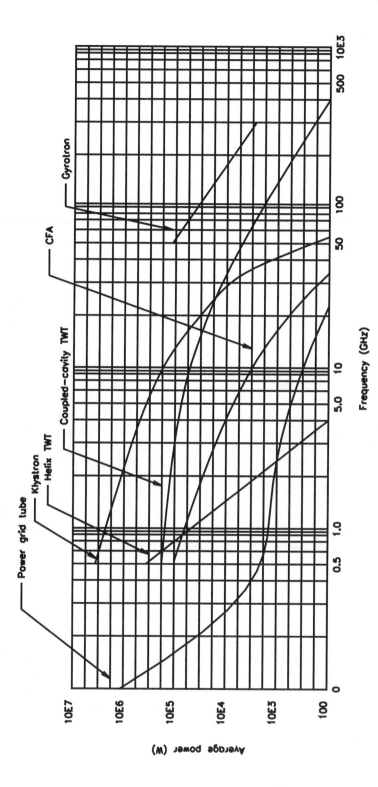

Figure 1.2 Vacuum tube operating parameters based on power and frequency.

Band A 49 meters	5.950–6.200 MHz
Band B 32 meters	9.500–9.775 MHz
Band C 25 meters	11.700–11.975 MHz
Band D 19 meters	15.100–15.450 MHz
Band E 16 meters	17.700–17.900 MHz
Band F 14 meters	21.450–21.750 MHz
Band G 11 meters	25.600–26.100 MHz

Figure 1.3 Operating frequency bands for shortwave broadcasting.

1.2.2 AM Radio Broadcasting

AM radio stations operate on 10 kHz channels spaced evenly from 540 to 1600 kHz. Various classes of stations have been established by the Federal Communications Commission (FCC) and agencies in other countries to allocate the available spectrum to given regions and communities. In the United States, the basic classes are *clear, regional,* and *local.* Current practice uses the CCIR (international) designations as class A, B, and C, respectively. Operating power levels range from 50 kW for a clear channel station to as little as 250 W for a local station. AM stations choosing to do so may operate in stereo using the *C-QUAM system.* (C-QUAM is a registered trademark of Motorola.) To receive a stereo AM broadcast, consumers must purchase a new stereo radio. The C-QUAM system transmits the stereo *sum signal* (in others words, the monophonic signal) in the usual manner and places the stereo *difference signal* on a phase-modulated subchannel. Decoder circuits in the receiver reconstruct the stereo signals.

1.2.3 Shortwave Broadcasting

The technologies used in commercial and government-sponsored shortwave broadcasting are closely allied with those used in AM radio. However, shortwave stations usually operate at significantly higher powers than AM stations.

International broadcast stations use frequencies ranging from 5.95 to 26.1 MHz. The transmissions are intended for reception by the general public in foreign countries. Figure 1.3 shows the frequencies assigned by the FCC for international broadcast shortwave service. The minimum output power is 50 kW. Assignments are made for specific hours of operation at specific frequencies.

High-power shortwave transmitters have been installed to serve large geographical areas and to overcome jamming efforts by foreign governments. Systems rated for power outputs of 500 kW and more are common. RF circuits and vacuum tube devices designed specifically for high-power operation are utilized.

Most shortwave transmitters have the unique requirement for automatic tuning to one of several preset operating frequencies. A variety of schemes exist to accomplish this task, including multiple exciters (each set to the desired operating frequency) and motor-controlled variable inductors and capacitors. Tune-up at each frequency is performed by the transmitter manufacturer. The settings of all tuning controls are stored in a memory device or as a set of trim potentiometer adjustments. Automatic retuning of a high-power shortwave transmitter can be accomplished in less than 30 s in most cases.

1.2.4 FM Radio Broadcasting

FM radio stations operate on 200 kHz channels spaced evenly from 88.1 to 107.9 MHz. In the United States, channels below 92.1 MHz are reserved for noncommercial, educational stations. The FCC has established three classifications for FM stations operating east of the Mississippi River and four classifications for stations west of the Mississippi. Power levels range from a high of 100 kW *effective radiated power* (ERP) to 3 kW or less for a lower classification. The ERP of a station is a function of transmitter power output (TPO) and antenna gain. ERP is determined by multiplying these two quantities together and allowing for line loss.

A transmitting antenna is said to have "gain" if, by design, it concentrates useful energy at low radiation angles, rather than allowing a substantial amount of energy to be radiated above the horizon (and be lost in space). FM and TV transmitting antennas are designed to provide gain through vertically stacking individual radiating elements.

Stereo broadcasting is used almost universally in FM radio today. Introduced in the mid-1960s, stereo has contributed in large part to the success of FM radio. The left and right sum (monophonic) information is transmitted as a standard frequency-modulated signal. Filters restrict this *main channel* signal to a maximum bandwidth of approximately 17 kHz. A pilot signal is transmitted at low amplitude at 19 kHz to enable decoding at the receiver. The left and right difference signal is transmitted as an amplitude-modulated subcarrier that frequency-modulates the main FM carrier. The center frequency of the subcarrier is 38 kHz. Decoder circuits in the FM receiver matrix the sum and difference signals to reproduce the left and right audio channels. Figure 1.4 illustrates the baseband signal of a stereo FM station.

Auxiliary Services

Modern FM broadcast stations are capable of broadcasting not only stereo programming, but one or more subsidiary channels as well. These signals, referred to by the FCC as *subsidiary communications authorization* (SCA) services, are used for the transmission of stock market data, background music, control signals, and other information not normally part of the station's main programming. Although these services do not provide the same range of coverage or audio fidelity as the main stereo

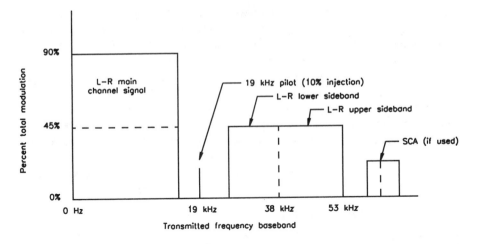

Figure 1.4 Composite baseband stereo FM signal. A full left-only or right-only signal will modulate the main (L+R) channel to a maximum of 45 percent. The stereophonic subchannel is composed of upper sideband (USB) and lower sideband (LSB) components.

program, they perform a public service and can be a valuable source of income for the broadcaster.

SCA systems provide efficient use of the available spectrum. The most common subcarrier frequency is 67 kHz, although higher subcarrier frequencies may be used. Stations that operate subcarrier systems are permitted by the FCC to exceed (by a small amount) the maximum 75 kHz deviation limit, under certain conditions. The subcarriers utilize low modulation levels, and the energy produced is maintained essentially within the 200 kHz bandwidth limitation of FM channel radiation.

1.2.5 TV Broadcasting

TV transmitters in the United States operate in three frequency bands:

- *Low-band VHF*—channels 2 through 6 (54 to 72 MHz and 76 to 88 MHz).

- *High-band VHF*—channels 7 through 13 (174 to 216 MHz).

- *UHF*—channels 14 through 69 (470 to 806 MHz). UHF channels 70 through 83 (806 to 890 MHz) currently are assigned to land mobile radio services. Certain TV translators may continue to operate on these frequencies on a secondary basis.

Because of the wide variety of operating parameters for TV stations outside the United States, this section will focus primarily on TV transmission as it relates to the United States. Table 1.1 lists the frequencies used by TV broadcasting. Maximum power output limits are specified by the FCC for each type of service. The maximum ef-

Table 1.1 Channel Designations for VHF- and UHF-TV Stations in the United States

Channel	Frequency (MHz)	Channel	Frequency (MHz)	Channel	Frequency (MHz)
2	54 – 60	30	566 – 572	58	734 – 740
3	60 – 66	31	572 – 578	59	740 – 746
4	66 – 72	32	578 – 584	60	746 – 752
5	76 – 82	33	584 – 590	61	752 – 758
6	82 – 88	34	590 – 596	62	758 – 764
7	174 – 180	35	596 – 602	63	764 – 770
8	180 – 186	36	602 – 608	64	770 – 776
9	186 – 192	37	608 – 614	65	776 – 782
10	192 – 198	38	614 – 620	66	782 – 788
11	198 – 204	39	620 – 626	67	788 – 794
12	204 – 210	40	626 – 632	68	794 – 800
13	210 – 216	41	632 – 638	69	800 – 806
14	470 – 476	42	638 – 644	70	806 – 812
15	476 – 482	43	644 – 650	71	812 – 818
16	482 – 488	44	650 – 656	72	818 – 824
17	488 – 494	45	656 – 662	73	824 – 830
18	494 – 500	46	662 – 668	74	830 – 836
19	500 – 506	47	668 – 674	75	836 – 842
20	506 – 512	48	674 – 680	76	842 – 848
21	512 – 518	49	680 – 686	77	848 – 854
22	518 – 524	50	686 – 692	78	854 – 860
23	524 – 530	51	692 – 698	79	860 – 866
24	530 – 536	52	698 – 704	80	866 – 872
25	536 – 542	53	704 – 710	81	872 – 878
26	542 – 548	54	710 – 716	82	878 – 884
27	548 – 554	55	716 – 722	83	884 – 890
28	554 – 560	56	722 – 728		
29	560 – 566	57	728 – 734		

fective radiated power for low-band VHF is 100 kW; for high-band VHF it is 316 kW; and for UHF it is 5 MW.

The second major factor that affects the coverage area of a TV station is antenna height, known in the industry as *height above average terrain* (HAAT). HAAT takes into consideration the effects of the geography in the vicinity of the transmitting tower. The maximum HAAT permitted by the FCC for a low- or high-band VHF station is 1000 ft (305 m) east of the Mississippi River and 2000 ft (610 m) west of the Mississippi. UHF stations are permitted to operate with a maximum HAAT of 2000 ft (610 m) anywhere in the United States (including Alaska and Hawaii).

The ratio of visual output power to aural power may vary from one installation to another, but the aural typically is operated at between 10 to 20 percent of the visual power. This difference is the result of the reception characteristics of the two signals. Much greater signal strength is required at the consumer's receiver to recover the visual portion of the transmission than the aural portion. The aural power output is intended to be sufficient for good reception at the fringe of the station's coverage area, but not beyond. It is pointless for a consumer to be able to receive a TV station's audio signal, but not the video.

1.2.6 Satellite Transmission

Commercial satellite communication began on July 10, 1962, when TV pictures were first beamed across the Atlantic Ocean through the Telstar 1 satellite. Three years later, the Intelsat system of *geostationary* relay satellites saw its initial craft, Early Bird 1, launched into a rapidly growing communications industry. In the same year, the U.S.S.R. inaugurated the Molnya series of satellites, traveling in an elliptical orbit to better meet the needs of that nation. The Molnya satellites were placed in an orbit inclined about 64°, relative to the equator, with an orbital period half that of the earth.

All commercial satellites in use today operate in a geostationary orbit. A geostationary satellite is one that maintains a fixed position in space relative to earth because of its altitude, roughly 22,300 miles above the earth. Two primary frequency bands are used: the *C-band* (4 to 6 GHz) and the *Ku-band* (11 to 14 GHz). Any satellite relay system involves three basic sections:

- An *uplink* transmitting station, which beams signals toward the satellite in its equatorial geostationary orbit

- The satellite (the space segment of the system), which receives, amplifies, and retransmits the signals back to earth

- The *downlink* receiving station, which completes the relay path

Because of the frequencies involved, satellite communications is designated as a microwave radio service. As such, certain requirements are placed upon the system. As with terrestrial microwave, the path between transmitter and receiver must be line of sight. Meteorological conditions, such as rain and fog, result in detrimental attenuation of the signal. Arrangements must be made to shield satellite receive antennas from terrestrial interference. Because received signal strength is based upon the inverse square law, highly directional transmit and receive parabolic antennas are used, in turn requiring a high degree of aiming accuracy. To counteract the effects of galactic and thermal noise sources on low-level signals, amplifiers are designed for exceptional low noise characteristics. Figure 1.5 shows the primary elements of a satellite relay system.

Satellite Link

Like other relay stations, the communications spacecraft contains antennas for receiving and retransmission. From the receive antenna, signals pass through a low-noise

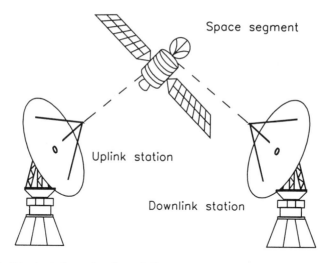

Figure 1.5 Principal elements of a satellite communications link.

amplifier before frequency conversion to the transmit band. A high-power amplifier (HPA) feeds the received signal to a directional antenna, which beams the information to a predetermined area of the earth to be served by the satellite, as illustrated in Figure 1.6.

Power to operate the electronics hardware is generated by solar cells. Inside the satellite, storage batteries, kept recharged by the solar cell arrays, carry the electronic load, particularly when the satellite is eclipsed by the earth. Power to the electronics on the craft requires protective regulation to maintain consistent signal levels. Most of the equipment operates at low voltages, but the final stage of each transponder chain ends in a high-power amplifier. The HPA of C-band satellite channels may include a traveling wave tube (TWT) or a solid-state power amplifier (SSPA). Ku-band systems rely primarily on TWT devices at this writing. Klystrons and TWTs require multiple voltage levels. The filaments operate at low voltages, but beam focus and electron collection electrodes require potentials in the hundreds and thousands of volts. To develop such a range of voltages, the satellite power supply includes voltage converters.

From these potentials, the klystron or TWT produces output powers in the range of 8.5 to 20 W. Most systems are operated at the lower end of the range to increase reliability and life expectancy. In general, the lifetime of the spacecraft is assumed to be 7 years.

A guidance system is included to stabilize the attitude of the craft as it rotates around the earth. Small rocket engines are provided for maintaining an exact position in the assigned geostationary arc. This work is known as *station-keeping*.

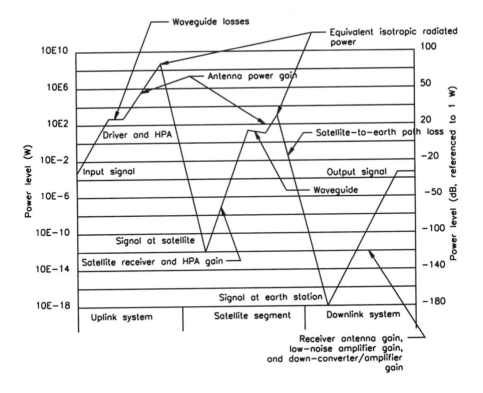

Figure 1.6 The power levels in transmission of a video signal via satellite.

1.2.7 Radar

The word radar is an acronym for *radio detection and ranging*. The name accurately spells out the basic function of a radar system. Measurement of target angles is an additional function of most radar equipment. Doppler velocity also may be measured as an important parameter. A block diagram of a typical pulsed radar system is shown in Figure 1.7. Any system can be divided into six basic subsections:

- Exciter and synchronizer: controls the sequence of transmission and reception functions.

- Transmitter: generates a high-power RF pulse of specified frequency and shape.

- Microwave network: couples the transmitter and receiver sections to the antenna.

- Antenna system: consists of a radiating/receiving structure mounted on a mechanically steered servo-driven pedestal. A *stationary array*, which uses electri-

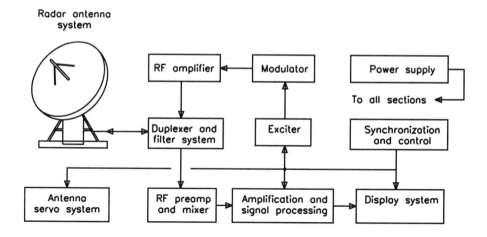

Figure 1.7 Simplified block diagram of a pulsed radar system.

cal steering of the antenna system, may be used in place of the mechanical system shown in the figure.

- Receiver: selects and amplifies the return pulse picked up by the antenna.

- Signal processor and display: integrates the detected echo pulse, synchronizer data, and antenna pointing data for presentation to an operator.

Radar technology is used for countless applications. Table 1.2 lists some of the more common uses.

Operating Parameters

Because radar systems have many diverse applications, the parameters of frequency, power, and transmission format also vary widely. There are no fundamental bounds on the operating frequencies of radar. In fact, any system that locates objects by detecting echoes scattered from a target that has been illuminated with electromagnetic energy can be considered radar. Although the principles of operation are similar regardless of the frequency, the functions and circuit parameters of most radar systems can be divided into specific operating bands. Table 1.3 shows the primary bands in use today. As shown in the table, letter designations have been developed for most of the operating bands.

Radar frequencies have been selected to minimize atmospheric attenuation by rain and snow, clouds and fog, and (at some frequencies) electrons in the air. The frequency bands must also support wide bandwidth radiation and high antenna gain.

Table 1.2 Typical Radar Applications

Air surveillance	Long-range early warning Ground-controlled intercept Target acquisition Height finding and three-dimensional analysis Airport and air-route management
Space and missile surveillance	Ballistic missile warning Missile acquisition Satellite surveillance
Surface-search and military surveillance	Sea search and navigation Ground mapping Artillery location Airport taxiway control
Weather forecasting/tracking	Observation and prediction Aircraft weather avoidance Cloud-visibility determination
Tracking and guidance	Antiaircraft fire control Surface fire control Missile guidance Satellite instrumentation Aircraft approach and landing

Table 1.3 Standard Radar Frequency Bands

Frequency Band	Frequency Range	Radiolocation Bands Based on ITU Assignments in Region II
VHF	30 – 300 MHz	137 – 134 MHz
UHF	300 – 1000 MHz	216 – 255 MHz
L-band	1.0 – 2.0 GHz	1.215 – 1.4 GHz
S-band	2.0 – 4.0 GHz	2.3 – 2.55 GHz, 2.7 – 3.7 GHz
C-band	4.0 – 8.0 GHz	5.255 – 5.925 GHz
X-band	8.0 – 12.5 GHz	8.5 – 10.7 GHz
Ku-band	12.5 – 18.0 GHz	13.4 – 14.4 GHz, 15.7 – 17.7 GHz
K-band	18.0 – 26.5 GHz	23 – 24.25 GHz
Ka-band	26.5 – 40 GHz	33.4 – 36 GHz
Millimeter	Above 40 GHz	

Transmission Equipment

The operating parameters of a radar transmitter are entirely different from those of the other transmitters discussed so far. Broadcast and satellite systems are characterized by medium-power, continuous-duty applications. Radar, on the other hand, is characterized by high-power pulsed transmissions of relatively low duty cycle. The unique requirements of radar have led to the development of technology that is foreign to most communications systems.

Improvements in semiconductor design and fabrication have made solid-state radar sets possible. Systems producing several hundred watts of output power at frequencies up to 2 GHz have been installed. Higher operating powers are achieved by using parallel amplification.

Despite inroads made by solid-state devices, vacuum tubes continue to be the mainstay of radar technology. Tube-based systems consist of the following stages:

- Exciter: generates the necessary RF and local-oscillator frequencies for the system.

- Power supply: provides the needed operating voltages for the system.

- Modulator: triggers the power output tube into operation. Pulse-shaping of the transmitted signal is performed in the modulator stage.

- RF amplifier: converts the dc input from the power supply and the trigger signals from the modulator into a high-energy, short-duration pulse.

1.2.8 Electronic Navigation

Navigation systems based on radio transmissions are used every day by commercial airlines, general aviation aircraft, ships, and the military. Electronic position-fixing systems also are used in surveying work. Although the known speed of propagation of radio waves allows good accuracies to be obtained in free space, multipath effects along the surface of the earth are the primary enemies of practical airborne and shipborne systems. A number of different navigation tools, therefore, have evolved to obtain the needed accuracy and coverage area.

Electronic navigation systems can be divided into three primary categories:

- *Long-range*, useful for distances of greater than 200 miles. Long-range systems are used primarily for transoceanic navigation.

- *Medium-range*, useful for distances of 20 to 200 miles. Medium-range systems are used mainly in coastal areas and above populated land masses.

- *Short-range*, useful for distances of less than 20 miles. Short-range systems are used for approach, docking, or landing applications.

Electronic navigation systems can be divided further into *cooperative* or *self-contained* systems. Cooperative systems depend on one- or two-way transmission between one or more ground stations and the vehicle. Such systems are capable of providing the vehicle with a location fix, independent of its previous position. Self-contained systems, housed entirely within the vehicle, may be radiating or nonradiating. They typically are used to measure the distance traveled, but have errors that increase with distance and/or time. The type of system chosen for a particular application depends upon a number of considerations, including how often the location of the vehicle must be determined and how much accuracy is required.

Because aircraft and ships may travel to any part of the world, many electronic navigation systems have received standardization on an international scale.

Virtually all radio frequencies have been used in navigation at one point or another. Systems operating at low frequencies typically use high-power transmitters with massive antenna systems. With a few exceptions, frequencies and technologies have been chosen to avoid dependence on ionospheric reflection. Such reflections can be valuable in communications systems, but are usually unpredictable. Table 1.4 lists the principal frequency bands used for radionavigation.

Direction Finding

Direction finding (DF) is a simple and widely used navigation aid. The position of a moving transmitter may be determined by comparing the arrival coordinates of the radiated energy at two or more fixed (known) points. Conversely, the position of a receiving point may be determined by comparing the direction coordinates from two or more known transmitters.

The weakness of this system is its susceptibility to site errors. This shortcoming may be reduced through the use of a large DF antenna aperture. In many cases, a multiplicity of antennas, suitably combined, can be made to favor the direct path and discriminate against indirect paths, as illustrated in Figure 1.8.

Ship navigation is a common application of DF. Coastal beacons operate in the 285 to 325 kHz band specifically for ship navigation. This low frequency provides ground wave coverage over seawater to about 1000 miles. Operating powers vary from 100 W to 10 kW. A well-designed shipboard DF system can provide accuracies of about ±2° under typical conditions.

Two-Way Distance Ranging

Automatic distance measuring can be accomplished through placement of a transponder on a given target, as illustrated in Figure 1.9. The system receives an interrogator pulse and replies to it with another pulse, usually on a different frequency. Various codes can be employed to limit responses to a single target or class of target.

Distance-measuring equipment (DME) systems are one application of two-way distance ranging. An airborne interrogator transmits 1 kW pulses at a 30 Hz rate on one of 126 channels spaced 1 MHz apart. (The operating band is 1.025 to 1.150 GHz.) A ground transponder responds with similar pulses on a different channel (63 MHz above or below the interrogating channel).

In the airborne set, the received signal is compared with the transmitted signal. By determining the time difference of the two pulses, a direct reading of miles may be found. Typical accuracy of this system is ±0.2 miles.

Ground transponders usually are configured to handle interrogation from up to 100 aircraft simultaneously.

Table 1.4 Radio Frequencies Used for Electronic Navigation

System	Frequency Band
Omega	10 – 13 kHz
VLF	16 – 24 kHz
Decca	70 – 130 kHz
Loran C/D	100 kHz
ADF/NDB	200 – 1600 kHz
Coastal DF	285 – 325 kHz
Consol	250 – 350 kHz
Marker beacon	75 MHz
ILS localizer	108 – 112 MHz
VOR	108 – 118 MHz
ILS glide slope	329 – 335 MHz
DME, Tacan	960 – 1210 MHz
ATCRBS	1.03 – 1.09 GHz
GPS	1.227 – 1.575 GHz
Altimeter	4.2 GHz
Talking beacon	9.0 GHz
MLS	5.0 GHz

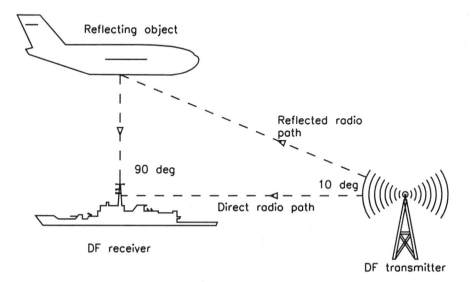

Figure 1.8 Direction-finding error resulting from beacon reflections.

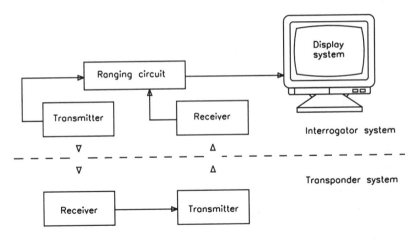

Figure 1.9 The concept of two-way distance ranging.

Differential Distance Ranging

The differential distance ranging system requires two widely-spaced transmitters, one at each end of a predefined link. By placing two transmitters on the ground, it is unnecessary to carry a transmitter in the vehicle. One transmitter is the master and the other is a slave repeating the master (see Figure 1.10). The mobile receiver measures the time difference in arrival of the two signals. For each time difference, there is a *hyperbolic line of position* that defines the target location. (Such systems are known as *hyperbolic systems*.) The transmissions may be either pulsed or continuous wave using different carrier frequencies. A minimum of two stations is required to produce a fix.

If both stations in a differential distance ranging system are provided with stable, synchronized clocks, distance measurements can be accomplished through one-way transmissions whose elapsed time is measured with reference to the clocks. This mode of operation is referred to as *one-way distance ranging* and is illustrated in Figure 1.11.

Loran C

Hyperbolic positioning is used in the *Loran C* navigation system. (Loran is named after its intended application, *long-range navigation*.) Chains of transmitters, located along coastal waters, radiate pulses at a carrier frequency of 100 kHz. Because all stations operate on the same frequency, discrimination among chains is accomplished by different pulse-repetition frequencies. A typical chain consists of a master station and two slaves, about 600 miles from the master. Each antenna is approximately 1300 ft high and is fed 5 MW pulses of defined rise and decay characteristics. Shaping of the rise and decay times is necessary to keep the radiated spectrum within the assigned band limits of 90 to 100 kHz.

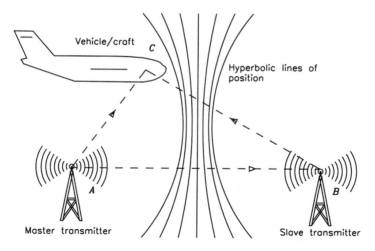

Figure 1.10 The concept of differential distance ranging (hyperbolic).

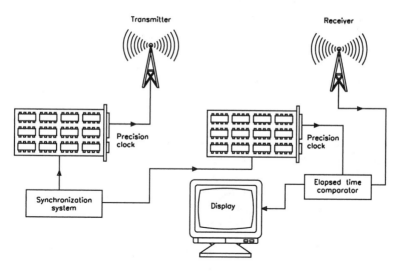

Figure 1.11 The concept of one-way distance ranging.

Pulsed transmissions are used to obtain greater average power at the receiver without resorting to higher peak power at the transmitters. The master station transmits groups of nine pulses 1 ms apart, and the slaves transmit groups of eight pulses 1 ms apart. These groups are repeated at a rate of 10 to 25 per second. Within each pulse, the phase of the RF carrier can be varied for communications purposes.

Coverage of Loran C extends to all U.S. coastal areas, plus certain areas of the North Pacific, North Atlantic, and Mediterranean. Currently, more than 17 chains employ about 50 transmitters.

Omega

Omega is another navigation system based on the hyperbolic concept. The system is designed to provide worldwide coverage from just eight stations. Omega operates on the VLF band, from 10 to 13 kHz. Skywave propagation is relatively stable for these frequencies, providing an overall system accuracy on the order of 1 mile, even at ranges of 5000 miles.

There are no masters or slaves in the Omega system; stations transmit according to their own standard. Each station has its particular operating code and transmits on one frequency at a time for a minimum of about 1 s. The cycle is repeated every 10 s. These slow rates are necessary because of the high Qs of the transmitting antennas.

A simple Omega receiver monitors for signals at 10.2 kHz and compares emissions from one station to those of another by use of an internal oscillator. The phase difference data is transferred to a map with hyperbolic coordinates.

Most Omega receivers also are able to use VLF communications stations for navigation. There are about 10 such facilities operating from 16 to 24 kHz. Output powers range from 50 kW to 1 MW. Frequency stability is maintained to 1 part in 10^{12}. This allows one-way DME to be accomplished with a high degree of accuracy.

GPS

The Global Positioning System (NAVSTAR GPS) is a U.S. Department of Defense (DoD) developed, worldwide, satellite-based radionavigation system. The constellation consists of 24 operational satellites. GPS full operational capability was declared on July 17, 1995.

GPS provides two levels of service—a Standard Positioning Service (SPS) and a Precise Positioning Service (PPS). SPS is a positioning and timing service that is available to all GPS users on a continuous, worldwide basis. SPS is provided on the GPS L1 frequency, which contains a coarse acquisition (C/A) code and a navigation data message. SPS provides, on a daily basis, the capability to obtain horizontal positioning accuracy within 100 meters (95 percent probability) and 300 meters (99.99 percent probability), vertical positioning accuracy within 140 meters (95 percent probability), and timing accuracy within 340 ns (95 percent probability).

The GPS L1 frequency also contains a precision (P) code that is not a part of the SPS. PPS is a highly accurate military positioning, velocity, and timing service that is available on a continuous, worldwide basis to users authorized by the DoD. PPS is the data transmitted on GPS L1 and L2 frequencies. PPS is designed primarily for U.S. military use. P-code-capable military user equipment provides a predictable positioning accuracy of at least 22 meters horizontally and 27.7 meters vertically, and timing/time interval accuracy within 90 ns (95 percent probability).

The GPS satellites transmit on two L-band frequencies: L1 = 1575.42 MHz and L2 = 1227.6 MHz. Three pseudo-random noise (PRN) ranging codes are used. The coarse/acquisition (C/A) code has a 1.023 MHz chip rate, a period of 1 ms, and is used by civilian users for ranging and by military users to acquire the P-code. Bipolar-phase shift key (BPSK) modulation is utilized. The transmitted PRN code sequence is actually the modulo-2 addition of a 50 Hz navigation message and the C/A code. The SPS

Table 1.5 Common-Carrier Microwave Frequencies Used in the United States

Band (GHz)	Allotted frequencies	Bandwidth	Applications
2	2.11 – 2.13 Hz 2.16 – 2.18 GHz	20 MHz	General purpose, limited use
4	3.7 – 4.2 GHz	20 MHz	Long-haul point-to-point relay
6	5.925 – 6.425 GHz	500 MHz	Long and short haul
11	10.7 – 11.7 GHz	500 MHz	Short haul
18	17.7 – 19.7 GHz	1.0 GHz	Short haul, limited use

receiver demodulates the received code from the L1 carrier, and detects the differences between the transmitted and the receiver-generated code. The SPS receiver uses an exclusive-OR truth table to reconstruct the navigation data, based upon the detected differences in the two codes.

The precision (P) code has a 10.23 MHz rate, a period of seven days and is the principle navigation ranging code for military users. The Y-code is used in place of the P-code whenever the *anti-spoofing* (A-S) mode of operation is activated. The C/A code is available on the L1 frequency and the P-code is available on both L1 and L2. The various satellites all transmit on the same frequencies, L1 and L2, but with individual code assignments.

Each satellite transmits a navigation message containing its orbital elements, clock behavior, system time, and status messages. In addition, an almanac is provided that gives the approximate data for each active satellite. This allows the user set to find all satellites after the first has been acquired.

1.2.9 Microwave Radio

Microwave radio relay systems carry a significant portion of long-haul telecommunications in the United States and other countries. Table 1.5 lists the major common-carrier bands and their typical uses. The goal of microwave relay technology has been to increase channel capacity and lower costs. Solid-state devices have provided the means to accomplish this goal. Current efforts focus on the use of fiber optic landlines for terrestrial long-haul communications systems. Satellite circuits also have been used extensively for long-distance common-carrier applications.

Single-sideband amplitude modulation is used for microwave systems because of its spectrum efficiency. Single-sideband systems, however, require a high degree of linearity in amplifying circuits. Several techniques have been used to provide the needed channel linearity. The most popular is *amplitude predistortion* to cancel the inherent nonlinearity of the power amplifier.

Microwave relay systems typically use *frequency-division multiplexing* (FDM) to combine signals for more efficient transmission. For example, three 600-channel master groups may be multiplexed into a single baseband signal. This multiplexing often is done some distance (a mile or more) from the transmitter site. A coaxial wireline is

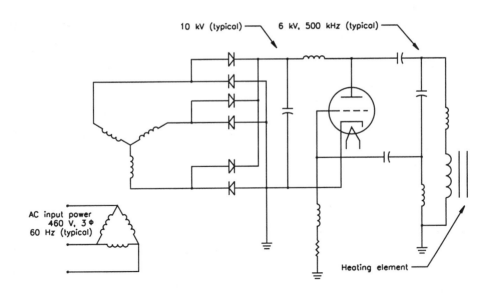

Figure 1.12 Basic schematic of a 20 kW induction heater circuit.

used to bring the baseband signal to the microwave equipment. Depending on the distance, intermediate repeaters may be used. The baseband signal is applied to an FM terminal (FMT) that frequency-modulates a carrier of typically 70 MHz. This IF signal then modulates a 20-MHz-wide channel in the 4 GHz band.

1.2.10 Induction Heating

Induction heating is achieved by placing a coil carrying alternating current adjacent to a metal workpiece so that the magnetic flux produced induces a voltage in the workpiece. This causes current flow and heats the workpiece. Power sources for induction heating include:

- Motor-generator sets, which operate at low frequencies and provide outputs from 1 kW to more than 1 MW.

- Vacuum tube oscillators, which operate at 3 kHz to several hundred megahertz at power levels of 1 kW to several hundred kilowatts. Figure 1.12 shows a 20 kW induction heater using a vacuum tube as the power generating device.

- Inverters, which operate at 10 kHz or more at power levels of as much as several megawatts. Inverters using thyristors (silicon controlled rectifiers) are replacing motor-generator sets in most high-power applications.

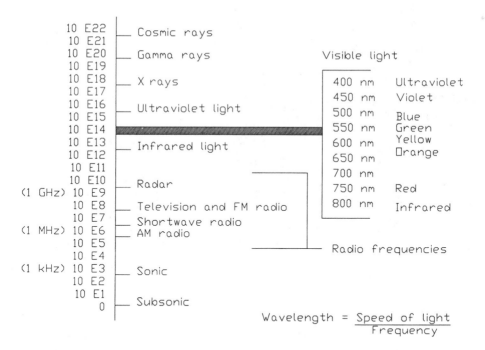

Figure 1.13 The electromagnetic spectrum.

Dielectric Heating

Dielectric heating is a related application for RF technology. Instead of heating a conductor, as in induction heating, dielectric heating relies on the capacitor principle to heat an insulating material. The material to be heated forms the dielectric of a capacitor, to which power is applied. The heat generated is proportional to the *loss factor* (the product of the dielectric constant and the power factor) of the material. Because the power factor of most dielectrics is low at low frequencies, the range of frequencies employed for dielectric heating is higher than for induction heating. Frequencies of a few megahertz to several gigahertz are common.

1.2.11 Electromagnetic Radiation Spectrum

The usable spectrum of electromagnetic radiation frequencies extends over a range from below 100 Hz for power distribution to 10^{20} Hz for the shortest X rays (see Figure 1.13). The lower frequencies are used primarily for terrestrial broadcasting and communications. The higher frequencies include visible and near-visible infrared and ultraviolet light, and X rays. The frequencies of interest to RF engineers range from 30 kHz to 30 GHz. (See Table 1.6.)

Table 1.6 Frequency Band Designations

Description	Designation	Frequency
Extremely Low Frequency	ELF (1) Band	3 Hz up to 30 Hz
Super Low Frequency	SLF (2) Band	30 Hz up to 300 Hz
Ultra Low Frequency	ULF (3) Band	300 Hz up to 3 kHz
Very Low Frequency	VLF (4) Band	3 kHz up to 30 kHz
Low Frequency	LF (5) Band	30 kHz up to 300 kHz
Medium Frequency	MF (6) Band	300 kHz up to 3 MHz
High Frequency	HF (7) Band	3 MHz up to 30 MHz
Very High Frequency	VHF (8) Band	30 MHz up to 300 MHz
Ultra High Frequency	UHF (9) Band	300 MHz up to 3 GHz
Super High Frequency	SHF (10) Band	3 GHz up to 30 GHz
Extremely High Frequency	EHF (11) Band	30 GHz up to 300 GHz
—	— (12) Band	300 GHz up to 3 THz

Low Frequency (LF): 30 to 300 kHz

The LF band is used for around-the-clock communications services over long distances and where adequate power is available to overcome high levels of atmospheric noise. Applications include:

- Radionavigation
- Fixed/maritime communications and navigation
- Aeronautical radionavigation
- Low-frequency broadcasting (Europe)
- Underwater submarine communications (10 to 30 kHz)

Medium Frequency (MF): 300 kHz to 3 MHz

The low-frequency portion of this band is used for around-the-clock communication services over moderately long distances. The upper portion of the MF band is used principally for moderate-distance voice communications. Applications in this band include:

- AM radio broadcasting (535.5 to 1605.5 kHz)
- Radionavigation
- Fixed/maritime communications
- Aeronautical radionavigation
- Fixed and mobile commercial communications

- Amateur radio

- Standard time and frequency services

High Frequency (HF): 3 to 30 MHz

This band provides reliable medium-range coverage during daylight and, when the transmission path is in total darkness, worldwide long-distance service. The reliability and signal quality of long-distance service depends to a large degree upon ionospheric conditions and related long-term variations in sunspot activity affecting sky-wave propagation. Applications include:

- Shortwave broadcasting

- Fixed and mobile service

- Telemetry

- Amateur radio

- Fixed/maritime mobile

- Standard time and frequency services

- Radio astronomy

- Aeronautical fixed and mobile

Very High Frequency (VHF): 30 to 300 MHz

The VHF band is characterized by reliable transmission over medium distances. At the higher portion of the VHF band, communication is limited by the horizon. Applications include:

- FM radio broadcasting (88 to 108 MHz)

- Low-band VHF-TV broadcasting (54 to 72 MHz and 76 to 88 MHz)

- High-band VHF-TV broadcasting (174 to 216 MHz)

- Commercial fixed and mobile radio

- Aeronautical radionavigation

- Space research

- Fixed/maritime mobile

- Amateur radio

- Radiolocation

Ultrahigh Frequency (UHF): 300 MHz to 3 GHz

Transmissions in this band are typically line of sight. Short wavelengths at the upper end of the band permit the use of highly directional parabolic or multielement antennas. Applications include:

- UHF terrestrial television (470 to 806 MHz)
- Fixed and mobile communications
- Telemetry
- Meteorological aids
- Space operations
- Radio astronomy
- Radionavigation
- Satellite communications
- Point-to-point microwave relay

Superhigh Frequency (SHF): 3 to 30 GHz

Communication in this band is strictly line of sight. Very short wavelengths permit the use of parabolic transmit and receive antennas of exceptional gain. Applications include:

- Satellite communications
- Point-to-point wideband relay
- Radar
- Specialized wideband communications
- Developmental research
- Military support systems
- Radiolocation
- Radionavigation
- Space research

1.3 Bibliography

Battison, John, "Making History," *Broadcast Engineering*, Intertec Publishing, Overland Park, KS, June 1986.

Benson, K. B., and J. C. Whitaker (eds.), *Television Engineering Handbook*, revised ed., McGraw-Hill, New York, 1991.

Benson, K. B., and J. C. Whitaker, *Television and Audio Handbook for Technicians and Engineers*, McGraw-Hill, New York, 1989.

Brittain, James E., "Scanning the Past: Guglielmo Marconi," *Proceedings of the IEEE*, Vol. 80, no. 8, August 1992.

Burke, William, "WLW: The Nation's Station," *Broadcast Engineering*, Intertec Publishing, Overland Park, KS, November 1967.

Clerc, Guy, and William R. House, "The Case for Tubes," *Broadcast Engineering*, Intertec Publishing, Overland Park, KS, pp. 67 - 82, May 1991.

Dorschug, Harold, "The Good Old Days of Radio," *Broadcast Engineering*, Intertec Publishing, Overland Park, KS, May 1971.

Fink, D., and D. Christiansen (eds.), *Electronics Engineers' Handbook*, 3rd ed., McGraw-Hill, New York, 1989.

Fink, D., and D. Christiansen (eds.), *Electronics Engineer's Handbook*, 2nd ed., McGraw-Hill, New York, 1982.

Goldstein, Irving, "Communications Satellites: A Revolution in International Broadcasting," *Broadcast Engineering*, Intertec Publishing, Overland Park, KS, May 1979.

Jordan, Edward C. (ed.), *Reference Data for Engineers: Radio, Electronics, Computer and Communications*, 7th Ed., Howard W. Sams Company, Indianapolis, IN, 1985.

McCroskey, Donald, "Setting Standards for the Future," *Broadcast Engineering*, Intertec Publishing, Overland Park, KS, May 1989.

Nelson, Cindy, "RCA Demolishes Last Antenna Tower at Historic Radio Central," *Broadcast Engineering*, Intertec Publishing, Overland Park, KS, February 1978.

Paulson, Robert, "The House That Radio Built," *Broadcast Engineering*, Intertec Publishing, Overland Park, KS, April 1989.

Peterson, Benjamin B., "Electronic Navigation Systems," in *The Electronics Handbook*, Jerry C. Whitaker (ed.), CRC Press, Boca Raton, Fla., pp. 1710–1733, 1996.

Pond, N. H., and C. G. Lob, "Fifty Years Ago Today or On Choosing a Microwave Tube," *Microwave Journal*, September 1988.

Riggins, George, "The Real Story on WLW's Long History," *Radio World*, Falls Church, VA, March 8, 1989.

Schow, Edison, "A Review of Television Systems and the Systems for Recording Television," *Sound and Video Contractor*, Intertec Publishing, Overland Park, KS, May 1989.

Schubin, Mark, "From Tiny Tubes to Giant Screens," *Video Review*, April 1989.

Stokes, John W., *Seventy Years of Radio Tubes and Valves*, Vestal Press, New York, 1982.

Symons, Robert S., "Tubes—Still Vital After All These Years," *IEEE Spectrum*, IEEE, New York, pp. 53–63, April 1998.

Terman, Frederick E., *Radio Engineering*, 3rd ed., McGraw-Hill, New York, 1947.

Varian Associates, "An Early History," Varian, Palo Alto, CA.

Whitaker, J. C., *Maintaining Electronic Systems*, CRC Press, Boca Raton, FL, 1992.

Whitaker, J. C., *Radio Frequency Transmission Systems: Design and Operation*, McGraw-Hill, New York, 1991.

2

Modulation Systems and Characteristics

2.1 Introduction

Radio frequency (RF) power amplifiers are used in countless applications at tens of thousands of facilities around the world. The wide variety of applications, however, stem from a few basic concepts of conveying energy and information by means of an RF signal. Furthermore, the devices used to produce RF energy have many similarities, regardless of the final application. Although communications systems represent the most obvious use of high-power RF generators, numerous other common applications exist, including:

- Induction heating and process control systems

- Radar (ground, air, and shipboard)

- Satellite communications

- Atomic science research

- Medical research, diagnosis, and treatment

The process of generating high-power RF signals has been refined over the years to an exact science. Advancements in devices and circuit design continue to be made each year, pushing ahead the barriers of efficiency and maximum operating frequency. Although different applications place unique demands on the RF design engineer, the fundamental concepts of RF amplification are applicable to virtually any system.

2.1.1 Modulation Systems

The primary purpose of most communications and signaling systems is to transfer information from one location to another. The message signals used in communication and control systems usually must be limited in frequency to provide for efficient transfer. This frequency may range from a few hertz for control systems to a few megahertz for video signals to many megahertz for multiplexed data signals. To facilitate efficient and controlled distribution of these components, an *encoder* generally is required between the source and the transmission channel. The encoder acts to *modulate* the signal, producing at its output the *modulated waveform*. Modulation is a process whereby the characteristics of a wave (the *carrier*) are varied in accordance with a message signal, the modulating waveform. Frequency translation is usually a by-product of this process. Modulation may be continuous, where the modulated wave is always present, or pulsed, where no signal is present between pulses.

There are a number of reasons for producing modulated waves, including:

- *Frequency translation.* The modulation process provides a vehicle to perform the necessary frequency translation required for distribution of information. An input signal may be translated to its assigned frequency band for transmission or radiation.

- *Signal processing.* It is often easier to amplify or process a signal in one frequency range as opposed to another.

- *Antenna efficiency.* Generally speaking, for an antenna to be efficient, it must be large compared with the signal wavelength. Frequency translation provided by modulation allows antenna gain and beamwidth to become part of the system design considerations. The use of higher frequencies permits antenna structures of reasonable size and cost.

- *Bandwidth modification.* The modulation process permits the bandwidth of the input signal to be increased or decreased as required by the application. Bandwidth reduction permits more efficient use of the spectrum, at the cost of signal fidelity. Increased bandwidth, on the other hand, provides increased immunity to transmission channel disturbances.

- *Signal multiplexing.* In a given transmission system, it may be necessary or desirable to combine several different signals into one baseband waveform for distribution. Modulation provides the vehicle for such *multiplexing*. Various modulation schemes allow separate signals to be combined at the transmission end and separated (*demultiplexed*) at the receiving end. Multiplexing may be accomplished by using, among other systems, *frequency-domain multiplexing* (FDM) or *time-domain multiplexing* (TDM).

Modulation of a signal does not come without the possible introduction of undesirable attributes. Bandwidth restriction or the addition of noise or other disturbances are the two primary problems faced by the transmission system designer.

2.1.2 Principles of Resonance

All RF generations rely on the principles of resonance for operation. Three basic systems exist:

- Series resonance circuits

- Parallel resonance circuits

- Cavity resonators

Series Resonant Circuits

When a constant voltage of varying frequency is applied to a circuit consisting of an inductance, capacitance, and resistance (all in series), the current that flows depends upon frequency in the manner shown in Figure 2.1. At low frequencies, the capacitive reactance of the circuit is large and the inductive reactance is small, so that most of the voltage drop is across the capacitor, while the current is small and leads the applied voltage by nearly 90°. At high frequencies, the inductive reactance is large and the capacitive reactance is low, resulting in a small current that lags nearly 90° behind the applied voltage; most of the voltage drop is across the inductance. Between these two extremes is the *resonant frequency*, at which the capacitive and inductive reactances are equal and, consequently, neutralize each other, leaving only the resistance of the circuit to oppose the flow of current. The current at this resonant frequency is, accordingly, equal to the applied voltage divided by the circuit resistance, and it is very large if the resistance is low.

The characteristics of a series resonant circuit depend primarily upon the ratio of inductive reactance ωL to circuit resistance R, known as the circuit Q:

$$Q = \frac{\omega L}{R} \qquad\qquad (2.1)$$

The circuit Q also may be defined by:

$$Q = 2\pi \left(\frac{E_s}{E_d} \right) \qquad\qquad (2.2)$$

Where:
E_s = energy stored in the circuit
E_d = energy dissipated in the circuit during one cycle

Most of the loss in a resonant circuit is the result of coil resistance; the losses in a properly constructed capacitor are usually small in comparison with those of the coil.

The general effect of different circuit resistances (different values of Q) is shown in Figure 2.1. As illustrated, when the frequency differs appreciably from the resonant frequency, the actual current is practically independent of circuit resistance and is nearly the current that would be obtained with no losses. On the other hand, the current at the resonant frequency is determined solely by the resistance. The effect of increasing the resistance of a series circuit is, accordingly, to flatten the resonance curve by reducing the current at resonance. This broadens the top of the curve, giving a more uniform current over a band of frequencies near the resonant point. This broadening is achieved, however, by reducing the selectivity of the tuned circuit.

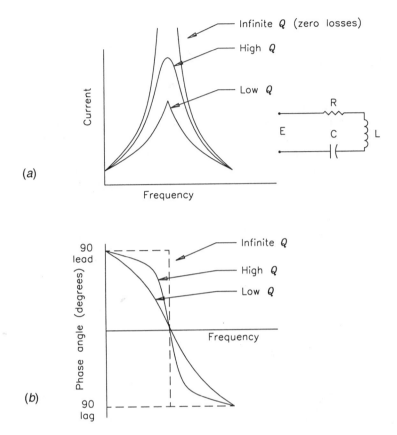

Figure 2.1 Characteristics of a series resonant circuit as a function of frequency for a constant applied voltage and different circuit Qs: (a) magnitude, (b) phase angle.

Parallel Resonant Circuits

A parallel circuit consisting of an inductance branch in parallel with a capacitance branch offers an impedance of the character shown in Figure 2.2. At low frequencies, the inductive branch draws a large lagging current while the leading current of the capacitive branch is small, resulting in a large lagging line current and a low lagging circuit impedance. At high frequencies, the inductance has a high reactance compared with the capacitance, resulting in a large leading line current and a corresponding low circuit impedance that is leading in phase. Between these two extremes is a frequency at which the lagging current taken by the inductive branch and the leading current entering the capacitive branch are equal. Being 180° out of phase, they neutralize, leaving only a small resultant in-phase current flowing in the line; the impedance of the parallel circuit is, therefore, high.

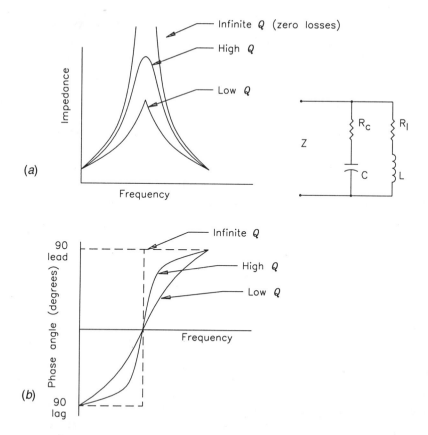

Figure 2.2 Characteristics of a parallel resonant circuit as a function of frequency for different circuit Qs: (a) magnitude, (b) phase angle.

The effect of circuit resistance on the impedance of the parallel circuit is similar to the influence that resistance has on the current flowing in a series resonant circuit, as is evident when Figures 2.1 and 2.2 are compared. Increasing the resistance of a parallel circuit lowers and flattens the peak of the impedance curve without appreciably altering the sides, which are relatively independent of the circuit resistance.

The resonant frequency F_o of a parallel circuit can be taken as the same frequency at which the same circuit is in series resonance:

$$F_o = \frac{1}{2\pi \sqrt{LC}}$$

(2.3)

Where:
L = inductance in the circuit
C = capacitance in the circuit

When the circuit Q is large, the frequencies corresponding to the maximum impedance of the circuit and to unity power factor of this impedance coincide, for all practical purposes, with the resonant frequency defined in this way. When the circuit Q is low, however, this rule does not necessarily apply.

Cavity Resonators

Any space completely enclosed with conducting walls may contain oscillating electromagnetic fields. Such a space also exhibits certain resonant frequencies when excited by electrical oscillations. Resonators of this type, commonly termed *cavity resonators*, find extensive use as resonant circuits at very high frequencies and above. For such applications, cavity resonators have a number of advantages, including:

- Mechanical and electrical simplicity

- High Q

- Stable parameters over a wide range of operating conditions

If desired, a cavity resonator can be configured to develop high shunt impedance.

The simplest cavity resonators are sections of waveguides short-circuited at each end and $\lambda_g/2$ wavelengths long, where λ_g is the guide wavelength. This arrangement results in a resonance analogous to that of a $\frac{1}{2}$-wavelength transmission line short-circuited at the receiving end.

Any particular cavity is resonant at a number of frequencies, corresponding to different possible field configurations that exist within the enclosure. The resonance having the longest wavelength (lowest frequency) is termed the *dominant* or *fundamental* resonance. In the case of cavities that are resonant sections of cylindrical or rectangular waveguides, most of the possible resonances correspond to various modes that exist in the corresponding waveguides.

The resonant wavelength is proportional in all cases to the size of the resonator. If the dimensions are doubled, the wavelength corresponding to resonance will likewise be doubled. This fact simplifies the construction of resonators of unusual shapes whose proper dimensions cannot be calculated easily.

The resonant frequency of a cavity can be changed through one or more of the following actions:

- Altering the mechanical dimensions of the cavity

- Coupling reactance into the cavity

- Inserting a copper paddle inside the cavity and moving it to achieve the desired resonant frequency

Small changes in mechanical dimensions can be achieved by flexing walls, but large changes typically require some type of sliding member. Reactance can be coupled into the resonator through a coupling loop, thus affecting the resonant frequency. A copper paddle placed inside the resonator affects the normal distribution of flux and tends to raise the resonant frequency by an amount determined by the orientation of the paddle. This technique is similar to the way in which copper can be used to produce small variations in the inductance of a coil.

Coupling to a cavity resonator can be achieved by means of a coupling loop or a coupling electrode. Magnetic coupling is achieved by means of a small coupling loop oriented so as to enclose magnetic flux lines existing in the desired mode of operation. A current passed through such a loop then excites oscillations of this mode. Conversely, oscillations existing in the resonator induce a voltage in such a coupling loop. The combination of a coupling loop and cavity resonator is equivalent to the ordinary coupled circuit shown in Figure 2.3. The magnitude of magnetic coupling can be readily controlled by rotating the loop. The coupling is reduced to zero when the plane of the loop is parallel to the magnetic flux.

The practical application of cavity resonators for VHF frequencies is discussed in Section 4.3; resonators for microwave devices are discussed in Section 7.3.6.

2.1.3 Frequency Source

Every RF amplifier requires a stable frequency reference. At the heart of most systems is a quartz crystal. Quartz acts as a stable high Q mechanical resonator. Crystal resonators are available for operation at frequencies ranging from 1 kHz to 300 MHz and beyond.

The operating characteristics of a crystal are determined by the *cut* of the device from a bulk "mother" crystal. The behavior of the device depends heavily on the size and shape of the crystal, and the angle of the cut. To provide for operation at a wide range of frequencies, different cuts, vibrating in one or more selected modes, are used.

The properties of a piezoelectric quartz crystal can be expressed in terms of three sets of axes. (See Figure 2.4.) The axis joining the points at the ends of the crystal is known as the *optical axis*, and electrical stresses applied in this direction exhibit no piezoelectric properties. The three axes X', X'', and X''', passing through the corners of the hexagon that forms the section perpendicular to the optical axes, are known as the *electrical axes*. The three axes Y', Y'', and Y''', which are perpendicular to the faces of the crystal, are the *mechanical axes*.

If a flat section is cut from a quartz crystal in such a way that the flat sides are perpendicular to an electrical axis, as shown in Figure 2.4 (b) (*X-cut*), the mechanical stresses along the Y-axis of such a section produce electric charges on the flat sides of the crystal. If the direction of these stresses is changed from tension to compression, or vice versa, the polarity of the charges on the crystal surfaces is reversed. Conversely, if electric charges are placed on the flat sides of the crystal by applying a voltage across the faces, a mechanical stress is produced in the direction of the Y-axis. This property by which mechanical and electrical properties are interconnected in a crystal is known as the *piezoelectric effect* and is exhibited by all sections cut from a piezoelectric crystal. Thus, if mechanical forces are applied across the faces of a crystal section having its flat sides perpendicular to a Y-axis, as shown in Figure 2.4 (a), piezoelectric charges will be produced because forces and potentials developed in such a crystal have components across at least one of the Y- or X-axes, respectively.

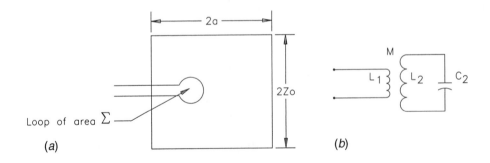

Figure 2.3 Cylindrical resonator incorporating a coupling loop: (*a*) orientation of loop with respect to cavity, (*b*) equivalent coupled circuit.

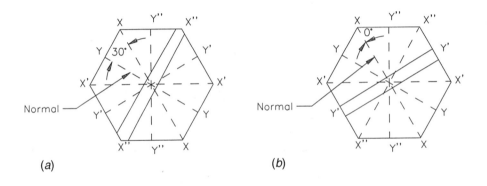

Figure 2.4 Cross section of a quartz crystal taken in the plane perpendicular to the optical axis: (*a*) *Y*-cut plate, (*b*) *X*-cut plate.

An alternating voltage applied across a quartz crystal will cause the crystal to vibrate and, if the frequency of the applied alternating voltage approximates a frequency at which mechanical resonance can exist in the crystal, the amplitude vibrations will be large. Any crystal has a number of such resonant frequencies that depend upon the crystal dimensions, the type of mechanical oscillation involved, and the orientation of the plate cut from the natural crystal.

Crystals are temperature-sensitive, as shown in Figure 2.5. The extent to which a device is affected by changes in temperature is determined by its cut and packaging. Crystals also exhibit changes in frequency with time. Such *aging* is caused by one or both of the following:

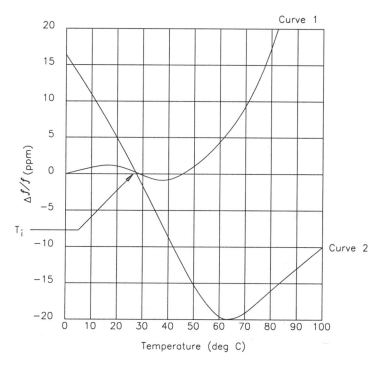

Figure 2.5 The effects of temperature on two types of AT-cut crystals.

- Mass transfer to or from the resonator surface
- Stress relief within the device itself
- Crystal aging is most pronounced when the device is new. As stress within the internal structure is relieved, the aging process slows.

Frequency Stabilization

The stability of a quartz crystal is inadequate for most commercial and industrial applications. Two common methods are used to provide the required long-term frequency stability:

- *Oven-controlled crystal oscillator*—The crystal is installed in a temperature-controlled box. Because the temperature in the box is constant, controlled by a thermostat, the crystal remains on-frequency. The temperature of the enclosure usually is set to the *turnover temperature* of the crystal. (The turnover point is illustrated in Figure 2.5.)

- *Temperature-compensated crystal oscillator* (TCXO)—The frequency vs. temperature changes of the crystal are compensated by varying a load capacitor. A thermistor network typically is used to generate a correction voltage that feeds a

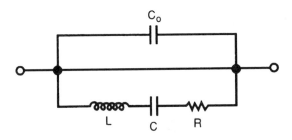

Figure 2.6 Equivalent network of a piezoelectric resonator. (*From* [1]. *Used with permission.*)

voltage-controlled capacitor (*varactor*) to retune the crystal to the desired on-frequency value.

Equivalent Circuit of a Quartz Resonator

Because of the piezoelectric effect, the quartz resonator behaves physically as a vibrating object when driven by a periodic electric signal near a resonant frequency of the cavity [1,2]. This resonant behavior may be characterized by equivalent electrical characteristics, which may then be used to accurately determine the response and performance of the electromechanical system. Dye and Butterworth (1926) [3] developed the equivalent lump electrical network that describes an electrostatic signal driving a mechanically vibrating system (an *electromechanical transducer*). The equivalent electric network that models the piezo resonator was later developed by Van Dyke (1928), Butterworth, and others, and is shown in Figure 2.6. In the equivalent network, the circuit parameters R_1, L_1, and C_1 represent parameters associated with the quartz crystal, while C_0 represents the shunt capacitance of the resonator electrodes in parallel with the container that packages the crystal.

The inductance L_1 is associated with the mass of the quartz slab, and the capacitance C_1 is associated with the stiffness of the slab. Finally, the resistance, R_1 is determined by the loss processes associated with the operation of the crystal. Each of the equivalent parameters can be measured using standard network measurement techniques. Typical parameters are shown in Table 2.1.

Application of an electric field across the electrodes of the crystal results in both face and thickness shear waves developing within the crystal resonator. For a Y-cut crystal, face shear modes occur in the X-Z plane, while thickness shear modes develop in the X-Y plane.

The vibration is enhanced for electric signals near a resonant frequency of the resonator. Low frequency resonances tend to occur in the face shear modes, while high frequency resonances occur in thickness shear modes. In either case, when the applied electric signal is near a resonant frequency, the energy required to maintain the vibrational motion is small.

Table 2.1 Typical Equivalent Circuit Parameter Values for Quartz Crystals (*After*[1].)

Parameter	200 kHz Fundamental	2 MHz	30 Mhz 3rd overtone	90 Mhz 5th overtone
R_1	2 kS	100 Ω	20 Ω	40 Ω
L_1	27 H	520 mH	11 mH	6 mH
C_1	0.024 pF	0.012 pF	0.0026 pF	0.0005 pF
C_0	9 pF	4 pF	6 pF	4 pF
Q	18×10^3	18×10^3	18×10^3	18×10^3

Temperature Compensation

Compensation for temperature-related frequency shifts can be accomplished by connecting a reactance, such as a variable voltage capacitor (a *varactor*) in series with the resonator [1]. The effect of this reactance is to pull the frequency to cancel the effect of frequency drift caused by temperature changes. An example of a frequency versus temperature characteristic both before and after compensation is shown in Figure 2.7.

Several techniques for temperature compensation are possible. They include:

- Analog temperature compensation

- Hybrid analog-digital compensation

- Digital temperature compensation

For analog compensation, a varactor is controlled by a voltage divider network that contains both thermistors and resistors to manage the compensation. Stability values are in the range 0.5 to 10 ppm. The lowest value requires that individual components in the voltage divider network be adjusted under controlled-temperature conditions. This is difficult over large temperature ranges for analog networks because of the tight component value tolerances required and because of interactions between different analog segments. Analog networks can be segmented in order to increase the independence of the component adjustments required to improve stability.

These problems can be alleviated through the use of some form of hybrid or digital compensation. In this scheme, the crystal oscillator is connected to a varactor in series, as in the analog case. This capacitor provides the coarse compensation (about ±4 parts in 10^7). The fine compensation is provided by the digital network to further improve the stability. The hybrid analog-digital method is then capable of producing stability values of ±5 parts in 10^8 over the range –40°C to +80°C.

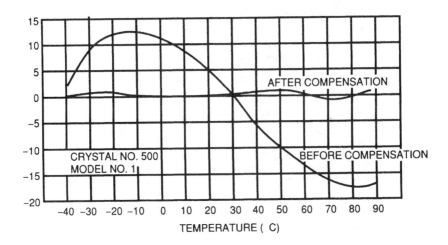

Figure 2.7 Frequency vs. temperature characteristics for a typical temperature-compensated crystal oscillator. (*After* [2].)

Stress Compensation

Stress compensation is employed in order to stabilize the crystal resonator frequency against shifts caused by mechanical stresses [1]. These result from temperature changes or from electrode changes, due to differences in the expansion coefficients, which can occur over time. The frequency variations are caused by alteration of the crystal elastic coefficients with stress bias. The method developed for stress compensation is a special crystal cut called the *stress-compensated* or SC-cut. This cut was developed to obtain a resonator whose frequency would be independent of specified stresses.

The SC-cut has some advantages over the more common AT- and BT-cuts. These include a flatter frequency-temperature response, especially at the turning points, and less susceptibility to aging because of changes in the mounting stresses. The lower sensitivity to frequency changes means that SC-cut resonators can be frequency-temperature tested more rapidly than their AT- and BT-cut counterparts. A disadvantage of the SC-cut is that it is sensitive to air-damping and must be operated in a vacuum environment to obtain optimum Q factors. The frequency-temperature curve is also roughly 70° higher for the SC-cut over the more common AT-cut, which could also be a disadvantage in some applications.

Aging Effects

Aging in crystal oscillators generally refers to any change over time that will affect the frequency characteristics of the oscillator, or the physical parameters that describe the device; such as motional time constant and equivalent circuit parameters [1]. Some factors that influence aging include the following:

- Surface deterioration

- Surface contamination

- Electrode composition

- Environmental conditions

Deterioration on the surface contributes to strain in the crystal, and this type of strain can impact the operating frequency of the oscillator. More importantly, the surface strain is sensitive to humidity and temperature early in the life of the crystal. One method of combating this surface degradation mechanism is to etch away the strained surface layers. The initial slope of the frequency change versus etch time curve is exponential. However, after the strained surface layers have been removed, the slope becomes linear.

Another surface issue is contamination from foreign materials. The presence of foreign substances will load the oscillator with an additional mass, and this can be reflected in the equivalent circuit parameters. The contaminants can be introduced during fabrication processes, from out-gassing of other materials after the device has been packaged, or from contaminants in the environment during storage. The contamination process is often increased in higher temperature environments, and temperature cycling will lead to stability problems for the device.

Aging problems associated with the device electrodes includes delamination, leading to increased resistance or corrosion effects. Typical electrode materials include gold, silver, and aluminum. Of these, silver is the most popular material.

2.1.4 Operating Class

Amplifier stage operating efficiency is a key element in the design and application of an electron tube device or system. As the power level of an RF generator increases, the overall efficiency of the system becomes more important. Increased efficiency translates into lower operating costs and, usually, improved reliability of the system. Put another way, for a given device dissipation, greater operating efficiency translates into higher power output. The operating mode of the final stage, or stages, is the primary determining element in the maximum possible efficiency of the system.

An electron amplifying device is classified by its individual *class of operation*. Four primary class divisions apply to vacuum tube devices:

- *Class A*—A mode wherein the power amplifying device is operated over its linear transfer characteristic. This mode provides the lowest waveform distortion, but also the lowest efficiency. The basic operating efficiency of a class A stage is 50 percent. Class A amplifiers exhibit low intermodulation distortion, making them well suited to linear RF amplifier applications.

- *Class B*—A mode wherein the power amplifying device is operated just outside its linear transfer characteristic. This mode provides improved efficiency at the expense of some waveform distortion. *Class AB* is a variation on class B operation. The transfer characteristic for an amplifying device operating in this mode is, predictably, between class A and class B.

- *Class C*—A mode wherein the power amplifying device is operated significantly outside its linear transfer characteristic, resulting in a pulsed output waveform. High efficiency (up to 90 percent) can be realized with class C operation, but significant distortion of the waveform will occur. Class C is used extensively as an efficient RF power generator.

- *Class D*—A mode that essentially results in a switched device state. The power amplifying device is either *on* or *off*. This is the most efficient mode of operation. It is also the mode that produces the greatest waveform distortion.

The angle of current flow determines the class of operation for a power amplifying device. Typically, the following generalizations regarding conduction angle apply:

- Class A: 360°
- Class AB: between 180 and 360°
- Class B: 180°
- Class C: less than 180°

Subscripts also may be used to denote grid current flow. The subscript "1" means that no grid current flows in the stage; the subscript "2" denotes grid current flow. Figure 2.8 charts operating efficiency as a function of conduction angle for an RF amplifier.

The class of operation is not directly related to the type of amplifying circuit. Vacuum tube stages may be grid- or cathode-driven without regard to the operating class.

Operating Efficiency

The design goal of all RF amplifiers is to convert input power into an RF signal at the greatest possible efficiency. Direct current input power that is not converted into a useful output signal is, for the most part, converted to heat. This heat is wasted energy, which must be removed from the amplifying device. Removal of heat is a problem common to all high-power RF amplifiers. Cooling methods include:

- Natural convection
- Radiation
- Forced convection
- Liquid
- Conduction
- Evaporation

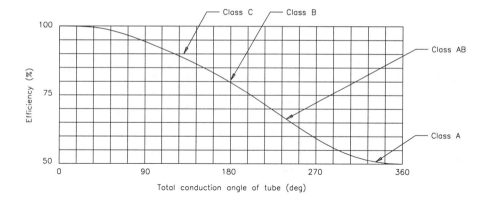

Figure 2.8 Plate efficiency as a function of conduction angle for an amplifier with a tuned load.

The type of cooling method chosen is dictated in large part by the type of active device used and the power level involved. For example, liquid cooling is used almost exclusively for high-power (100 kW) vacuum tubes; conduction is used most often for low-power (10 kW) devices. (Power-device cooling techniques are discussed in detail in Section 8.2.)

2.1.5 Broadband Amplifier Design

RF design engineers face a continuing challenge to provide adequate bandwidth for the signals to be transmitted while preserving as much efficiency from the overall system as possible. These two parameters, although not mutually exclusive, often involve tradeoffs for both designers and operators.

An ideal RF amplifier will operate over a wide band of frequencies with minimum variations in output power, phase, distortion, and efficiency. The bandwidth of the amplifier depends to a great extent on the type of active device used, the frequency range required, and the operating power. As a general rule, a bandwidth of 20 percent or greater at frequencies above 100 MHz can be considered *broadband*. Below 100 MHz, broadband amplifiers typically have a bandwidth of one octave or more.

Stagger Tuning

Several stages with narrowband response (relative to the desired system bandwidth) can be cascaded and, through the use of *stagger tuning*, made broadband. Although there is an efficiency penalty for this approach, it has been used for years in all types of equipment. The concept is simple: Offset the center operating frequencies (and, therefore, peak amplitude response) of the cascaded amplifiers so that the resulting passband is flat and broad.

For example, the first stage in a three-stage amplifier is adjusted for peak response at the center operating frequency of the system. The second stage is adjusted above the center frequency, and the third stage is adjusted below center. The resulting composite response curve yields a broadband trace. The efficiency penalty for this scheme varies depending on the power level of each stage, the amount of stagger tuning required to achieve the desired bandwidth, and the number of individual stages.

Matching Circuits

The individual stages of an RF generator must be coupled together. Rarely do the output impedance and power level of one stage precisely match the input impedance and signal-handling level of the next stage. There is a requirement, therefore, for broadband matching circuits. Matching at radio frequencies can be accomplished with several different techniques, including:

- A $\frac{1}{4}$-wave transformer: A matching technique using simply a length of transmission line $\frac{1}{4}$-wave long, with a characteristic impedance of $Z_{line} = \sqrt{Z_{in} \times Z_{out}}$, where Z_{in} and Z_{out} = the terminating impedances. Multiple $\frac{1}{4}$-wave transformers can be cascaded to achieve more favorable matching characteristics. Cascaded transformers permit small matching ratios for each individual section.

- A *balun* transformer: A transmission-line transformer in which the turns are arranged physically to include the interwinding capacitance as a component of the characteristic impedance of the transmission line. This technique permits wide bandwidths to be achieved in the device without unwanted resonances.

Balun transformers usually are made of twisted wire pairs or twisted coaxial lines. Ferrite toroids may be used as the core material.

Power Combining

The two most common methods of extending the operating power of active devices are *direct paralleling* of components and *hybrid splitting/combining*. Direct paralleling has been used for tube designs, but application of this simple approach is limited by variations in device operating parameters. Two identical devices in parallel do not necessarily draw the same amount of current (supply the same amount of power to the load). Paralleling at UHF and above can be difficult because of the restrictions of operating wavelength.

The preferred approach involves the use of identical stages driven in parallel from a *hybrid coupler*. The coupler provides a constant-source impedance and directs any reflected energy from the driven stages to a *reject port* for dissipation. A hybrid coupler offers a standing-wave-canceling effect that improves system performance. Hybrids also provide a high degree of isolation between active devices in a system.

2.1.6 Thermal and Circuit Noise

All electronic circuits are affected by any number of factors that cause their performance to be degraded from the ideal assumed in simple component models, in ways that can be controlled but not eliminated entirely [4]. One limitation is the failure of the model to account properly for the real behavior of components, either the result of an oversimplified model description or variations in manufacture. Usually, by careful design, the engineer can work around the limitations of the model and produce a device or circuit whose operation is very close to predictions. One source of performance degradation that cannot be easily overcome, however, is noise.

When vacuum tubes first came into use in the early part of the twentieth century, they were extensively used in radios as signal amplifiers and for other related signal conditioning functions. Thus, the measure of performance that was of greatest importance to circuit designers was the quality of the sound produced at the radio speaker. It was immediately noticed that the sound coming from speakers not only consisted of the transmitted signal but also of popping, crackling, and hissing sounds, which seemed to have no pattern and were distinctly different from the sounds that resulted from other interfering signal sources, such as other radio stations using neighboring frequencies. This patternless or random signal was labeled "noise," and has become the standard term for signals that are random and are combined with the circuit signal to affect the overall performance of the system.

As the study of noise has progressed, engineers have come to realize that there are many sources of noise in circuits. The following definitions are commonly used in discussions of circuit noise:

- *White noise*: Noise whose energy is evenly distributed over the entire frequency spectrum, within the frequency range of interest (typically below frequencies in the infrared range). Because noise is totally random it may seem inappropriate to refer to its frequency range, because it is not really periodic in the ordinary sense. Nevertheless, by examining an oscilloscope trace of white noise, one can verify that every trace is different—as the noise never repeats itself—and yet each trace looks the same. Similarly, a TV set tuned to an unused frequency displays never-ending "snow" that always looks the same, yet clearly is always changing. There is a strong theoretical foundation to represent the frequency content of such signals as covering the frequency spectrum evenly. In this way the impact on other periodic signals can be analyzed. The term "white noise" arises from the fact that—similar to white light, which has equal amounts of all light frequencies—white noise has equal amounts of noise at all frequencies within circuit operating range.

- *Interference*: The name given to any predictable, periodic signal that occurs in an electronic circuit in addition to the signal the circuit is designed to process. This component is distinguished from a noise signal by the fact that it occupies a relatively small frequency range, and because it is predictable, it can often be filtered out. Interference usually results from another electronic system, such as an interfering radio source.

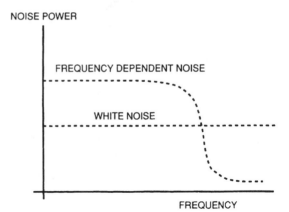

Figure 2.9 A plot of signal power vs frequency for white noise and frequency limited noise. (*From* [4]. *Used with permission.*)

- *Thermal noise*: Any temperature-dependent noise generated within a circuit. This signal usually is the result of the influence of temperature directly on the operating characteristics of circuit elements, which—because of the random motion of molecules as a result of temperature—in turn creates a random fluctuation of the signal being processed.

- *Shot noise*: A type of circuit noise that is not temperature-dependent, and is not white noise in the sense that it tends to diminish at higher frequencies. (See Figure 2.9.) This noise usually occurs in components whose operation depends on a mean particle residence time for the active electrons within the device. The cutoff frequency above which noise disappears is closely related to the inverse of this characteristic particle residence time. It is called "shot noise" because in a radio it can make a sound similar to buckshot hitting a drum, as opposed to white noise, which tends to sound more like hissing (because of the higher frequency content).

Thermal Noise

By definition, thermal noise is internally generated noise that is temperature-dependent [4]. While first observed in vacuum tube devices (because their amplifying capabilities tend to bring out thermal noise), it is also observed in semiconductor devices. It is a phenomenon resulting from the ohmic resistance of devices that dissipate the energy lost in them as heat. Heat consists of random motion of molecules, which are more energetic as temperature increases. Because the motion is random, it is to be expected that as electrons pass through the ohmic device incurring resistance, there should be some random deviations in the rate of energy loss. This fluctuation has the effect of causing variation in the resulting current, which is noise. As the temperature of the device increases, the random motion of the molecules increases, and so does the corresponding noise level.

Noise in Systems of Cascaded Stages

The noise level produced by thermal noise sources is not necessarily large, however because source signal power may also be low, it is usually necessary to amplify the source signal. Because noise is combined with the source signal, and both are then amplified, with more noise added at each successive stage of amplification, noise can become a noticeable phenomenon [4].

2.2 Amplitude Modulation

In the simplest form of amplitude modulation, an analog carrier is controlled by an analog modulating signal. The desired result is an RF waveform whose amplitude is varied by the magnitude of the applied modulating signal and at a rate equal to the frequency of the applied signal. The resulting waveform consists of a carrier wave plus two additional signals:

- An upper-sideband signal, which is equal in frequency to the carrier *plus* the frequency of the modulating signal

- A lower-sideband signal, which is equal in frequency to the carrier *minus* the frequency of the modulating signal

This type of modulation system is referred to as *double-sideband amplitude modulation* (DSAM).

The radio carrier wave signal onto which the analog amplitude variations are to be impressed is expressed as:

$$e(t) = AE_c \cos(\omega_c t) \tag{2.4}$$

Where:
$e(t)$ = instantaneous amplitude of carrier wave as a function of time (t)
A = a factor of amplitude modulation of the carrier wave
ω_c = angular frequency of carrier wave (radians per second)
E_c = peak amplitude of carrier wave

If A is a constant, the peak amplitude of the carrier wave is constant, and no modulation exists. Periodic modulation of the carrier wave results if the amplitude of A is caused to vary with respect to time, as in the case of a sinusoidal wave:

$$A = 1 + (E_m/E_c) \cos(\omega_m t) \tag{2.5}$$

Where:
E_m/E_c = the ratio of modulation amplitude to carrier amplitude

The foregoing relationship leads to:

$$e(t) = E_c [1 + (E_m/E_c) \cos(\omega_m t) \cos(\omega_c t)] \tag{2.6}$$

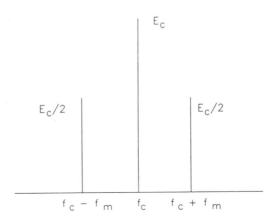

Figure 2.10 Frequency-domain representation of an amplitude-modulated signal at 100 percent modulation. E_c = carrier power, F_c = frequency of the carrier, and F_m = frequency of the modulating signal.

This is the basic equation for periodic (sinusoidal) amplitude modulation. When all multiplications and a simple trigonometric identity are performed, the result is:

$$e(t) = E_c \cos(\omega_c t) + (M/2) \cos(\omega_c t + \omega_m t) + (M/2) \cos(\omega_c t - \omega_m t) \qquad (2.7)$$

Where:
M = the amplitude modulation factor E_m/E_c

Amplitude modulation is, essentially, a multiplication process in which the time functions that describe the modulating signal and the carrier are multiplied to produce a modulated wave containing *intelligence* (information or data of some kind). The frequency components of the modulating signal are translated in this process to occupy a different position in the spectrum.

The bandwidth of an AM transmission is determined by the modulating frequency. The bandwidth required for full-fidelity reproduction in a receiver is equal to twice the applied modulating frequency.

The magnitude of the upper sideband and lower sideband will not normally exceed 50 percent of the carrier amplitude during modulation. This results in an upper-sideband power of one-fourth the carrier power. The same power exists in the lower sideband. As a result, up to one-half of the actual carrier power appears additionally in the sum of the sidebands of the modulated signal. A representation of the AM carrier and its sidebands is shown in Figure 2.10. The actual occupied bandwidth, assuming pure sinusoidal modulating signals and no distortion during the modulation process, is equal to twice the frequency of the modulating signal.

The extent of the amplitude variations in a modulated wave is expressed in terms of the *degree of modulation* or *percentage of modulation*. For sinusoidal variation, the degree of modulation m is determined from the following:

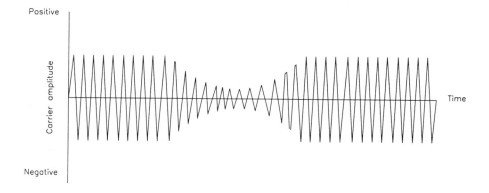

Figure 2.11 Time-domain representation of an amplitude-modulated signal. Modulation at 100 percent is defined as the point at which the peak of the waveform reaches twice the carrier level, and the minimum point of the waveform is zero.

$$m = \frac{E_{avg} - E_{min}}{E_{avg}} \qquad (2.8)$$

Where:

E_{avg} = average envelope amplitude
E_{min} = minimum envelope amplitude

Full (100 percent) modulation occurs when the peak value of the modulated envelope reaches twice the value of the unmodulated carrier, and the minimum value of the envelope is zero. The envelope of a modulated AM signal in the time domain is shown in Figure 2.11.

When the envelope variation is not sinusoidal, it is necessary to define the degree of modulation separately for the peaks and troughs of the envelope:

$$m_{pp} = \frac{E_{max} - E_{avg}}{E_{avg}} \times 100 \qquad (2.9)$$

$$m_{np} = \frac{E_{avg} - E_{min}}{E_{avg}} \times 100 \qquad (2.10)$$

Where:

m_{pp} = positive peak modulation (percent)
E_{max} = peak value of modulation envelope
m_{np} = negative peak modulation (percent)

E_{avg} = average envelope amplitude
E_{min} = minimum envelope amplitude

When modulation exceeds 100 percent on the negative swing of the carrier, spurious signals are emitted. It is possible to modulate an AM carrier asymmetrically; that is, to restrict modulation in the negative direction to 100 percent, but to allow modulation in the positive direction to exceed 100 percent without a significant loss of fidelity. In fact, many modulating signals normally exhibit asymmetry, most notably human speech waveforms.

The carrier wave represents the average amplitude of the envelope and, because it is the same regardless of the presence or absence of modulation, the carrier transmits no information. The information is carried by the sideband frequencies. The amplitude of the modulated envelope may be expressed as follows [5]:

$$E = E_0 + E_1 \sin (2\pi f_1 t + \Phi_1) + E_2 \sin (2\pi f_2 t + \Phi_2) \qquad (2.11)$$

Where:
E = envelope amplitude
E_0 = carrier wave crest value (volts)
E_1 = 2 × first sideband crest amplitude (volts)
f_1 = frequency difference between the carrier and the first upper/lower sidebands
E_2 = 2 × second sideband crest amplitude (volts)
f_2 = frequency difference between the carrier and the second upper/lower sidebands
Φ_1 = phase of the first sideband component
Φ_2 = phase of the second sideband component

2.2.1 High-Level AM Modulation

High-level anode modulation is the oldest and simplest method of generating a high-power AM signal. In this system, the modulating signal is amplified and combined with the dc supply source to the anode of the final RF amplifier stage. The RF amplifier is normally operated class C. The final stage of the modulator usually consists of a pair of tubes operating class B in a push-pull configuration. A basic high-level modulator is shown in Figure 2.12.

The RF signal normally is generated in a low-level transistorized oscillator. It is then amplified by one or more solid-state or vacuum tube stages to provide final RF drive at the appropriate frequency to the grid of the final class C amplifier. The modulation input is applied to an intermediate power amplifier (usually solid-state) and used to drive two class B (or class AB) push-pull output devices. The final amplifiers provide the necessary modulating power to drive the final RF stage. For 100 percent modulation, this modulating power is equal to 50 percent of the actual carrier power.

The modulation transformer shown in Figure 2.12 does not usually carry the dc supply current for the final RF amplifier. The modulation reactor and capacitor shown provide a means to combine the signal voltage from the modulator with the dc supply to the final RF amplifier. This arrangement eliminates the necessity of having direct current

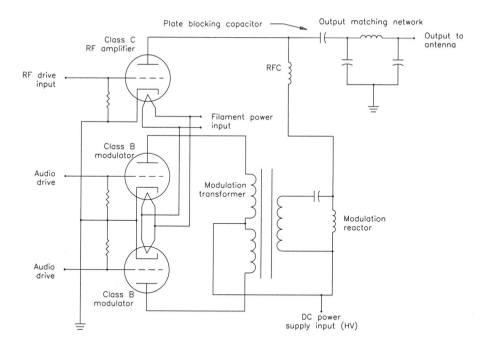

Figure 2.12 Simplified diagram of a high-level amplitude-modulated amplifier.

flow through the secondary of the modulation transformer, which would result in magnetic losses and saturation effects. In some transmitter designs, the modulation reactor is eliminated from the system, thanks to improvements in transformer technology.

The RF amplifier normally operates class C with grid current drawn during positive peaks of the cycle. Typical stage efficiency is 75 to 83 percent. An RF tank following the amplifier resonates the output signal at the operating frequency and, with the assistance of a low-pass filter, eliminates harmonics of the amplifier caused by class C operation.

This type of system was popular in AM applications for many years, primarily because of its simplicity. The primary drawback is low overall system efficiency. The class B modulator tubes cannot operate with greater than 50 percent efficiency. Still, with inexpensive electricity, this was not considered to be a significant problem. As energy costs increased, however, more efficient methods of generating high-power AM signals were developed. Increased efficiency normally came at the expense of added technical complexity.

2.2.2 Vestigial-Sideband Amplitude Modulation

Because the intelligence (modulating signal) of conventional AM transmission is identical in the upper *and* lower sidebands, it is possible to eliminate one sideband and

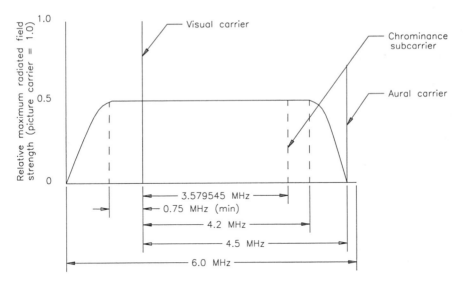

Figure 2.13 Idealized amplitude characteristics of the FCC standard waveform for monochrome and color TV transmission. (*Adapted from*: FCC Rules, Sec. 73.699.)

still convey the required information. This scheme is implemented in *vestigial-sideband AM* (VSBAM). Complete elimination of one sideband (for example, the lower sideband) requires an ideal high-pass filter with infinitely sharp cutoff. Such a filter is difficult to implement in any practical design. VSBAM is a compromise technique wherein one sideband (typically the lower sideband) is attenuated significantly. The result is a savings in occupied bandwidth and transmitter power.

VSBAM is used for television broadcast transmission and other applications. A typical bandwidth trace for a VSBAM TV transmitter is shown in Figure 2.13.

2.2.3 Single-Sideband Amplitude Modulation

The carrier in an AM signal does not convey any intelligence. All of the modulating information is in the sidebands. It is possible, therefore, to suppress the carrier upon transmission, radiating only one or both sidebands of the AM signal. The result is much greater efficiency at the transmitter (that is, a reduction in the required transmitter power). Suppression of the carrier may be accomplished with DSAM and SSBAM signals. *Single-sideband suppressed carrier* AM (SSB-SC) is the most spectrum- and energy-efficient mode of AM transmission. Figure 2.14 shows representative waveforms for suppressed carrier transmissions.

A waveform with carrier suppression differs from a modulated wave containing a carrier primarily in that the envelope varies at twice the modulating frequency. In addition, it will be noted that the SSB-SC wave has an apparent phase that reverses every time the modulating signal passes through zero. The wave representing a single sideband consists of a number of frequency components, one for each component in the

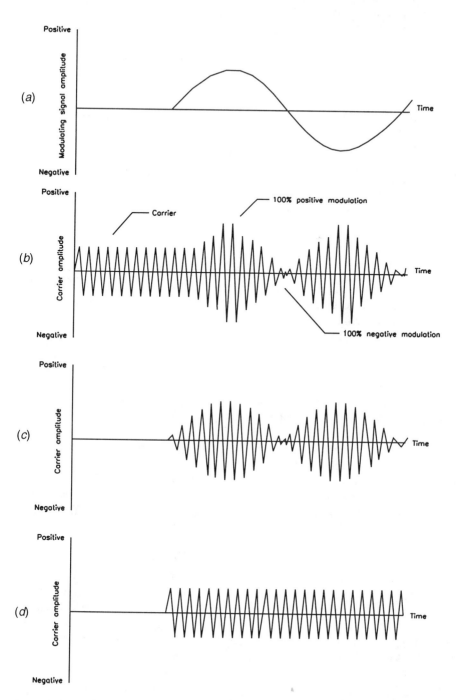

Figure 2.14 Types of suppressed carrier amplitude modulation: (*a*) the modulating signal, (*b*) double-sideband AM signal, (*c*) double-sideband suppressed carrier AM, (*d*) single-sideband suppressed carrier AM.

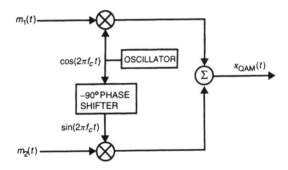

Figure 2.15 Simplified QAM modulator. (*From* [6]. *Used with permission.*)

original signal. Each of these components has an amplitude proportional to the amplitude of the corresponding modulating component and a frequency differing from that of the carrier by the modulating frequency. The result is that, in general, the envelope amplitude of the single sideband signal increases with the degree of modulation, and the envelope varies in amplitude in accordance with the difference frequencies formed by the various frequency components of the single sideband interacting with each other.

An SSB-SC system is capable of transmitting a given intelligence within a frequency band only half as wide as that required by a DSAM waveform. Furthermore, the SSB system saves more than two-thirds of the transmission power because of the elimination of one sideband and the carrier.

The drawback to suppressed carrier systems is the requirement for a more complicated receiver. The carrier must be regenerated at the receiver to permit demodulation of the signal. Also, in the case of SSBAM transmitters, it is usually necessary to generate the SSB signal in a low-power stage and then amplify the signal with a linear power amplifier to drive the antenna. Linear amplifiers generally exhibit low efficiency.

2.2.4 Quadrature Amplitude Modulation (QAM)

Single sideband transmission makes very efficient use of the spectrum; for example, two SSB signals can be transmitted within the bandwidth normally required for a single DSB signal. However, DSB signals can achieve the same efficiency by means of *quadrature amplitude modulation* (QAM), which permits two DSB signals to be transmitted and received simultaneously using the same carrier frequency [6]. A basic QAM DSB modulator is shown schematically in Figure 2.15.

Two DSB signals coexist separately within the same bandwidth by virtue of the 90° phase shift between them. The signals are, thus, said to be in *quadrature*. Demodulation uses two local oscillator signals that are also in quadrature, i.e., a sine and a cosine signal, as illustrated in Figure 2.16.

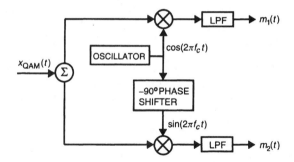

Figure 2.16 Simplified QAM demodulator. (*From* [6]. *Used with permission.*)

The chief disadvantage of QAM is the need for a coherent local oscillator at the receiver exactly in phase with the transmitter oscillator signal. Slight errors in phase or frequency can cause both loss of signal and interference between the two signals (cochannel interference or crosstalk).

The relative merits of the various AM systems are summarized in Table 2.2

2.3 Frequency Modulation

Frequency modulation is a technique whereby the phase angle or phase shift of a carrier is varied by an applied modulating signal. The *magnitude* of frequency change of the carrier is a direct function of the *magnitude* of the modulating signal. The *rate* at which the frequency of the carrier is changed is a direct function of the *frequency* of the modulating signal. In FM modulation, multiple pairs of sidebands are produced. The actual number of sidebands that make up the modulated wave is determined by the *modulation index* (MI) of the system.

2.3.1 Modulation Index

The modulation index is a function of the frequency deviation of the system and the applied modulating signal:

$$MI = \frac{F_d}{M_f} \tag{2.12}$$

Where:
MI = the modulation index
F_d = frequency deviation
M_f = modulating frequency

Table 2.2 Comparison of Amplitude Modulation Techniques (*After* [6].)

Modulation Scheme	Advantages	Disadvantages	Comments
DSB SC	Good power efficiency; good low-frequency response.	More difficult to generate than DSB+C; detection requires coherent local oscillator, pilot, or phase-locked loop; poor spectrum efficiency.	
DSB+C	Easier to generate than DSB SC, especially at high-power levels; inexpensive receivers using envelope detection.	Poor power efficiency; poor spectrum efficiency; poor low-frequency response; exhibits threshold effect in noise.	Used in commercial AM broadcasting.
SSB SC	Excellent spectrum efficiency.	Complex transmitter design; complex receiver design (same as DSB SC); poor low-frequency response.	Used in military communication systems, and to multiplex multiple phone calls onto long-haul microwave links.
SSB+C	Good spectrum efficiency; low receiver complexity.	Poor power efficiency; complex transmitters; poor low-frequency response; poor noise performance.	
VSB SC	Good spectrum efficiency; excellent low-frequency response; transmitter easier to build than for SSB.	Complex receivers (same as DSB SC).	
VSB+C	Good spectrum efficiency; good low-frequency response; inexpensive receivers using envelope detection.	Poor power efficiency; poor performance in noise.	Used in commercial TV broadcasting.
QAM	Good low-frequency response; good spectrum efficiency.	Complex receivers; sensitive to frequency and phase errors.	Two SSB signals may be preferable.

The higher the MI, the more sidebands produced. It follows that the higher the modulating frequency for a given deviation, the fewer number of sidebands produced, but the greater their spacing.

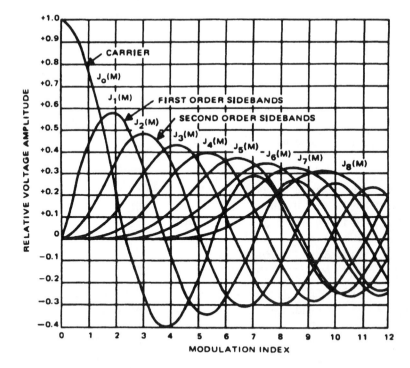

Figure 2.17 Plot of Bessel functions of the first kind as a function of modulation index.

To determine the frequency spectrum of a transmitted FM waveform, it is necessary to compute a Fourier series or Fourier expansion to show the actual signal components involved. This work is difficult for a waveform of this type, because the integrals that must be performed in the Fourier expansion or Fourier series are not easily solved. The result, however, is that the integral produces a particular class of solution that is identified as the *Bessel function*, illustrated in Figure 2.17.

The carrier amplitude and phase, plus the sidebands, can be expressed mathematically by making the modulation index the argument of a simplified Bessel function. The general expression is given from the following equations:

RF output voltage $= E_t = E_c + S_{1u} - S_{1l} + S_{2u} - S_{2l} + S_{3u} - S_{3l} + S_{nu} - S_{nl}$

Carrier amplitude $= E_c = A\,[J_0\,(M)\,\sin\,\omega c(t)]$

First-order upper sideband $= S_{1u} = J_1\,(M)\,\sin\,(\omega c + \omega m)\,t$

First-order lower sideband $= S_{1l} = J_1\,(M)\,\sin\,(\omega c - \omega m)\,t$

Second-order upper sideband $= S_{2u} = J_2\,(M)\,\sin\,(\omega c + 2\omega m)\,t$

Second-order lower sideband = $S_{2l} = J_2 (M) \sin (\omega c - 2\omega m) t$

Third-order upper sideband = $S_{3u} = J_3 (M) \sin (\omega c + 3\omega m) t$

Third-order lower sideband = $S_{3l} = J_3 (M) \sin (\omega c - 3\omega m) t$

Nth-order upper sideband = $S_{nu} = J_n (M) \sin (\omega c + n\omega m) t$

Nth-order lower sideband = $S_{nl} = J_n (M) \sin (\omega c - n\omega m) t$

Where:
A = the unmodulated carrier amplitude constant
J_0 = modulated carrier amplitude
$J_1, J_2, J_3 ... J_n$ = amplitudes of the nth-order sidebands
M = modulation index
$\omega c = 2\pi F_c$, the carrier frequency
$\omega m = 2\pi F_m$, the modulating frequency

Further supporting mathematics will show that an FM signal using the modulation indices that occur in a wideband system will have a multitude of sidebands. From the purist point of view, *all* sidebands would have to be transmitted, received, and demodulated to reconstruct the modulating signal with complete accuracy. In practice, however, the channel bandwidths permitted FM systems usually are sufficient to reconstruct the modulating signal with little discernible loss in fidelity, or at least an acceptable loss in fidelity.

Figure 2.18 illustrates the frequency components present for a modulation index of 5. Figure 2.19 shows the components for an index of 15. Note that the number of significant sideband components becomes quite large with a high MI. This simple representation of a single-tone frequency-modulated spectrum is useful for understanding the general nature of FM, and for making tests and measurements. When typical modulation signals are applied, however, many more sideband components are generated. These components vary to the extent that sideband energy becomes distributed over the entire occupied bandwidth, rather than appearing at discrete frequencies.

Although complex modulation of an FM carrier greatly increases the number of frequency components present in the frequency-modulated wave, it does not, in general, widen the frequency band occupied by the energy of the wave. To a first approximation, this band is still roughly twice the sum of the maximum frequency deviation at the peak of the modulation cycle plus the highest modulating frequency involved.

FM is not a simple frequency translation, as with AM, but involves the generation of entirely new frequency components. In general, the new spectrum is much wider than the original modulating signal. This greater bandwidth may be used to improve the *signal-to-noise ratio* (S/N) of the transmission system. FM thereby makes it possible to exchange bandwidth for S/N enhancement.

The power in an FM system is constant throughout the modulation process. The output power is increased in the amplitude modulation system by the modulation process,

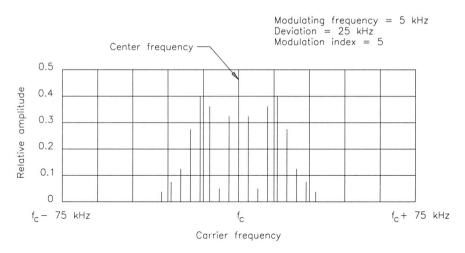

Figure 2.18 RF spectrum of a frequency-modulated signal with a modulation index of 5 and other operating parameters as shown.

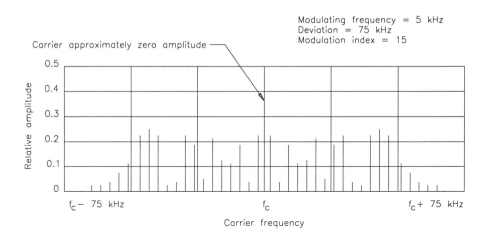

Figure 2.19 RF spectrum of a frequency-modulated signal with a modulation index of 15 and operating parameters as shown.

but the FM system simply distributes the power throughout the various frequency components that are produced by modulation. During modulation, a wideband FM system does not have a high amount of energy present in the carrier. Most of the energy will be found in the sum of the sidebands.

The constant-amplitude characteristic of FM greatly assists in capitalizing on the low noise advantage of FM reception. Upon being received and amplified, the FM signal normally is clipped to eliminate all amplitude variations above a certain threshold. This removes noise picked up by the receiver as a result of man-made or atmospheric signals. It is not possible (generally speaking) for these random noise sources to change the frequency of the desired signal; they can affect only its amplitude. The use of *hard limiting* in the receiver will strip off such interference.

2.3.2 Phase Modulation

In a phase modulation (PM) system, intelligence is conveyed by varying the phase of the RF wave. Phase modulation is similar in many respects to frequency modulation, except in the interpretation of the modulation index. In the case of PM, the modulation index depends only on the amplitude of the modulation; MI is independent of the frequency of the modulating signal. It is apparent, therefore, that the phase-modulated wave contains the same sideband components as the FM wave and, if the modulation indices in the two cases are the same, the relative amplitudes of these different components also will be the same.

The modulation parameters of a PM system relate as follows:

$$\Delta f = m_p \times f_m \tag{2.13}$$

Where:
Δf = frequency deviation of the carrier
m_p = phase shift of the carrier
f_m = modulating frequency

In a phase-modulated wave, the phase shift m_p is independent of the modulating frequency; the frequency deviation Δf is proportional to the modulating frequency. In contrast, with a frequency-modulated wave, the frequency deviation is independent of modulating frequency. Therefore, a frequency-modulated wave can be obtained from a phase modulator by making the modulating voltage applied to the phase modulator inversely proportional to frequency. This can be readily achieved in hardware.

2.3.3 Modifying FM Waves

When a frequency-modulated wave is passed through a harmonic generator, the effect is to increase the modulation index by a factor equal to the frequency multiplication involved. Similarly, if the frequency-modulated wave is passed through a frequency divider, the effect is to reduce the modulation index by the factor of frequency division. Thus, the frequency components contained in the wave—and, consequently, the bandwidth of the wave—will be increased or decreased, respectively, by frequency multiplication or division. No distortion in the nature of the modulation is introduced by the frequency change.

When an FM wave is translated in the frequency spectrum by heterodyne action, the modulation index—hence the relative positions of the sideband frequencies and the bandwidths occupied by them—remains unchanged.

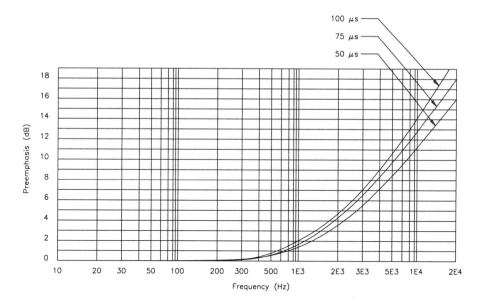

Figure 2.20 Preemphasis curves for time constants of 50, 75, and 100 μs.

2.3.4 Preemphasis and Deemphasis

The FM transmission/reception system offers significantly better noise-rejection characteristics than AM. However, FM noise rejection is more favorable at low modulating frequencies than at high frequencies because of the reduction in the number of sidebands at higher frequencies. To offset this problem, the input signal to the FM transmitter may be *preemphasized* to increase the amplitude of higher-frequency signal components in normal program material. FM receivers utilize complementary *deemphasis* to produce flat overall system frequency response.

FM broadcasting, for example, uses a 75 μs preemphasis curve, meaning that the time constant of the resistance-inductance (*RL*) or resistance-capacitance (*RC*) circuit used to provide the boost of high frequencies is 75 μs. Other values of preemphasis are used in different types of FM communications systems. Figure 2.20 shows three common preemphasis curves.

2.3.5 Modulation Circuits

Early FM transmitters used *reactance modulators* that operated at a low frequency. The output of the modulator then was multiplied to reach the desired output frequency. This approach was acceptable for some applications and unsuitable for others. Modern FM systems utilize what is referred to as *direct modulation*; that is, the

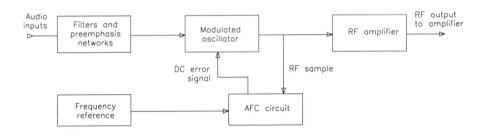

Figure 2.21 Block diagram of an FM exciter.

frequency modulation occurs in a modulated oscillator that operates on a center frequency equal to the desired transmitter output frequency.

Various techniques have been developed to generate the direct-FM signal. One of the most popular uses a variable-capacity diode as the reactive element in the oscillator. The modulating signal is applied to the diode, which causes the capacitance of the device to vary as a function of the magnitude of the modulating signal. Variations in the capacitance cause the frequency of the oscillator to change. The magnitude of the frequency shift is proportional to the amplitude of the modulating signal, and the rate of frequency shift is equal to the frequency of the modulating signal.

The direct-FM modulator is one element of an FM transmitter *exciter*, which generates the composite FM waveform. A block diagram of a complete FM exciter is shown in Figure 2.21. Input signals are buffered, filtered, and preemphasized before being summed to feed the modulated oscillator. Note that the oscillator is not normally coupled directly to a crystal, but to a free-running oscillator adjusted as closely as possible to the carrier frequency of the transmitter. The final operating frequency is maintained carefully by an automatic frequency control system employing a *phase locked loop* (PLL) tied to a reference crystal oscillator or frequency synthesizer.

A solid-state class C amplifier typically follows the modulated oscillator and raises the operating power of the FM signal to 20 to 30 W. One or more subsequent amplifiers in the transmitter raise the signal power to several hundred watts for application to the final power amplifier stage. Nearly all high-power FM broadcast transmitters use solid-state amplifiers up to the final RF stage, which is generally a vacuum tube for operating powers of 15 kW and above. All stages operate in the class C mode. In contrast to AM systems, each stage in an FM power amplifier can operate class C; no information is lost from the frequency-modulated signal because of amplitude changes. As mentioned previously, FM is a constant-power system.

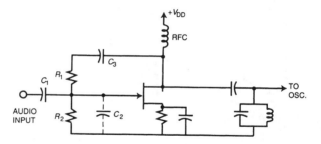

Figure 2.22 Simplified reactance modulator. (*From* [7]. *Used with permission.*)

Direct-FM Modulator

Many types of circuits have been used to produce a direct-FM signal [7]. In each case, a reactance device is used to shunt capacitive or inductive reactance across an oscillator. The value of capacitive or inductive reactance is made to vary as the amplitude of the modulating signal varies. Because the reactive load is placed across an oscillator's tuned circuit, the frequency of the oscillator will therefore shift by a predetermined amount, thereby creating an FM signal.

A typical example of a reactance modulator is illustrated in Figure 2.22. The circuit uses a field-effect transistor (FET), where the modulating signal is applied to the modulator through C_1. The actual components that affect the overall reactance consist of R_1 and C_2. Typically, the value of C_2 is small as this is the input capacitance to the FET, which may only be a few picofarads. However, this capacitance will generally be much larger by a significant amount as a result of the *Miller effect*. Capacitor C_3 has no significant effect on the reactance of the modulator; it is strictly a blocking capacitor that keeps dc from changing the gate bias of the FET.

To further understand the performance of the reactance modulator, the equivalent circuit of Figure 2.22 is represented in Figure 2.23. The FET is represented as a current source, gmV_g, with the internal drain resistance, r_d. The impedances Z_1 and Z_2 are a combination of resistance and capacitive reactance, which are designed to provide a 90° phase shift.

Using vector diagrams, we can analyze the phase relationship of the reactance modulator. Referring to Figure 2.22, the resistance of R_1 is typically high compared to the capacitive reactance of C_2. The R_1C_2 circuit is then resistive. Because this circuit is resistive, the current, I_{AB}, that flows through it, is in phase with the voltage V_{AB}. Voltage V_{AB} is also across R_1C_2 (or Z_1Z_2 in Figure 2.23). This is true because current and voltage tend to be in phase in a resistive network. However, voltage V_{C2}, which is across C_2, is out of phase with I_{AB}. This is because the voltage that is across a capacitor lags behind its current by 90°. (See Figure 2.24.)

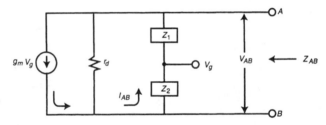

Figure 2.23 Equivalent circuit of the reactance modulator. (*From* [7]. *Used with permission.*)

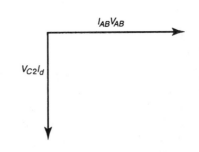

Figure 2.24 Vector diagram of a reactance modulator producing FM. Note: $V_g = V_{c2}$. (*From* [7]. *Used with permission.*)

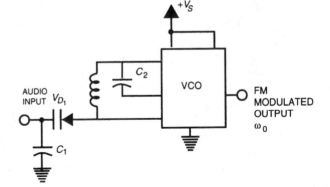

Figure 2.25 Voltage-controlled direct-FM modulator. (*From* [7]. *Used with permission.*)

VCO Direct-FM Modulator

One of the more common direct-FM modulation techniques uses an analog *voltage controlled oscillator* (VCO) in a phase locked loop arrangement [7]. In this configuration, shown in Figure 2.25, a VCO produces the desired carrier frequency that is—in turn—modulated by applying the input signal to the VCO input via a variactor diode. The variactor is used to vary the capacitance of an oscillator tank circuit. Therefore, the variactor behaves as a variable capacitor whose capacitance changes as the signal voltage across it changes. As the input capacitance of the VCO is varied by

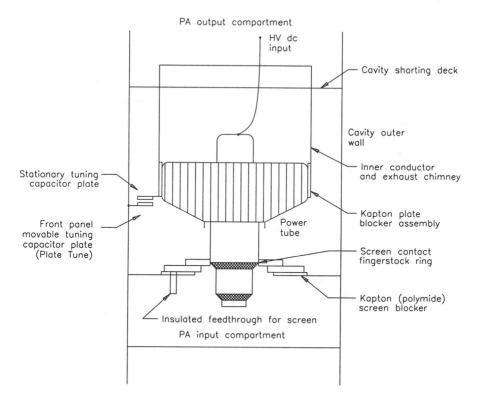

PA output compartment

HV dc
input

Cavity shorting deck

Cavity outer
wall

Inner conductor
and exhaust chimney

Stationary tuning
capacitor plate

Kapton plate
blocker assembly

Front panel
movable tuning
capacitor plate
(Plate Tune)

Power
tube

Screen contact
fingerstock ring

Kapton (polymide)
screen blocker

Insulated feedthrough for screen

PA input compartment

Figure 2.26 Mechanical layout of a common type of ¼-wave PA cavity for FM service.

the variactor, the output frequency of the VCO is shifted, which produces a direct-FM modulated signal.

FM Power Amplifiers

Nearly all high-power FM transmitters employ cavity designs. The ¼-wavelength cavity is the most common. The design is simple and straightforward. A number of variations can be found in different transmitters, but the underlying theory of operation is the same. The goal of any cavity amplifier is to simulate a resonant tank circuit at the operating frequency and provide a means to couple the energy in the cavity to the transmission line.

A typical ¼-wave cavity is shown in Figure 2.26. The plate of the tube connects directly to the inner section (tube) of the plate-blocking capacitor. The blocking capacitor can be formed in one of several ways. In at least one design, it is made by wrapping the outside surface of the inner tube conductor with multiple layers of insulating film. The exhaust chimney/inner conductor forms the other element of the blocking capacitor. The cavity walls form the outer conductor of the ¼-wave transmission line circuit. The dc plate voltage is applied to the PA tube by a cable routed inside the exhaust chimney and inner tube conductor. In this design, the screen-contact fingerstock ring mounts on a metal plate that is insulated from the grounded cavity deck by a blocking capacitor. This hardware makes up the screen-blocker assembly. The dc screen voltage feeds to the fingerstock ring from underneath the cavity deck using an insulated feedthrough.

Some transmitters that employ the $\frac{1}{4}$-wave cavity design use a grounded-screen configuration in which the screen-contact fingerstock ring is connected directly to the grounded cavity deck. The PA cathode then operates at below ground potential (in other words, at a negative voltage), establishing the required screen voltage for the tube.

Coarse-tuning of the cavity is accomplished by adjusting the cavity length. The top of the cavity (the *cavity shorting deck*) is fastened by screws or clamps and can be raised or lowered to set the length of the assembly for the particular operating frequency. Fine-tuning is accomplished by a variable-capacity plate-tuning control built into the cavity. In the example shown in Figure 2.26, one plate of this capacitor—the stationary plate—is fastened to the inner conductor just above the plate-blocking capacitor. The movable tuning plate is fastened to the cavity box, the outer conductor, and is linked mechanically to the front-panel tuning control. This capacity shunts the inner conductor to the outer conductor and varies the electrical length and resonant frequency of the cavity.

The theory of operation of cavity amplifier systems is discussed in Section 4.3.

2.4 Pulse Modulation

The growth of digital processing and communications has led to the development of modulation systems tailor-made for high-speed, spectrum-efficient transmission. In a *pulse modulation* system, the unmodulated carrier usually consists of a series of recurrent pulses. Information is conveyed by modulating some parameter of the pulses, such as amplitude, duration, time of occurrence, or shape. Pulse modulation is based on the *sampling principle*, which states that a message waveform with a spectrum of finite width can be recovered from a set of discrete samples if the sampling rate is higher than twice the highest sampled frequency (the Nyquist criteria). The samples of the input signal are used to modulate some characteristic of the carrier pulses.

2.4.1 Digital Modulation Systems

Because of the nature of digital signals (on or off), it follows that the amplitude of the signal in a pulse modulation system should be one of two heights (present or absent/positive or negative) for maximum efficiency. Noise immunity is a significant advantage of such a system. It is necessary for the receiving system to detect only the presence or absence (or polarity) of each transmitted pulse to allow complete reconstruction of the original intelligence. The pulse shape and noise level have minimal effect (to a point). Furthermore, if the waveform is to be transmitted over long distances, it is possible to regenerate the original signal exactly for retransmission to the next relay point. This feature is in striking contrast to analog modulation systems in which each modulation step introduces some amount of noise and signal corruption.

In any practical digital data system, some corruption of the intelligence is likely to occur over a sufficiently large span of time. Data encoding and manipulation schemes have been developed to detect and correct or conceal such errors. The addition of error-correction features comes at the expense of increased system overhead and (usually) slightly lower intelligence throughput.

2.4.2 Pulse Amplitude Modulation

Pulse amplitude modulation (PAM) is one of the simplest forms of data modulation. PAM departs from conventional modulation systems in that the carrier exists as a series of pulses, rather than as a continuous waveform. The amplitude of the pulse train is modified in accordance with the applied modulating signal to convey intelligence, as illustrated in Figure 2.27. There are two primary forms of PAM sampling:

- *Natural sampling* (or *top sampling*), where the modulated pulses follow the amplitude variation of the sampled time function during the sampling interval.

- *Instantaneous sampling* (or *square-topped sampling*), where the amplitude of the pulses is determined by the instantaneous value of the sampled time function corresponding to a single instant of the sampling interval. This "single instant" may be the center or edge of the sampling interval.

There are two common methods of generating a PAM signal:

- Variation of the amplitude of a pulse sequence about a fixed nonzero value (or *pedestal*). This approach constitutes double-sideband amplitude modulation.

- Double-polarity modulated pulses with no pedestal. This approach constitutes double-sideband suppressed carrier modulation.

2.4.3 Pulse Time Modulation (PTM)

A number of modulating schemes have been developed to take advantage of the noise immunity afforded by a constant amplitude modulating system. *Pulse time modulation* (PTM) is one of those systems. In a PTM system, instantaneous samples of the intelligence are used to vary the time of occurrence of some parameter of the pulsed carrier. Subsets of the PTM process include:

- *Pulse duration modulation* (PDM), where the time of occurrence of either the leading or trailing edge of each pulse (or both pulses) is varied from its unmodulated position by samples of the input modulating waveform. PDM also may be described as *pulse length* or *pulse width* modulation (PWM).

- *Pulse position modulation* (PPM), where samples of the modulating input signal are used to vary the position in time of pulses, relative to the unmodulated waveform. Several types of pulse time modulation waveforms are shown in Figure 2.28.

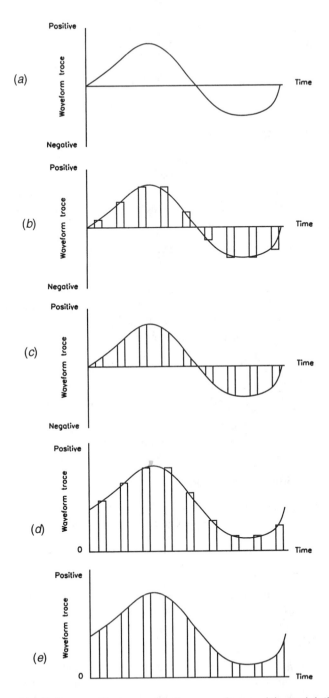

Figure 2.27 Pulse amplitude modulation waveforms: (*a*) modulating signal; (*b*) square-topped sampling, bipolar pulse train; (*c*) topped sampling, bipolar pulse train; (*d*) square-topped sampling, unipolar pulse train; (*e*) top sampling, unipolar pulse train.

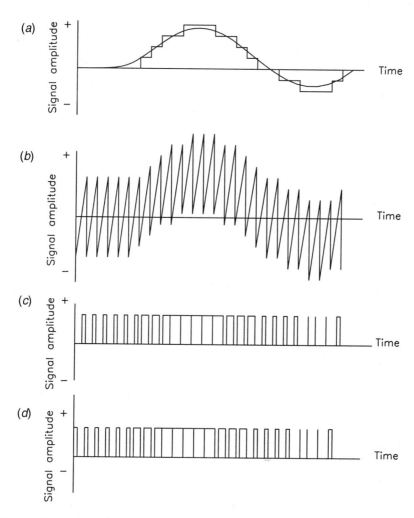

Figure 2.28 Pulse time modulation waveforms: (*a*) modulating signal and sample-and-hold (S/H) waveforms, (*b*) sawtooth waveform added to S/H, (*c*) leading-edge PTM, (*d*) trailing-edge PTM.

- *Pulse frequency modulation* (PFM), where samples of the input signal are used to modulate the frequency of a series of carrier pulses. The PFM process is illustrated in Figure 2.29.

It should be emphasized that all of the pulse modulation systems discussed thus far may be used with both analog and digital input signals. Conversion is required for either signal into a form that can be accepted by the pulse modulator.

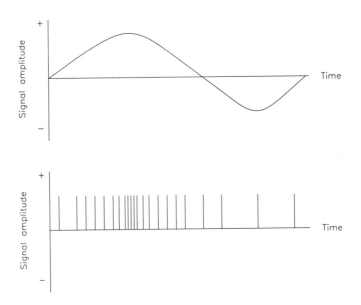

Figure 2.29 Pulse frequency modulation.

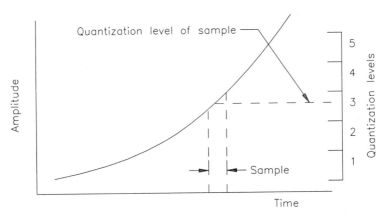

Figure 2.30 The quantization process.

2.4.4 Pulse Code Modulation

The pulse modulation systems discussed previously are *unencoded* systems. *Pulse code modulation* (PCM) is a scheme wherein the input signal is *quantized* into discrete steps and then sampled at regular intervals (as in conventional pulse modulation). In the *quantization* process, the input signal is sampled to produce a code representing the instantaneous value of the input within a predetermined range of values. Figure 2.30 illustrates the concept. Only certain discrete levels are allowed in the quantization process. The code is then transmitted over the communications system as a pattern of pulses.

Quantization inherently introduces an initial error in the amplitude of the samples taken. This *quantization error* is reduced as the number of quantization steps is increased. In system design, tradeoffs must be made regarding low quantization error, hardware complexity, and occupied bandwidth. The greater the number of quantization steps, the wider the bandwidth required to transmit the intelligence or, in the case of some signal sources, the slower the intelligence must be transmitted.

In the classic design of a PCM encoder, the quantization steps are equal. The quantization error (or *quantization noise*) usually can be reduced, however, through the use of nonuniform spacing of levels. Smaller quantization steps are provided for weaker signals, and larger steps are provided near the peak of large signals. Quantization noise is reduced by providing an encoder that is matched to the *level distribution* (*probability density*) of the input signal.

Nonuniform quantization typically is realized in an encoder through processing of the input (analog) signal to compress it to match the desired nonuniformity. After compression, the signal is fed to a uniform quantization stage.

2.4.5 Delta Modulation

Delta modulation (DM) is a coding system that measures changes in the direction of the input waveform, rather than the instantaneous value of the wave itself. Figure 2.31 illustrates the concept. The clock rate is assumed to be constant. Transmitted pulses from the pulse generator are positive if the signal is changing in a positive direction; they are negative if the signal is changing in a negative direction.

As with the PCM encoding system, quantization noise is a parameter of concern for DM. Quantization noise can be reduced by increasing the sampling frequency (the pulse generator frequency). The DM system has no fixed maximum (or minimum) signal amplitude. The limiting factor is the slope of the sampled signal, which must not change by more than one level or step during each pulse interval.

2.4.6 Digital Coding Systems

A number of methods exist to transmit digital signals over long distances in analog transmission channels. Some of the more common systems include:

- *Binary on-off keying* (BOOK), a method by which a high-frequency sinusoidal signal is switched on and off corresponding to 1 and 0 (on and off) periods in the input digital data stream. In practice, the transmitted sinusoidal waveform does not start or stop abruptly, but follows a predefined ramp up or down.

- *Binary frequency-shift keying* (BFSK), a modulation method in which a continuous wave is transmitted that is shifted between two frequencies, representing 1s and 0s in the input data stream. The BFSK signal may be generated by switching between two oscillators (set to different operating frequencies) or by applying a binary baseband signal to the input of a voltage-controlled oscillator (VCO). The

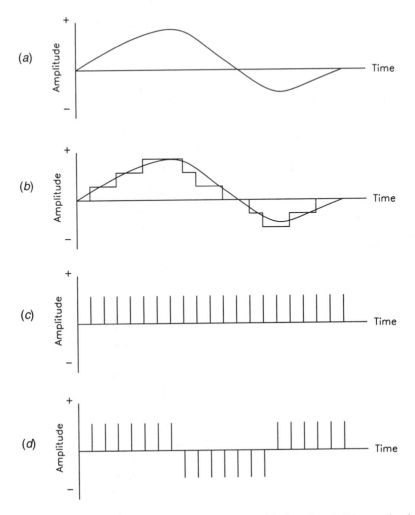

Figure 2.31 Delta modulation waveforms: (*a*) modulating signal, (*b*) quantized modulating signal, (*c*) pulse train, (*d*) resulting delta modulation waveform.

transmitted signals often are referred to as a *mark* (binary digit 1) or a *space* (binary digit 0). Figure 2.32 illustrates the transmitted waveform of a BFSK system.

- *Binary phase-shift keying* (BPSK), a modulating method in which the phase of the transmitted wave is shifted 180° in synchronism with the input digital signal. The phase of the RF carrier is shifted by $+\frac{\pi}{2}$ radians or $-\frac{\pi}{2}$ radians, depending upon whether the data bit is a 0 or a 1. Figure 2.33 shows the BPSK transmitted waveform.

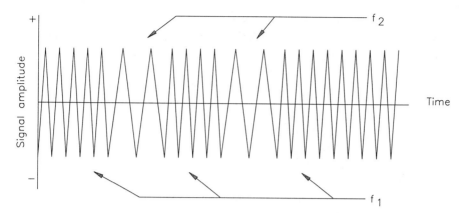

Figure 2.32 Binary FSK waveform.

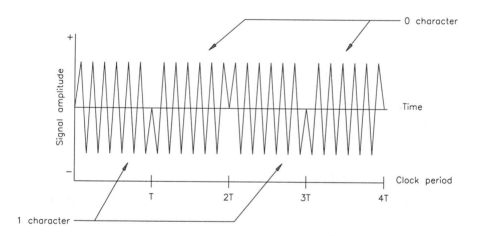

Figure 2.33 Binary PSK waveform.

- *Quadriphase-shift keying* (QPSK), a modulation scheme similar to BPSK except that quaternary modulation is employed, rather than binary modulation. QPSK requires half the bandwidth of BPSK for the same transmitted data rate.

2.4.7 Baseband Digital Pulse Modulation

After the input samples have been quantized, they are transmitted through a channel, received, and converted back to their approximate original form [8].The format (modulation scheme) applied to the quantized samples is determined by a number of factors, not the least of which is the channel through which the signal passes. A number of different formats are possible and practical.

Several common digital modulation formats are shown in Figure 2.34. The first (*a*) is referred to as *non-return-to-zero* (NRZ) polar because the waveform does not return to zero during each signaling interval, but switches from $+V$ to $-V$, or vice versa, at the end of each signaling interval (NRZ unipolar uses the levels V and 0). On the other hand, the *unipolar return-to-zero* (RZ) format, shown in (*b*) returns to zero in each signaling interval. Because bandwidth is inversely proportional to pulse duration, it is apparent that RZ requires twice the bandwidth that NRZ does. Also, RZ has a nonzero dc component, whereas NRZ does not necessarily have a nonzero component (unless there are more 1s than 0s or vice versa). An advantage of RZ over NRZ is that a pulse transition is guaranteed in each signaling interval, whereas this is not the case for NRZ. Thus, in cases where there are long strings of 1s or 0s, it may be difficult to synchronize the receiver to the start and stop times of each pulse in NRZ-based systems. A very important modulation format from the standpoint of synchronization considerations is NRZ-*mark*, also known as *differential encoding*, where an initial reference bit is chosen and a subsequent 1 is encoded as a change from the reference and a 0 is encoded as no change. After the initial reference bit, the current bit serves as a reference for the next bit, and so on. An example of this modulation format is shown in (*c*).

Manchester is another baseband data modulation format that guarantees a transition in each signaling interval and does not have a dc component. Also known as *biphase* or *split phase*, this scheme is illustrated in (*d*). The format is produced by *OR*ing the data clock with an NRZ-formatted signal. The result is a + to − transition for a logic 1, and a − to + zero crossing for a logic 0.

A number of other data formats have been proposed and employed in the past, but further discussion is beyond the scope of this chapter.

2.4.8   Spread Spectrum Systems

As the name implies, a *spread spectrum* system requires a frequency range substantially greater than the basic information-bearing signal. Spread spectrum systems have some or all of the following properties:

- Low interference to other communications systems

- Ability to reject high levels of external interference

- Immunity to jamming by hostile forces

- Provision for secure communications paths

- Operability over multiple RF paths

Spread spectrum systems operate with an entirely different set of requirements than transmission systems discussed previously. Conventional modulation methods are designed to provide for the easiest possible reception and demodulation of the transmitted intelligence. The goals of spread spectrum systems, on the other hand, are secure and reliable communications that cannot be intercepted by unauthorized persons. The most common modulating and encoding techniques used in spread spectrum communications include:

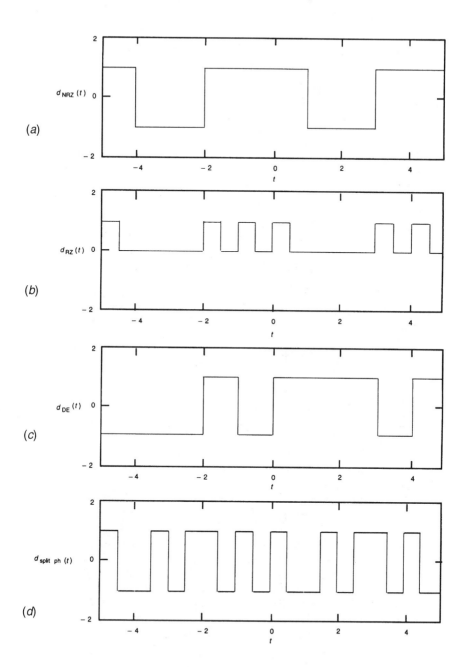

Figure 2.34 Various baseband modulation formats: (*a*) non-return-to zero, (*b*) unipolar return-to-zero, (*c*) differential encoded (NRZ-mark), (*d*) split phase. (*From* [8]. *Used with permission.*)

- *Frequency hopping*, where a random or *pseudorandom* number (PN) sequence is used to change the carrier frequency of the transmitter. This approach has two basic variations: *slow frequency hopping*, where the hopping rate is smaller than the data rate, and *fast frequency hopping*, where the hopping rate is larger than the data rate. In a fast frequency-hopping system, the transmission of a single piece of data occupies more than one frequency. Frequency-hopping systems permit multiple-access capability to a given band of frequencies because each transmitted signal occupies only a fraction of the total transmitted bandwidth.

- *Time hopping*, where a PN sequence is used to switch the position of a message-carrying pulse within a series of frames.

- *Message corruption*, where a PN sequence is added to the message before modulation.

- *Chirp spread spectrum*, where linear frequency modulation of the main carrier is used to spread the transmitted spectrum. This technique is commonly used in radar and also has been applied to communications systems.

In a spread spectrum system, the signal power is divided over a large bandwidth. The signal, therefore, has a small average power in any single narrowband slot. This means that a spread spectrum system can share a given frequency band with one or more narrowband systems. Furthermore, because of the low energy in any particular band, detection or interception of the transmission is difficult.

2.5   References

1. Tate, Jeffrey P., and Patricia F. Mead, "Crystal Oscillators," in *The Electronics Handbook*, Jerry C. Whitaker (ed.), CRC Press, Boca Raton, FL, pp. 185–199, 1996.
2. Frerking, M. E., *Crystal Oscillator Design and Temperature Compensation*, Van Nostrand Reinhold, New York, 1978.
3. Dye, D. W., *Proc. Phys. Soc.*, Vol. 38, pp. 399–457, 1926.
4. Douglass, Barry G., "Thermal Noise and Other Circuit Noise," in *The Electronics Handbook*, Jerry C. Whitaker (ed.), pp. 30–36, 1996.
5. Terman, F. E., *Radio Engineering*, 3rd ed., McGraw-Hill, New York, pg. 468, 1947.
6. Kubichek, Robert, "Amplitude Modulation," in *The Electronics Handbook*, Jerry C. Whitaker (ed.), CRC Press, Boca Raton, FL, pp. 1175–1187, 1996.
7. Seymour, Ken, "Frequency Modulation," in *The Electronics Handbook*, Jerry C. Whitaker (ed.), CRC Press, Boca Raton, FL, pp. 1188–1200, 1996.
8. Ziemer, Rodger E., "Pulse Modulation," in *The Electronics Handbook*, Jerry C. Whitaker (ed.), CRC Press, Boca Raton, FL, pp. 1201–1212, 1996.

2.6   Bibliography

Benson, K. B., and Jerry. C. Whitaker (eds.), *Television Engineering Handbook*, McGraw-Hill, New York, 1986.

Benson, K. B., and Jerry. C. Whitaker, *Television and Audio Handbook for Technicians and Engineers*, McGraw-Hill, New York, 1989.

Crutchfield, E. B. (ed.), *NAB Engineering Handbook*, 8th Ed., National Association of Broadcasters, Washington, DC, 1991.

Fink, D., and D. Christiansen (eds.), *Electronics Engineers' Handbook*, 3rd Ed., McGraw-Hill, New York, 1989.

Fink, D., and D. Christiansen (eds.), *Electronics Engineers' Handbook*, 2nd Ed., McGraw-Hill, New York, 1982.

Hulick, Timothy P., "Using Tetrodes for High Power UHF," *Proceedings of the Society of Broadcast Engineers*, Vol. 4, SBE, Indianapolis, IN, pp. 52-57, 1989.

Jordan, Edward C., *Reference Data for Engineers: Radio, Electronics, Computer and Communications*, 7th Ed., Howard W. Sams, Indianapolis, IN, 1985.

Laboratory Staff, *The Care and Feeding of Power Grid Tubes*, Varian Eimac, San Carlos, CA, 1984.

Mendenhall, G. N., "Fine Tuning FM Final Stages," *Broadcast Engineering*, Intertec Publishing, Overland Park, KS, May 1987.

Whitaker, Jerry. C., *Maintaining Electronic Systems*, CRC Press, Boca Raton, FL, 1992.

Whitaker, Jerry. C., *Radio Frequency Transmission Systems: Design and Operation*, McGraw-Hill, New York, 1991.

3

Vacuum Tube Principles

3.1 Introduction

A power grid tube is a device using the flow of free electrons in a vacuum to produce useful work [1]. It has an emitting surface (the cathode), one or more grids that control the flow of electrons, and an element that collects the electrons (the anode). Power tubes can be separated into groups according to the number of electrodes (grids) they contain. The physical shape and location of the grids relative to the plate and cathode are the main factors that determine the *amplification factor* (μ) and other parameters of the device (see Section 3.2.5). The physical size and types of material used to construct the individual elements determine the power capability of the tube. A wide variety of tube designs are available to commercial and industrial users. By far the most common are triodes and tetrodes.

3.2 Characteristics of Electrons

Electrons are minute, negatively charged particles that are constituents of all matter. They have a mass of 9×10^{-28} g ($\frac{1}{1840}$ that of a hydrogen atom) and a charge of 1.59×10^{-19} coulomb. Electrons are always identical, irrespective of their source. Atoms are composed of one or more such electrons associated with a much heavier nucleus, which has a positive charge equal to the number of the negatively charged electrons contained in the atom; an atom with a full quota of electrons is electrically neutral. The differences in chemical elements arise from differences in the nucleus and in the number of associated electrons.

Free electrons can be produced in a number of ways. *Thermonic emission* is the method normally employed in vacuum tubes (discussed in Section 3.2.2). The principle of thermonic emission states that if a solid body is heated sufficiently, some of the electrons that it contains will escape from the surface into the surrounding space. Electrons also are ejected from solid materials as a result of the impact of rapidly moving electrons or ions. This phenomenon is referred to as *secondary electron emission*, because it is necessary to have a primary source of electrons (or ions) before the secondary emission can be obtained (Section 3.2.3). Finally, it is possible to pull electrons directly out of solid substances by an intense electrostatic field at the surface of the material.

Positive ions represent atoms or molecules that have lost one or more electrons and so have become charged bodies having the weight of the atom or molecule concerned, and a positive charge equal to the negative charge of the lost electrons. Unlike electrons, positive ions are not all alike and may differ in charge or weight, or both. They are much heavier than electrons and resemble the molecule or atom from which they are derived. Ions are designated according to their origin, such as mercury ions or hydrogen ions.

3.2.1 Electron Optics

Electrons and ions are charged particles and, as such, have forces exerted upon them by an electrostatic field in the same way as other charged bodies. Electrons, being negatively charged, tend to travel toward the positive or anode electrode, while the positively charged ions travel in the opposite direction (toward the negative or cathode electrode). The force F exerted upon a charged particle by an electrostatic field is proportional to the product of the charge e of the particle and the voltage gradient G of the electrostatic field [1]:

$$F = G \times e \times 10^7 \qquad (3.1)$$

Where:
F = force in dynes
G = voltage gradient in volts per centimeter
e = charge in coulombs

This force upon the ion or electron is exerted in the direction of the electrostatic flux lines at the point where the charge is located. The force acts toward or away from the positive terminal, depending upon whether a negative or positive charge, respectively, is involved.

The force that the field exerts on the charged particle causes an acceleration in the direction of the field at a rate that can be calculated by the laws of mechanics where the velocity does not approach that of light:

$$A = \frac{F}{M} \qquad (3.2)$$

Where:
A = acceleration in centimeters per second per second
F = force in dynes
m = mass in grams

The velocity an electron or ion acquires in being acted upon by an electrostatic field can be expressed in terms of the voltage through which the electron (or ion) has fallen in acquiring the velocity. For velocities well below the speed of light, the relationship between velocity and the acceleration voltage is:

$$v = \sqrt{\frac{2 \times V \times e \times 10^7}{m}} \qquad\qquad (3.3)$$

Where:
v = velocity in centimeters per second corresponding to V
V = accelerating voltage
e = charge in coulombs
m = mass in grams

Electrons and ions move at great velocities in even moderate-strength fields. For example, an electron dropping through a potential difference of 2500 V will achieve a velocity of approximately one-tenth the speed of light.

Electron optics, as discussed in this section, relies on the principles of classical physics. While modern tube design uses computer simulation almost exclusively, the preceding information is still valid and provides a basis for the understanding of electron motion within a vacuum tube device.

Magnetic Field Effects

An electron in motion represents an electric current of magnitude ev, where e is the magnitude of the charge on the electron and v is its velocity. A magnetic field accordingly exerts a force on a moving electron exactly as it exerts a force on an electric current in a wire. The magnitude of the force is proportional to the product of the equivalent current ev represented by the moving electron and the strength of the component of the magnetic field in a direction at right angles to the motion of the electron. The resulting force is, then, in a direction at right angles both to the direction of motion of the electron and to the component of the magnetic field that is producing the force. As a result, an electron entering a magnetic field with a high velocity will follow a curved path. Because the acceleration of the electron that the force of the magnetic field produces is always at right angles to the direction in which the electron is traveling, an electron moving in a uniform magnetic field will follow a circular path. The radius of this circle is determined by the strength of the magnetic field and the speed of the electron moving through the field.

When an electron is subjected to the simultaneous action of both electric and magnetic fields, the resulting force acting on the electron is the vector sum of the force resulting from the electric field and the force resulting from the magnetic field, each considered separately.

Magnetic fields are not used for conventional power grid tubes. Microwave power tubes, on the other hand, use magnetic fields to confine and focus the electron stream.

3.2.2 Thermal Emission From Metals

Thermonic emission is the phenomenon of an electric current leaving the surface of a material as the result of thermal activation. Electrons with sufficient thermal energy to overcome the surface-potential barrier escape from the surface of the material. This thermally emitted electron current increases with temperature because more electrons have sufficient energy to leave the material.

The number of electrons released per unit area of an emitting surface is related to the absolute temperature of the emitting material and a quantity b that is a measure of the work an electron must perform in escaping through the surface, according to the equation [1]:

$$I = A T^2 \varepsilon^{-b/T} \tag{3.4}$$

Where:
T = absolute temperature of the emitting material
b = the work an electron must perform in escaping the emitter surface
I = electron current in amperes per square centimeter
A = a constant (value varies with type of emitter)

The exponential term in the equation accounts for most of the variation in emission with temperature. The temperature at which the electron current becomes appreciable is accordingly determined almost solely by the quantity b. Figure 3.1 plots the emission resulting from a cathode operated at various temperatures.

Thermal electron emission can be increased by applying an electric field to the cathode. This field lowers the surface-potential barrier, enabling more electrons to escape. This field-assisted emission is known as the *Schottky effect*.

Figure 3.2 illustrates common heater-cathode structures for power tubes.

3.2.3 Secondary Emission

Almost all metals and some insulators will emit low-energy electrons (secondary electrons) when bombarded by other energetic electrons. The number of secondary electrons emitted per primary electron is determined by the velocity of the primary bombarding electrons and the nature and condition of the material composing the surface being bombarded. Figure 3.3 illustrates a typical relationship for two types of surfaces. As shown in the figure, no secondary electrons are produced when the primary velocity is low. However, with increasing potential (and consequently higher velocity), the ratio of secondary to primary electrons increases, reaching a maximum and then decreasing. With pure metal surfaces, the maximum ratio of secondary to primary electrons ranges from less than 1 to approximately 3. Some complex surfaces based on alkali metal compounds yield ratios of secondary to primary electrons as high as 5 to 10.

The majority of secondary electrons emitted from a conductive surface have relatively low velocity. However, a few secondary electrons usually are emitted with a velocity nearly equal to the velocity of the bombarding primary electrons.

For insulators, the ratio of secondary to primary electrons as a function of primary electron potential follows along the same lines as for metals. The net potential of the insulating surface being bombarded is affected by the bombardment. If the ratio of secondary to primary current is less than unity, the insulator acquires a net negative charge

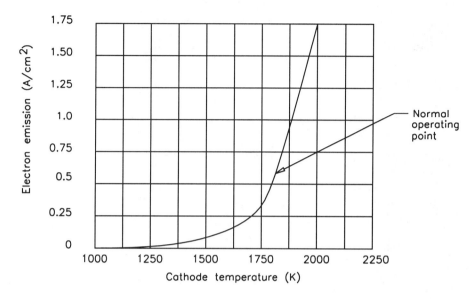

Figure 3.1 Variation of electron emission as a function of absolute temperature for a thoriated-tungsten emitter.

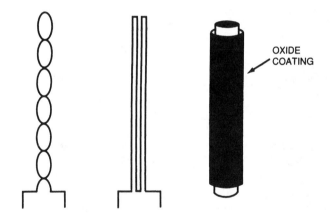

Figure 3.2 Common types of heater and cathode structures. (*From* [2]. *Used with permission.*)

because more electrons arrive than depart. This causes the insulator to become more negative and, finally, to repel most of the primary electrons, resulting in a blocking action. In the opposite extreme, when the ratio of secondary to primary electrons exceeds unity, the insulating surface loses electrons through secondary emission faster than

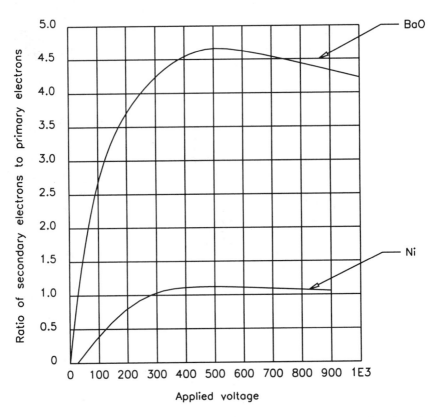

Figure 3.3 Ratio of secondary emission current to primary current as a function of primary electron velocity.

they arrive; the surface becomes increasingly positive. This action continues until the surface is sufficiently positive that the ratio of secondary to primary electrons decreases to unity as a result of the increase in the velocity of the bombarding electrons, or until the surface is sufficiently positive that it attracts back into itself a significant number of secondary electrons. This process makes the number of electrons gained from all sources equal to the number of secondary electrons emitted.

3.2.4 Diode

A diode is a two-electrode vacuum tube containing a cathode, which emits electrons by thermonic emission, surrounded by an anode (or plate). (See Figure 3.4) Such a tube is inherently a rectifier because when the anode is positive, it attracts electrons; current, therefore, passes through the tube. When the anode is negative, it repels the electrons and no current flows.

The typical relationship between anode voltage and current flowing to the positive anode is shown in Figure 3.5. When the anode voltage is sufficiently high, electrons are

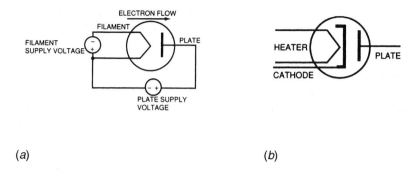

Figure 3.4 Vacuum diode: (*a*) directly heated cathode, (*b*) indirectly heated cathode. (*From* [2]. *Used with permission.*)

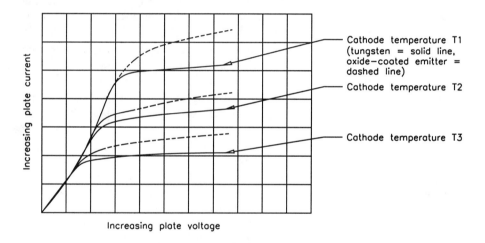

Figure 3.5 Anode current as a function of anode voltage in a two-electrode tube for three cathode temperatures.

drawn from the cathode as rapidly as they are emitted. The anode current is then limited by the electron emission of the cathode and, therefore, depends upon cathode temperature rather than anode voltage.

At low anode voltages, however, plate current is less than the emission of which the cathode is capable. This occurs because the number of electrons in transit between the cathode and plate at any instant cannot exceed the number that will produce a negative *space charge*, which completely neutralizes the attraction of the positive plate upon the electrons just leaving the cathode. All electrons in excess of the number necessary to neutralize the effects of the plate voltage are repelled into the cathode by the negative space charge of the electrons in transit; this situation applies irrespective of how many excess electrons the cathode emits. When the plate current is limited in this way by space charge, plate current is determined by plate potential and is substantially independent of the electron emission of the cathode.

Detailed examination of the space-charge situation will reveal that the negative charge of the electrons in transit between the cathode and the plate is sufficient to give the space in the immediate vicinity of the cathode a slight negative potential with respect to the cathode. The electrons emitted from the cathode are projected out into this field with varying emission velocities. The negative field next to the cathode causes the emitted electrons to slow as they move away from the cathode, and those having a low velocity of emission are driven back into the cathode. Only those electrons having the highest velocities of emission will penetrate the negative field near the cathode and reach the region where they are drawn toward the positive plate. The remainder (those electrons having low emission velocities) will be brought to a stop by the negative field adjacent to the cathode and will fall back into the cathode.

The energy that is delivered to the tube by the source of anode voltage is first expended in accelerating the electrons traveling from the cathode to the anode; it is converted into kinetic energy. When these swiftly moving electrons strike the anode, this kinetic energy is then transformed into heat as a result of the impact and appears at the anode in the form of heat that must be radiated to the walls of the tube.

The basic function of a vacuum tube diode—to rectify an ac voltage—has been superseded by solid-state devices. An understanding of how the diode operates, however, is important in understanding the operation of triodes, tetrodes, and pentodes.

3.2.5 Triode

The power triode is a three-element device commonly used in a wide variety of RF generators. Triodes have three internal elements: the cathode, control grid, and plate. Most tubes are cylindrically symmetrical. The filament or cathode structure, the grid, and the anode all are cylindrical in shape and are mounted with the axis of each cylinder along the center line of the tube, as illustrated in Figure 3.6.

The grid normally is operated at a negative potential with respect to the cathode, and so attracts no electrons. However, the extent to which it is negative affects the electrostatic field in the vicinity of the cathode and, therefore, controls the number of electrons that pass between the grid and the plate. The grid, in effect, functions as an imperfect electrostatic shield. It allows some, but not all, of the electrostatic flux from the anode to leak between its wires. The number of electrons that reach the anode in a triode tube under space-charge-limited conditions is determined almost solely by the electrostatic field near the cathode; the field in the rest of the interelectrode space has little effect. This phenomenon results because the electrons near the cathode are moving slowly compared with the electrons that have traveled some distance toward the plate. The result of this condition is that the volume density of electrons in proportion to the rate of flow is large near the cathode and low in the remainder of the interelectrode space. The total space charge of the electrons in transit toward the plate, therefore, consists almost

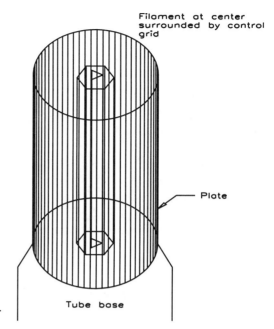

Filoment ot center
surrounded by control
grid

Plate

Tube bose

Figure 3.6 Mechanical configuration of a power triode.

solely of the electrons in the immediate vicinity of the cathode. After an electron has traveled beyond this region, it reaches the plate so quickly as to contribute to the space charge for only a brief additional time interval. The result is that the space current in a three-electrode vacuum tube is, for all practical purposes, determined by the electrostatic field that the combined action of the grid and plate potentials produces near the cathode.

When the grid structure is symmetrical, the field E at the surface of the cathode is proportional to the quantity:

$$\frac{E_c + E_b}{\mu} \tag{3.5}$$

Where:
E_c = control grid voltage (with respect to cathode)
E_b = anode voltage (with respect to cathode)
μ = a constant determined by the geometry of the tube

The constant μ, the amplification factor, is independent of the grid and plate voltages. It is a measure of the relative effectiveness of grid and plate voltages in producing electrostatic fields at the surfaces of the cathode. Placement of the control grid relative to the cathode and plate determines the amplification factor. The μ values of triodes generally range from 5 to 200. Key mathematical relationships include the following:

$$\mu = \frac{\Delta E_b}{\Delta E_{c1}}$$ (3.6)

$$R_p = \frac{\Delta E_b}{\Delta I_b}$$ (3.7)

$$S_m = \frac{\Delta I_b}{\Delta E_{c1}}$$ (3.8)

Where:
μ = amplification factor (with plate current held constant)
R_p = dynamic plate resistance
S_m = transconductance (also may be denoted G_m)
E_b = total instantaneous plate voltage
E_{c1} = total instantaneous control grid voltage
I_b = total instantaneous plate current

The total cathode current of an ideal triode can be determined from the equation:

$$I_k = \left\{ E_c + \frac{E_b}{\mu} \right\}^{\frac{3}{2}}$$ (3.9)

Where:
I_k = cathode current
K = a constant determined by tube dimensions
E_c = grid voltage
E_b = plate voltage
μ = amplification factor

Figure 3.7 plots plate and grid current as a function of plate voltage at various grid voltages for a triode with a μ of 12. The tube, a 304TL, is a classic design and, while not used in new equipment, provides a common example of the relationship between the parameters plotted. Figure 3.8 plots the same parameters for a tube with a μ of 20. Observe how much more plate current at a given plate voltage can be obtained from the 304TL ($\mu = 12$) without driving the grid into the positive grid region. Note also how much more bias voltage is required for the 304TL to cut the plate current off at some given plate voltage. With this increased bias, there is a corresponding increase in grid voltage swing to drive up the zero grid voltage point on the curve. Low-μ tubes have lower voltage gain by definition. This fact can be seen by comparing Figures 3.7 and 3.8.

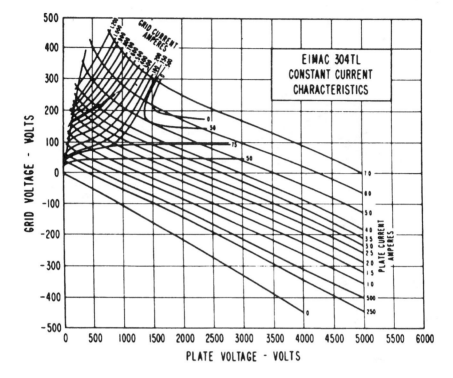

Figure 3.7 Constant-current characteristics for a triode with a μ of 12.

Triodes with a μ of 20 to 50 generally are used in conventional RF amplifiers and oscillators. High-μ triodes (200 or more) may be designed so that the operating bias is zero, as depicted in Figure 3.9. These *zero-bias triodes* are available with plate dissipation ratings of 400 W to 10 kW or more. The zero-bias triode commonly is used in grounded-grid amplification. The tube offers good power gain and circuit simplicity. No bias power source is required. Furthermore, no protection circuits for loss of bias or drive are needed. Despite these attributes, present-day use of the zero-bias triode is limited.

Low- and medium-μ devices usually are preferred for induction heating applications because of the wide variations in load that an induction or dielectric heating oscillator normally works into. Such tubes exhibit lower grid-current variation with a changing load. The grid current of a triode with a μ of 20 will rise substantially less under a light- or no-load condition than a triode with a μ of 40. High-μ triode oscillators can be designed for heating applications, but extra considerations must be given to current rise under no-load conditions.

Vacuum tubes specifically designed for induction heating are available, intended for operation under adverse conditions. The grid structure is ruggedized with ample dissi-

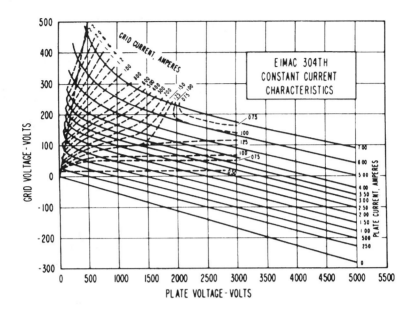

Figure 3.8 Constant-current characteristics for a triode with a μ of 20.

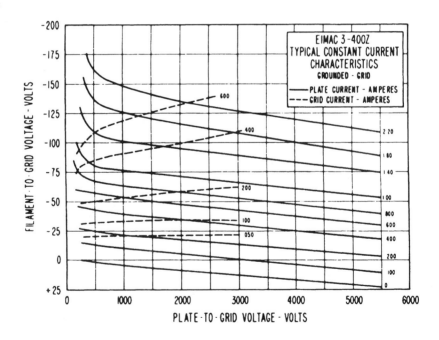

Figure 3.9 Grounded-grid constant-current characteristics for a zero-bias triode with a μ of 200.

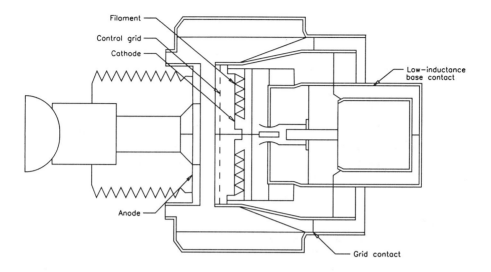

Figure 3.10 Internal configuration of a planar triode.

pation capability to deal with wide variations in load. As the load decreases, grid dissipation increases.

Triodes also are manufactured with the cathode, grid, and anode in the shape of a flat surface, as shown in Figure 3.10. Tubes so constructed are called *planar triodes*. This construction technique permits operation at high frequencies. The close spacing reduces electron *transit time*, allowing the tube to be used at high frequencies (up to 3 GHz or so). The physical construction of planar triodes results in short lead lengths, which reduces lead inductance. Planar triodes are used in both *continuous wave* (CW) and pulsed modes. The contacting surfaces of the planar triode are arranged for easy integration into coaxial and waveguide resonators.

3.2.6 Tetrode

The tetrode is a four-element tube with two grids. The control grid serves the same purpose as the grid in a triode, while a second (*screen*) grid with the same number of vertical elements (bars) as the control grid is mounted between the control grid and the anode. The grid bars of the screen grid are mounted directly behind the control-grid bars, as observed from the cathode surface, and serve as a shield or screen between the input circuit and the output circuit of the tetrode. The principal advantages of a tetrode over a triode include:

• Lower internal plate-to-grid feedback.

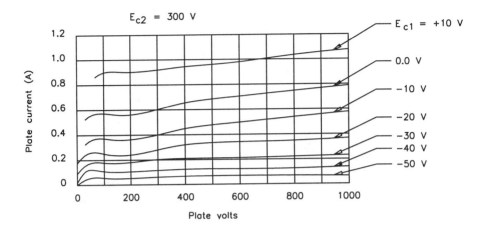

Figure 3.11 Tetrode plate current characteristics. Plate current is plotted as a function of plate voltage, with grid voltages as shown.

- Lower drive power requirements. In most cases, the driving circuit need supply only 1 percent of the output power.
- More efficient operation. Tetrodes allow the design of compact, simple, flexible equipment with little spurious radiation and low intermodulation distortion.

Plate current is almost independent of plate voltage in a tetrode. Figure 3.11 plots plate current as a function of plate voltage at a fixed screen voltage and various grid voltages. In an ideal tetrode, a change in plate current does not cause a change in plate voltage. The tetrode, therefore, can be considered a *constant-current device*. The voltages on the screen and control grids determine the amount of plate current.

The total cathode current of an ideal tetrode is determined by the equation:

$$I_k = K \left\{ E_{c1} + \frac{E_{c2}}{\mu_s} + \frac{E_b}{\mu_p} \right\}^{\frac{3}{2}} \tag{3.10}$$

Where:
I_k = cathode current
K = a constant determined by tube dimensions
E_{c1} = control grid voltage
E_{c2} = screen grid voltage
μ_s = screen amplification factor
μ_p = plate amplification factor
E_b = plate voltage

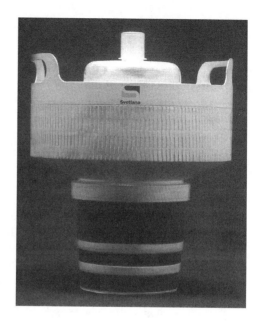

Figure 3.12 Radial beam power tetrode (4CX15000A).

The arithmetic value of the screen μ generally is not used in the design of an RF amplifier. In most tetrode applications, the screen amplification factor is useful to categorize roughly the performance to be expected.

Application Example

Figure 3.12 shows a radial beam power tetrode (4CX15000A) designed for class AB_1 or class C power amplification. The device is particularly well suited for RF linear power amplifier service. The tube has a directly heated thoriated-tungsten mesh filament for mechanical ruggedness and high efficiency. The maximum rated plate dissipation of the tube is 15 kW using air cooling.

The tube must be protected from damage that may result from an internal arc occurring at high plate voltage. A protective resistance typically is inserted in series with the tube anode to help absorb stored power supply energy in case an internal arc occurs.

The maximum control grid dissipation is 200 W, determined (approximately) by the product of the dc grid current and the peak positive grid voltage.

Screen grid maximum dissipation is 450 W. With no ac applied to the screen grid, dissipation is the product of dc screen voltage and dc screen current. Plate voltage, plate loading, and/or bias voltage must never be removed while filament and screen voltages are present.

The 4CX15000A must be mounted vertically, base up or down. The tube requires forced-air cooling in all applications. The tube socket is mounted in a pressurized compartment where cooling air passes through the socket and is guided to the anode cooling fins by an air chimney. Adequate movement of cooling air around the base of the tube

Table 3.1 Minimum Cooling Airflow Requirements for the 4CX15000A Power Tetrode at Sea Level

Plate Dissipation (W)	Airflow (CFM)	Pressure Drop (inches of water)
7500	230	0.7
12,500	490	2.7
15,000	645	4.6

keeps the base and socket contact fingers at a safe operating temperature. Although the maximum temperature rating for seals and the anode is 250°C, good engineering practice dictates that a safety factor be provided. Table 3.1 lists cooling parameters for the tube with the cooling air at 50°C and a maximum anode temperature of 225°C. The figures given in the table apply to designs in which air passes in the base-to-anode direction. Pressure drop values shown are approximate and apply to the tube/socket/chimney combination.

At altitudes significantly above sea level, the flow rate must be increased for equivalent cooling. At 5000 ft above sea level, both the flow rate and the pressure drop is increased by a factor of 1.20; at 10,000 ft, both flow rate and pressure drop are increased by 1.46.

Anode and base cooling is applied before or simultaneously with filament voltage turn-on, and normally should continue for a brief period after shutdown to allow the tube to cool properly.

An outline of the principal tube dimensions is given in Figure 3.13. General specifications are listed in Table 3.2.

3.2.7 Pentode

The pentode is a five-electrode tube incorporating three grids. The control and screen grids perform the same function as in a tetrode. The third grid, the *suppressor grid*, is mounted in the region between the screen grid and the anode. The suppressor grid produces a *potential minimum*, which prevents secondary electrons from being interchanged between the screen and plate. The pentode's main advantages over the tetrode include:

- Reduced secondary emission effects.

- Good linearity.

- Ability to let plate voltage swing below the screen voltage without excessive screen dissipation. This allows slightly higher power output for a given operating plate voltage.

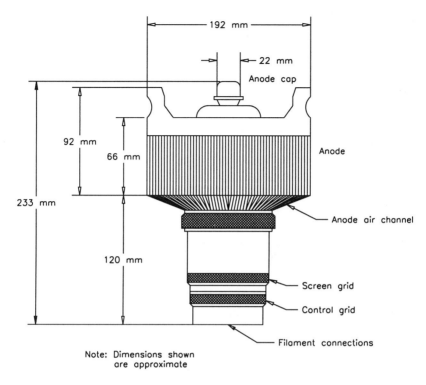

Figure 3.13 Principal dimensions of 4CX15000A tetrode.

Because of the design of the pentode, plate voltage has even less effect on plate current than in the tetrode. The same total space-current equation applies to the pentode as with the tetrode:

$$I_k = K \left\{ E_{c1} + \frac{E_{c2}}{\mu_s} + \frac{E_b}{\mu_p} \right\}^{3/2} \qquad (3.11)$$

Where:
I_k = cathode current
K = a constant determined by tube dimensions
E_{c1} = control grid voltage
E_{c2} = screen grid voltage
μ_s = screen amplification factor
μ_p = plate amplification factor
E_b = plate voltage

The suppressor grid may be operated negative or positive with respect to the cathode. It also may be operated at cathode potential. It is possible to control plate current

Table 3.2 General Characteristics of the 4CX15000A Power Tetrode (*Courtesy of Svetlana Electron Devices.*)

Electrical Characteristics		
Filament type		Thoriated-tungsten mesh
Filament voltage		6.3 ± 0.3 V
Filament current		164 A (at 6.3 V)
Amplification factor (average), grid to screen		4.5
Direct interelectrode capacitance (grounded cathode)	C_{in}	158 pF
	C_{out}	25.8 pF
	C_{pk}	1.3 pF
Direct interelectrode capacitance (grounded grid)	C_{in}	67 pF
	C_{out}	25.6 pF
	C_{gk}	0.21 pF
Maximum frequency for full ratings (CW)		110 MHz
Mechanical Characteristics:		
Length		238 mm (9.38 in)
Diameter		193 mm (7.58 in)
Net weight		5.8 kg (12.8 lb)
Operating position		Axis vertical, base up or down
Maximum operating temperature (seals/envelope)		250°C
Cooling method		Forced air
Base type		Coaxial
Radio Frequency Power Amplifier (class C FM) (absolute maximum ratings)		
DC plate voltage		10,000 V
DC screen voltage		2000 V
DC grid voltage		−750 V
DC plate current		5.0 A
Plate dissipation		15 kW
Screen dissipation		450 W
Grid dissipation		200 W
Typical Operation (frequencies up to 110 MHz)		
DC plate voltage	7.5 kV dc	10.0 kV dc
DC screen voltage	750 V dc	750 V dc
DC grid voltage	−510 V dc	−550 V dc
DC plate current	4.65 A dc	4.55 A dc
DC screen current	0.59 A dc	0.54 A dc
DC grid current	0.30 A dc	0.27 A dc
Peak RF grid voltage	730 V	790 V
Calculated driving power	220 W	220 W
Plate dissipation	8.1	9.0 kW
Plate output power	26.7	36.5 kW

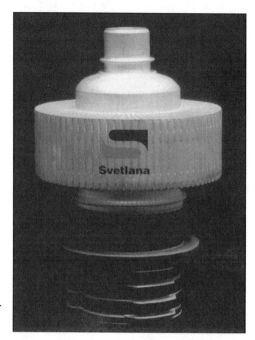

Figure 3.14 5CX1500B power pentode device.

by varying the potential on the suppressor grid. Because of this ability, a modulating voltage can be applied to the suppressor to achieve amplitude modulation. The required modulating power is low because of the low electron interception of the suppressor.

Application Example

The 5CX1500B is a ceramic/metal power pentode designed for use as a class AB_1 linear amplifier or FM VHF amplifying device. The exterior of the device is shown in Figure 3.14. Basic specifications for the tube are given in Table 3.3.

The filament of the 5CX1500B is rated for 5.0 V. Filament voltage, measured at the socket, is maintained to ±5 percent of the rated value for maximum tube life.

The rated dissipation of the control grid is 25 W. This value is approximately the product of dc grid current and peak positive grid voltage. Operation at bias and drive levels near those listed in Table 3.3 will ensure safe operation.

The power dissipated by the screen grid must not exceed 75 W. Screen dissipation, in cases where there is no ac applied to the screen, is the simple product of screen voltage and screen current. If the screen voltage is modulated, screen dissipation will depend on the rms screen current and voltage. Screen dissipation is likely to rise to excessive levels if the plate voltage, bias voltage, or plate load are removed with filament and screen voltages applied. Suitable protective means must be provided to limit screen dissipation to 75 W in the event of a circuit failure.

The rated dissipation of the suppressor grid is 25 W. Suppressor current is zero or nearly zero for all typical operating conditions specified in Table 3.3. The 5CX1500B

Table 3.3 General Characteristics of the 5CX1500B Power Pentode (*Courtesy of Svetlana Electron Devices.*)

Electrical Characteristics		
Filament type		Thoriated-tungsten mesh
Filament voltage		5.0 ± 0.25 V
Filament current		38.5 A (at 5.0 V)
Transconductance (avg.), I_b = 1.0 A dc, E_{c2} = 500 V dc		24,000 μmhos
Amplification factor (average), grid to screen		5.5
Direct interelectrode capacitance (grounded cathode)	Input	75 pF
	Output	17.8 pF
	Feedback	0.20 pF
Frequency (maximum) CW		110 MHz
Mechanical Characteristics		
Cooling method		Forced air
Base		Ring and breechblock
Recommended air system socket		SK-840 series
Recommended (air) chimney		SK-806
Operating position		Axis vertical, base down or up
Maximum operating temperature		250°C
Maximum dimensions	Length	130 mm (5.2 in)
	Diameter	85.6 mm (3.37 in)
Net weight		850 gm (30 oz)
Radio Frequency Linear Amplifier (class C, CW conditions) Maximum Ratings		
Plate voltage		5000 V
Screen voltage		750 V
Plate dissipation		1500 W
Suppressor dissipation		25 W
Screen dissipation		75 W
Grid dissipation		25 W
Typical Operation (frequencies to 30 MHz)		
Plate voltage	3000 V dc	4500 V dc
Suppressor voltage	0	0
Screen voltage	500 V dc	500 V dc
Grid voltage	–200 V dc	–200 V dc
Plate current	900 mA dc	900 mA dc
Screen current	94 mA dc	88 mA dc
Grid current	35 mA dc	34 mA dc
Peak RF grid voltage	255 V	255 V
Calculated driving power	9.0 W	9.0 W
Plate input power	2700 W	4050 W
Plate dissipation	720 W	870 W
Plate output power	1980 W	3180 W

Table 3.4 Minimum Cooling Airflow Requirements for the 5CX1500B Power Tetrode

Plate Dissipation (W)		Airflow (CFM)	Pressure Drop (inches of water)
At sea level	1000	27	0.33
	1550	47	0.76
At 6000 ft	1000	33	0.40
	1550	58	0.95

has been designed for zero voltage operation of the suppressor grid for most applications.

The plate dissipation rating of the 5CX1500B is 1500 W. The tube and associated circuitry should be protected against surge current in the event of an arc inside the tube. A current-limiting resistance of 10 to 25 Ω between the power supply and plate is sufficient. The resistor should be capable of withstanding the surge currents that may develop; it should not be used as a fuse.

Cooling requirements for the 5CX1500B follow along the same lines as for the 4CX15000A discussed in the previous section. Table 3.4 lists the airflow requirements for the pentode. Note that the power dissipated by the device is equal to the power dissipated by the plate, heater, and grids (control, screen, and suppressor) combined. In the case of the 5CX1500B, the heater and three grids account for approximately 350 W dissipation.

Mechanical dimensions for the 5CX1500B are given in Figure 3.15.

3.2.8 High-Frequency Operating Limits

As with most active devices, performance of a given vacuum tube deteriorates as the operating frequency is increased beyond its designed limit. Electron *transit time* is a significant factor in the upper-frequency limitation of electron tubes. A finite time is taken by electrons to traverse the space from the cathode, through the grid, and travel on to the plate. As the operating frequency increases, a point is reached at which the electron transit-time effects become significant. This point is a function of the accelerating voltages at the grid and anode and their respective spacings. Tubes with reduced spacing in the grid-to-cathode region exhibit reduced transit-time effects.

A power limitation also is interrelated with the high-frequency limit of a device. As the operating frequency is increased, closer spacing and smaller-sized electrodes must be used. This reduces the power-handling capability of the tube. Figure 3.16 illustrates the relationship.

Gridded tubes at all power levels for frequencies up to about 1 GHz are invariably cylindrical in form. At higher frequencies, planar construction is almost universal. As the operating frequency is increased beyond design limits, output power and efficiency both decrease. Figure 3.17 illustrates the relationship.

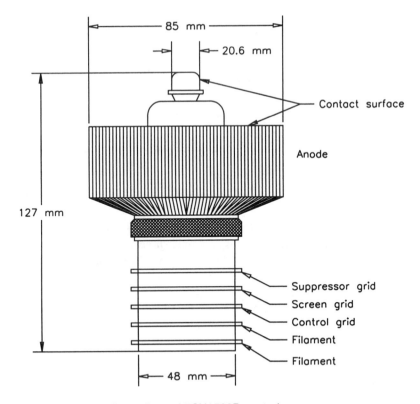

Figure 3.15 Principal dimensions of 5CX1500B pentode.

Transit time typically is not a problem for power grid tubes operating below 30 MHz. Depending on the application, power grid tubes can be used at 100 MHz and above without serious consideration of transit time effects. Klystrons and other micro-wave tubes actually take advantage of transit time, as discussed in Chapter 6.

Transit-Time Effects

When class C, class B, or similar amplifier operations are carried out at frequencies sufficiently high that the transit time of the electrons is not a negligible fraction of the waveform cycle, the following complications are observed in grid-based vacuum tubes:

- *Back-heating* of the cathode

- Loading of the control grid circuit as a result of energy transferred to electrons that do not necessarily reach the grid to produce a dc grid current

- Debunching of plate current pulses

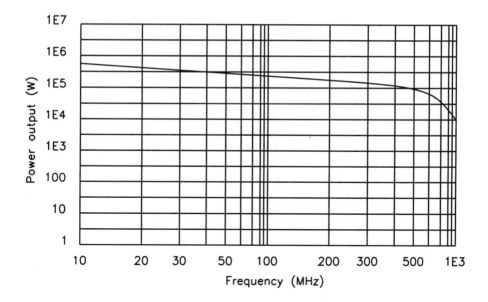

Figure 3.16 Continuous wave output power capability of a gridded vacuum tube.

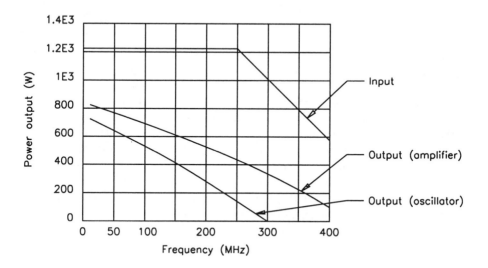

Figure 3.17 Performance of a class C amplifier as the operating frequency is increased beyond the design limits of the vacuum tube.

- Phase differences between the plate current and the exciting voltage applied to the control grid

Back-heating of the cathode occurs when the transit time in the grid-cathode space is sufficiently great to cause an appreciable number of electrons to be in transit at the instant the plate current pulse would be cut off in the case of low-frequency operation. A considerable fraction of the electrons thus trapped in the interelectrode space are returned to the cathode by the negative field existing in the grid-cathode space during the cutoff portion of the cycle. These returning electrons act to heat the cathode. At very high frequencies, this back-heating is sufficient to supply a considerable fraction of the total cathode heating required for normal operation. Back-heating may reduce the life of the cathode as a result of electron bombardment of the emitting surface. It also causes the required filament current to depend upon the conditions of operation within the tube.

Energy absorbed by the control grid as a result of input loading is transferred directly to the electron stream in the tube. Part of this stream acts to produce back-heating of the cathode. The remainder affects the velocity of the electrons as they arrive at the anode of the tube. This portion of the energy is not necessarily all wasted. In fact, a considerable percentage of it may, under favorable conditions, appear as useful output in the tube. To the extent that this is the case, the energy supplied by the exciting voltage to the electron stream is simply transferred directly from the control grid to the output circuits of the tube without amplification.

An examination of the total time required by electrons to travel from the cathode to the anode in a triode, tetrode, or pentode operated as a class C amplifier reveals that the resulting transit times for electrons at the beginning, middle, and end of the current pulse will differ as the operating frequency is increased. In general, electrons traversing the distance during the first segment of the pulse will have the shortest transit time, while those near the middle and end of the pulse will have the longest transit times, as illustrated in Figure 3.18. The first electrons in the pulse have a short transit time because they approach the plate before the plate potential is at its minimum value. Electrons near the middle of the pulse approach the plate with the instantaneous plate potential at or near minimum and, consequently, travel less rapidly in the grid-plate space. Finally, those electrons that leave the cathode late in the current pulse (those just able to escape being trapped in the control grid-cathode space and returned toward the cathode) will be slowed as they approach the grid, and so have a large transit time. The net effect is to cause the pulse of plate current to be longer than it would be in operation at a low frequency. This causes the efficiency of the amplifier to drop at high frequencies, because a longer plate current pulse increases plate losses.

3.3 Vacuum Tube Design

Any particular power vacuum tube may be designed to meet a number of operating parameters, the most important usually being high operating efficiency and high gain/bandwidth properties. Above all, the tube must be reliable and provide long operating life. The design of a new power tube is a lengthy process that involves

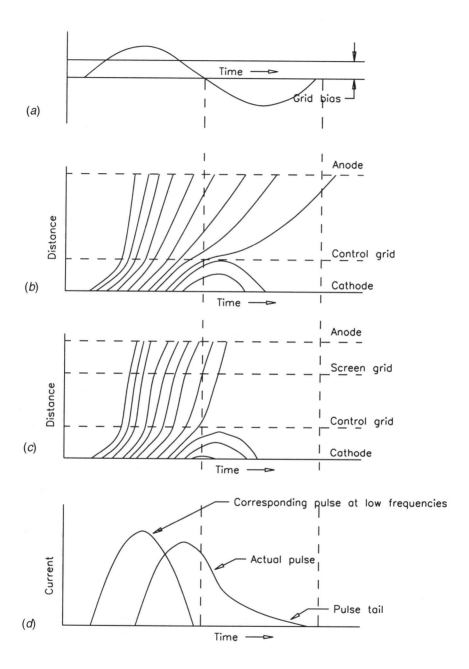

Figure 3.18 Transit-time effects in a class C amplifier: (*a*) control grid voltage, (*b*) electron position as a function of time (triode case), (*c*) electron position as a function of time (tetrode case), (*d*) plate current (triode case).

computer-aided calculations and modeling. The design engineers must examine a laundry list of items, including:

- **Cooling**: how the tube will dissipate heat generated during normal operation. A high-performance tube is of little value if it will not provide long life in typical applications. Design questions include whether the tube will be air-cooled or water-cooled, the number of fins the device will have, and the thickness and spacing of the fins.

- **Electro-optics**: how the internal elements line up to achieve the desired performance. A careful analysis must be made of what happens to the electrons in their paths from the cathode to the anode, including the expected power gain of the tube.

- **Operational parameters**: what the typical interelectrode capacitances will be, and the manufacturing tolerances that can be expected. This analysis includes: spacing variations among elements within the tube, the types of materials used in construction, the long-term stability of the internal elements, and the effects of thermal cycling.

3.3.1 Device Cooling

The main factor that separates tube types is the method of cooling used: air, water, or vapor. Air-cooled tubes are common at power levels below 50 kW. A water cooling system, although more complicated, is more effective than air cooling—by a factor of 5 to 10 or more—in transferring heat from the device. Air cooling at the 100 kW level is virtually impossible because it is difficult to physically move enough air through the device (if the tube is to be of reasonable size) to keep the anode sufficiently cool. Vapor cooling provides an even more efficient method of cooling a power amplifier (PA) tube than water cooling, for a given water flow and a given power dissipation. Naturally, the complexity of the external blowers, fans, ducts, plumbing, heat exchangers, and other hardware must be taken into consideration in the selection of a cooling method. Figure 3.19 shows how the choice of cooling method is related to anode dissipation.

Air Cooling

A typical air cooling system for a transmitter is shown in Figure 3.20. Cooling system performance for an air-cooled device is not necessarily related to airflow volume. The cooling capability of air is a function of its mass, not its volume. An appropriate airflow rate within the equipment is established by the manufacturer, resulting in a given resistance to air movement.

The altitude of operation is also a consideration in cooling system design. As altitude increases, the density (and cooling capability) of air decreases. To maintain the same cooling effectiveness, increased airflow must be provided. (Air cooling systems are discussed in detail in Section 8.2.1.)

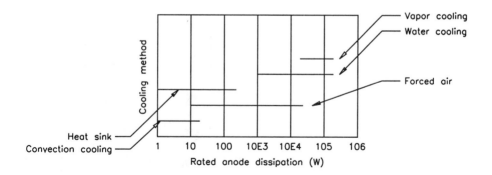

Figure 3.19 The relationship between anode dissipation and cooling method.

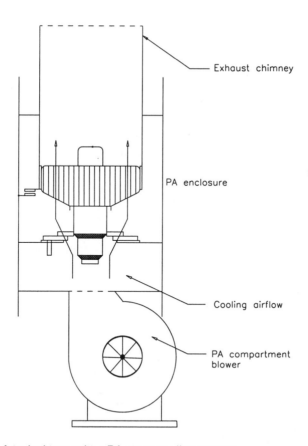

Figure 3.20 A typical transmitter PA stage cooling system.

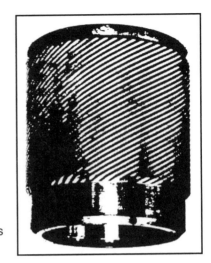

Figure 3.21 Water-cooled anode with grooves for controlled water flow.

Water Cooling

Water cooling usually is preferred over air cooling for power outputs above about 50 kW. Multiple grooves on the outside of the anode, in conjunction with a cylindrical jacket, force the cooling water to flow over the surface of the anode, as illustrated in Figure 3.21.

Because the water is in contact with the outer surface of the anode, a high degree of purity must be maintained. A resistivity of 1 mΩ-cm (at 25°C) typically is specified by tube manufacturers. Circulating water can remove about 1 kW/cm, of effective internal anode area. In practice, the temperature of water leaving the tube must be limited to 70°C to prevent the possibility of spot boiling.

After leaving the anode, the heated water is passed through a heat exchanger where it is cooled to 30 to 40°C before being pumped back to the tube.

In typical modern water-cooled tubes, the dissipation ranges from 500 W to 1 kW per square centimeter. Water-cooled anodes have deen designed that can operate at 10 kW per square centimeter and above. However, with the exception of "big science" applications, dissipations this high are seldom required. (Water cooling systems are discussed in detail in Section 8.2.2.)

Vapor-Phase Cooling

Vapor cooling allows the permissible output temperature of the water to rise to the boiling point, giving higher cooling efficiency compared with water cooling. The benefits of vapor-phase cooling are the result of the physics of boiling water. Increasing the temperature of one gram of water from 40 to 70°C requires 30 calories of energy. However, transforming 1 g of water at 100°C into steam vapor requires 540 calories. Thus, a vapor-phase cooling system permits essentially the same cooling capacity as water cooling, but with greatly reduced water flow. Viewed from another per-

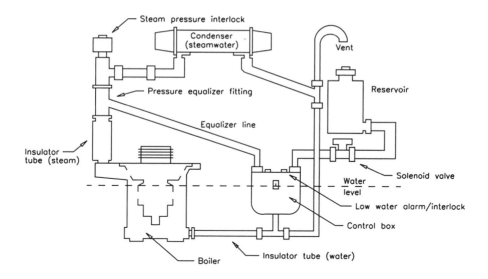

Figure 3.22 Typical vapor-phase cooling system.

spective, for the same water flow, the dissipation of the tube may be increased significantly (all other considerations being the same).

A typical vapor-phase cooling system is shown in Figure 3.22. A tube incorporating a specially designed anode is immersed in a boiler filled with distilled water. When power is applied to the tube, anode dissipation heats the water to the boiling point, converting the water to steam vapor. The vapor passes to a condenser, where it gives up its energy and reverts to a liquid state. The condensate then is returned to the boiler, completing the cycle. Electric valves and interlocks are included in the system to provide for operating safety and maintenance. A vapor-phase cooling system for a transmitter with multiple PA tubes is shown in Figure 3.23. (Vapor-phase cooling systems are discussed in detail in Section 8.2.3.)

Special Applications

Power devices used for research applications must be designed for transient overloading, requiring special considerations with regard to cooling. Oil, heat pipes, refrigerants (such as Freon), and, where high-voltage hold-off is a problem, gases (such as sulfahexafluoride) are sometimes used to cool the anode of a power tube.

3.3.2 Cathode Assembly

The ultimate performance of any vacuum tube is determined by the accuracy of design and construction of the internal elements. The requirements for a successful tube

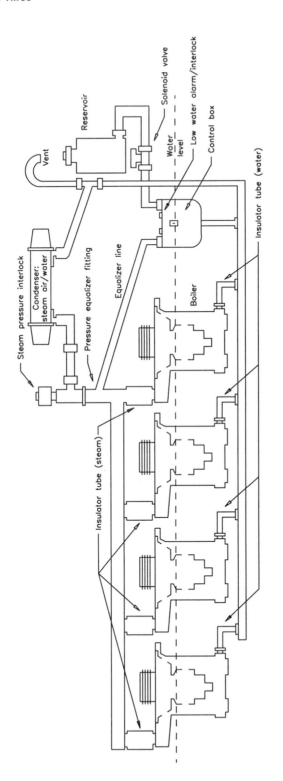

Figure 3.23 Vapor-phase cooling system for a 4-tube transmitter using a common water supply.

Table 3.5 Characteristics of Common Thermonic Emitters

Emitter	Heating Method[1]	Operating Temp. (°C)	Emission Density (A/cm²)	
			Average	Peak
Oxide	Direct and indirect	700 to 820	0.10 to 0.50	0.10 to 20
Thoriated tungsten	Direct	1600 to 1800	0.04 to 0.43	0.04 to 10
Impregnated tungsten	Direct and indirect	900 to 1175	0.5 to 8.0	0.5 to 12
[1] Directly heated refers to a filament-type cathode.				

include the ability to operate at high temperatures and withstand physical shock. Each element is critical to this objective.

The cathode used in a power tube obtains the energy required for electron emission from heat. The cathode may be directly heated (filament type) or indirectly heated. The three types of emitting surfaces most commonly used are:

- Thoriated-tungsten

- Alkaline-earth oxides

- Tungsten-barium-aluminate-impregnated emitters

The thoriated-tungsten and tungsten-impregnated cathodes are preferred in power tube applications because they are more tolerant to *ion bombardment*. The characteristics of the three emitting surfaces are summarized in Table 3.5.

A variety of materials may be used as a source of electrons in a vacuum tube. Certain combinations of materials are preferred, however, for reasons of performance and economics.

Oxide Cathode

The conventional production-type oxide cathode consists of a coating of barium and strontium oxides on a base metal such as nickel. Nickel alloys, in general, are stronger, tougher, and harder than most nonferrous alloys and many steels. The most important property of nickel alloys is their ability to retain strength and toughness at elevated temperatures. The oxide layer is formed by first coating a nickel structure (a can or disk) with a mixture of barium and strontium carbonates, suspended in a binder material. The mixture is approximately 60 percent barium carbonate and 40 percent strontium carbonate.

During vacuum processing of the tube, these elements are *baked* at high temperatures. As the binder is burned away, the carbonates subsequently are reduced to oxides. The cathode is then said to be *activated* and will emit electrons.

An oxide cathode operates CW at 700 to 820°C and is capable of an average emission density of 100 to 500 mA/cm². High emission current capability is one of the main

Figure 3.24 Photo of a typical oxide cathode. (*Courtesy of Varian.*)

advantages of the oxide cathode. Other advantages include high peak emission for short pulses and low operating temperature. As shown in Table 3.5, peak emission of up to 20 A/cm^2 is possible from an oxide cathode. A typical device is shown in Figure 3.24.

Although oxide-coated emitters provide more peak emission per watt of heating power than any other type, they are not without their drawbacks. Oxide emitters are more easily damaged or "poisoned" than other emitters and also deteriorate more rapidly when subjected to bombardment from high-energy particles.

The oxide cathode material will evaporate during the life of the tube, causing free barium to migrate to other areas within the device. This evaporation can be minimized in the design stage by means of a high-efficiency cathode that runs as cool as possible but still is not emission-limited at the desired heater voltage. In the field, the heater voltage must not exceed the design value. An oxide cathode that is overheated produces little, if any, additional emission. The life of the tube under such operation, however, is shortened significantly.

Thoriated-Tungsten Cathode

The thoriated-tungsten filament is another form of atomic-film emitter commonly used in power grid tubes. Tungsten is stronger than any other common metal at temperatures of more than 3500°F. The melting point of tungsten is 6170°F, higher than that of any other metal. The electrical conductivity of tungsten is approximately one-third that of copper, but much better than the conductivity of nickel, platinum, or iron-based alloys. The resistivity of tungsten in wire form is exploited in filament applications. The thoriated-tungsten filament (or cathode) is created in a high-temperature gaseous atmosphere to produce a layer of *ditungsten carbide* on the surface of the cathode element(s). Thorium is added to tungsten in the process of making tungsten wire. The thorium concentration is typically about 1.5 percent, in the form of thoria. By proper processing during vacuum pumping of the tube envelope, the metallic thorium is brought to the surface of the filament wire. The result is an increase in emission of approximately 1000 times over a conventional cathode.

At a typical operating temperature of 1600 to 1800°C, a thoriated-tungsten filament will produce an average emission of 40 to 430 mA/cm^2. Peak current ranges up to 10 A/cm^2 or more.

One of the advantages of a thoriated-tungsten cathode over an oxide cathode is the ability to operate the plate at higher voltages. Because oxide cathodes are susceptible to deterioration caused by ion bombardment, plate voltage must be limited. A thoriated-tungsten cathode is more tolerant of ion bombardment, so higher plate voltages can be safely applied.

The end of useful life for a thoriated-tungsten tube occurs when most of the carbon has evaporated or has combined with residual gas, depleting the carbide surface layer. Theoretically, a 3 percent increase in filament voltage will result in a 20 degree Kelvin (K) increase in cathode temperature, a 20 percent increase in peak emission, and a 50 percent decrease in tube life because of carbon loss. This cycle works in reverse, too. For a small decrease in temperature and peak emission, the life of the carbide layer—hence, the tube—may be increased.

Tungsten-Impregnated Cathode

The tungsten-impregnated cathode typically operates at 900 to 1175°C and provides the highest average emission density of the three types of cathodes discussed in this section (500 mA/cm^2 to 8 A/cm^2). Peak power performance ranges up to 12 A/cm^2.

Tungsten, as an element, is better able than other emitters to withstand bombardment by high-energy positive ions without suffering emission impairment. These positive ions are always present in small numbers in vacuum tubes as a result of ionization by collision with the residual gas.

Cathode Construction

Power tube filaments can be assembled in several different configurations. Figure 3.25 shows a spiral-type filament, and Figure 3.26 shows a bar-type design. The spiral filament is used extensively in low-power tubes. As the size of the tube increases, mechanical considerations dictate a bar-type filament with spring loading to compensate for thermal expansion. A mesh filament can be used for both small and large tubes. It is more rugged than other designs and less subject to damage from shock and vibration. The rigidity of a cylindrical mesh cathode is determined by the following parameters:

- The diameter of the cathode

- The number, thickness, and length of the wires forming the cathode

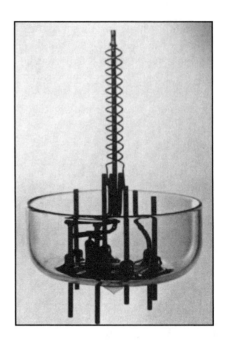

Figure 3.25 Spiral-type tungsten filament. (*Courtesy of Varian.*)

Figure 3.26 Bar-type tungsten filament. (*Courtesy of Varian.*)

- The ratio of welded to total wire crossings

A mesh cathode is shown in Figure 3.27.

Most power grid tubes are designed as a series of electron gun structures arranged in a cylinder around a centerline. This construction allows large amounts of plate current

Figure 3.27 Mesh tungsten filament. (*Courtesy of Varian.*)

to flow and to be controlled with a minimum of grid interception. With reduced grid interception, less power is dissipated in the grid structures. In the case of the control grid, less driving power is required for the tube.

In certain applications, the construction of the filament assembly may affect the performance of the tube, and that of the RF system as a whole. For example, filaments built in a basket-weave mesh arrangement usually offer lower distortion in critical high-level AM modulation circuits.

Velocity of Emission

The electrons emitted from a hot cathode depart with a velocity that represents the difference between the kinetic energy possessed by the electron just before emission and the energy that must be given up to escape. Because the energy of different electrons within the emitter is not the same, the velocity of emission will vary as well, ranging from zero to a maximum value determined by the type and style of emitter.

3.3.3 Grid Structures

The type of grid used for a power tube is determined principally by the power level and operating frequency required. For most medium-power tubes (5 to 25 kW dissipation) welded-wire construction is common. At higher power levels, laser-cut *pyrolytic graphite* grids may be found. The grid structures of a power tube must maintain their shape and spacing at elevated temperatures. They also must withstand shock and vibration.

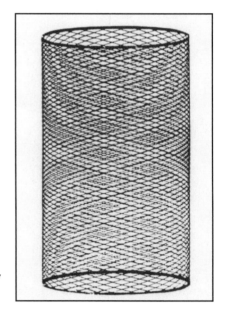

Figure 3.28 Mesh grid structure. (*Courtesy of Varian.*)

Wire Grids

Conventional wire grids are prepared by operators that wind the assemblies using special *mandrels* (forms) that include the required outline of the finished grid. The operators spot weld the wires at intersecting points, as shown in Figure 3.28. Most grids of this type are made with tungsten or *molybdenum*, which exhibit stable physical properties at elevated temperatures. On a strength basis, pure molybdenum generally is considered the most suitable of all refractory metals at temperatures of 1600 to 3000°F. The thermal conductivity of molybdenum is more than three times that of iron and almost half that of copper.

Grids for higher-power tubes typically are built using a bar-cage type of construction. A number of vertical supports are fastened to a metal ring at the top and to a base cone at the bottom. The lower end of the assembly is bonded to a contact ring. The construction of the ring, metal base cone, and cylindrical metal base give the assembly low lead inductance and low RF resistance.

The external loading of a grid during operation and the proximity of the grid to the hot cathode impose severe demands on both the mechanical stability of the structure and the physical characteristics of its surface. The grid absorbs a high proportion of the heat radiated by the cathode. It also intercepts the electron beam, converting part of its kinetic energy into heat. Furthermore, high-frequency capacitive currents flowing in the grid create additional heat.

The result is that grids are forced to work at temperatures as high as 1500°C. Their primary and secondary emissions, however, must be low. To prevent grid emission, high electron affinity must be ensured throughout the life of the tube, even though it is impossible to prevent material that evaporates from the cathode from contaminating the grid surface.

In tubes with oxide cathodes, grids made of tungsten or molybdenum wire are coated with gold to reduce primary emission caused by deposition. The maximum safe operating temperature for gold plating, however, is limited (about 550°C). Special coatings developed for high-temperature applications are effective in reducing grid emission. In tubes with thoriated-tungsten cathodes, grids made of tungsten or molybdenum are coated with proprietary compounds to reduce primary emission.

Primary grid emission is usually low in a thoriated tungsten cathode device. In the case of an oxide cathode, however, free barium may evaporate from the cathode coating material and find its way to the control and screen grids. The rate of evaporation is a function of cathode temperature and time. A grid contaminated with barium will become another emitting surface. The hotter the grid, the greater the emissions.

K-Grid

To permit operation at higher powers (and, therefore, higher temperatures) the *K-grid* has been developed (Philips). The K-grid is a spot-welded structure of molybdenum wire doped to prevent brittleness and recrystallization. To eliminate mechanical stresses, the grid is annealed at 1800 K. It is then baked under vacuum at 2300 K to degas the molybdenum wire. The grid structure next is coated with zirconium carbide (sintered at 2300 K) and, finally, with a thin layer of platinum.

Despite high grid temperatures, the platinum coating keeps primary emissions to a minimum. The surface structure maintains low secondary emissions.

Pyrolytic Grid

Pyrolytic grids are a high-performance alternative to wire or bar grid assemblies. Used primarily for high-power devices, pyrolytic grids are formed by laser-cutting a graphite cup of the proper dimensions. The computer-controlled laser cuts numerous holes in the cup to simulate a conventional-style grid. Figure 3.29 shows a typical pyrolytic-type grid before and after laser processing.

Pyrolytic (or oriented) graphite is a form of crystallized carbon produced by the decomposition of a hydrocarbon gas at high temperatures in a controlled environment. A layer of pyrolytic graphite is deposited on a grid form. The thickness of the layer is proportional to the amount of time deposition is allowed to continue. The structural and mechanical properties of the deposited graphite depend upon the imposed conditions.

Pyrolytic grids are ideal vacuum tube elements because they do not expand like metal. Their small coefficient of expansion prevents movement of the grids inside the tube at elevated temperatures. This preserves the desired electrical characteristics of the device. Because tighter tolerances can be maintained, pyrolytic grids can be spaced more closely than conventional wire grids. Additional benefits are that the grid:

- Is a single structure having no weld points

- Has a thermal conductivity in two of the three planes nearly that of copper

(a)   (b)

Figure 3.29 Pyrolytic graphite grid: (*a*) before laser processing, (*b*) completed assembly. (*Courtesy of Varian.*)

- Can operate at high temperatures with low vapor pressure

Grid Physical Structure

The control, screen, and suppressor grids are cylindrical and concentric. Each is slightly larger than the previous grid, as viewed from the cathode. Each is fastened to a metal base cone, the lower end of which is bonded to a contact ring. Figure 3.30 shows the construction of a typical screen grid assembly. Figure 3.31 provides a cutaway view of a tetrode power tube.

The shape of the control grid and its spacing from the cathode define, in large part, the operating characteristics of the tube. For best performance, the grid must be essentially transparent to the electron path from the cathode to the plate. In a tetrode, the control and screen grids must be precisely aligned to minimize grid current. For pentode tubes, these two conventions apply, in addition to the requirement for precise alignment and minimum beam interception for the suppressor grid.

Secondary Emission Considerations

The relationship of the properties of secondary electrons to the grid structures and other elements of a vacuum tube must be considered carefully in any design. As the

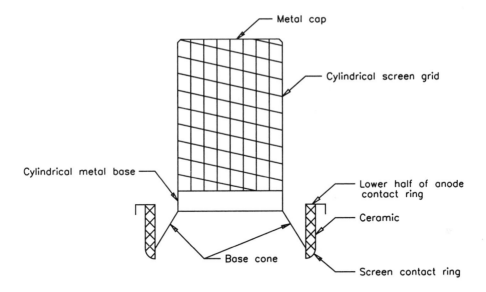

Figure 3.30 The screen grid assembly of a typical tetrode PA tube.

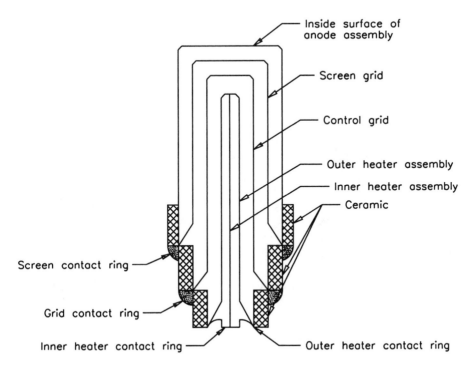

Figure 3.31 The internal arrangement of the anode, screen, control grid, and cathode assemblies of a tetrode power tube.

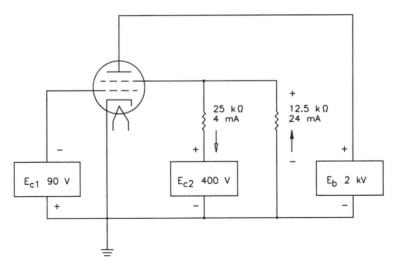

Figure 3.32 Circuit for providing a low-impedance screen supply for a tetrode.

power capability of a tube increases, the physical size of the elements also increases. This raises the potential for secondary emission from the control, screen, and suppressor grids. Secondary emission may occur regardless of the type of cathode used. The yield of secondary electrons may be reduced through the application of surface treatments.

In a tetrode, the screen is operated at a relatively low potential, necessary to accelerate the electrons emitted from the cathode. Not all electrons pass through the screen on their way to the plate. Some are intercepted by the screen grid. As the electrons strike the screen, other low-energy electrons are emitted. If these electrons have a stronger attraction to the screen, they will fall back to that element. If, however, they pass into the region between the screen and the plate, the much higher anode potential will attract them. The result is electron flow from screen to plate.

Because of the physical construction of a tetrode, the control grid will have virtually no control over screen-to-plate current flow as a result of secondary electrons. During a portion of the operating cycle of the device, it is possible that more electrons will leave the screen grid than will arrive. The result will be a reverse electron flow on the screen element, a condition common to high-power tetrodes. A low-impedance path for reverse electron flow must be provided.

Tube manufacturers typically specify the recommended value of bleeding current from the screen power supply to counteract the emission current. Two common approaches are illustrated in Figures 3.32 and 3.33. If the screen power supply impedance is excessively high in the direction of reverse electron flow, the screen voltage will attempt to rise to the plate voltage. Note the emphasis on low impedance in the reverse electron flow direction. Most regulated power supplies are low impedance in the forward electron flow direction only. If the supply is not well bled, the reverse electrons will try to flow from anode to cathode in the regulator series pass element. As the screen

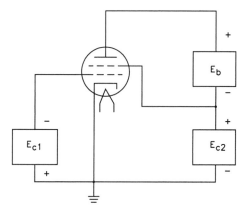

Figure 3.33 Series power supply configuration for swamping the screen circuit of a tetrode.

voltage rises, the secondary and plate currents will increase, and the tube will enter a runaway condition.

As shown in Figure 3.33, the addition of a 12.5 kΩ resistor from screen to ground provides a path for screen grid emission (20 mA for the circuit shown). In the circuit of Figure 3.33, plate current flows through the screen power supply, swamping the screen supply. The screen power supply must, obviously, carry the normal screen and plate currents. This scheme is used extensively in circuits where the screen is operated at dc ground potential. The plate-to-cathode voltage is then the sum of the E_b and E_{c2} power supplies.

The suppressor grid of a pentode lessens the effects of secondary emission in the device, reducing the requirement to provide a reverse electron flow path for the screen grid power supply. The screen current requirement for a pentode, however, may be somewhat higher than for a tetrode of the same general characteristics.

The designer also must consider the impedance of the control grid circuit. Primary grid emission can result in operational problems if the grid circuit impedance is excessively high. Primary grid emission, in the case of an oxide cathode tube, will increase with tube life.

The size and power of gridded tubes dictate certain characteristics of electrical potential. As this geometry increases in electrical terms, significant secondary emission from the control grid can occur. Control grid secondary emission can exist whether the cathode is a thoriated-tungsten or oxide emitter, and it can occur in a triode, tetrode, or pentode. A typical curve of grid current as a function of grid voltage for a high-power thoriated-tungsten filament tetrode is shown in Figure 3.34. As shown in the figure, grid current decreases and eventually takes a reverse direction as the grid voltage increases. This reduction and reversal of grid current can be explained by the normal secondary emission characteristics of the metals used in the grid structure.

The secondary emission characteristics of common metals are presented in curve form in Figure 3.35. The ratio of secondary-to-primary electron current is given as a

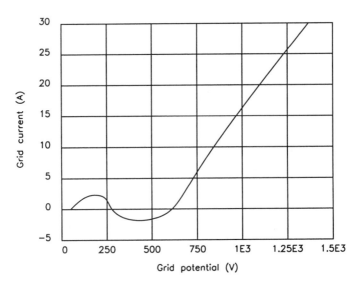

Figure 3.34 Typical curve of grid current as a function of control grid voltage for a high-power thoriated-tungsten filament tetrode.

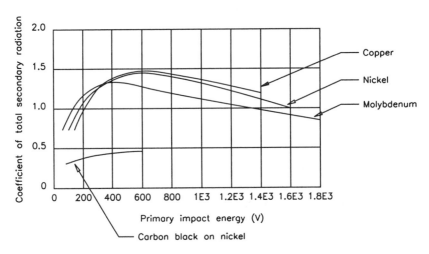

Figure 3.35 Secondary emission characteristics of various metals under ordinary conditions.

function of primary electron potential. An examination of the chart will show the region between 200 and 600 V to be rather critical as far as secondary emission is concerned. Any power grid tube that normally operates with 200 to 600 V on the grid can exhibit the negative resistance characteristic of decreasing grid current with increasing grid voltage when another electrode—such as the anode in a triode or the screen grid in a tetrode—is at a sufficiently high potential to attract the emitted electrons from the control grid. A driver stage that works into such a nonlinear load normally must be designed in such a manner as to tolerate this condition. One technique involves swamping the driver so that changes in the load resulting from secondary grid emission represent only a small percentage of the total load the driver works into.

3.3.4 Plate Assembly

The plate assembly of a power tube is typically a collection of many smaller parts that are machined and assembled to tight specifications. Copper generally is used to construct the anode. It is an excellent material for this purpose because it has the highest electrical conductivity of any metal except pure silver. Furthermore, copper is easily fabricated and ideally suited to cold-forming operations such as deep drawing, bending, and stamping.

The anode and cooling fins (in the case of an air-cooled device) begin as flat sheets of copper. They are stamped by the tube manufacturer into the necessary sizes and shapes. After all of the parts have been machined, the anode and cooling fins are stacked in their proper positions, clamped, and brazed into one piece in a brazing furnace.

The plate of a power tube resembles a copper cup with the upper half of a plate contact ring welded to the mouth and cooling fins silver-soldered or welded to the outside of the assembly. The lower half of the anode contact ring is bonded to a base ceramic spacer. At the time of assembly, the two halves of the ring are welded together to form a complete unit, as shown in Figure 3.36.

In most power tubes, the anode is a part of the envelope and, because the outer surface is external to the vacuum, it can be cooled directly. Glass envelopes were used in older power tubes. Most have been replaced, however, with devices that use ceramic as the envelope material.

3.3.5 Ceramic Elements

Ceramics are an integral part of modern power vacuum tubes. Three types of ceramics are in common usage in the production of vacuum devices:

- *Aluminum oxide*-based ceramics

- *Beryllium oxide* (BeO)-based ceramics

- *Aluminum nitride* (AlNi)-based ceramics

Aluminum Oxide Ceramics

Aluminum oxide-based ceramic insulators are a common construction material for a wide variety of electric components, including vacuum tubes. Aluminum oxide is 20 times higher in thermal conductivity than most oxides. The flexure strength of commercial high-alumina ceramics is two to four times greater than that of most oxide ceramics. There are drawbacks, however, to the use of alumina ceramics, including:

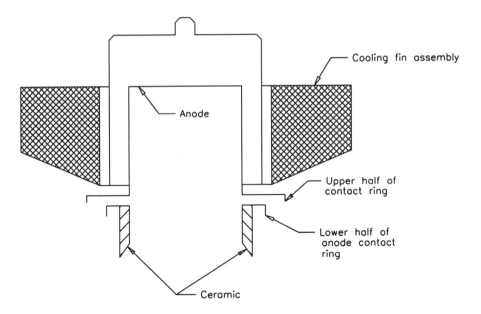

Figure 3.36 A cutaway view of the anode structure of an RF power amplifier tube.

- Relatively high thermal expansion (approximately 7 ppm/°C), compared to other ceramic materials, such as BeO

- Moderately high dielectric constant (approximately 10)

Aluminas are fabricated from aluminum oxide powders with various percentages of *sintering promoters*, which produce the *glassy phase*. The later additives reduce the densification temperatures to between 1500 and 1600°C. Based on the final application, the powders may be pressed, extruded, or prepared in slurries for slip casting or tape casting. The surface finish of aluminas is typically 3 to 25 μm/in as a result of normal processing. For very smooth finishes (2 μm/in), the surfaces may be lapped or polished.

Alumina ceramics for vacuum tube applications rarely are used apart from being bonded to metals. The means by which this is accomplished frequently dictate the processing technique. Metallization of aluminas usually is accomplished by either high-temperature firing or low-temperature thick-film processing. Fired, shaped aluminas usually are postmetallized by refiring the formed article after it has been coated with a slurry of molybdenum and manganese powder or tungsten metal powder. Based on the purity of the alumina, glass powder may be added to the metal powder. The mechanism of metallization requires an internal glassy phase or the added glassy phase for proper bonding. This is accomplished by firing in slightly reducing and moist atmospheres at temperatures above 1500°C. The resulting metallization usually is plated electrochemically with nickel, copper, or both.

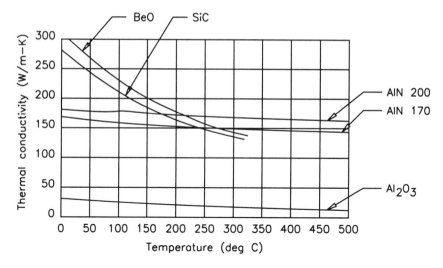

Figure 3.37 Thermal conductivity of common ceramics.

Thick-film processing of alumina ceramics traditionally has been performed in oxidizing atmospheres at moderate temperatures (800 to 1000°C). Precious metals are used, such as gold, silver, platinum, palladium, and their combinations. A glassy phase usually is incorporated in the thick-film paste.

Each approach to metallization has advantages and disadvantages. The advantage of the high-temperature metallization schemes with molybdenum or tungsten is their moderate cost. The disadvantage is the high resistivity of the resulting films. These attributes are reversed for thick-film materials. Another advantage of thick-film metallization is that the process often is applied by the device fabricator, not the ceramic vendor. This can allow for greater efficiency in design modification.

In tubes designed for operation at 100 MHz and below, alumina typically is the ceramic of choice.

Beryllium Oxide Ceramics

Beryllium oxide-based ceramics are in many ways superior to alumina-based ceramics. The major drawback is the toxicity of BeO. Beryllium and its compounds are a group of materials that are potentially hazardous and must be handled properly. With the necessary safeguards, BeO has been used successfully in many tube designs.

Beryllium oxide materials are particularly attractive for use in power vacuum tubes because of their electrical, physical, and chemical properties. The thermal conductivity of BeO is approximately 10 times higher than that of alumina-based materials. Figure 3.37 compares the thermal conductivity of BeO with that of alumina and some alternative materials. As the chart illustrates, BeO has a lower dielectric constant and coefficient of thermal expansion than alumina. It is, however, also slightly lower in strength.

Beryllia materials are fabricated in much the same way as alumina compounds, although the toxic properties of the powders mandate that they be processed in laboratories equipped to handle them safely. Simple cutting, drilling, and postmetallization also are handled by vendors with special equipment. Reliable thick-film systems may be applied to BeO substrates. Such coatings usually necessitate less elaborate safety precautions and are often applied by the device fabricator. General safety precautions relating to BeO ceramics are given in Section 11.3.2.

Other Ceramics

Aluminum nitride (AlNi)-based ceramics have been developed as an alternative to the toxicity concerns of BeO-based materials. As shown in Figure 3.37, the thermal conductivity of AlNi is comparable to that of BeO but deteriorates less with temperature. The dielectric constant of AlNi is comparable to that of alumina (a drawback), but its thermal expansion is low (4 ppm/°C).

Other ceramic materials that find some use in vacuum devices include silicon carbide (SiC)-based substances and boron nitride ceramics.

3.3.6 Tube Construction

Each type of power grid tube is unique insofar as its operating characteristics are concerned. The basic physical construction, however, is common to most devices. A vacuum tube is built in essentially two parts:

- The base, which includes the filament and supporting stem, control grid, screen grid, and suppressor grid (if used)

- The anode, which includes the heat-dissipating fins made in various machining steps

The base subassembly is welded using a *tungsten-inert gas* (TIG) process in an oxygen-free atmosphere (a process sometimes referred to as *Heliarc welding*) to produce a finished base unit.

The ceramic elements used in a vacuum tube are critical parts of the device. Assembled in sections, each element builds upon the previous one to form the base of the tube. The ceramic-to-metal seals are created using a material that is *painted* onto the ceramic and then heated in a brazing oven. After preparation in a high-temperature oven, the painted area provides a metallic structure that is molecularly bonded to the ceramic and provides a surface suitable for brazing.

This process requires temperature sequences that dictate completion of the highest-temperature stages first. As the assembly takes form, lower oven temperatures are used so that completed bonds will not be compromised.

Despite all the advantages that make ceramics one of the best insulators for a tube envelope, their brittleness is a potential cause of failure. The smallest cracks in the ceramic, not in themselves damaging, can cause the ceramic to break when mechanically stressed by temperature changes.

After the base assembly has been matched with the anode, the completed tube is brazed into a single unit. The device then goes through a *bake-out* procedure. Baking stations are used to evacuate the tube and bake out any oxygen or other gases from the copper parts of the assembly. Although oxygen-free copper is used in tube construction, some residual oxygen exists in the metal and must be driven out for long component life. A typical vacuum reading of 10^{-8} Torr is specified for most power tubes (formerly expressed as 10^{-8} mm of mercury). For comparison, this is the degree of vacuum in outer space about 200 miles above the earth.

A vacuum offers excellent electrical insulation characteristics. This property is essential for reliable operation of a vacuum tube, the elements of which typically operate at high potentials with respect to each other and to the surrounding environment. An electrode containing absorbed gases, however, will exhibit reduced breakdown voltage because the gas will form on the electrode surface, increasing the surface gas pressure and lowering the breakdown voltage in the vicinity of the gas pocket.

To maintain a high vacuum during the life of the component, power tubes contain a *getter* device. The name comes from the function of the element: to "get" or trap and hold gases that may exist inside the tube. Materials used for getters include zirconium, cerium, barium, and titanium.

The operation of a vacuum tube is an evolving chemical process. End of life in a vacuum tube generally is caused by loss of emission.

3.3.7 Connection Points

The high power levels and frequencies at which vacuum tubes operate place stringent demands on the connectors used to tie the outside world to the inside elements. Tubes are designed to be mounted vertically on their electrical connectors. The connectors provide a broad contact surface and mechanical support for the device.

The cathode and grids typically are mounted on ring-shaped *Kovar* bases, which also serve as contact rings for the external connections. Kovar is an iron-nickel-cobalt alloy whose coefficient of thermal expansion is comparable to that of aluminum oxide ceramic. The different diameters of the various contact rings allow them to be grouped coaxially. The concentric tube/connector design provides for operation at high frequencies. Conductivity is improved by silver plating.

3.3.8 Tube Sockets

Any one tube design may have several possible socket configurations, depending upon the frequency of operation. If the tube terminals are large cylindrical surfaces, the contacting portions of the socket consist of either spring *collets* or an assembly of spring fingerstock. Usually, these multiple-contacting surfaces are made of beryllium copper to preserve spring tension at the high temperatures present at the tube terminals. The fingers are silver-plated to reduce RF resistance. Figure 3.38 shows a cutaway view of the base of a tetrode socket.

If the connecting fingers of a power tube socket fail to provide adequate contact with the tube element rings, a concentration of RF currents will result. Depending on the ex-

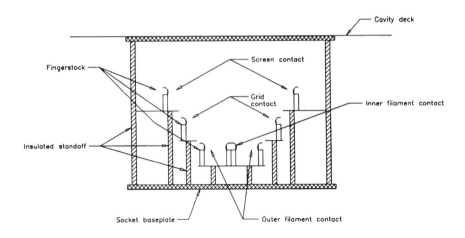

Figure 3.38 Cross section of the base of a tetrode socket showing the connection points.

tent of this concentration, damage may result to the socket. After a connecting finger loses its spring action, the heating effect is aggravated, and tube damage is possible.

A series of specialized power tubes is available with no sockets at all. Intended for cathode-driven service, the grid assembly is formed into a flange that is bolted to the chassis. The filament leads are connected via studs on the bottom of the tube. Such a configuration eliminates the requirement for a socket. This type of device is useful for low-frequency applications, such as induction heating.

3.4 Neutralization

An RF power amplifier must be properly neutralized to provide acceptable perform-ance in most applications. The means to accomplish this end vary considerably from one design to another. An RF amplifier is neutralized when two operating conditions are met:

- The interelectrode capacitance between the input and output circuits is canceled.

- The inductance of the screen grid and cathode assemblies (in a tetrode) is can-celed.

Cancellation of these common forms of coupling between the input and output cir-cuits of vacuum tube amplifiers prevents self-oscillation and the generation of spuri-ous products.

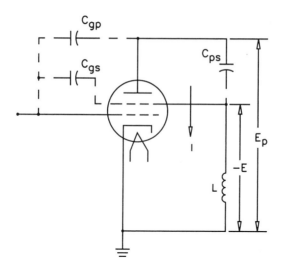

Figure 3.39 The elements involved in the neutralization of a tetrode PA stage.

3.4.1 Circuit Analysis

Figure 3.39 illustrates the primary elements that affect neutralization of a vacuum tube RF amplifier operating in the VHF band. (Many of the following principles also apply to lower frequencies.) The feedback elements include the residual grid-to-plate capacitance (C_{gp}), plate-to-screen capacitance (C_{ps}), and screen grid lead inductance (L_s). The RF energy developed in the plate circuit (E_p) causes a current (I) to flow through the plate-to-screen capacitance and the screen lead inductance. The current through the screen inductance develops a voltage ($-E$) with a polarity opposite that of the plate voltage (E_p). The $-E$ potential often is used as a method of neutralizing tetrode and pentode tubes operating in the VHF band.

Figure 3.40 graphically illustrates the electrical properties at work. The circuit elements of the previous figure have been arranged so that the height above or below the zero potential line represents the magnitude and polarity of the RF voltage for each part of the circuit with respect to ground (zero). For the purposes of this illustration, assume that all of the circuit elements involved are pure reactances. The voltages represented by each, therefore, are either in phase or out of phase and can be treated as positive or negative with respect to each other.

The voltages plotted in the figure represent those generated as a result of the RF output circuit voltage (E_p). No attempt is made to illustrate the typical driving current on the grid of the tube. The plate (P) has a high positive potential above the zero line, established at the ground point. Keep in mind that the distance above the baseline represents increasing positive potential. The effect of the out-of-phase screen potential developed as a result of inductance L_s is shown, resulting in the generation of $-E$.

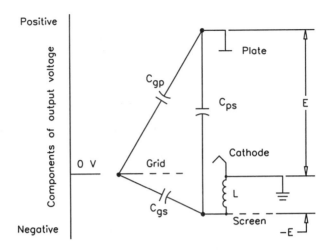

Figure 3.40 A graphical representation of the elements involved in the self-neutralization of a tetrode RF stage.

As depicted, the figure constitutes a perfectly neutralized circuit. The grid potential rests at the zero baseline. The grid operates at filament potential insofar as any action of the output circuit on the input circuit is concerned.

The total RF voltage between plate and screen is made up of the plate potential and screen lead inductance voltage, $-E$. This total voltage is applied across a divider circuit that consists of the grid-to-plate capacitance and grid-to-screen capacitance (C_{gp} and C_{gs}). When this potential divider is properly matched for the values of plate RF voltage (Ep) and screen lead inductance voltage ($-E$), the control grid will exhibit zero voltage difference with respect to the filament as a result of E_p.

3.4.2 Circuit Design

A variety of methods may be used to neutralize a vacuum tube amplifier. Generally speaking, a grounded-grid, cathode-driven triode can be operated into the VHF band without external neutralization components. The grounded-grid element is sufficient to prevent spurious oscillations. Tetrode amplifiers generally will operate through the MF band without neutralization. However, as the gain of the stage increases, the need to cancel feedback voltages caused by tube interelectrode capacitances and external connection inductances becomes more important. At VHF and above, it is generally necessary to provide some form of stage neutralization.

Below VHF

For operation at frequencies below the VHF band, neutralization for a tetrode typically employs a capacitance bridge circuit to balance out the RF feedback caused by

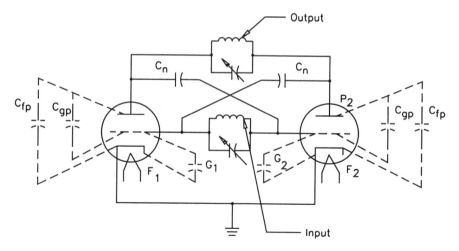

Figure 3.41 Push-pull grid neutralization.

residual plate-to-grid capacitance. This method assumes that the screen is well by-passed to ground, providing the expected screening action inside the tube.

Neutralization of low-power push-pull tetrode or pentode tubes can be accomplished with cross-neutralization of the devices, as shown in Figure 3.41. Small-value neutralization capacitors are used. In some cases, neutralization can be accomplished with a simple wire connected to each side of the grid circuit and brought through the chassis deck. Each wire is positioned to "look at" the plate of the tube on the opposite half of the circuit. Typically, the wire (or a short rod) is spaced a short distance from the plate of each tube. Fine adjustment is accomplished by moving the conductor in or out from its respective tube.

A similar method of neutralization can be used for a cathode-driven symmetrical stage, as shown in Figure 3.42. Note that the neutralization capacitors (C_n) are connected from the cathode of one tube to the plate of the opposite tube. The neutralizing capacitors have a value equal to the internal cathode-to-plate capacitance of the PA tubes.

In the case of a single-ended amplifier, neutralization can be accomplished using either a push-pull output or push-pull input circuit. Figure 3.43 shows a basic push-pull grid neutralization scheme that provides the out-of-phase voltage necessary for proper neutralization. It is usually simpler to create a push-pull network in the grid circuit than in the plate because of the lower voltages present. The neutralizing capacitor, C_n, is small and may consist of a simple feedthrough wire (described previously). A padding capacitor in parallel with C_1 often is added to maintain the balance of the input circuit while tuning. The padding capacitor generally is equal in size to the input capacitance of the tube.

Single-ended tetrode and pentode stages also can be neutralized using the method shown in Figure 3.44. The input resonant circuit is placed above ground by a small amount because of the addition of capacitor C. The voltage to ground that develops

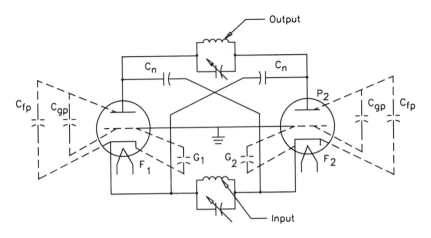

Figure 3.42 Symmetrical stage neutralization for a grounded-grid circuit.

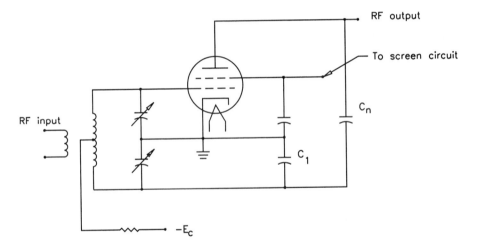

Figure 3.43 Push-pull grid neutralization in a single-ended tetrode stage.

across C upon the application of RF drive is out of phase with the grid voltage, and is fed back to the plate through C_n to provide neutralization. In such a design, C_n is considerably larger in value than the grid-to-plate interelectrode capacitance.

The single-ended grid neutralization circuit is redrawn in Figure 3.45 to show the capacitance bridge that makes the design work. Balance is obtained when the following condition is met:

$$\frac{C_n}{C} = \frac{C_{gp}}{C_{gf}} \tag{3.12}$$

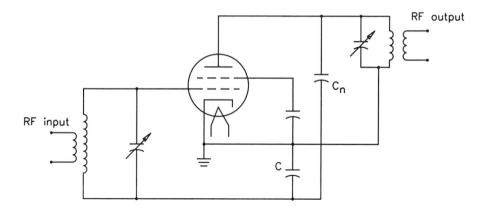

Figure 3.44 Single-ended grid neutralization for a tetrode.

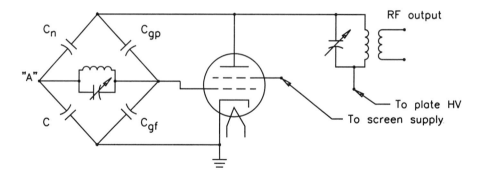

Figure 3.45 The previous figure redrawn to show the elements involved in neutralization.

Where:
C_n = neutralization capacitance
C = input circuit bypass capacitor
C_{gp} = grid-to-plate interelectrode capacitance
C_{gf} = total input capacitance, including tube and stray capacitance

A single-ended amplifier also can be neutralized by taking the plate circuit slightly above ground and using the tube capacitances as part of the neutralizing bridge. This circuit differs from the usual RF amplifier design in that the plate bypass capacitor is returned to the screen side of the screen bypass capacitor, as shown in Figure 3.46. The size of screen bypass capacitor C_s and the amount of stray capacitances in C_p are chosen

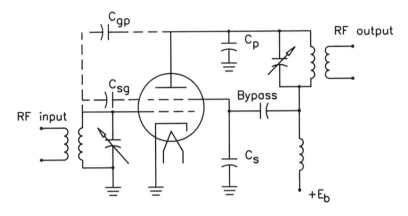

Figure 3.46 Single-ended plate neutralization.

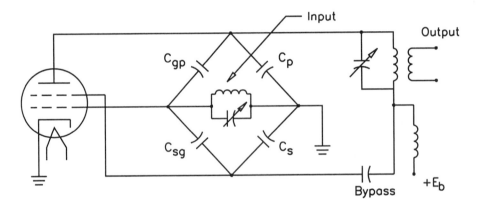

Figure 3.47 Single-ended plate neutralization showing the capacitance bridge present.

to balance the voltages induced in the input by the internal tube capacitances, grid-to-plate (C_{gp}) and screen-to-grid (C_{sg}). This circuit is redrawn in Figure 3.47 in the usual bridge form. Balance is obtained when the following condition is met:

$$\frac{C_p}{C_s} = \frac{C_{gp}}{C_{sg}} \tag{3.13}$$

In usual tetrode and pentode structures, the screen-to-grid capacitance is approximately half the published tube input capacitance. The tube input capacitance is primarily the sum of the grid-to-screen capacitance and the grid-to-cathode capacitance.

Note that in the examples given, it is assumed that the frequency of operation is low enough that inductances in the socket and connecting leads can be ignored. This is basically true in MF applications and below. At higher bands, however, the effects of stray inductances must be considered, especially in single-ended tetrode and pentode stages.

VHF and Above

Neutralization of power grid tubes operating at very high frequencies provides special challenges and opportunities to the design engineer. At VHF and above, significant RF voltages may develop in the residual inductance of the screen, grid, and cathode elements. If managed properly, these inductances can be used to accomplish neutralization in a simple, straightforward manner.

At VHF and above, neutralization is required to make the tube input and output circuits independent of each other with respect to reactive currents. Isolation is necessary to ensure independent tuning of the input and output. If variations in the output voltage of the stage produce variations of phase angle of the input impedance, phase modulation will result.

As noted previously, a circuit exhibiting independence between the input and output circuits is only half of the equation required for proper operation at radio frequencies. The effects of incidental inductance of the control grid also must be canceled for complete stability. This condition is required because the suppression of coupling by capacitive currents between the input and output circuits is not, by itself, sufficient to negate the effects of the output signal on the cathode-to-grid circuit. Both conditions—input and output circuit independence and compensation for control grid lead inductance—must be met for complete stage stability at VHF and above.

Figure 3.48 shows a PA stage employing stray inductance of the screen grid to achieve neutralization. In this grounded-screen application, the screen is bonded to the cavity deck using six short connecting straps. Two additional adjustable ground straps are set to achieve neutralization.

3.4.3 Grounded-Grid Amplifier Neutralization

Grounded-grid amplifiers offer an attractive alternative to the more common grid-driven circuit. The control grid is operated at RF ground and serves as a shield to capacitive currents from the output to the input circuit. Generally, neutralization is not required until the control grid lead inductive reactance becomes significant. The feedback from the output to the input circuit is no longer the result of plate-to-filament capacitance. The physical size of the tube and the operating frequency determine when neutralization is required.

Two methods of neutralization commonly are used with grounded-grid amplifiers. In the first technique, the grids of a push-pull amplifier are connected to a point having zero impedance to ground, and a bridge of neutralizing capacitances is used that is equal to the plate-filament capacitances of the tubes.

The second method of neutralization requires an inductance between the grid and ground, or between the grids of a push-pull amplifier, of a value that will compensate for the coupling between input and output circuits resulting from the internal capacitances of the tubes.

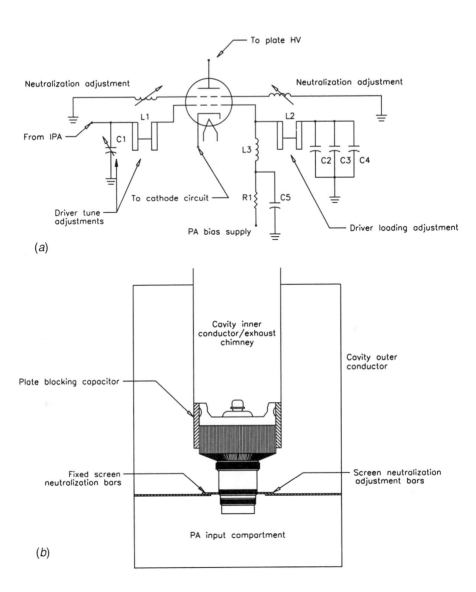

Figure 3.48 A grounded-screen PA stage neutralized through the use of stray inductance between the screen connector and the cavity deck: (*a*) schematic diagram, (*b*) mechanical layout of cavity.

Behavior of these two circuits is quite different. They may be considered as special forms of the more general case in which the neutralizing capacitors have values differing from the internal capacitances of the tubes, and in which an appropriate reactance is connected between the grids. Under these conditions, the value of neutralizing capaci-

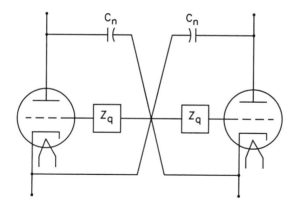

Figure 3.49 Circuit of grounded-grid amplifier having grid impedance and neutralized for reactive currents.

tance permits continuous variation of power amplification, stability, and negative feedback.

Grid Impedance

In the special case of a grounded-grid amplifier having a grid impedance and the reactive currents neutralized, the following equations apply (see Figure 3.49):

$$C_n = C_{fp} - \frac{C_{fg}}{\mu} \tag{3.14}$$

$$Z_g = -\frac{1}{j\omega C_{fg} + C_{gp}(1+\mu)} \tag{3.15}$$

If in solving the equation for C_n the sign is negative, this indicates that in-phase neutralization is required. Conversely, if the sign of C_n is positive, then out-of-phase neutralization is needed. A negative value of Z_g indicates capacitive reactance required, and a positive value indicates that inductive reactance is to be used.

Application Example

If the grids of a push-pull cathode-driven amplifier are not at ground potential because the inductance of the leads is not negligible, coupling may exist between the input and output circuits through the plate-grid capacitances, cathode-grid capacitances, and grid-to-grid inductance. One method of reducing this coupling is to insert between the grids a series-tuned circuit that has zero reactance at the operating frequency. This technique is illustrated in Figure 3.50. This neutralization scheme is use-

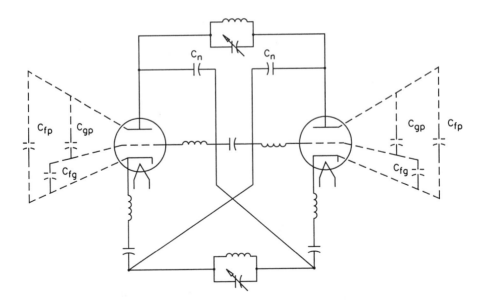

Figure 3.50 Neutralization by cross-connected capacitors of a symmetrical cathode-excited amplifier wth compensation of lead inductance.

ful only for the case in which no grid current flows. If grid current flows, a grid resistance will appear in parallel with the grid-to-filament capacitance. If the resistance is small in comparison with the reactance of this grid-to-filament capacitance, phase modulation will result.

Another important property of the preceding neutralization scheme is that power amplification is a function of the neutralizing capacitance, while the independence of cathode and plate circuits from the viewpoint of reactive currents may be obtained with any value of neutralizing capacitance. If the neutralizing capacitance is less than the plate-to-filament capacitance of the tube, the stage will operate with low excitation power and high power amplification.

If the neutralizing capacitance is greater than the plate-to-filament capacitance, the power amplification will be quite low, but the total output power possible will be increased.

3.4.4 Self-Neutralizing Frequency

The voltage-dividing action between the plate-to-grid capacitance (C_{pg}) and the grid-to-screen capacitance (C_{gs}) will not change with variations in operating frequency. The voltage division between the plate and screen, and screen and ground caused by the charging current (I) will, however, vary significantly with frequency. There will be a particular frequency, therefore, at which this potential-dividing circuit

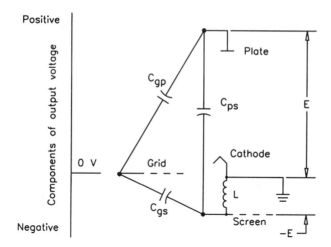

Figure 3.51 Graphical representation of the elements of a tetrode when self-neutralized.

will effectively place the grid at filament potential insofar as the plate is concerned. This point is known as the *self-neutralizing frequency*, illustrated in Figure 3.51.

At the self-neutralizing frequency, the tetrode or pentode is inherently neutralized by the circuit elements within the device itself, and external screen inductance to ground. When a device is operated below its self-neutralizing frequency, the normal cross-neutralization circuits apply. When the operating frequency is above the self-neutralizing frequency, the voltage developed in the screen lead inductance is too large to give the proper voltage division between the internal capacitances of the device. One approach to neutralization in this case involves adjusting the inductive reactance of the screen lead to ground to lower the total reactance. In the example shown in Figure 3.52, this is accomplished with a series variable capacitor.

Another approach is shown in Figure 3.53, in which the potential divider network made up of the tube capacitance is changed. In the example, additional plate-to-grid capacitance is added external to the tube. The external capacitance (C_{ext}) can take the form of a small wire or rod positioned adjacent to the plate of the tube. This approach is similar to the one described in Section 3.4.2, except that in this case the neutralizing probe is connected to the grid of the tube, rather than to an opposite polarity in the circuit.

If the RF power amplifier is operating above the self-neutralizing frequency of the tube and must be tuned over a range of frequencies, it is probably easier to use the screen series-tuning capacitor method and make this control available to the operator. If operation is desired over a range of frequencies including the self-neutralizing frequency of the tube, this circuit is also desirable because the incidental lead inductance in the variable capacitor lowers the self-neutralizing frequency of the circuit so that the neutralizing series capacitor can be made to operate over the total desired frequency range. If this range is too great, switching of neutralizing circuits will be required. A small 50 to 100 pF variable capacitor in the screen lead often has been found to be satisfactory.

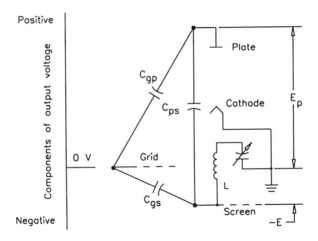

Figure 3.52 Components of the output voltage of a tetrode when neutralized by adding series screen-lead capacitance.

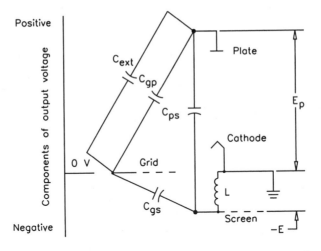

Figure 3.53 Components of the output voltage of a tetrode when neutralized by adding external grid-to-plate capacitance.

Another method of changing the self-neutralizing frequency of a tetrode or pentode can be fashioned from the general bypassing arrangement of the screen and filament shown in Figure 3.54. The screen lead is bypassed with minimum inductance to the filament terminal of the tube. Some inductance is introduced in the common filament and screen ground leads. The grid is shown below the zero voltage or chassis potential, indicating that the voltage developed in the screen lead inductance to chassis is excessive. If the filament is tapped on this inductance, a point can be found where the voltage

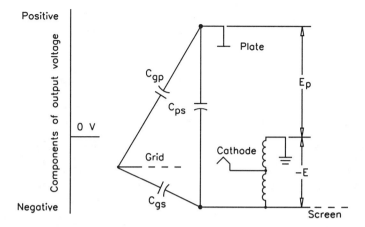

Figure 3.54 Components of the output voltage of a tetrode neutralized by adding inductance common to the screen and cathode return.

difference between the grid and filament is zero, as far as the components of plate voltage are concerned. This arrangement will be found to self-neutralize at a higher frequency than if the filament and screen were bypassed separately to the chassis. Thus, increasing the self-neutralizing frequency of the tube and screen bypass arrangement reduces the tendency of the VHF parasitic to occur.

If the frequency of the VHF parasitic is reduced by increasing the inductance of the plate lead (presuming this is the principal frequency-defining circuit), the circuit can be made to approach the self-neutralizing frequency of the tube and, therefore, suppress the parasitic.

3.4.5 Neutralization Adjustment

Most neutralization circuits must be adjusted for operation at a given frequency. The exact procedure followed to make these adjustments varies from one circuit to the next. The following generalizations, however, apply to most systems.

The first step in the process of neutralization is to break the dc connections to the plate voltage and screen voltage supplies, leaving the RF circuits intact. If the direct current path is not broken, some current can flow in either of these circuits even though the voltages are zero. The presence of this current causes the amplifier to work in the normal manner, generating RF power in the plate circuit. It would then be incorrect to adjust for zero power in the plate circuit. Sufficient RF grid drive must be applied to provide some grid current or to cause a sensitive RF meter coupled to the plate to give an indication of feedthrough power. When the plate circuit is tuned through resonance, the grid current will dip when the circuit is out of neutralization, or the RF meter will peak. The neutralization adjustments are made until the indication is minimum.

Another powerful tool for roughly neutralizing an RF amplifier is to feed the power output from a signal generator into the grid circuit. A sensitive RF detector is inserted between the output connector and the load. Neutralization can then be adjusted for minimum feedthrough. This technique is useful in working with prototype equipment. Actual qualitative measurements can be made. If the insertion loss of the amplifier is less than the expected gain, oscillation will occur. Circuit modifications can be made until the isolation is sufficient to warrant a test with high voltages applied. The advantages of this "cold" system test are that:

- No components are subjected to unusual stress if the amplifier is unstable.

- Circuit adjustments can be made safely because no high voltages are present.

For final trimming of the neutralization adjustment, the stage should be returned to operating condition at reduced power (similar to that used when testing for parasitic oscillations), or under the final loaded operating conditions. At higher frequencies, particularly in the VHF region, it will be found that a small additional trimming adjustment of the neutralization circuit is usually required. When the plate circuit is tuned through resonance, minimum plate current and maximum control grid current should occur simultaneously. In the case of the tetrode and pentode, the dc screen current should be maximum at the same time.

These neutralizing procedures apply not only to the HF radio frequencies, but also to the VHF and UHF regions.

3.5 References

1. Terman, F. E., *Radio Engineering*, 3rd Ed., McGraw-Hill, New York, 1947.
2. Ferris, Clifford D., "Electron Tube Fundamentals," in *The Electronics Handbook*, Jerry C. Whitaker (ed.), CRC Press, Boca Raton, FL, pp. 295–305, 1996.

3.6 Bibliography

Birdsall, C. K., *Plasma Physics via Computer Simulation*, Adam Hilger, 1991.
Block, R., "CPS Microengineers New Breed of Materials," *Ceramic Ind.*, pp. 51–53, April 1988.
Buchanan, R. C., *Ceramic Materials for Electronics*, Marcel Dekker, New York, 1986.
"Ceramic Products," Brush Wellman, Cleveland, OH.
Chaffee, E. L., *Theory of Thermonic Vacuum Tubes*, McGraw-Hill, New York, 1939.
"Combat Boron Nitride, Solids, Powders, Coatings," Carborundum Product Literature, form A-14, 011, September 1984.
"Coors Ceramics—Materials for Tough Jobs," Coors Data Sheet K.P.G.-2500-2/87 6429.
Cote, R. E., and R. J. Bouchard, "Thick Film Technology," in *Electronic Ceramics*, L. M. Levinson (ed.), Marcel Dekker, New York, pp. 307–370, 1988.
Dettmer, E. S., and H. K. Charles, Jr., "AlNi and SiC Substrate Properties and Processing Characteristics," *Advances in Ceramics*, Vol. 31, American Ceramic Society, Columbus, OH, 1989.

Dettmer, E. S., H. K. Charles, Jr., S. J. Mobley, and B. M. Romenesko, "Hybrid Design and Processing Using Aluminum Nitride Substrates," *ISHM 88 Proc.*, pp. 545–553, 1988.

Eastman, Austin V., *Fundamentals of Vacuum Tubes*, McGraw-Hill, 1941.

Fink, D., and D. Christiansen (eds.), *Electronics Engineers' Handbook*, 3rd Ed., McGraw-Hill, New York, 1989.

Floyd, J. R., "How to Tailor High-Alumina Ceramics for Electrical Applications," *Ceramic Ind.*, pp. 44–47, February 1969; pp. 46–49, March 1969.

Gray, Truman S., *Applied Electronics*, John Wiley & Sons, New York, 1954.

Harper, C. A., *Electronic Packaging and Interconnection Handbook*, McGraw-Hill, New York, 1991.

"High Power Transmitting Tubes for Broadcasting and Research," Philips Technical Publication, Eindhoven, the Netherlands, 1988.

Iwase, N., and K. Shinozaki, "Aluminum Nitride Substrates Having High Thermal Conductivity," *Solid State Technol.*, pp. 135–137, October 1986.

Jordan, Edward C., (ed.), *Reference Data for Engineers: Radio, Electronics, Computer and Communications*, 7th Ed., Howard W. Sams, Indianapolis, IN, 1985.

Kingery, W. D., H. K. Bowen, and D. R. Uhlmann, *Introduction to Ceramics*, John Wiley & Sons, New York, pp. 637, 1976.

Kohl, Walter, *Materials Technology for Electron Tubes*, Reinhold, New York.

Laboratory Staff, *The Care and Feeding of Power Grid Tubes*, Varian Eimac, San Carlos, CA, 1984.

Lafferty, J. M., *Vacuum Arcs*, John Wiley & Sons, New York, 1980.

Mattox, D. M., and H. D. Smith, "The Role of Manganese in the Metallization of High Alumina Ceramics," *J. Am. Ceram. Soc.*, Vol. 64, pp. 1363–1369, 1985.

Mistler, R. E., D. J. Shanefield, and R. B. Runk, "Tape Casting of Ceramics," in G. Y. Onoda, Jr., and L. L. Hench (eds.), *Ceramic Processing Before Firing*, John Wiley & Sons, New York, pp. 411–448, 1978.

Muller, J. J., "Cathode Excited Linear Amplifiers," *Electrical Communications*, Vol. 23, 1946.

Powers, M. B., "Potential Beryllium Exposure While Processing Beryllia Ceramics for Electronics Applications," Brush Wellman, Cleveland, OH.

Reich, Herbert J., *Theory and Application of Electronic Tubes*, McGraw-Hill, New York, 1939.

Roth, A., *Vacuum Technology*, 3rd Ed., Elsevier Science Publishers B. V., 1990.

Sawhill, H. T., A. L. Eustice, S. J. Horowitz, J. Gar-El, and A. R. Travis, "Low Temperature Co-Fireable Ceramics with Co-Fired Resistors," in *Proc. Int. Symp. on Microelectronics*, pp. 173–180, 1986.

Schwartz, B., "Ceramic Packaging of Integrated Circuits," in *Electronic Ceramics*, L. M. Levinson (ed.), Marcel Dekker, New York, pp. 1–44, 1988.

Strong, C. E., "The Inverted Amplifier," *Electrical Communications*, Vol. 19, no. 3, 1941.

Whitaker, J. C., *Radio Frequency Transmission Systems: Design and Operation*, McGraw-Hill, New York, 1991.

Designing Vacuum Tube Circuits

4.1   Introduction

Any number of configurations may be used to generate RF signals using vacuum tubes. Circuit design is dictated primarily by the operating frequency, output power, type of modulation, duty cycle, and available power supply. Tube circuits can be divided generally by their operating class and type of modulation employed. As discussed in Section 2.1.4, the angle of plate current flow determines the class of operation:

- Class A = 360° conduction angle
- Class B = 180° conduction angle
- Class C = conduction angle less than 180°
- Class AB = conduction angle between 180 and 360°

The class of operation has nothing to do with whether the amplifier is grid-driven or cathode-driven. A cathode-driven amplifier, for example, can be operated in any desired class. The class of operation is only a function of the plate current conduction angle. The efficiency of an amplifier is also a function of the plate current conduction angle.

The efficiency of conversion of dc to RF power is one of the more important characteristics of a vacuum tube amplifier circuit. The dc power that is not converted into useful output energy is, for the most part, converted to heat. This heat represents wasted power; the result of low efficiency is increased operating cost for energy. Low efficiency also compounds itself. This wasted power must be dissipated, requiring increased cooling capacity. The efficiency of the amplifier must, therefore, be carefully considered, consistent with the other requirements of the system. Figure 4.1 shows the theoretical efficiency attainable with a tuned or resistive load assuming that the peak ac plate voltage is equal to the plate supply voltage.

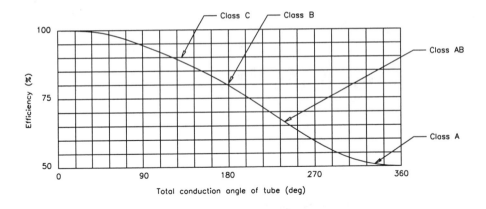

Figure 4.1 Plate efficiency as a function of conduction angle for an amplifier with a tuned load.

4.1.1 Class A Amplifier

A class A amplifier is used in applications requiring low harmonic distortion in the output signal. A class A amplifier can be operated with low intermodulation distortion in linear RF amplifier service. Typical plate efficiency for a class A amplifier is about 30 percent. Power gain is high because of the low drive power required. Gains as high as 30 dB are typical.

4.1.2 Class B and AB Amplifiers

A class AB power amplifier is capable of generating more power—using the same tube—than the class A amplifier, but more intermodulation distortion also will be generated. A class B RF linear amplifier will generate still more intermodulation distortion, but is acceptable in certain applications. The plate efficiency is typically 66 percent, and stage gain is about 20 to 25 dB.

4.1.3 Class C Amplifier

A class C power amplifier is used where large amounts of RF energy need to be generated with high efficiency. Class C RF amplifiers must be used in conjunction with tuned circuits or cavities, which restore the amplified waveform through the *flywheel effect*.

The grounded-cathode class C amplifier is the building block of RF technology. It is the simplest method of amplifying CW, pulsed, and FM signals. The basic configuration is shown in Figure 4.2. Tuned input and output circuits are used for impedance

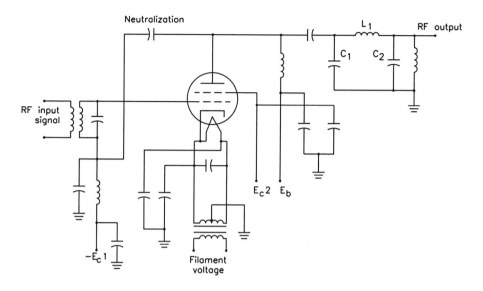

Figure 4.2 Basic grounded-cathode RF amplifier circuit.

matching and to resonate the stage at the desired operating frequency. The cathode is bypassed to ground using low-value capacitors. Bias is applied to the grid as shown. The bias power supply may be eliminated if a self-bias configuration is used. The typical operating efficiency of a class C stage ranges from 65 to 85 percent.

Figure 4.3 illustrates the application of a zero-bias triode in a grounded-grid arrangement. Because the grid operates at RF ground potential, this circuit offers stable performance without the need for neutralization (at MF and below). The input signal is coupled to the cathode through a matching network. The output of the triode feeds a *pi* network through a blocking capacitor.

4.2   Principles of RF Power Amplification

In an RF power amplifier, a varying voltage is applied to the control grid (or cathode, in the case of a grounded-grid circuit) from a driver stage whose output is usually one of the following:

- Carrier-frequency signal only

- Modulation (intelligence) signal only

- Modulated carrier signal

Simultaneous with the varying control grid signal, the plate voltage will vary in a similar manner, resulting from the action of the amplified current flowing in the plate circuit. In RF applications with resonant circuits, these voltage changes are smooth sine

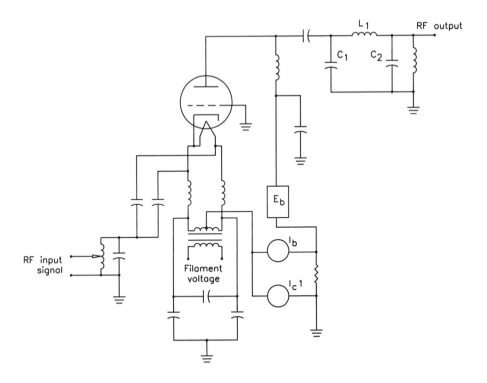

Figure 4.3 Typical amplifier circuit using a zero-bias triode. Grid current is measured in the return lead from ground to the filament.

wave variations, 180° out of phase with the input. The relationship is illustrated in Figure 4.4. Note how these variations center about the dc plate voltage and the dc control grid bias. In Figure 4.5, the variations have been indicated next to the plate voltage and grid voltage scales of a typical constant-current curve. At some instant in time, shown as t on the time scales, the grid voltage has a value denoted e_g on the grid voltage sine wave.

Any point on the operating line (when drawn on constant-current curves as illustrated in Figure 4.5) tells the instantaneous values of plate current, screen current, and grid current that must flow when these particular values of grid and plate voltage are applied to the tube. Thus, by plotting the values of plate and grid current as a function of time t, it is possible to produce a curve of instantaneous values of plate and grid current. Such plots are shown in Figure 4.6.

By analyzing the plate and grid current values, it is possible to predict with accuracy the effect on the plate circuit of a change at the grid. It follows that if a properly loaded resonant circuit is connected to the plate, a certain amount of RF power will be delivered to that circuit. If the resonant circuit is tuned to the fundamental frequency (the same frequency as the RF grid voltage), the power delivered will be that of the fundamental or principal RF component of plate current. If the circuit is tuned to a harmonic

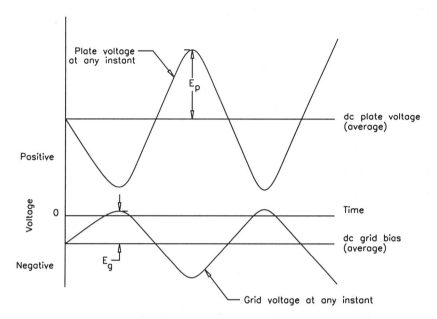

Figure 4.4 Variation of plate voltage as a function of grid voltage.

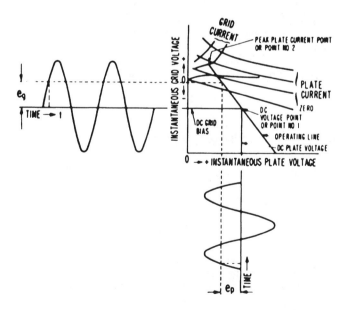

Figure 4.5 Relationship between grid voltage and plate voltage plotted on a constant-current curve.

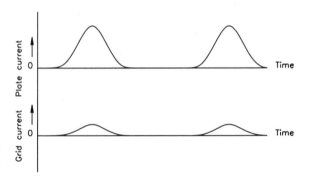

Figure 4.6 Instantaneous values of plate and grid current as a function of time.

of the grid voltage frequency, the power delivered will be the result of a harmonic component of the plate current.

4.2.1 Drive Power Requirements

The technical data sheet for a given tube type lists the approximate drive power required for various operating modes. As the frequency of operation increases and the physical size of the tube structure becomes large with respect to this frequency, the drive power requirements also will increase.

The drive power requirements of a typical grounded-cathode amplifier consist of six major elements:

- The power consumed by the bias source, given by:

$$P_1 = I_{c1} \times E_{c1} \tag{4.1}$$

- The power dissipated in the grid as a result of rectified grid current:

$$P_2 = I_{c1} \times e_{cmp} \tag{4.2}$$

- The power consumed in the tuned grid circuit:

$$P_3 = I_{c,rms}^2 \times R_{rf} \tag{4.3}$$

- The power lost as a result of transit-time effects:

$$P_4 = \left\{ \frac{e_{c,rms}}{R_t} \right\}^2 \tag{4.4}$$

R_t is that part of the resistive component of the tube input impedance resulting from transit-time effects, and is given by:

$$R_t = \frac{1}{Kg_m f^2 T^2} \qquad (4.5)$$

- The power consumed in that part of the resistive component of the input impedance resulting from cathode lead inductance:

$$P_5 = \frac{e_g^{\;2}}{R_s} \qquad (4.6)$$

Input resistance resulting from the inductance of the cathode leads is found from the following:

$$R_s = \frac{1}{\omega^2 \, g_m \, L_k \, C_{gk}} \qquad (4.7)$$

- The power dissipated in the tube envelope because of dielectric loss:

$$P_6 = 1.41 f E_1^{\;2} \, \varepsilon \qquad (4.8)$$

Where:
I_{c1} = dc grid current
E_{c1} = dc grid voltage
E_{cmp} = maximum positive grid voltage
$I_{c,rms}$ = rms value of RF grid current
R_{rf} = RF resistance of grid circuit
$e_{c,rms}$ = rms value of RF grid voltage
R_t = resistance resulting from transit-time loading
K = a constant (function of the tube geometry)
g_m = transconductance
f = frequency (Hz)
T = transit-time, cathode to grid
R_s = cathode lead inductance input resistance loading
$\omega = 2\pi f$
L_k = cathode lead inductance (Henrys)
C_{gk} = grid-to-cathode capacitance (Farads)
E_1 = voltage gradient (kilovolt per inch, rms)
ε = loss factor of dielectric materials

The total drive power P_t is, then, equal to:

$$P_t = P_1 + P_2 + P_3 + P_4 + P_5 + P_6 \qquad (4.9)$$

Particular attention must be given to grid dissipation when a tube is operated in the VHF and UHF regions. The total driving power required for a given output may be greater than the grid dissipation capability of the device.

Operational Considerations for VHF and UHF

For operation of a tube in the VHF and UHF regions, several techniques may be applied to minimize the driving power without appreciably affecting plate conversion efficiency. The most common techniques are:

- Use the minimum dc control bias. Frequently, it is advisable to bring the bias down to approximately cutoff.

- Maintain a high value of dc screen voltage, even though it appears to increase the fraction of the cycle during which plate current flows.

- Use the minimum RF excitation voltage necessary to obtain the desired plate circuit performance, even though the dc grid current is considerably lower than would be expected at lower frequencies.

- Keep the cathode lead inductance to the output and input circuits as low as possible. This can be accomplished by 1) using short and wide straps, 2) using two separate return paths for the input and output circuits, or 3) properly choosing a cathode bypass capacitor(s).

These techniques do not necessarily decrease the plate efficiency significantly when the circuit is operated at VHF and UHF. The steps should be tried experimentally to determine whether the plate circuit efficiency is appreciably affected. It is usually acceptable, and even preferable, to sacrifice some plate efficiency for improved tube life when operating at VHF and UHF.

Optimum power output at these frequencies is obtained when the loading is greater than would be used at lower frequencies. Fortunately, the same condition reduces driving power and screen current (for the tetrode and pentode), and improves tube life expectancy in the process.

4.2.2 Mechanical and Electrical Considerations

To maintain proper isolation of the output and input circuits, careful consideration must be given to the location of the component parts of the amplifier. All elements of the grid or input circuit and any earlier stages must be kept out of the plate circuit compartment. Similarly, plate circuit elements must be kept out of the input compartment. Note, however, that in the case of the tetrode and pentode, the screen lead of the tube and connections via the socket are common to both the output and input resonant circuits. Because of the plate-to-screen capacitance of a tetrode or pentode, the RF plate voltage (developed in the output circuit) causes an RF current to flow out of the screen lead to the chassis. In the case of a push-pull stage, this current may flow from the screen terminal of one tube to the screen terminal of the other tube. Similarly, because of the grid-to-screen capacitance of the tube, the RF voltage in the input circuit will cause an RF current to flow in this same screen lead to the chassis, or to the opposite tube of the push-pull circuit.

The inductance of the lead common to both the output and input circuits has the desirable feature of providing voltage of opposite polarity to neutralize the feedback voltage of the residual plate to control grid capacitance of the tube. (The properties of neutralization are discussed in Section 3.4.) Note, however, that the mutual coupling from the screen lead to the input resonant circuit may be a possible source of trouble, depending on the design.

Lead Lengths

The interconnecting lead wires close to the tube should be designed with low inductance to minimize the generation of VHF parasitic oscillations. If two or more tubes are used in a given circuit, they should be placed reasonably close together to facilitate short interconnecting leads. The lead lengths of HF circuits usually can be much longer; the length depends, to a large extent, upon the frequency of the fundamental. All of the dc, keying, modulation, and control circuit wires can be relatively long if properly filtered and arranged physically apart from the active RF circuits. The following interconnecting lead wires in a tetrode or pentode power amplifier should have low inductance:

- Filament and screen bypass leads

- Suppressor bypass leads

- Leads from the grid and the plate to the tuning capacitor of the RF circuit (and return)

- Interconnections from tube to tube in push-pull or parallel arrangements (except for parasitic suppressors in the plate)

For a lead to have low inductance, it must have a large surface and be short in length, such as a strap or a ribbon. This consideration also applies to that portion of a lead inside a bypass capacitor.

Power Supply Considerations

The power supply requirements for a triode are straightforward. The degree of regulation and ripple depends upon the requirements of the system. In the case of a linear RF amplifier, it is important to have good plate power supply regulation. Without tight regulation, the plate voltage will drop during the time the plate is conducting current heavily. This drop will cause *flat topping* and will appear as distortion in the output. In push-pull applications where grid current flows, it is important to keep the grid circuit resistance to a minimum. If this is not done, positive peak clipping will occur.

In the case of the tetrode and pentode, the need for screen voltage introduces some new considerations and provides some new operating possibilities. Voltage for the screen grid of a low-power tetrode or pentode can readily be taken from the power supply used for the plate of the tube. In this case, a series resistor or potential-dividing resistor is chosen so that, with the intended screen current flowing, the voltage drop through the resistor is adequate to give the desired screen voltage. The potential-dividing resistor is the preferred technique for those tubes with significant secondary screen emission.

For high-power tubes, screen voltage commonly is taken from a separate power supply. A combination scheme also may be employed, where a dropping resistor is used in conjunction with a low-voltage or intermediate-voltage supply. Often, a combination of series resistor and voltage source can be chosen so that the rated screen dissipation will not be exceeded, regardless of variations in screen current. With a fixed screen supply, there are advantages in using an appreciable amount of fixed grid bias so as to provide protection against the loss of excitation, or for cases where the driver stage is being keyed.

If the screen voltage is taken through a dropping resistor from the plate supply, there is usually little point in using a fixed grid bias because an unreasonable amount of bias would be required to protect the tube if the excitation failed. Under operating conditions with normal screen voltage, the cutoff bias is low (screen voltage divided by the screen μ). When a stage loses excitation and runs statically, the screen current falls to nearly zero. (See the static curves of the tube in question.) If the screen voltage is obtained through a simple dropping resistor from the plate supply, the screen voltage will then rise close to full plate voltage. Because the cutoff bias required is proportional to screen voltage, the grid bias required will be much greater than the amount of bias desired under normal operating conditions. When a screen dropping resistor is used, most of the bias normally is supplied through a grid resistor, and other means are used for tube protection.

The power output from a tetrode or pentode is sensitive to screen voltage. For this reason, any application requiring a high degree of linearity through the amplifier requires a well-regulated screen power supply. A screen dropping resistor from the plate supply is not recommended in such cases.

The suppressor grid power supply requirements are similar to those of the control grid power supply. The suppressor grid intercepts little current, so a low-power supply may be used. Any variation in suppressor voltage as a result of ripple or lack of regulation will appear at the output of the amplifier because of suppressor grid modulation of the plate current.

4.2.3 Bypassing Tube Elements

Operation at high frequencies requires attention to bypassing of the tube elements. Areas of concern include:

- Filament circuit
- Screen grid circuit
- Suppressor grid circuit

Filament Bypassing

Low-inductance bypass capacitors should be used at the filament in an RF amplifier. Good engineering practice calls for placement of the capacitor directly between the filament socket terminals. If the circuit design allows it, strap one filament directly to the chassis; if not, use a second bypass capacitor from one terminal to the chassis.

If two or more tubes are used in a push-pull or parallel circuit, a short strap interconnecting one of the filament terminals of each socket can be used. The midpoint of the interconnecting strap is then bypassed or grounded directly.

For tubes equipped with an isolating screen cone terminal, the general circuit arrangement is usually different. The filament or cathode should go directly, or through bypass capacitors, to the cavity wall or chassis to which the screen terminal is bypassed.

Screen and Suppressor Grid Bypassing

Low-inductance leads are generally advisable for screen and suppressor grid terminal connections. For all frequencies, good engineering practice calls for routing of the screen bypass capacitors directly from the screen to one filament terminal. The suppressor grid is bypassed in the same manner when the suppressor is operated with a potential other than cathode potential. With the suppressor operating at cathode potential, the suppressor should be grounded to the chassis directly in a circuit where the cathode is at chassis potential. This applies to tubes in push-pull as well as tubes in single-ended stages.

In the VHF region, the connection to the screen terminals—for those tubes with two screen pins—should be made to the midpoint of a strap placed between the two screen terminals of the socket. This arrangement provides for equal division of the RF currents in the screen leads and minimizes heating effects.

At operating frequencies above the self-neutralizing frequency of the tetrode or pentode being used, the screen bypass capacitors are sometimes variable. By proper adjustment of this variable capacitor, the amount and phase of the screen RF voltage can be made to cancel the effects of the feedback capacitance within the tube. Thus, neutralization is accomplished. The screen lead inductance and the variable capacitor are not series resonant. The variable capacitor is adjusted so that a net inductive reactance remains to provide the proper voltage and phase for neutralization.

The preceding guidelines apply directly to tubes having the screen and suppressor grids mounted on internal supporting lead rods. Tube types having isolating screen cone terminals tend to work best when the screen or suppressor bypass capacitor is a flat sandwich type built directly onto the peripheral screen-contacting collet of the socket. The size of the bypass capacitor is a function of the operating frequency. The dielectric material may be of Teflon, mica, isomica, or a similar material. (Teflon is a trademark of DuPont.)

Application Examples

The use of bypassing capacitors for a grounded-cathode RF amplifier is illustrated in Figure 4.7. Bypassing for a grounded-grid amplifier is shown in Figure 4.8. The con-

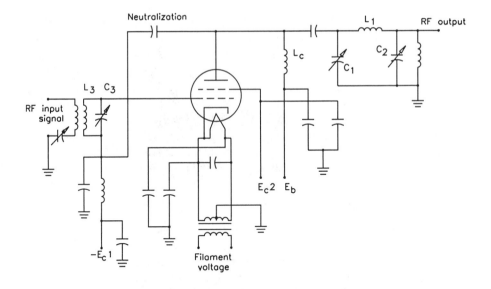

Figure 4.7 Schematic of a tetrode RF power amplifier showing tube element bypass capacitors.

trol grid may be operated at dc ground potential, shown in Figure 4.8*a*, or bypassed directly at the socket, shown in Figure 4.8*b*. For the sake of simplicity, operating the grid at dc ground potential is preferred. Functionally, these two circuits are identical. They vary only in the method used to measure grid current. In Figure 4.8*a*, grid current is measured in the return lead from ground to filament. In Figure 4.8*b*, the grid is raised 1 Ω above dc ground to allow grid current to be measured.

4.2.4 Parasitic Oscillations

Most self-oscillations in RF power amplifiers using gridded tubes have been found to fall within the following three classes:

- Oscillation at VHF from about 40 to 200 MHz, regardless of the operating frequency of the amplifier

- Self-oscillation on the fundamental frequency of the amplifier

- Oscillation at a low radio frequency below the normal frequency of the amplifier

Low-frequency oscillation in an amplifier usually involves the RF chokes, especially when chokes are used in both the output and input circuits. Oscillation near the fundamental frequency involves the normal resonant circuits and brings up the question of neutralization. (Neutralization is discussed in Section 3.4.) When a parasitic

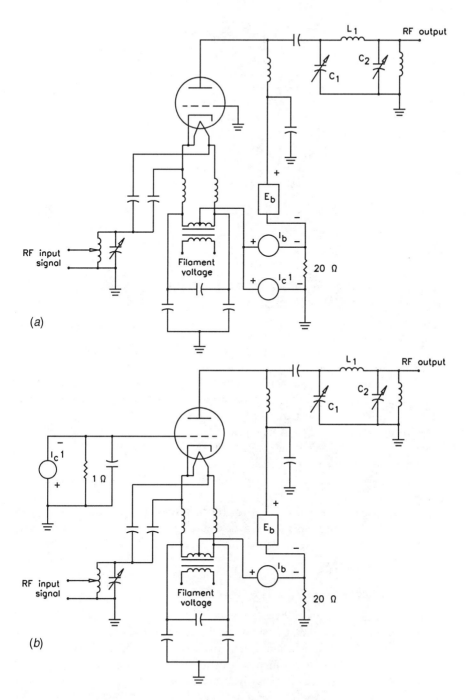

Figure 4.8 Schematic of a triode RF power amplifier showing tube element bypass capacitors: (*a*) control grid at dc ground potential, (*b*) control grid bypassed to ground.

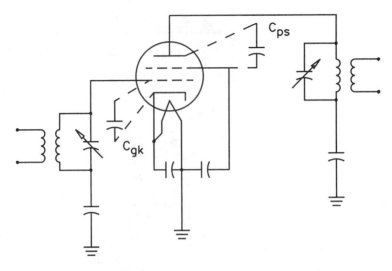

Figure 4.9 Interelectrode capacitances supporting VHF parasitic oscillation in a high-frequency RF amplfier.

self-oscillation is found on a very high frequency, the interconnecting leads of the tube, tuning capacitor, and bypass capacitors typically are involved.

VHF oscillation occurs commonly in amplifiers where the RF circuits consist of coils and capacitors, as opposed to cavity designs. As illustrated in Figure 4.9, the tube capacitances effectively form a tuned-plate, tuned-grid oscillator.

The frequency of a VHF parasitic typically is well above the self-neutralizing frequency of the tube. However, if the self-neutralizing frequency of the device can be increased and the frequency of the parasitic lowered, complete suppression of the parasitic may result, or its suppression by resistor-inductor parasitic suppressors may be made easier.

It is possible to predict, with the use of a grid dip wavemeter, the parasitic frequency to be expected in a given circuit. The circuit should be complete and have no voltages applied to the tube. Couple the meter to the plate or screen lead, and determine the resonant frequency.

Elimination of the VHF parasitic oscillation may be accomplished using one of the following techniques:

- Place a small coil and resistor combination in the plate lead between the plate of the tube and the tank circuit, as shown in Figure 4.10*a*. The resistor-coil combination usually is made up of a noninductive resistor of about 25 to 100 Ω, shunted by three or four turns of approximately $\frac{1}{2}$ in diameter and frequently wound around the resistor. In some cases, it may be necessary to use such a suppressor in both the plate and grid leads. The resistor-coil combination operates on the principle that the resistor loads the VHF circuit but is shunted by the coil for the lower fundamental frequency. In the process of adjusting the resistor-coil combination, it is

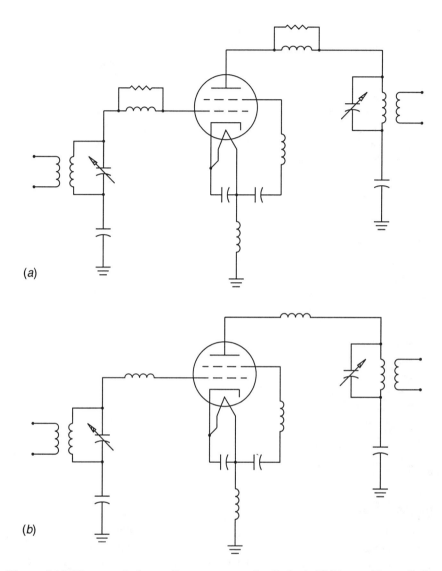

(a)

(b)

Figure 4.10 Placement of parasitic suppressors to eliminate VHF parasitic oscillations in a high-frequency RF amplifier: (a) resistor-coil combination, (b) parasitic choke.

often found that the resistor runs hot. This heat usually is caused by the dissipation of fundamental power in the resistor, and it is an indication of too many turns in the suppressor coil. Use only enough turns to suppress the parasitic, and no more. Once the parasitic has been suppressed, there will be no parasitic voltage or current present. Therefore, there is no parasitic power to be dissipated.

- Use small parasitic chokes in the plate lead, as shown in Figure 4.10*b*. The size of the coil will vary considerably depending upon the tube and circuit layout. A coil of four to ten turns with a diameter of approximately ½ in is typical. The presence of the choke in the frequency-determining part of the circuit lowers the frequency of a possible VHF parasitic so that it falls near the self-neutralizing frequency of the tube and bypass leads. In addition to variation in the size of the suppressor choke, the amount of inductance common to the screen and filament in the filament grounding strap may be a factor. This parameter can be varied simultaneously with the suppressor choke.

Of the two methods outlined for suppressing VHF parasitic oscillations, the first is probably the simpler and is widely employed.

Dynatron Oscillation

Another form of commonly encountered self-oscillation is known as *dynatron oscillation*, caused when an electrode in a vacuum tube has *negative resistance*. At times, there may be more electrons leaving the screen grid than arriving. If the screen voltage is allowed to increase under these conditions, even more electrons will leave the grid. This phenomenon implies a negative resistance characteristic. If there is high alternating current impedance in the circuit from the screen grid through the screen grid power supply, and from the plate power supply to the plate, dynatron oscillation may be sustained.

Dynatron oscillation typically occurs in the region of 1 to 20 Hz. This low-frequency oscillation usually is accompanied by another oscillation in the 1 to 2 kHz region. Suppression of these oscillations can be accomplished by placing a large bypass capacitor (1000 μF) across the output of the screen grid power supply. The circuit supporting the oscillation also can be detuned by a large inductor. Increasing the circuit losses at the frequency of oscillation is also effective.

Harmonic Energy

It is generally not appreciated that the pulse of grid current contains energy on harmonic frequencies and that control of these harmonic energies may be important. The ability of the tetrode and pentode to isolate the output circuit from the input over a wide range of frequencies is significant in avoiding feedthrough of harmonic voltages from the grid circuit. Properly designed tetrode and pentode amplifiers provide for complete shielding in the amplifier layout so that coupling external to the tube is prevented.

In RF amplifiers operating either on the fundamental or on a desired harmonic frequency, the control of unwanted harmonics is important. The following steps permit reduction of the unwanted harmonic energies present in the output circuit:

- Keep the circuit impedance between the plate and cathode low for the high harmonic frequencies. This requirement may be achieved by having some or all of the tuning capacitance of the resonant circuit close to the tube.

- Completely shield the input and output compartments.

- Use inductive output coupling from the resonant plate circuit and possibly a capacitive or Faraday shield between the coupling coil and the tank coil, or a high-frequency attenuating circuit such as a *pi* or *pi-L* network.

- Use low-pass filters on all supply leads and wires coming into the output and input compartments.

- Use resonant traps for particular frequencies.

- Use a low-pass filter in series with the output transmission line.

4.2.5 Shielding

In an RF amplifier, shielding between the input and output circuits must be considered. Triode amplifiers are more tolerant of poor shielding because power gain is relatively low. If the circuit layout is reasonable and no inductive coupling is allowed to exist, a triode amplifier usually can be built without extensive shielding. Even if shielding is not necessary to prevent fundamental-frequency oscillation, it will aid in eliminating any tendency toward parasitic oscillation. The higher the gain of an amplifier, the more important the shielding.

Pierced Shields

Tetrode and pentode power amplifiers require comprehensive shielding to prevent input-to-output circuit coupling. It is advisable to use nonmagnetic materials such as copper, aluminum, or brass in the RF fields to provide the shielding. Quite often, a shield must have holes through it to allow the passage of cooling air. In the LF and part of the HF range, the presence of small holes will not impair shield effectiveness. As the frequency is increased, however, the RF currents flowing around the hole in one compartment cause fields to pass through the hole into another compartment. Currents, therefore, are induced on the shield in the second compartment. This problem can be eliminated by using holes that have significant length. A piece of pipe with a favorable length-to-diameter ratio as compared to the frequency of operation will act as a *waveguide-beyond-cutoff* attenuator.

If more than one hole is required to pass air, a material resembling a honeycomb may be used. This material is available commercially and provides excellent isolation with a minimum of air pressure drop. A section of honeycomb shielding is shown in Figure 4.11. Some tube sockets incorporate a waveguide-beyond-cutoff air path. These sockets allow the tube in the amplifier to operate at high gain and up through VHF.

Metal Base Shells and Submounted Sockets

Some tetrodes and pentodes are supplied with metal base shells. The shell is typically grounded by the clips provided with the socket. This completes the shielding between

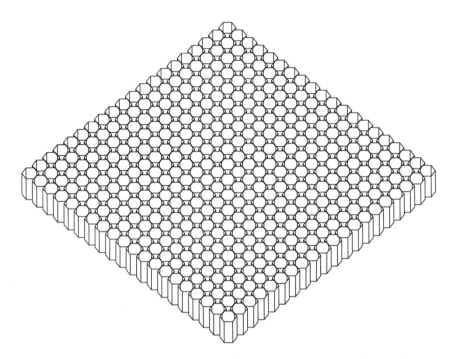

Figure 4.11 A section of honeycomb shielding material used in an RF amplifier.

the output and input circuits because the base shell of the tube comes up opposite the screen shield within the tube itself.

Some pentodes use this metal base shell as the terminal for the suppressor grid. If the suppressor is to be at some potential other than ground, then the base shell must not be dc grounded. The base shell is bypassed to ground for RF and insulated from ground for dc.

There is a family of tetrodes and pentodes without the metal base shell. For this type of tube structure, it is good practice to submount the socket so that the internal screen shield is at the same level as the chassis deck. This technique will improve the input-to-output circuit shielding. In submounting a tube, it is important that adequate clearance be provided around the base of the device for the passage of cooling air.

Compartments

By placing the tube and related circuits in completely enclosed compartments and properly filtering incoming supply wires, it is possible to prevent coupling of RF energy out of the circuit by means other than the desired output coupling. Such filtering prevents the coupling of energy that may be radiated or fed back to the input section or to earlier stages in the amplifier chain. Energy fed back to the input circuit causes undesirable interaction in tuning and/or self-oscillation. If energy is fed back to ear-

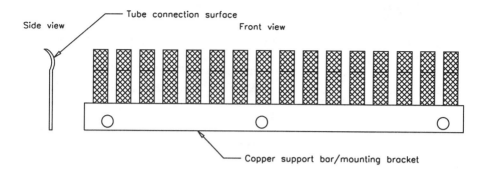

Figure 4.12 A selection of common fingerstock.

lier stages, significant operational problems may result because of the large power gain over several stages.

In the design of an RF amplifier, doors or removable panels typically must be used. The requirement for making a good, low-resistance joint at the discontinuity must be met. There are several materials available commercially for this purpose. Fingerstock, shown in Figure 4.12, has been used for many years. *Teknit*[1] is also a practical solution. Sometimes, after the wiring has been completed, further shielding of a particular conductor is required. Various types of shielding tapes[2] can be wound on a conductor as a temporary or permanent solution.

4.2.6 Protection Measures

Power grid tubes are designed to withstand considerable abuse. The maximum ratings for most devices are conservative. For example, the excess anode dissipation resulting from detuning the plate circuit will have no ill effects on most tubes if it is not applied long enough to overheat the envelope and the seal structure.

Similarly, the control, screen, and suppressor grids will stand some excess dissipation. Typically, the maximum dissipation for each grid indicated on the data sheet should not be exceeded except for time intervals of less than 1 s. The maximum dissipation rating for each grid structure is usually considerably above typical values used for maximum output so that ample operating reserve is provided. The time of duration of overload on a grid structure is necessarily short because of the small heat-storage capacity of the grid wires. Furthermore, grid temperatures cannot be measured or seen, so no warning of accidental overload is apparent.

---

1 Technical Wire Products, Inc., Cranford, NJ
2 Perfection Mica Co., Magnetic Shield Division, Chicago, IL

Table 4.1 Protection Guidelines for Tetrode and Pentode Devices

Circuit Failure Type	Fixed Screen Supply		Screen Voltage Through Dropping Resistor	
	Fixed Grid Bias	**Resistor Grid Bias**	**Fixed Grid Bias**	**Resistor Grid Bias**
Loss of excitation	No protection required	Plate current relay	Plate current relay	Plate current relay or screen control circuit
Loss of antenna loading	Screen current relay	Screen current relay	Grid current relay	No protection required
Excess antenna loading	Screen under-current relay	Screen under-current relay	Plate current relay	Plate current relay
Failure of plate supply	Screen current relay	Screen current relay	Grid current relay	No protection required
Failure of screen supply	Grid current relay	No protection required	Does not apply	Does not apply
Failure of grid bias supply	Plate current relay or screen current relay	Does not apply	Plate current relay or grid current relay	Does not apply

The type and degree of protection required in an RF amplifier against circuit failure varies with the type of screen and grid voltage supply. Table 4.1 lists protection criteria for tetrode and pentode devices. The table provides guidelines on the location of a suitable relay that should act to remove the principal supply voltage from the stage or transmitter to prevent damage to the tube.

For designs where screen voltage is taken through a dropping resistor from the plate supply, a plate relay provides almost universal protection. For a fixed screen supply, a relay provides protection in most cases. For protection against excessive antenna loading and subsequent high plate dissipation, a screen undercurrent relay also may be used in some services.

The plate, screen, and bias voltages may be applied simultaneously to a tetrode. The same holds true for a pentode, plus the application of the suppressor voltage. In a grid-driven amplifier, grid bias and excitation usually can be applied alone to the tube, especially if a grid leak resistor is used. Plate voltage can be applied to the tetrode and pentode before the screen voltage with or without excitation to the control grid. Never apply screen voltage before the plate voltage. The only exceptions are when the tube is cut off so that no space current (screen or plate current) will flow, or when the excitation and screen voltage are low. If screen voltage is applied before the plate voltage and screen current can flow, the maximum allowable screen dissipation will almost always be exceeded, and tube damage will result.

Table 4.2 lists protection guidelines for a triode. The table covers the grid-driven triode amplifier and the high-μ (zero-bias) cathode-driven triode amplifier. Drive voltage must never be applied to a zero-bias triode amplifier without plate voltage. The table indicates the recommended location of a suitable relay that should act to remove the prin-

Table 4.2 Protection Guidelines for a Triode

Circuit Failure Type	Triode		Zero-Bias Triode
	Fixed Grid Bias	**Resistor Grid Bias**	
Loss of excitation	No protection required	Plate overcurrent relay	No protection required
Loss of antenna loading	RF output detector and relay	RF output detector and relay	Grid overcurrent relay
Excess antenna loading	RF output detector and relay	RF output detector and relay	RF output detector and relay
Failure of plate supply	No protection required	No protection required	Grid overcurrent relay
Failure of grid bias supply	Plate overcurrent relay	Does not apply	Does not apply

cipal supply voltage from the stage or transmitter to prevent damage to the tube or transmitter.

4.3 Cavity Amplifier Systems

Power grid tubes are ideally suited for use as the power generating element in a cavity amplifier. Because of the physical dimensions involved, cavity designs typically are limited to VHF and above. Lower-frequency RF generators utilize discrete L and C devices to create the needed tank circuit. Two types of cavity amplifiers commonly are used: $\frac{1}{4}$-wavelength and $\frac{1}{2}$-wavelength systems.

In a cavity amplifier, the tube becomes part of a resonant transmission line. The stray interelectrode and distributed capacity and inductance of the tube are used to advantage as part of the resonant line. This resonant line is physically larger than the equivalent lumped constant LRC resonant circuit operating at the same frequency, and this larger physical size aids in solving the challenges of high-power operation, skin effect losses, and high-voltage-standoff concerns.

A shorted $\frac{1}{4}$-wavelength transmission line has a high, purely resistive input impedance. Electrically, it appears as a parallel resonant circuit, as shown in Figure 4.13. When the physical length of the line is less than $\frac{1}{4}$ wavelength, the impedance will be lower and the line will appear inductive, as illustrated in Figure 4.14. This inductance is used to resonate with the capacitive reactance in the tube and the surrounding circuit.

4.3.1 Bandwidth and Efficiency

Power amplifier bandwidth has a significant effect on modulation performance. Available bandwidth determines the amplitude response, phase response, and group-delay response. Performance tradeoffs often must be considered in the design of a cavity amplifier, including bandwidth, gain, and efficiency.

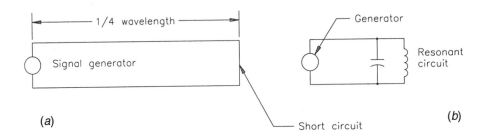

Figure 4.13 Shorted ¼-wavelength line: (a) physical circuit, (b) electrical equivalent.

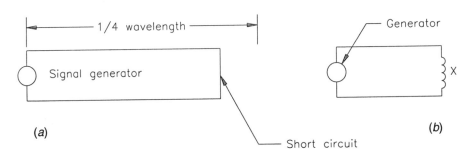

Figure 4.14 Shorted line less than ¼ wavelength: (a) physical circuit, (b) electrical equivalent.

Power amplifier bandwidth is restricted by the equivalent load resistance across the parallel tuned circuits in the stage. Tuned circuits are necessary to cancel low reactive impedance presented by the relatively high input and output capacitances of the amplifying device. The bandwidth for a single tuned circuit is proportional to the ratio of capacitive reactance X_c to load resistance R_l appearing across the tuned circuit:

$$BW \approx \frac{K}{2\pi f_c \, C \, R_l} \approx \frac{K \, (X_c)}{R_l} \tag{4.10}$$

Where:
BW = bandwidth between half-power (- 3 dB) points
K = proportionality constant
R_l = load resistance appearing across tuned circuit
C = total capacitance of tuned circuit (includes stray capacitances and output or input capacitances of the tube)
X_c = capacitive reactance of C
f_c = carrier frequency

The RF voltage swing across the tuned circuit also depends on the load resistance. For the same power and efficiency, the bandwidth can be increased if the capacitance is reduced.

The efficiency of a cavity PA depends primarily on the RF plate voltage swing, the plate current conduction angle, and the cavity efficiency. Cavity efficiency is related to the ratio of loaded to unloaded Q as follows:

$$N = 1 - \frac{Q_l}{Q_u} \times 100 \tag{4.11}$$

Where:
N = cavity efficiency in percent
Q_l = cavity loaded Q
Q_u = cavity unloaded Q

The loaded Q is determined by the plate load impedance and output circuit capacitance. Unloaded Q is determined by the cavity volume and the RF resistivity of the conductors resulting from the skin effect. (Skin effect is discussed in Chapter 7.) For best cavity efficiency, the following conditions are desirable:

- High unloaded Q

- Low loaded Q

As the loaded Q increases, the bandwidth decreases. For a given tube output capacitance and power level, loaded Q decreases with decreasing plate voltage or with increasing plate current. The increase in bandwidth at reduced plate voltage occurs because the load resistance is directly related to the RF voltage swing on the tube anode. For the same power and efficiency, the bandwidth also can be increased if the output capacitance is reduced. Power tube selection and minimization of stray capacitance are important considerations when designing for maximum bandwidth.

4.3.2 Current Paths

The operation of a cavity amplifier is an extension of the current paths inside the tube. Two elements must be examined in this discussion:

- The input circulating currents

- The output circulating currents

Input Circuit

The grid/cathode assembly resembles a transmission line whose termination is the RF resistance of the electron stream within the tube. Figure 4.15 shows the current path of an RF generator (the RF driver stage output) feeding a signal into the grid/cathode circuit.

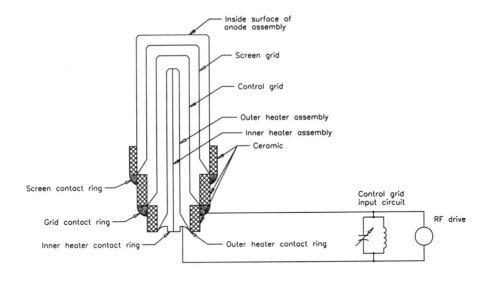

Figure 4.15 A simplified representation of the grid input circuit of a PA tube.

The outer contact ring of the cathode heater assembly makes up the inner conductor of a transmission line formed by the cathode and control grid assemblies. The filament wires are returned down the center of the cathode. For the input circuit to work correctly, the cathode must have a low RF impedance to ground. This cathode bypassing may be accomplished in one of several ways.

Below 30 MHz, the cathode can be grounded to RF voltages by simply bypassing the filament connections with capacitors, as shown in Figure 4.16a. Above 30 MHz, this technique does not work well because of the stray inductance of the filament leads. Notice that in (b), the filament leads appear as RF chokes, preventing the cathode from being placed at RF ground potential. This causes negative feedback and reduces the efficiency of the input and output circuits.

In Figure 4.16c, the cathode circuit is configured to simulate a ½-wave transmission line. The line is bypassed to ground with large-value capacitors ½ wavelength from the center of the filament (at the filament voltage feed point). This transmission line RF short circuit is repeated ½ wavelength away at the cathode (heater assembly), effectively placing it at ground potential.

Because ½-wavelength bypassing is usually bulky at VHF (and may be expensive), RF generators often are designed using certain values of inductance and capacitance in the filament/cathode circuit to create an artificial transmission line that will simulate a ½-wavelength shorted transmission line. As illustrated in the figure, the inductance and capacitance of the filament circuit can resemble an artificial transmission line of ½ wavelength, if the values of L and C are properly selected.

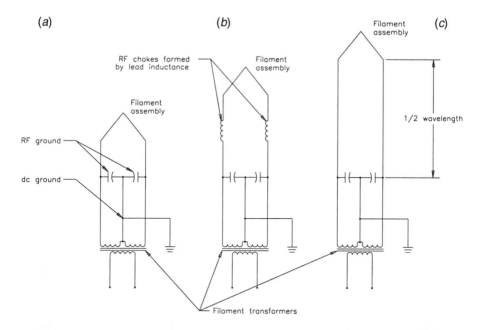

Figure 4.16 Three common methods for RF bypassing of the cathode of a tetrode PA tube: (*a*) grounding the cathode below 30 MHz, (*b*) grounding the cathode above 30 MHz, (*c*) grounding the cathode via a ½-wave transmission line.

Output Circuit

The plate-to-screen circulating current of the tetrode is shown in Figure 4.17. For the purposes of example, consider that the output RF current is generated by an imaginary current generator located between the plate and screen grid. The RF current travels along the inside surface of the plate structure (because of the skin effect), through the ceramic at the lower half of the anode contact ring, across the bottom of the fins, and to the band around the outside of the fins. The RF current then flows through the plate bypass capacitor to the RF tuned circuit and load, and returns to the screen grid.

The return current travels through the screen bypass capacitor (if used) and screen contact ring, up the screen base cone to the screen grid, and back to the imaginary generator.

The screen grid has RF current returning to it, but because of the assembly's low impedance, the screen grid is effectively at RF ground potential. The RF current generator, therefore, appears to be feeding an open-ended transmission line consisting of the anode (plate) assembly and the screen assembly. The RF voltage developed by the anode is determined by the plate impedance (Z_p) presented to the anode by the resonant circuit and its load.

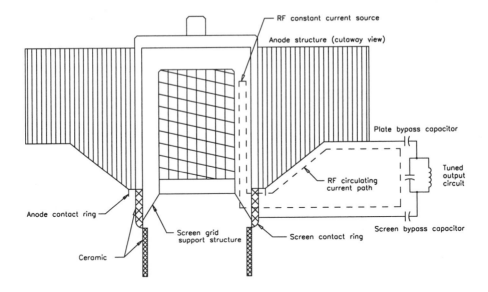

Figure 4.17 RF circulating current path between the plate and screen in a tetrode PA tube.

When current flows on one conductor of a transmission line cavity circuit, an equal-magnitude current flows in the opposite direction on the other conductor. This means that a large value of RF circulating current is flowing in the cavity amplifier outer conductor (the cavity box). All of the outer conductor circulating currents start at and return to the screen grid (in a tetrode-based system). The front or back access panel (door) of the cavity is part of the outer conductor, and large values of circulating current flow into it, through it, and out of it. A mesh contact strap generally is used to electrically connect the access panel to the rest of the cavity.

4.3.3 The 1/4-Wavelength Cavity

The ¼-wavelength PA cavity is common in transmitting equipment. The design is simple and straightforward. A number of variations can be found in different RF generators, but the underlying theory of operation is the same.

A typical ¼-wave cavity is shown in Figure 4.18. The plate of the tube connects directly to the inner section (tube) of the plate-blocking capacitor. The exhaust chimney/inner conductor forms the other plate of the blocking capacitor. The cavity walls form the outer conductor of the ¼-wave transmission line circuit. The dc plate voltage is applied to the PA tube by a cable routed inside the exhaust chimney and inner tube conductor. In the design shown in the figure, the screen-contact fingerstock ring mounts on a metal plate that is insulated from the grounded cavity deck by an insulating material. This hardware makes up the screen-blocking capacitor assembly. The dc screen voltage

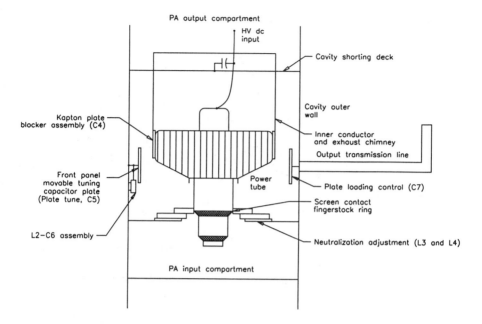

Figure 4.18 The physical layout of a common type of ¼-wavelength PA cavity designed for VHF operation.

feeds to the fingerstock ring from underneath the cavity deck through an insulated feedthrough assembly.

A grounded screen configuration also may be used in this design in which the screen-contact fingerstock ring is connected directly to the grounded cavity deck. The PA cathode then operates at below ground potential (in other words, at a negative voltage), establishing the required screen voltage to the tube.

The cavity design shown in the figure is set up to be slightly shorter than a full ¼ wavelength at the operating frequency. This makes the load inductive and resonates the tube's output capacity. Thus, the physically foreshortened shorted transmission line is resonated and electrically lengthened to ¼ wavelength.

Figure 4.19 illustrates the paths taken by the RF circulating currents in the circuit. RF energy flows from the plate, through the plate-blocking capacitor, along the inside surface of the chimney/inner conductor (because of the skin effect), across the top of the cavity, down the inside surface of the cavity box, across the cavity deck, through the screen-blocking capacitor, over the screen-contact fingerstock, and into the screen grid.

Figure 4.20 shows a graph of RF current, voltage, and impedance for a ¼-wavelength coaxial transmission line. The plot shows that infinite impedance, zero RF current, and maximum RF voltage occur at the feed point. This would not be suitable for a

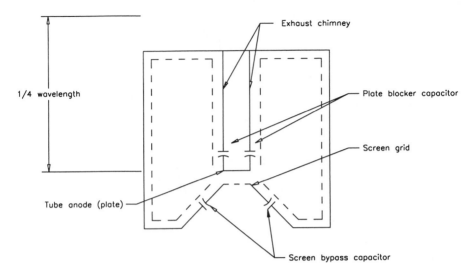

Figure 4.19 The RF circulating current paths for the ¼-wavelength cavity shown in Figure 4.18.

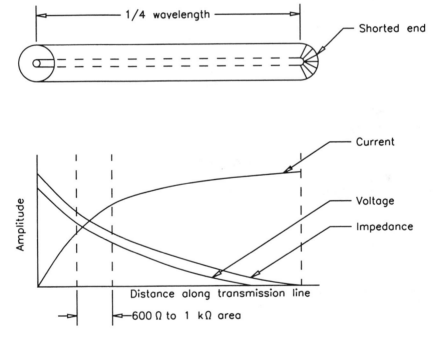

Figure 4.20 Graph of the RF current, RF voltage, and RF impedance for a ¼-wavelength shorted transmission line. At the feed point, RF current is zero, RF voltage is maximum, and RF impedance is infinite.

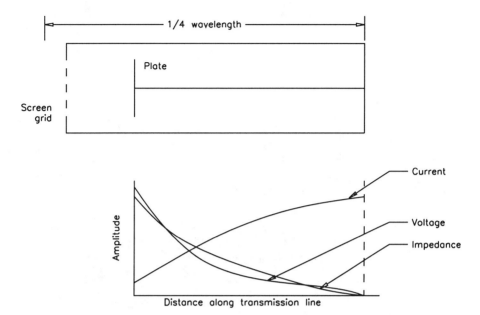

Figure 4.21 Graph of the RF current, RF voltage, and RF impedance produced by the physically foreshortened coaxial transmission line cavity.

practical PA circuit because arcing would result from the high RF voltage, and poor efficiency would be caused by the mismatch between the tube and the load.

Notice, however, the point on the graph marked at slightly less than ¼ wavelength. This length yields an impedance of 600 to 800 Ω and is ideal for the PA plate circuit. It is necessary, therefore, to physically foreshorten the shorted coaxial transmission line cavity to provide the correct plate impedance. Shortening the line also is a requirement for resonating the tube's output capacity, because the capacity shunts the transmission line and electrically lengthens it.

Figure 4.21 shows a graph of the RF current, voltage, and impedance presented to the plate of the tube as a result of the physically foreshortened line. This plate impedance is now closer to the ideal 600 to 800 Ω value required by the tube anode.

Tuning the Cavity

Coarse-tuning of the cavity is accomplished by adjusting the cavity length. The top of the cavity (the cavity shorting deck) is fastened by screws or clamps and can be raised or lowered to set the length of the cavity for the particular operating frequency. Fine-tuning is accomplished by a variable-capacity plate-tuning control built into the cavity. In a typical design, one plate of the tuning capacitor—the stationary plate—is fastened to the inner conductor just above the plate-blocking capacitor. The movable

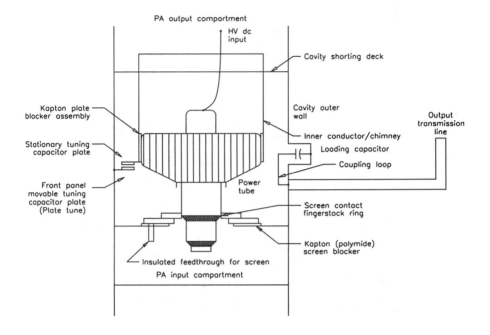

Figure 4.22 Basic loading mechanism for a VHF ¼-wave cavity.

tuning plate is fastened to the cavity box—the outer conductor—and mechanically linked to the front-panel tuning control. This capacity shunts the inner conductor to the outer conductor and is used to vary the electrical length and resonant frequency of the cavity.

The ¼-wavelength cavity is inductively coupled to the output port. This coupling is usually on the side opposite the cavity access door. The inductive pickup loop can take one of several forms. In one design it consists of a half-loop of flat copper bar stock that terminates in the loading capacitor at one end and feeds the output transmission line inner conductor at the other end. This configuration is shown in Figure 4.22. The inductive pickup ideally would be placed at the maximum current point in the ¼-wavelength cavity. However, this point is located at the cavity shorting deck, and, when the deck is moved for coarse-tuning, the magnetic coupling will be changed. A compromise in positioning, therefore, must be made. The use of a broad, flat copper bar for the coupling loop adds some capacitive coupling to augment the reduced magnetic coupling.

Adjustment of the loading capacitor couples the 50 Ω transmission line impedance to the impedance of the cavity. Heavy loading lowers the plate impedance presented to the tube by the cavity. Light loading reflects a higher load impedance to the amplifier plate.

Methods commonly used to increase the operating bandwidth of the cavity include minimizing added tuning capacitance. The ideal case would be to resonate the plate capacitance alone with a "perfect" inductor. Practical cavities, however, require either the

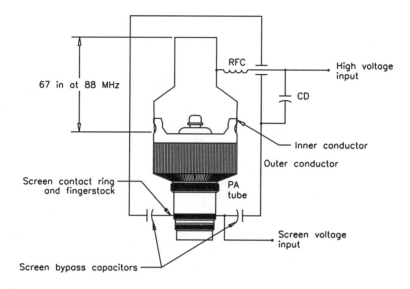

67 in at 88 MHz

RFC

High voltage
input

CD

Inner conductor

Outer conductor

Screen contact ring
and fingerstock

PA
tube

Screen voltage
input

Screen bypass capacitors

Figure 4.23 The ½-wavelength PA cavity in its basic form.

addition of a variable capacitor or a variable inductor using sliding contacts for tuning. Other inherent mechanical and electrical compromises include:

- The requirement for a plate dc-blocking capacitor.

- The presence of maximum RF current at the grounded end of the simulated transmission line where the conductor may be nonhomogeneous. This can result in accompanying losses in the contact resistance.

4.3.4 The 1/2-Wavelength Cavity

The operation of a ½-wavelength cavity follows the same basic idea as the ¼-wavelength design outlined previously. The actual construction of the system depends upon the power level used and the required bandwidth. An example will help to illustrate how a ½-wavelength cavity operates.

The design of a basic ½-wavelength PA cavity for operation in the region of 100 MHz is shown in Figure 4.23. The tube anode and a silver-plated brass pipe serve as the inner conductor of the ½-wave transmission line. The cavity box serves as the outer conductor. The transmission line is open at the far end and repeats this condition at the plate of the tube. The line is, in effect, a parallel resonant circuit for the PA tube.

The physical height of the circuit shown (67 in) was calculated for operation down to 88 MHz. To allow adequate clearance at the top of the transmission line and space for

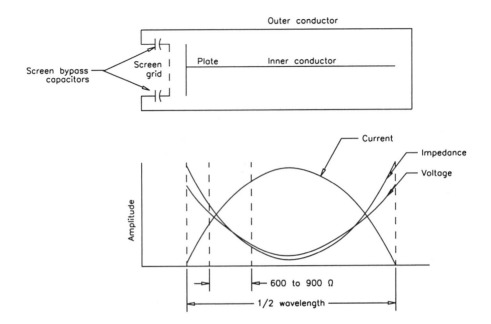

Figure 4.24 The distribution of RF current, voltage, and impedance along the inner conductor of a ½-wavelength cavity.

input circuitry at the bottom of the assembly, the complete cavity box would have to be almost 8 ft tall. This is too large for any practical transmitter.

Figure 4.24 shows RF voltage, current, and impedance for the inner conductor of the transmission line and the anode of the tube. The load impedance at the plate is thousands of ohms. Therefore, the RF current is extremely small, and the RF voltage is extremely large. In the application of such a circuit, arcing would become a problem. The high plate impedance also would make amplifier operation inefficient.

The figure also shows an area between the anode and the ¼-wavelength location where the impedance of the circuit is 600 to 800 Ω. As noted previously, this value is ideal for the anode of the PA tube. To achieve this plate impedance, the inner conductor must be less than a full ½ wavelength. The physically foreshortened transmission-line circuit must, however, be electrically resonated (lengthened) to ½ wavelength for proper operation.

If the line length were changed to operate at a different frequency, the plate impedance also would change because of the new distribution of RF voltage and current on the new length of line. The problem of frequency change, therefore, is twofold: The length of the line must be adjusted for resonance, and the plate impedance of the tube must be kept constant for good efficiency.

To accommodate operation of this system at different frequencies, while keeping the plate impedance constant, two forms of coarse-tuning and one form of fine-tuning

Cavity box

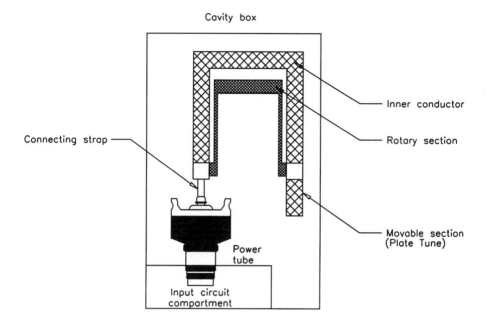

Inner conductor

Rotary section

Connecting strap

Movable section
(Plate Tune)

Power
tube

Input circuit
compartment

Figure 4.25 The configuration of a practical ½-wavelength PA cavity.

are built into the ½-wave PA cavity. Figure 4.25 shows the tube and its plate line (inner conductor). The inner conductor is U-shaped to reduce the cavity height. With the movable section (the plate tune control) fully extended, the inner conductor measures 38 in, and the anode strap measures 7 in. The RF path from the anode strap to the inside of the tube plate (along the surface because of the skin effect) is estimated to be about 8 in. This makes the inner conductor maximum length about 53 in. This is too short to be a ½ wavelength at 88 MHz, the target low-end operating frequency in this example. The full length of a ½-wave line is 67.1 in at 88 MHz.

The coarse-tuning and fine-tuning provisions of the cavity, coupled with the PA tube's output capacity, resonate the plate line to the exact operating frequency. In effect, they electrically lengthen the physically foreshortened line. This process, along with proper loading, determines the plate impedance and, therefore, the efficiency.

Lengthening the Plate Line

The output capacity of the tube is the first element that electrically lengthens the plate line. A ½-wave transmission line that is too short offers a high impedance that is both resistive and inductive. The output capacity of the tube resonates this inductance. The detrimental effects of the tube's output capacity, therefore, are eliminated. The anode strap and the cavity inner conductor rotary section provide two methods of coarse frequency adjustment for resonance.

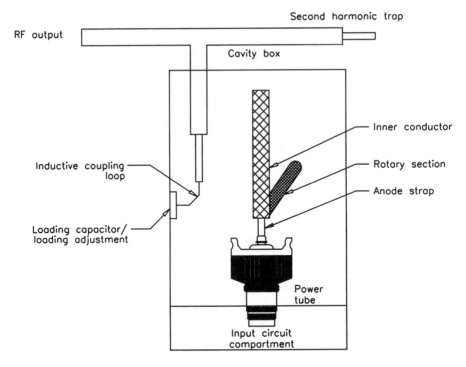

Figure 4.26 Coarse-tuning mechanisms for a ½-wave cavity.

The anode strap, shown in Figure 4.26, has a smaller cross-sectional area than the inner conductor of the transmission line. Therefore, it has more inductance than an equal length of the inner conductor. The anode-coupling strap acts as a series inductance and electrically lengthens the plate circuit.

At low frequencies, one narrow anode strap is used. At midband frequencies, one wide strap is used. The wide strap exhibits less inductance than the narrow strap and does not electrically lengthen the plate circuit as much. At the upper end of the operating band, two anode straps are used. The parallel arrangement lowers the total inductance of the strap connection and adds still less electrical length to the plate circuit.

The main section of the plate resonant line, together with the rotary section, functions as a parallel inductance. RF current flows in the same direction through the transmission line and the rotary section. Therefore, the magnetic fields of the two paths add. When the rotary section is at maximum height, the magnetic coupling between the main section of the transmission line and the rotary assembly is maximum. Because of the relatively large mutual inductance provided by this close coupling, the total inductance of these parallel inductors increases. This electrically lengthens the transmission line and lowers the resonant frequency. The concept is illustrated in Figure 4.27a.

When the rotary section is at minimum height, the magnetic coupling between the two parts of the inner conductor is minimum. This reduced coupling lowers the mutual inductance, which lowers the total inductance of the parallel combination. The reduced

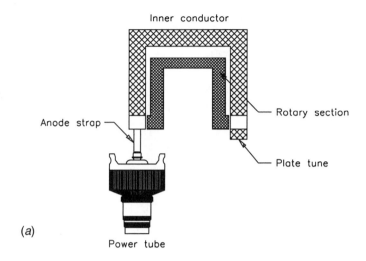

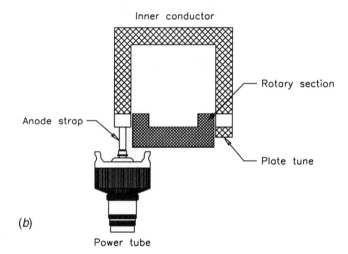

Figure 4.27 Using the cavity's movable section to adjust for resonance: (*a*) the rotary section at maximum height, (*b*) the rotary section at minimum height.

inductance allows operation at a higher resonant frequency. This condition is illustrated in Figure 4.27*b*. The rotary section provides an infinite number of coarse settings for various operating frequencies.

The movable plate-tune assembly is located at the end of the inner-plate transmission line. It can be moved up and down to change the physical length of the inner conductor by about 4-¾ in. This assembly is linked to the front-panel plate-tuning knob, providing a fine adjustment for cavity resonance.

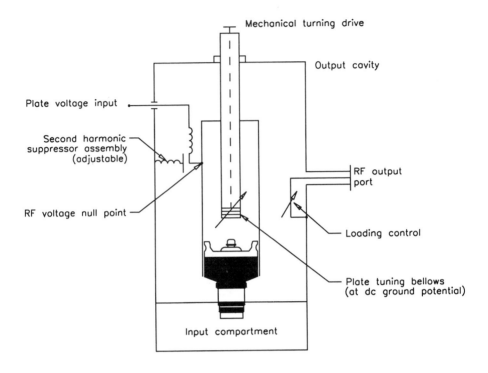

Figure 4.28 The basic design of a folded ½-wavelength cavity.

4.3.5 Folded 1/2-Wavelength Cavity

A special case of the ½-wavelength PA cavity is shown in Figure 4.28. The design employs a folded ½-wave resonator constructed with coaxial aluminum and copper tubing. This cavity arrangement eliminates the high-voltage blocking capacitor and high-current shorting contacts of conventional designs by connecting the main transmission-line resonant circuit conductor directly to the anode of the power tube. A grounded, concentric transmission-line center conductor tunes the cavity with a variable reentrant length inserted into the end of the main conductor opposite the tube.

The main conductor (the fixed portion of the plate line) is insulated from ground and carries the anode dc potential. High-voltage power is fed at the fundamental-frequency RF voltage null point, approximately ¼ wavelength from the anode, for easy RF decoupling. A large surface area without sliding contacts results in minimal loss.

Incorporated into the tank design is a second harmonic suppressor. Rather than attenuating the second harmonic after the signal has been generated and amplified, this design essentially prevents formation of second-harmonic energy by series-*LC* trapping the second-harmonic waveform at the point where the wave exhibits a high impedance (approximately $\frac{1}{4}$ wavelength from the anode). The second harmonic will peak in voltage at the same point that the dc-plate potential is applied.

Plate tuning is accomplished by an adjustable bellows on the center portion of the plate line, which is maintained at chassis ground potential. Output coupling is accomplished with an untuned loop intercepting the magnetic field concentration at the voltage null (maximum RF current) point of the main line. The PA loading control varies the angular position of the plane of the loop with respect to the plate line, changing the amount of magnetic field that it intercepts. Multiple phosphor-bronze leaves connect one side of the output loop to ground and the other side to the center conductor of the output transmission line. This allows for mechanical movement of the loop by the PA loading control without using sliding contacts. The grounded loop improves immunity to lightning and to static buildup on the antenna.

4.3.6 Wideband Cavity

The cavity systems discussed so far in this section provide adequate bandwidth for many applications. Some uses, however, dictate an operating bandwidth beyond what a conventional cavity amplifier can provide. One method of achieving wider operating bandwidth involves the use of a double-tuned overcoupled amplifier, shown in Figure 4.29. The system includes four controls to accomplish tuning:

- *Primary tune*, which resonates the plate circuit and tends to tilt the response and slide it up and down the bandpass.

- *Coupling*, which sets the bandwidth of the PA. Increased coupling increases the operating bandwidth and lowers the PA efficiency. When the coupling is adjusted, it can tilt the response and change the center of the bandpass, necessitating readjustment of the plate-tune control.

- *Secondary tune*, which resonates the secondary cavity and will tilt the response if so adjusted. The secondary tune control typically can be used to slide the response up and down the bandpass, similar to the primary tune control.

- *Loading*, which determines the value of ripple in the bandpass response. Adjustment of the loading control usually tilts the response and changes the bandwidth, necessitating readjustment of the secondary tune and coupling controls.

4.3.7 Output Coupling

Coupling is the process by which RF energy is transferred from the amplifier cavity to the output transmission line. Wideband cavity systems use coupling to transfer energy from the primary cavity to the secondary cavity. Coupling in tube-type power amplifiers usually transforms a high (plate or cavity) impedance to a lower output (transmission line) impedance. Both capacitive (electrostatic) and inductive (magnetic) coupling methods are used in cavity RF amplifiers. In some designs, combinations of the two methods are used.

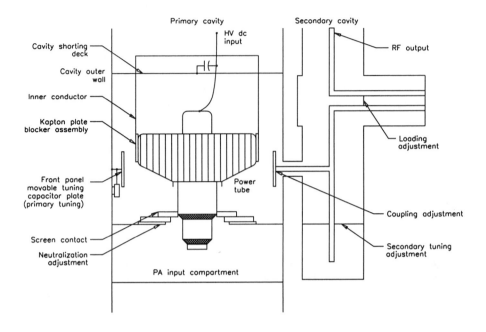

Figure 4.29 Double-tuned overcoupled broadband cavity amplifier.

Inductive Coupling

Inductive (magnetic) coupling employs the principles of transformer action. The efficiency of the coupling depends upon three conditions:

- The cross-sectional area under the coupling loop, compared to the cross-sectional area of the cavity (see Figure 4.30). This effect can be compared to the turns ratio principle of a transformer.

- The orientation of the coupling loop to the axis of the magnetic field. Coupling from the cavity is proportional to the cosine of the angle at which the coupling loop is rotated away from the axis of the magnetic field, as illustrated in Figure 4.31.

- The amount of magnetic field that the coupling loop intercepts (see Figure 4.32). The strongest magnetic field will be found at the point of maximum RF current in the cavity. This is the area where maximum inductive coupling is obtained. Greater magnetic field strength also is found closer to the center conductor of the cavity. Coupling, therefore, is inversely proportional to the distance of the coupling loop from the center conductor.

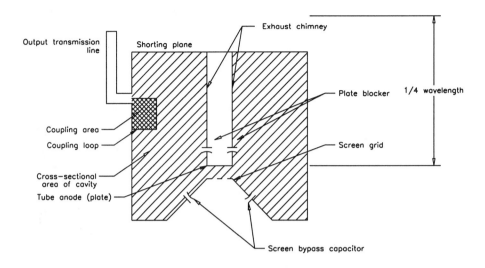

Figure 4.30 The use of inductive coupling in a ¼-wavelength PA stage.

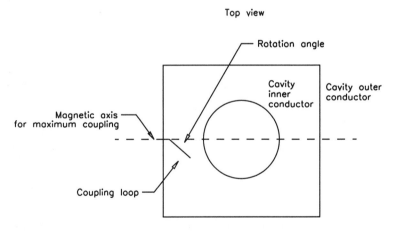

Figure 4.31 Top view of a cavity showing the inductive coupling loop rotated away from the axis of the magnetic field of the cavity.

In both ¼- and ½-wavelength cavities, the coupling loop generally feeds a 50 Ω transmission line (the load). The loop is in series with the load and has considerable inductance at very high frequencies. This inductance will reduce the RF current that flows into the load, thus reducing power output. This effect can be overcome by placing a variable capacitor in series with the output coupling loop. The load is connected to one end of the coupling loop, and the variable capacitor ties the other end of the loop to

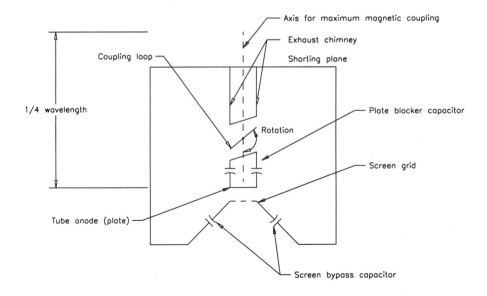

Figure 4.32 Cutaway view of the cavity showing the coupling loop rotated away from the axis of the magnetic field.

ground. The variable capacitor cancels some or all of the loop inductance. It functions as the PA-stage loading control.

Maximum loop current and output power occurs when the loading capacitor cancels all of the inductance of the loading loop. This lowers the plate impedance and results in heavier loading.

Light loading results if the loading capacitance does not cancel all of the loop inductance. The loop inductance that is not canceled causes a decrease in load current and power output, and an increase in plate impedance.

Capacitive Coupling

Capacitive (electrostatic) coupling, which physically appears to be straightforward, often baffles the applications engineer because of its unique characteristics. Figure 4.33 shows a cavity amplifier with a capacitive coupling plate positioned near its center conductor. This coupling plate is connected to the output load, which may be a transmission line or a secondary cavity (for wideband operation). The parameters that control the amount of capacitive coupling include:

- The area of the coupling capacitor plate (the larger the area, the greater the coupling)

- The distance from the coupling plate to the center conductor (the greater the distance, the lighter the coupling)

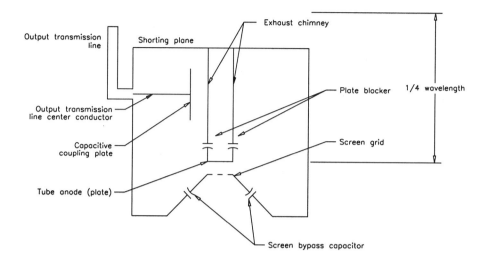

Figure 4.33 A ¼-wavelength cavity with capacitive coupling to the output load.

Maximum capacitive coupling occurs when the coupling plate is at the maximum voltage point on the cavity center conductor.

To understand the effects of capacitive coupling, observe the equivalent circuit of the cavity. Figure 4.34 shows the PA tube, cavity (functioning as a parallel resonant circuit), and output section. The plate-blocking capacitor isolates the tube's dc voltage from the cavity. The coupling capacitor and output load are physically in series, but electrically they appear to be in parallel, as shown in Figure 4.35. The resistive component of the equivalent parallel circuit is increased by the coupling reactance. The equivalent parallel coupling reactance is absorbed into the parallel resonant circuit. This explains the need to retune the plate after changing the PA stage coupling (loading). The coupling reactance may be a series capacitor or inductor.

The series-to-parallel transformations are explained by the following formula:

$$R_p = \frac{R_s^{\,2} + X_s^{\,2}}{R_s} \qquad (4.12)$$

$$X_p = \frac{R_s^{\,2} + X_s^{\,2}}{X_s} \qquad (4.13)$$

Where:
R_p = effective parallel resistance
R_s = actual series resistance

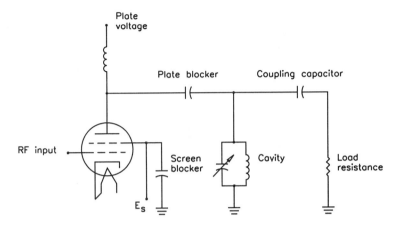

Figure 4.34 The equivalent circuit of a ¼-wavelength cavity amplifier with capacitive coupling.

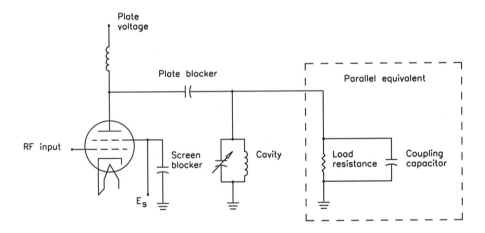

Figure 4.35 The equivalent circuit of a ¼-wavelength cavity amplifier showing how series capacitive coupling appears electrically as a parallel circuit to the PA tube.

X_s = actual series reactance
X_p = effective parallel reactance

PA Loading

Proper loading of a cavity PA stage to the output transmission line is critical to dependable operation and optimum efficiency of the overall system. Light coupling produces light loading and results in a high plate impedance; conversely, heavy coupling results in heavier loading and a lower plate impedance. Maximum output power, coinciding with maximum efficiency and acceptable dissipation, dictates a specific plate impedance for a cavity of given design. This plate impedance is also dependent upon the operating values of dc plate voltage (E_p) and plate current (I_p).

Plate impedance dictates the cavity parameters of loaded Q, RF circulating current, and bandwidth. The relationships can be expressed as follows:

- Loaded Q is directly proportional to the plate impedance and controls the other two cavity parameters. Loaded $Q = X_p / X_l$, where Z_p = cavity plate impedance and X_l = cavity inductive reactance.

- Circulating current in the cavity is much greater (by a factor of the loaded Q) than the RF current supplied by the tube. Circulating current $= Q \times I_p$, where I_p = the RF current supplied to the cavity by the tube.

- The cavity bandwidth is dependent on the loaded Q and operating frequency. Bandwidth $= F_r/Q$, where F_r = the cavity resonant frequency.

Heavy loading lowers the PA plate impedance and cavity Q. A lower Q reduces the cavity RF circulating currents. In some cavities, high circulating currents can cause cavity heating and premature failure of the plate or screen blocking capacitors. The effects of lower plate impedance—a by-product of heavy loading—are higher RF and dc plate currents and reduced RF plate voltage. The instantaneous plate voltage is the result of the RF plate voltage added to the dc plate voltage. The reduced swing of plate voltage causes less positive dc screen current to flow. Positive screen current flows only when the plate voltage swings close to or below the value of the positive screen grid.

Light loading raises the plate impedance and cavity Q. A higher Q will increase the cavity circulating currents, raising the possibility of component overheating and failure. The effects of higher plate impedance are reduced RF and dc plate current and increased RF and dc plate voltage excursions. The higher cavity RF or peak dc voltages may cause arcing in the cavity.

There is one value of plate impedance that will yield optimum output power, efficiency, dissipation, and dependable operation. It is dictated by cavity design and the values of the various dc and RF voltages and currents supplied to the stage.

Depending on the cavity design, light loading may seem deceptively attractive. The dc plate voltage is constant (set by the power supply), and the lower dc plate current resulting from light loading reduces the tube's overall dc input power. The RF output power may change with light loading, depending on the plate impedance and cavity design, while efficiency will probably increase or, at worst, remain constant. Caution must be exercised with light loading, however, because of the increased RF voltages and circulating currents that such operation creates. Possible adverse effects include

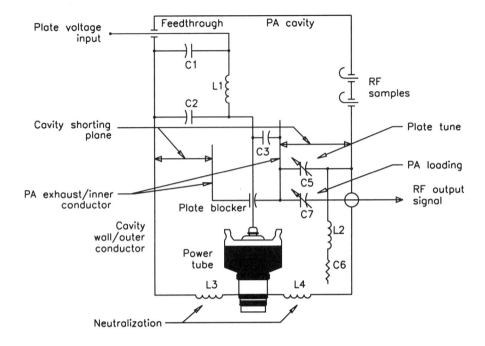

Figure 4.36 Typical VHF cavity amplifier.

cavity arcing and overheating of cavity components, such as capacitors. The manufacturer's recommendations on PA tube loading should, therefore, be carefully observed.

Despite the many similarities among various cavity designs, each imposes its own set of operational requirements and limitations. No two cavity systems will tune up in exactly the same fashion. Given proper maintenance, a cavity amplifier will provide years of reliable service.

4.3.8 Mechanical Design

The operation of a cavity amplifier can be puzzling because of the nature of the major component elements. It is often difficult to relate the electrical schematic diagram to the mechanical assembly that exists within the transmitter. Consider the PA cavity schematic diagram shown in Figure 4.36. The grounded screen stage is of conventional design. Decoupling of the high-voltage power supply is accomplished by C1, C2, C3, and L1. Capacitor C3 is located inside the PA chimney (cavity inner conductor). The RF sample lines provide two low-power RF outputs for a modulation monitor or other test instrument. Neutralization inductors L3 and L4 consist of adjustable grounding bars on the screen grid ring assembly.

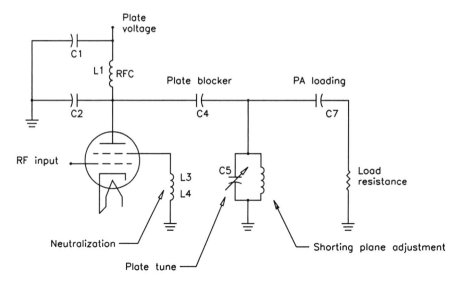

Figure 4.37 Electrical equivalent of the cavity amplifier shown in Figure 4.36.

Figure 4.37 shows the electrical equivalent of the PA cavity schematic diagram. The $\frac{1}{4}$-wavelength cavity acts as the resonant tank for the PA. Coarse-tuning of the cavity is accomplished by adjustment of the shorting plane. Fine-tuning is performed by the PA tuning control, which acts as a variable capacitor to bring the cavity into resonance. The PA loading control consists of a variable capacitor that matches the cavity to the load. The assembly made up of L2 and C6 prevents spurious oscillations within the cavity.

Blocking capacitor C4 is constructed of a roll of *Kapton* insulating material sandwiched between two circular sections of aluminum. (Kapton is a registered trademark of Du Pont.) PA plate-tuning control C5 consists of an aluminum plate of large surface area that can be moved in or out of the cavity to reach resonance. PA loading control C7 is constructed much the same as the PA tuning assembly, with a large-area paddle feeding the harmonic filter, located external to the cavity. The loading paddle may be moved toward the PA tube or away from it to achieve the required loading. The L2-C6 damper assembly actually consists of a 50 Ω noninductive resistor mounted on the side of the cavity wall. Component L2 is formed by the inductance of the connecting strap between the plate-tuning paddle and the resistor. Component C6 is the equivalent stray capacitance between the resistor and the surrounding cavity box.

It can be seen that cavity amplifiers involve as much mechanical engineering as they do electrical engineering. The photographs of Figure 4.38 show graphically the level of complexity that a cavity amplifier may involve. Figure 4.38a depicts a VHF power amplifier (Philips) with broadband input circuitry. Figure 4.38b shows a wideband VHF amplifier intended for television service (Varian). Figure 4.38c illustrates a VHF cavity amplifier designed for FM broadcast service (Varian).

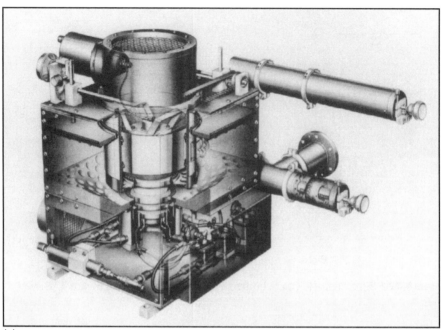

(a)

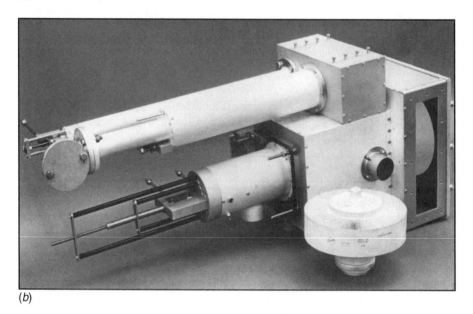

(b)

Figure 4.38 VHF cavity amplifiers: (a) cross-sectional view of a broadband design, (b) cavity designed for television service, (c, next page) FM broadcast cavity amplifier.

(c)

Figure 4.38c

4.4   High-Voltage Power Supplies

Virtually all power grid and microwave vacuum tubes require one or more sources of high voltage to operate. The usual source of operating power is a single-phase or multiphase supply operating from ac line current. Most supplies use silicon rectifiers as the primary ac-to-dc converting device.

4.4.1   Silicon Rectifiers

Rectifier parameters generally are expressed in terms of reverse-voltage ratings and mean forward-current ratings in a $\frac{1}{2}$-wave rectifier circuit operating from a 60 Hz supply and feeding a purely resistive load. The three principle reverse-voltage ratings are:

- *Peak transient reverse voltage* (V_{rm}). The maximum value of any nonrecurrent surge voltage. This value must never be exceeded.

- *Maximum repetitive reverse voltage* ($V_{RM(rep)}$). The maximum value of reverse voltage that may be applied recurrently (in every cycle of 60 Hz power). This includes oscillatory voltages that may appear on the sinusoidal supply.

- *Working peak reverse voltage* ($V_{RM(wkg)}$). The crest value of the sinusoidal voltage of the ac supply at its maximum limit. Rectifier manufacturers generally recommend a value that has a significant safety margin, relative to the peak transient reverse voltage (V_{rm}), to allow for transient overvoltages on the supply lines.

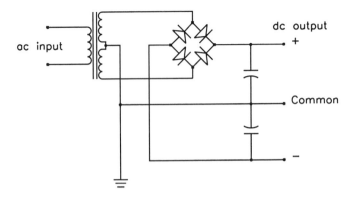

Figure 4.39 Conventional capacitor input filter full-wave bridge.

There are three forward-current ratings of similar importance in the application of silicon rectifiers:

- *Nonrecurrent surge current* ($I_{FM,surge}$). The maximum device transient current that must not be exceeded at any time. $I_{FM,surge}$ is sometimes given as a single value, but more often is presented in the form of a graph of permissible surge-current values vs. time. Because silicon diodes have a relatively small thermal mass, the potential for short-term current overloads must be given careful consideration.

- *Repetitive peak forward current* ($I_{FM,rep}$). The maximum value of forward-current reached in each cycle of the 60 Hz waveform. This value does not include random peaks caused by transient disturbances.

- *Average forward current* ($I_{FM,av}$). The upper limit for average load current through the device. This limit is always well below the repetitive peak forward-current rating to ensure an adequate margin of safety.

Rectifier manufacturers generally supply curves of the instantaneous forward voltage vs. instantaneous forward current at one or more specific operating temperatures. These curves establish the forward-mode upper operating parameters of the device.

Figure 4.39 shows a typical rectifier application in a bridge circuit.

4.4.2 Operating Rectifiers in Series

High-voltage power supplies (5 kV and greater) often require rectifier voltage ratings well beyond those typically available from the semiconductor industry. To meet the requirements of the application, manufacturers commonly use silicon diodes in a series configuration to give the required working peak reverse voltage. For such a configuration to work properly, the voltage across any one diode must not exceed the rated peak transient reverse voltage (V_{rm}) at any time. The dissimilarity commonly found between the reverse leakage current characteristics of different diodes of the same type number makes this objective difficult to achieve. The problem normally is

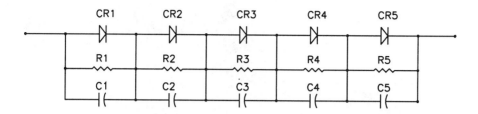

Figure 4.40 A portion of a high-voltage series-connected rectifier stack.

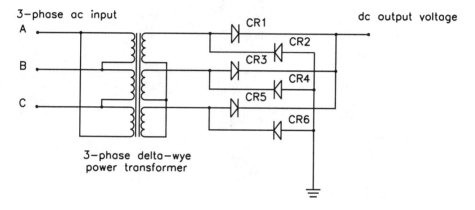

Figure 4.41 A 3-phase delta-connected high-voltage rectifier.

overcome by connecting shunt resistors across each rectifier in the chain, as shown in Figure 4.40. The resistors are chosen so that the current through the shunt elements (when the diodes are reverse-biased) will be several times greater than the leakage current of the diodes themselves.

The *carrier storage* effect also must be considered in the use of a series-connected rectifier stack. If precautions are not taken, different diode recovery times (caused by the carrier storage phenomenon) will effectively force the full applied reverse voltage across a small number of diodes, or even a single diode. This problem can be prevented by connecting small-value capacitors across each diode in the rectifier stack. The capacitors equalize the transient reverse voltages during the carrier storage recovery periods of the individual diodes.

Figure 4.41 illustrates a common circuit configuration for a high-voltage 3-phase rectifier bank. A photograph of a high-voltage series-connected 3-phase rectifier assembly is shown in Figure 4.42.

Figure 4.42 High-voltage rectifier assembly for a 3-phase delta-connected circuit.

4.4.3 Operating Rectifiers in Parallel

Silicon rectifiers are used in a parallel configuration when a large amount of current is required from the power supply. *Current sharing* is the major design problem with a parallel rectifier assembly because diodes of the same type number do not necessarily exhibit the same forward characteristics.

Semiconductor manufacturers often divide production runs of rectifiers into tolerance groups, matching forward characteristics of the various devices. When parallel diodes are used, devices from the same tolerance group must be used to avoid unequal sharing of the load current. As a margin of safety, designers allow a substantial derating factor for devices in a parallel assembly to ensure that the maximum operating limits of any one component are not exceeded.

The problems inherent in a parallel rectifier assembly can be reduced through the use of a resistance or reactance in series with each component, as shown in Figure 4.43. The buildout resistances (R1 through R4) force the diodes to share the load current equally. Such assemblies can, however, be difficult to construct and may be more expensive than simply adding diodes or going to higher-rated components.

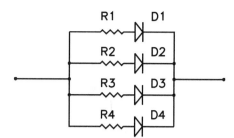

Figure 4.43 Using buildout resistances to force current sharing in a parallel rectifier assembly.

4.4.4 Silicon Avalanche Rectifiers

The *silicon avalanche diode* is a special type of rectifier that can withstand high reverse power dissipation. For example, an avalanche diode with a normal forward rating of 10 A can dissipate a reverse transient of 8 kW for 10 ms without damage. This characteristic of the device allows elimination of the surge-absorption capacitor and voltage-dividing resistor networks needed when conventional silicon diodes are used in a series rectifier assembly. Because fewer diodes are needed for a given applied reverse voltage, significant underrating of the device (to allow for reverse-voltage transient peaks) is not required.

When an extra-high-voltage rectifier stack is used, it is still advisable to install shunt capacitors—but not resistors—in an avalanche diode assembly. The capacitors are designed to compensate for the effects of carrier storage and stray capacitance in a long series assembly.

4.4.5 Thyristor Servo Systems

Thyristor control of ac power has become a common method of regulating high-voltage power supplies. The type of servo system employed depends on the application. Figure 4.44 shows a basic single-phase ac control circuit using discrete thyristors. The rms load current (I_{rms}) at any specific phase delay angle (α) is given in terms of the normal full-load rms current at a phase delay of zero ($I_{rms\text{-}0}$):

$$I_{rms} = I_{rms\text{-}\theta} \left\{ 1 - \frac{\alpha}{\pi} + (2\pi) - 1 \sin 2\alpha \right\}^{\frac{1}{2}} \tag{4.14}$$

The load rms voltage at any particular phase-delay angle bears the same relationship to the full-load rms voltage at zero phase delay as the previous equation illustrates for load current. An analysis of the mathematics shows that although the theoretical delay range for complete control of a resistive load is 0 to 180°, a practical span of 20 to 160° gives a power-control range of approximately 99 to 1 percent of maximum output to the load. Figure 4.45 illustrates typical phase-control waveforms.

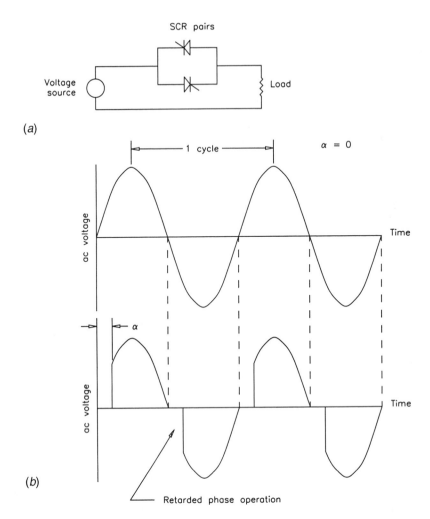

Figure 4.44 Inverse-parallel thyristor ac power control: (*a*) circuit diagram, (*b*) voltage and current waveforms for full and reduced thyristor control angles. The waveforms apply for a purely resistive load.

The circuit shown in Figure 4.44 requires a source of gate trigger pulses that must be isolated from each other by at least the peak value of the applied ac voltage. The two gate pulse trains must also be phased 180° with respect to each other. Also, the gate pulse trains must shift together with respect to the ac supply voltage phase when power throughput is adjusted.

Some power-control systems use two identical, but isolated, gate pulse trains operating at a frequency of twice the applied supply voltage (120 Hz for a 60 Hz system). Under such an arrangement, the forward-biased thyristor will fire when the gate pulses are applied to the silicon controlled rectifier (SCR) pair. The reverse-biased thyristor will

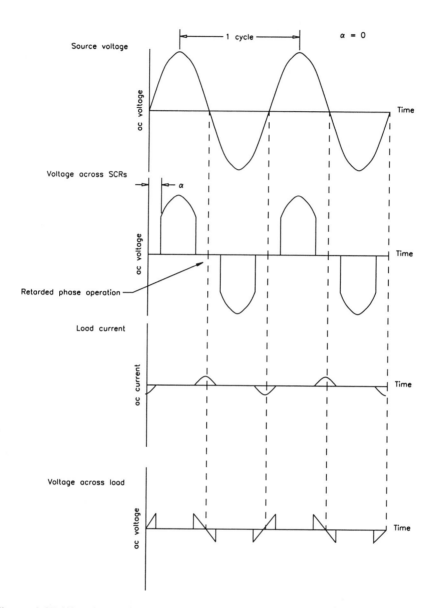

Figure 4.45 Waveforms in an ac circuit using thyristor power control.

not fire. Normally, it is considered unsafe to drive a thyristor gate positive while its anode-cathode is reverse-biased. In this case, however, it may be permissible because the thyristor that is fired immediately conducts and removes the reverse voltage from the other device. The gate of the reverse-biased device is then being triggered on a thyristor that essentially has no applied voltage.

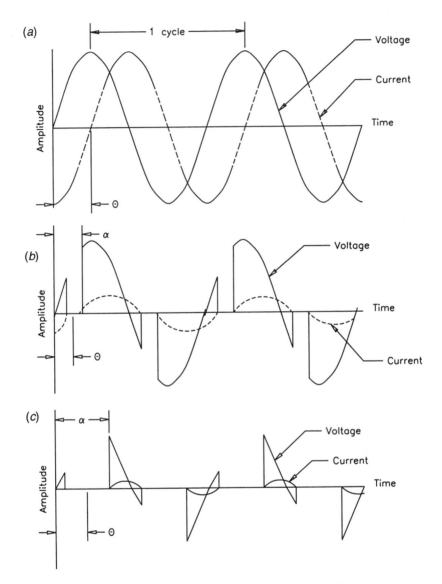

Figure 4.46 Voltage and current waveforms for inverse-parallel thyristor power control with an inductive load: (a) full conduction, (b) small-angle phase reduction, (c) large-angle phase reduction.

Inductive Loads

The waveforms shown in Figure 4.46 illustrate effects of phase control on an inductive load. When inductive loads are driven at a reduced conduction angle, a sharp transient change of load voltage occurs at the end of each current pulse (or loop). The

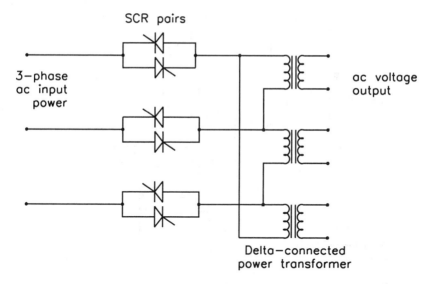

Figure 4.47 Modified full-thyristor 3-phase ac control of an inductive delta load.

transients generally have no effect on the load, but they can be dangerous to proper operation of the thyristors. When the conducting thyristor turns off, thereby disconnecting the load from the ac line supply, the voltage at the load rapidly drops to zero. This rapid voltage change, in effect, applies a sharply rising positive anode voltage to the thyristor opposing the device that has been conducting. If the thyristor dv/dt (change in voltage as a function of a change in time) rating is exceeded, the opposing device will turn on and conduction will take place, independent of any gate drive pulse.

A common protective approach involves the addition of a resistor-capacitor (RC) snubber circuit to control the rate of voltage change seen across the terminals of the thyristor pair. Whenever a thyristor pair is used to drive an inductive load, such as a power transformer, it is critically important that each device fires at a point in the applied waveform exactly 180° relative to the other. If proper timing is not achieved, the positive and negative current loops will differ in magnitude, causing a direct current to flow through the primary side of the transformer. A common trigger control circuit, therefore, should be used to determine gate timing for thyristor pairs.

Applications

Several approaches are possible for thyristor power control in a 3-phase ac system. The circuit shown in Figure 4.47 consists of essentially three independent, but interlocked, single-phase thyristor controllers. This circuit is probably the most common configuration found in commercial and industrial equipment.

In a typical application, the thyristor pairs feed a power transformer with multitap primary windings, thereby giving the user an adjustment range to compensate for varia-

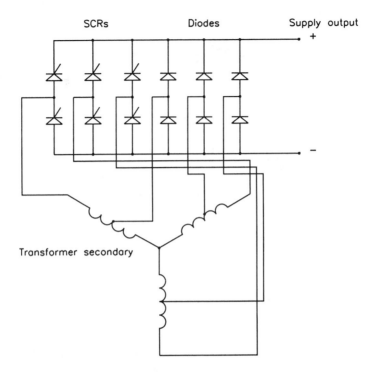

Figure 4.48 A 6-phase boost rectifier circuit.

tions in utility company line voltages from one location to another. A common procedure specifies selection of transformer tap positions that yield a power output of 105 percent when nominal utility company line voltages are present. The thyristor power-control system then is used to reduce the angle of conduction of the SCR pairs as necessary to cause a reduction in line voltage to the power transformer to yield 100 percent rated power output from the power supply. A servo loop from a sample point at the load may be used to compensate automatically for line-voltage variations. With such an arrangement, the thyristors are kept within a reasonable degree of retarded phase operation. Line voltages will be allowed to sag 5 percent or so without affecting the dc supply output. Utility supply voltage excursions above the nominal value simply will result in further delayed triggering of the SCR pairs.

Thyristor control of high-power loads (200 kW and above) typically uses transformers that provide 6- or 12-phase outputs. Although they are more complicated and expensive, such designs allow additional operational control, and filtering requirements are reduced significantly. Figure 4.48 shows a 6-phase *boost rectifier* circuit. The configuration consists basically of a full-wave 3-phase SCR bridge connected to a wye-configured transformer secondary. A second bridge, consisting of six diodes, is connected to low-voltage taps on the same transformer. When the SCRs are fully on, their output is at a higher voltage than the diode bridge. As a result, the diodes are reverse-biased and turned off. When the SCRs are partially on, the diodes are free to con-

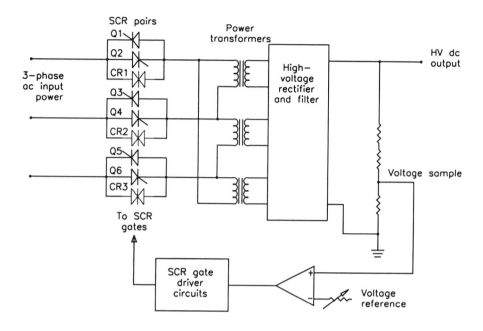

Figure 4.49 Thyristor-controlled high-voltage servo power supply.

duct. The diodes improve the quality of the output waveform during low-voltage (re-duced conduction angle) conditions. The minimum output level of the supply is determined by the transformer taps to which the diodes are connected.

A thyristor-driven 3-phase power-control circuit is shown in Figure 4.49. A single-phase power-control circuit is shown in Figure 4.50.

Triggering Circuits

Accurate, synchronized triggering of the gate pulses is a critical element in thyristor control of a 3-phase power supply. The gate signal must be synchronized properly with the phase of the ac line that it is controlling. The pulse also must properly match the phase-delay angle of the gates of other thyristors in the power-control system. Lack of proper synchronization of gate pulse signals between thyristor pairs can result in improper current sharing (*current hogging*) among individual legs of the 3-phase supply.

The gate circuit must be protected against electrical disturbances that could make proper operation of the power-control system difficult or unreliable. Electrical isolation of the gate is a common approach. Standard practice calls for the use of gate pulse transformers in thyristor servo system gating cards. Pulse transformers are fer-rite-cored devices with a single primary winding and (usually) multiple secondary

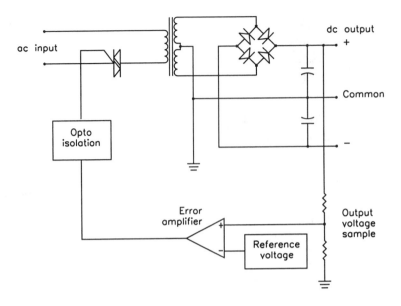

Figure 4.50 Phase-controlled power supply with primary regulation.

windings that feed, or at least control, the individual gates of a back-to-back thyristor pair. This concept is illustrated in Figure 4.51. Newer thyristor designs may use optocouplers to achieve the necessary electrical isolation between the trigger circuit and the gate.

It is common practice to tightly twist together the control leads from the gate and cathode of a thyristor to the gating card assembly. This practice provides a degree of immunity to high-energy pulses that might inadvertently trigger the thyristor gate. The gate circuit must be designed and configured carefully to reduce inductive and capacitive coupling that might occur between power and control circuits. Because of the high di/dt (change in current as a function of a change in time) conditions commonly found in thyristor-controlled power circuits, power wiring and control (gate) wiring must be separated physically as much as possible. Shielding of gating cards in a metal card cage is advisable.

Equipment manufacturers use various means to decrease gate sensitivity to transient sources, including placement of a series resistor in the gate circuit and/or a shunting capacitor between the gate and cathode. A series resistor has the effect of decreasing gate sensitivity, increasing the allowable dv/dt of the thyristor and reducing the turn-off time, which simultaneously increases the required holding and latching currents. The use of a shunt capacitor between the gate and cathode leads reduces high-frequency noise components that might be present on the gate lead and increases the dv/dt withstand capability of the thyristor.

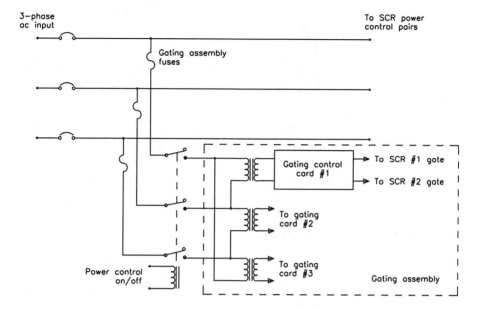

Figure 4.51 Simplified block diagram of the gating circuit for a phase-control system using back-to-back SCRs.

Fusing

Current limiting is a basic method of protection for any piece of equipment operated from the utility ac line. The device typically used for breaking fault currents is either a fuse or a circuit breaker. Some designs incorporate both components. *Semiconductor fuses* often are used in conjunction with a circuit breaker to provide added protection. Semiconductor fuses operate more rapidly (typically within 8.3 ms) and more predictably than common fuses or circuit breakers. Surge currents caused by a fault can destroy a semiconductor device, such as a power thyristor, before the ac line circuit breaker has time to act. Manufacturers of semiconductor fuses and thyristors usually specify in their data sheets the I^2t ratings of each device. Because the thyristor rating normally assumes that the device is operating at maximum rated current and maximum junction temperature (conditions that do not represent normal operation), a safety factor is ensured.

Control Flexibility

Thyristor servo control of a high-voltage power supply is beneficial to the user for a number of reasons, including:

- Wide control over ac input voltages

- Capability to compensate automatically for line-voltage variations

- Capability to soft-start the dc supply

Thyristor control circuits typically include a ramp generator that increases the ac line voltage to the power transformer from zero to full value within 2 to 5 s. This prevents high surge currents through rectifier stacks and filter capacitors during system startup.

Although thyristor servo systems are preferred over other power-control approaches from an operational standpoint, they are not without their drawbacks. The control system is complex and can be damaged by transient activity on the ac power line. Conventional power contactors are simple and straightforward. They either make contact or they do not. For reliable operation of the thyristor servo system, attention must be given to transient suppression at the incoming power lines.

4.4.6 Polyphase Rectifier Circuits

High-voltage power supplies typically used in vacuum tube circuits incorporate multiphase rectification of the ac line voltage. Common configurations include 3-, 6-, and 12-phase, with 3-phase rectification being the most common. Figure 4.52 illustrates four approaches to 3-phase rectification:

- *Three-phase half-wave wye*, Figure 4.52a. Three half-wave rectifiers are used in each leg of the secondary *wye* forming one phase. In such an arrangement, each diode carries current one-third of each cycle, and the output wave pulses at three times the frequency of the ac supply. In order to avoid direct-current saturation in the transformer, it is necessary to employ a 3-phase transformer rather than three single-phase transformers.

- *Three-phase full-wave bridge*, Figure 4.52b. Six diodes are used in this circuit to produce a low ripple output with a frequency of six times the input ac waveform. It is permissible in this configuration to use three single-phase transformers, if desired.

- *Six-phase star*, Figure 4.52c. This circuit, also known as a 3-phase diametric configuration, uses six diodes with a transformer secondary configured as a star, as illustrated in the figure. The output ripple frequency is six times the input ac waveform.

- *Three-phase double-wye*, Figure 4.52d. This circuit uses six diodes and a complicated configuration of transformer windings. Note the balance coil (*interphase transformer*) in the circuit.

The relative merits of these rectifier configurations are listed in Table 4.3.

Polyphase rectifiers are used when the dc power required is on the order of 2 kW or more. The main advantages of a polyphase power supply over a single-phase supply include the following:

- Division of the load current between three or more lines to reduce line losses.

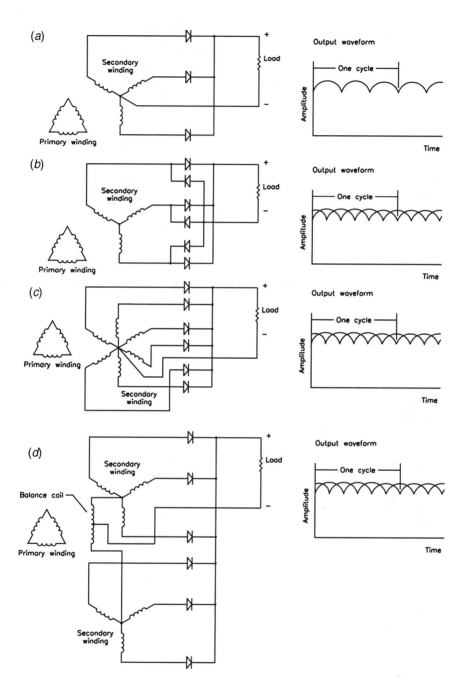

Figure 4.52 Basic 3-phase rectifier circuits: (a) half-wave wye, (b) full-wave bridge, (c) 6-phase star, (d) 3-phase double-wye.

Table 4.3 Operating Parameters of 3-Phase Rectifier Configurations

Parameter	3-Phase Star	3-Phase Bridge	6-Phase Star	3-Phase Double-Y	Multiplier[1]
Rectifier elements	3	6	6	6	
Rms dc voltage output	1.02	1.00	1.00	1.00	Average dc voltage output
Peak dc voltage output	1.21	1.05	1.05	1.05	Average dc voltage output
Peak reverse volts per rectifier	2.09	1.05	2.09	2.42	Average dc voltage output
	2.45	2.45	2.83	2.83	Rms secondary volts per transformer leg
	1.41	1.41	1.41	1.41	Rms secondary volts line-to-line
Average dc output current per rectifier	0.333	0.333	0.167	0.167	Average dc output current
Rms current per rectifier, resistive load	0.587	0.579	0.409	0.293	Average dc output current
Rms current per rectifier, inductive load	0.578	0.578	0.408	0.289	Average dc output current
Percent ripple	18.3	4.2	4.2	4.2	
Ripple frequency	3	6	6	6	Line frequency
AC line power factor	0.826	0.955	0.955	0.955	
Transformer secondary rms volts per leg[2]	0.855	0.428	0.740	0.855	Average dc voltage output
Transformer secondary rms volts line-to-line	1.48	0.740	1.48 (max)	1.71 (max, no load)	Average dc voltage output
Secondary line current	0.578	0.816	0.408	0.289	Average dc output current
Transformer secondary volt-amperes	1.48	1.05	1.81	1.48	DC watts output
Primary line current	0.817	1.41	0.817	0.707	(Avg. load I x secondary leg V) / primary line V

[1] To determine the value of a parameter in any column, multiply the factor shown by the value given in this column.
[2] For inductive load or large choke input filter

- Significantly reduced filtering requirements after rectification because of the low ripple output of a polyphase rectifier.

- Improved voltage regulation when using an inductive-input filter. Output voltage *soaring* is typically 6 percent or less from full-load to no-load conditions.

- Greater choice of output voltages from a given transformer by selection of either a *delta* or *wye* configuration.

The main disadvantage of a polyphase system is its susceptibility to phase imbalance. Resulting operational problems include increased ripple at the output of the supply and uneven sharing of the load current by the transformer windings.

4.4.7 Power Supply Filter Circuits

A filter network for a high-voltage power supply typically consists of a series inductance and one or more shunt capacitances. Bleeder resistors and various circuit protection devices are usually incorporated as well. Filter systems can be divided into two basic types:

- *Inductive input*: filter circuits that present a series inductance to the rectifier output

- *Capacitive input*: filter circuits that present a shunt capacitance to the rectifier output

Inductive Input Filter

The inductive input filter is the most common configuration found in high-power RF equipment. A common circuit is shown in Figure 4.53, along with typical current waveforms. When the input inductance is infinite, current through the inductance is constant and is carried at any moment by the rectifier anode that has the most positive voltage applied to it at that instant. As the alternating voltage being rectified passes through zero, the current suddenly transfers from one anode to another, producing square current waves through the individual rectifier devices.

When the input inductance is finite (but not too small), the situation changes to that shown by the solid lines of Figure 4.53. The current through the input inductance tends to increase when the output voltage of the rectifier exceeds the average or dc value, and to decrease when the rectifier output voltage is less than the dc value. This causes the current through the individual anodes to be modified as shown. If the input inductance is too small, the current decreases to zero during a portion of the time between the peaks of the rectifier output voltage. The conditions then correspond to a capacitor input filter system.

The output wave of the rectifier can be considered as consisting of a dc component upon which are superimposed ac voltages (ripple voltages). To a first approximation, the fluctuation in output current resulting from a finite input inductance can be considered as the current resulting from the lowest-frequency component of the ripple voltage acting against the impedance of the input inductance. This assumption is permissible

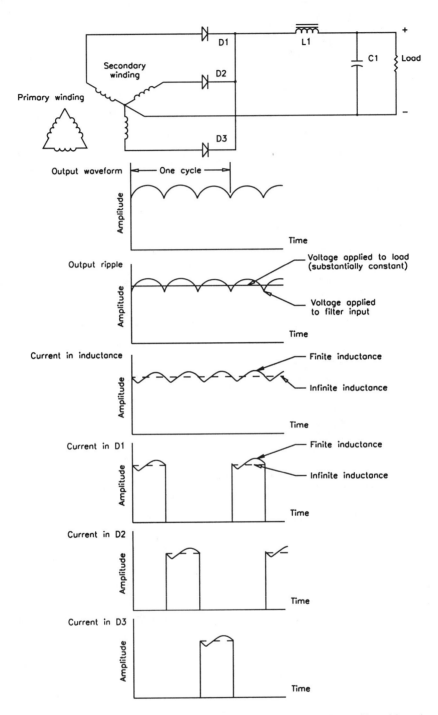

Figure 4.53 Voltage and current waveshapes for an inductive-input filter driven by a 3-phase supply.

because the higher-frequency components in the ripple voltage are smaller, and at the same time encounter higher impedance. Furthermore, in practical filters the shunting capacitor following the input inductance has a small impedance at the ripple frequency compared with the reactance of the input inductance. The peak current resulting from a finite input inductance, therefore, is given approximately by the relation [1]:

$$\frac{I_f}{I_i} = 1 + \frac{E_1}{E_0} \frac{R_{\textit{eff}}}{\omega L_1} \tag{4.15}$$

Where:
I_f = peak current with finite input inductance
I_i = peak current with infinite input inductance
E_1/E_0 = ratio of lowest-frequency ripple component to the dc voltage in the rectifier output
$R_{\textit{eff}}$ = effective load resistance
ωL_1 = reactance of the incremental value of the input inductance at the *lowest* ripple frequency

This equation is derived as follows:

- The peak alternating current through the input inductance is approximately $E_1 \big/ \omega L_1$
- The average or dc current is $E_0 \big/ R_{\textit{eff}}$
- The peak current with finite inductance is, therefore: $(E_1 \big/ \omega L_1) + (E_0 \big/ R_{\textit{eff}})$
- The current with infinite inductance is $E_0 \big/ R_{\textit{eff}}$

The effective load resistance value consists of the actual load resistance plus filter resistance plus equivalent diode and transformer resistances.

The normal operation of an inductive input filter requires that there be a continuous flow of current through the input inductance. The peak alternating current flowing through the input inductance, therefore, must be less than the dc output current of the rectifier. This condition is realized by satisfying the approximate relation:

$$\omega L_1 = R_{\textit{eff}} \frac{E_1}{E_0} \tag{4.16}$$

In the practical case of a 60 Hz single-phase full-wave rectifier circuit, the foregoing equation becomes:

$$L_1 = \frac{R_{\textit{eff}}}{1130} \tag{4.17}$$

In a polyphase system, the required value of L_1 is significantly less. The higher the load resistance (the lower the dc load current), the more difficult it is to maintain a continuous flow of current and, with a given L_1, the previous equation will not be satisfied when the load resistance exceeds a critical value.

The minimum allowable input inductance (ωL_1) is termed the *critical inductance*. When the inductance is less than the critical value, the filter acts as a capacitor input circuit. When the dc drawn from the rectifiers varies, it is still necessary to satisfy the ωL_1 equation at all times, particularly if good voltage regulation is to be maintained. To fulfill this requirement at small load currents without excessive inductance, it is necessary to place a bleeder resistance across the output of the filter system in order to limit R_{eff} to a value corresponding to a reasonable value of L_1.

Capacitive Input Filter

When a shunt capacitance rather than a series inductance is presented to the output of a rectifier, the behavior of the circuit is greatly modified. Each time the positive crest alternating voltage of the transformer is applied to one of the rectifier anodes, the input capacitor charges up to just slightly less than this peak voltage. The rectifier then ceases to deliver current to the filter until another anode approaches its peak positive potential, when the capacitor is charged again. During the interval when the voltage across the input capacitor is greater than the potential of any of the anodes, the voltage across the input capacitor drops off nearly linearly with time, because the first filter inductance draws a substantially constant current from the input capacitor. A typical set of voltage and current waves is illustrated in Figure 4.54.

The addition of a shunt capacitor to the input of a filter produces fundamental changes in behavior, including the following:

- The output voltage is appreciably higher than with an inductance input.

- The ripple voltage is less with a capacitance input filter than with an inductance input filter.

- The dc voltage across the filter input drops as the load current increases for the capacitive input case, instead of being substantially constant as for the inductive input case.

- The ratio of peak-to-average anode current at the rectifiers is higher in the capacitive case.

- The utilization factor of the transformer is less with a capacitive input configuration.

Filters incorporating shunt capacitor inputs generally are employed when the amount of dc power required is small. Inductance input filters are used when the amount of power involved is large; the higher utilization factor and lower peak current result in important savings in rectifier and transformer costs under these conditions. Inductance input systems are almost universally employed in polyphase rectifiers.

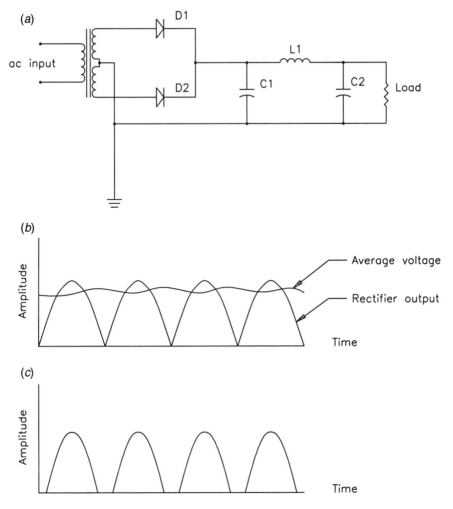

Figure 4.54 Characteristics of a capacitive input filter circuit: (*a*) schematic diagram, (*b*) voltage waveshape across the input capacitor, (*c*) waveshape of current flowing through the diodes.

4.5   Parameter Sampling Circuits

Most RF system parameters can be reduced to a *sample voltage* by using voltage dividers or current-sense resistors in the circuit to be measured [2]. Remote metering is accomplished by sending a signal representing the sample voltage to the remote metering point, and then displaying a value representing the original parameter. In practice, the actual sampling ratio (sample voltage/parameter value) is not particularly important as long as the sampling ratio is stable and the sample voltage is reasonable

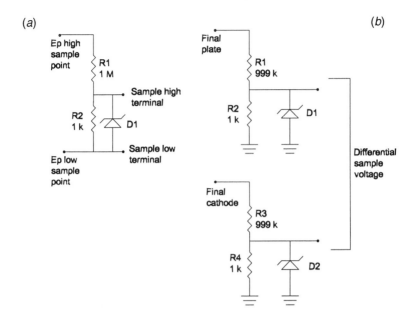

Figure 4.55 Plate voltage sampling circuits: (*a*) single-ended, (*b*) differential.

(i.e., low enough to be easily handled, yet high enough to be immune to noise). Figure 4.55 shows some typical voltage sampling circuits for high-power RF systems.

Figure 4.55*a* shows how a voltage divider can be used to sample plate voltage. The circuit is independent of the transmitter front panel plate voltage meter. A zener clamping diode is placed across the sample line for safety; if the shunt resistor opened, the full plate voltage would be present at the sampling output terminals (assuming minimal loading by the remote indicator). The zener will typically fail in a short-circuit mode, protecting the external sampling equipment.

The sampling circuit of Figure 4.55*a* is adequate for systems that measure plate voltage from a ground reference. In the event that the cathode operates at a potential other than ground, the sampling circuit of Figure 4.55*b* may be used. The sampling line is balanced, with both terminals floating above ground by 1 KΩ The differential output voltage is independent of the common mode input voltage if all the resistors are precisely matched.

Several techniques are available for dealing with differential sampling voltages (where neither side of the sample is grounded). Typical approaches include the following:

- *Isolation amplifier.* Isolation amplifiers provide isolation between the input and output by using magnetic, capacitive, optical, or thermal coupling between the in-

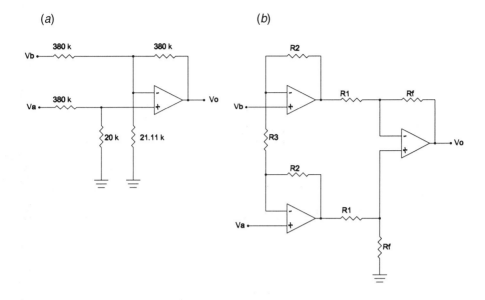

Figure 4.56 Differential amplifier circuits: (*a*) basic circuit, (*b*) instrumentation amplifier.

put and output. Some units include feedback techniques to reduce nonlinearities induced by the coupling method.

- *Differential amplifier.* In most cases, electrical isolation between the sampling circuit and the remote metering equipment is not required. Instead, a differential voltage must be changed to a *single ended* voltage (one side grounded). This can be accomplished using the differential amplifier shown in Figure 4.56*a* or the instrumentation amplifier shown in Figure 4.56*b*.

Measurement of plate and/or screen current is typically done by inserting a sampling resistor into the ground return lead of the respective supplies. This approach is shown in Figure 4.57. Notice that the remote plate current metering is taken across R1, with balancing resistors of 1.2 kΩ used to lift the sample above the ground reference. A bidirectional zener diode is placed across the output of the sample terminals for protection. The plate current meter on the transmitter is sampled across R2, and the over-current relay is sampled across R3. Device X1 is a crowbar that will shunt the plate current to ground in the event of a failure in the R1–R3 chain.

The initial tolerance of resistors used as current sense elements or in a voltage divider circuit is usually not particularly critical. An inexact resistance will cause the sample voltage to be something other than that expected. However, whatever sample voltage appears, the remote meter can be calibrated to agree with the local meter. The more important attribute is resistance changes with time and temperature.

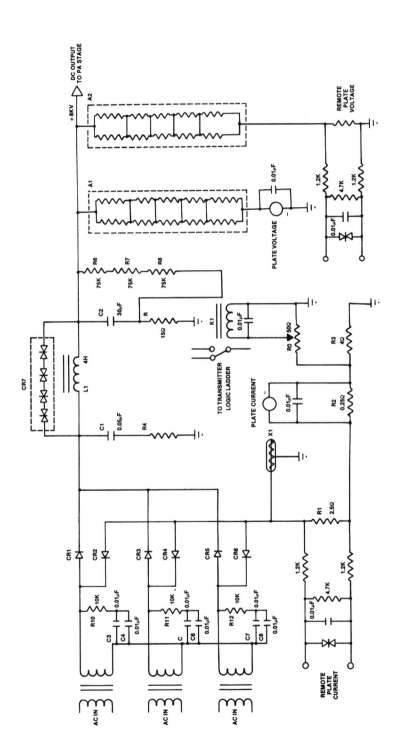

Figure 4.57 Transmitter high-voltage power supply showing remote plate current and voltage sensing circuits.

A carbon composition resistor can commonly have a temperature coefficient of about ±600 ppm/C° for resistors up to 1 KΩ. This increases to ±1875 ppm/C° for resistors up to 1 MΩ. If these resistors were used in a common plate voltage sampling circuit, such as than shown in Figure 4.57, the sample voltage could change as much as 20 percent over a 0 C° to 50 C° temperature range. Metal film resistors are available with temperature coefficients as low as 15 ppm/C°. Bulk metal resistors are available with temperature coefficients of 5 ppm/C°. If the resistors in a voltage divider are at the same temperature, the matching between the *temperature coefficient of resistance* (TCR) is more important than the actual TCR. Because the TCR matching is generally better than the TCR of an individual resistor, sampling systems should be designed to keep the resistors in the network at the same operating temperature.

In differential amplifiers, the matching of resistor values is critical. To insure the resistance values track with temperature, it is common practice to use a resistor network made using hybrid film techniques or integrated circuit techniques. Because the resistors are made at the same time and of the same materials, the TCRs track closely.

4.6   References

1. Terman, F. E., *Radio Engineering*, 3rd. Ed., McGraw-Hill, New York, pg. 560, 1947.
2. Hallikainen, Harold, "Transmitter Control Systems," in *NAB Engineering Handbook*, 9th Ed., Jerry C. Whitaker (ed.), National Association of Broadcasters, Washington, D.C., 1998.

4.7   Bibliography

Crutchfield, E. B. (ed.), *NAB Engineering Handbook*, 8th Ed., National Association of Broadcasters, Washington, DC, 1992.

Gray, T. S., *Applied Electronics*, Massachusetts Institute of Technology, 1954.

Heising, R. A., "Modulation in Radio Telephony," *Proceedings of the IRE*, Vol. 9, no. 3, pg. 305, June 1921.

Heising, R. A., "Transmission System," U.S. Patent no. 1,655,543, January 1928.

Jordan, Edward C. (ed.), *Reference Data for Engineers: Radio, Electronics, Computers, and Communications*, 7th Ed., Howard W. Sams, Indianapolis, IN, 1985.

Laboratory Staff, *The Care and Feeding of Power Grid Tubes*, Varian Eimac, San Carlos, CA, 1984.

Martin, T. L., Jr., *Electronic Circuits*, Prentice Hall, Englewood Cliffs, NJ, 1955.

Mendenhall, G. N., M. B. Shrestha, and E. J. Anthony, "FM Broadcast Transmitters," in *NAB Engineering Handbook*, 8th Ed., E. B. Crutchfield (ed.), National Association of Broadcasters, Washington, DC, pp. 429–474, 1992.

Shrestha, Mukunda B., "Design of Tube Power Amplifiers for Optimum FM Transmitter Performance," *IEEE Transactions on Broadcasting*, Vol. 36, no. 1, IEEE, New York, pp. 46–64, March 1990.

Terman, F. E., and J. R. Woodyard, "A High Efficiency Grid-Modulated Amplifier," *Proceedings of the IRE*, Vol. 26, no. 8, pg. 929, August 1938.

Terman, F. E., *Radio Engineering*, 3rd. Ed., McGraw-Hill, New York, 1947.

Weirather, R. R., G. L. Best, and R. Plonka, "TV Transmitters," in *NAB Engineering Handbook*, 8th Ed., E. B. Crutchfield (ed.), National Association of Broadcasters, Washington, DC, pp. 499–551, 1992.

Whitaker, Jerry C., *AC Power Systems*, 2nd Ed., CRC Press, Boca Raton, FL, 1998.

Whitaker, Jerry C., *Radio Frequency Transmission Systems: Design and Operation*, McGraw-Hill, New York, 1991.

Woodard, George W., "AM Transmitters," in *NAB Engineering Handbook*, 8th Ed., E. B. Crutchfield (ed.), National Association of Broadcasters, Washington, DC, pp. 353–381, 1992.

Applying Vacuum Tube Devices

5.1 Introduction

Numerous circuits have been developed to meet the varied needs of high-power RF amplification. Designers have adapted standard vacuum tube devices to specific applications that require unique performance criteria. Most of these circuits are built upon a few basic concepts.

5.2 AM Power Amplification Systems

Class C plate modulation is the classic method of producing an AM waveform [1]. Figure 5.1 shows the basic circuit. Triodes and tetrodes may be used as the modulator or carrier tube. Triodes offer the simplest and most common configuration.

Numerous variations on the basic design exist, including a combination of plate and screen grid modulation. The carrier signal is applied to the control grid, and the modulating signal is applied to the screen and plate. The plate is fully modulated, and the screen is modulated 70 to 100 percent to achieve 100 percent carrier modulation. Modulation of the screen can be accomplished using one of the following methods:

- Screen voltage supplied through a dropping resistor connected to the unmodulated dc plate supply, shown in Figure 5.2a

- An additional (third) winding on the modulation transformer, illustrated in Figure 5.2b

- A modulation choke placed in series with a low-voltage fixed screen supply, shown in Figure 5.2c

Depending on the design, screen/plate modulation also may require partial modulation of the control grid to achieve the desired performance characteristics.

During the portion of the modulation cycle when the plate voltage is increased, the screen current decreases. If the screen is supplied through an impedance, such as the screen dropping resistance of a modulation choke, the voltage drop in this series impedance becomes less and the screen voltage rises in the desired manner. During the part of the modulation cycle when the plate voltage is decreased, the screen current increases,

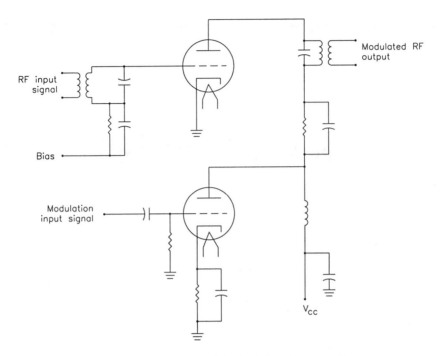

Figure 5.1 The classic plate-modulated PA circuit for AM applications.

causing a greater voltage drop in the screen series impedance, and lowering the voltage on the screen of the tube. The value of the screen bypass capacitor in the class C stage is a compromise between good RF bypassing and the shunting effect of this capacitance on the screen modulation circuit.

5.2.1 Control Grid Modulation

Although not as common as class C plate modulation, a class C RF amplifier may be modulated by varying the voltage on the control grid of a triode [1]. This approach, shown in Figure 5.3, produces a change in the magnitude of the plate current pulses and, therefore, variations in the output waveform taken from the plate tank. Both the carrier and modulation signals are applied to the grid.

The primary benefit of grid modulation is that a modulator is not required. Grid modulation, however, requires a fixed plate supply voltage that is twice the peak RF voltage without modulation. The result is higher plate dissipation at lower modulation levels, including carrier level. The typical plate efficiency of a grid-modulated stage may range from only 35 to 45 percent at carrier. Grid modulation typically is used only in systems where plate modulation transformers cannot provide adequate bandwidth for the intended application.

When grid modulation is used, the screen voltage and grid bias must be taken from sources with good regulation. This usually means separate low-voltage power supplies.

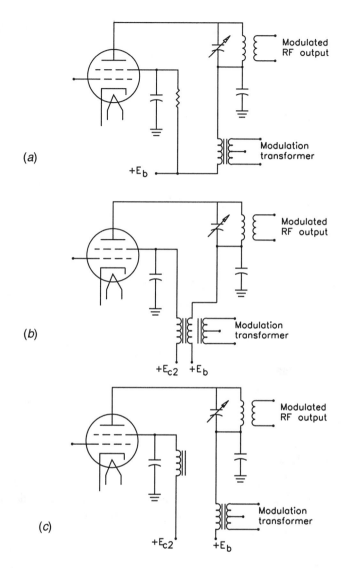

Figure 5.2 Basic methods of screen and plate modulation: (*a*) plate modulation, (*b*) plate and screen modulation, (*c*) plate modulation using separate plate/screen supplies.

5.2.2 Suppressor Grid Modulation

The output of a class C pentode amplifier may be controlled by applying a modulating voltage, superimposed on a suitable bias, to the suppressor grid to produce an AM waveform [1]. As the suppressor grid becomes more negative, the minimum instanta-

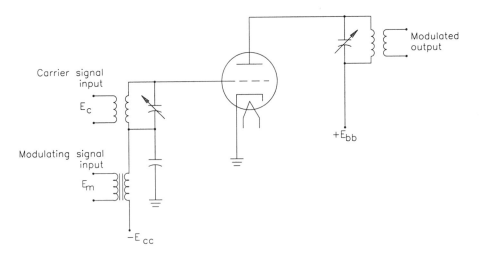

Figure 5.3 Grid modulation of a class C RF amplifier.

neous potential at which current can be drawn to the plate is increased. Thus, as modulation varies the suppressor grid potential, the output voltage changes.

This method of modulating an RF amplifier provides about the same plate efficiency as a grid-modulated stage. The overall operating efficiency, however, is slightly lower because of increased screen grid losses associated with the design. Linearity of the circuit is not usually high.

5.2.3 Cathode Modulation

Cathode modulation incorporates the principles of both control grid and plate modulation [1]. As shown in Figure 5.4, a modulation transformer in the cathode circuit varies the grid-cathode potential, as well as the plate-cathode potential. The ratio of grid modulation to plate modulation is set by adjustment of the tap shown in the figure. Grid leak bias typically is employed to improve linearity of the stage.

5.2.4 High-Level AM Amplification

The basic tuned-anode vacuum tube power amplifier is described in graphical form in Figure 5.5 [2]. The tube may be a triode, tetrode, or pentode. Tetrode final amplifiers are the most common. The RF excitation voltage is applied to the grid of the PA tube, and the ratio of dc grid bias voltage to peak RF excitation voltage (shown sinusoidal in Figure 5.5) determines the conduction angle θ_c of anode current:

$$\theta_c = 2arccos \left\{ \frac{E_{cc}}{Eg - E_{cc}} \right\}$$

(5.1)

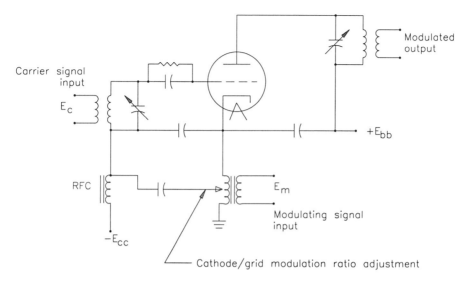

Figure 5.4 Cathode modulation of a class C RF amplifier.

Where:

E_g = exciting grid signal
E_{cc} = dc grid bias voltage

The shape of the anode current pulse is determined by the vacuum tube transfer characteristics and the input waveshape. The pulse of current thus generated is supplied by the dc power supply E_{bb} and passes through the resonant anode tank circuit. The tank is assumed to have sufficient operating Q to force anode voltage e_p to be essentially sinusoidal and of the same periodic frequency as the RF excitation voltage and resultant anode current pulse. The instantaneous anode dissipation is the product of instantaneous tube anode voltage drop and anode current. The tube transfer characteristic is a variable, dependent upon device geometry and other factors, including the maximum drive signal E_g. The exact shape and magnitude of the current waveform is normally obtained from a load-line plot on the constant-current characteristic tube curves supplied by the device manufacturer. The resonant anode load impedance is chosen and adjusted to allow $E_{p(min)}$ to be as low as possible without causing excessive screen grid (tetrode case) or control grid dissipation.

The overall anode efficiency of this circuit can be extended beyond the normal limits for typical class C amplification by employing a third-harmonic resonator between the output anode connection and the fundamental resonant circuit. In some cases, fifth-harmonic resonators are also employed. This technique has the effect of squaring up the anode voltage waveform e_p, thus causing the integral of the $e_p \times i_p$ product (anode dissipation) to be smaller, resulting in lower dissipation for a given RF power output. An amplifier employing the third-harmonic anode trap commonly is referred to as a class C-D stage, suggesting efficiency ranging between conventional class C opera-

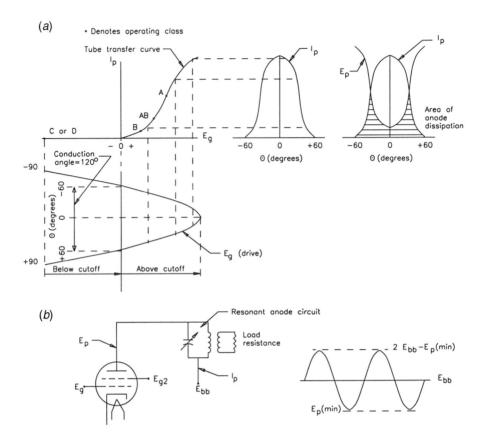

Figure 5.5 Classic vacuum tube class C amplifier: (*a*) representative waveforms, (*b*) schematic diagram. Conditions are as follows: sinusoidal grid drive, 120° anode current, resonant anode load. (*After* [2].)

tion and class D operation. The use of the third-harmonic technique permits efficiencies of 85 percent or more for transmitters of 10 kW power output and above.

Class B Modulator

The most common method of applying low-frequency (typically audio) intelligence to a high-level amplitude-modulated amplifier is the class B push-pull system illustrated in Figure 5.6 [1]. The vacuum tubes used in such a circuit may be triodes, tetrodes, or pentodes. The output circuit includes a modulation transformer, audio coupling capacitor, and dc shunt feed inductor. The capacitor and shunt inductor network is used to prevent unbalanced dc from magnetizing the modulation transformer core, which would result in poor low-frequency performance. Advanced core materials and improved transformer design have permitted elimination of the coupling ca-

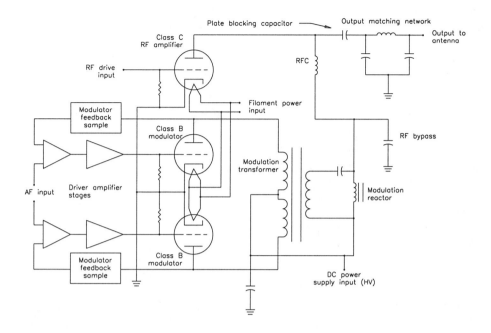

Figure 5.6 Class B push-pull modulator stage for a high-level AM amplifier.

pacitor and the shunt reactor in modern systems. The direct current to the modulated RF amplifier anode, therefore, flows directly through the secondary of the modulation transformer.

Elimination of the coupling capacitor and shunt reactor results in improved performance from the modulator circuit. The L and C components effectively form a three-pole high-pass filter that causes low-frequency transient distortion to be generated when driven with complex audio waveforms. The simplified circuit constitutes a single-pole high-pass filter, greatly reducing low-frequency transients.

Even in the improved design, the transient distortion performance of the class B push-pull modulator can be insufficient for some applications. Complex modulating waveforms with sharp transitions present special problems. Stray internal winding capacitances and leakage inductances effectively form a multipole low-pass filter at the high-frequency end of the audio spectrum. This filter can produce overshoots when driven with complex waveforms. Such distortion can be reduced by filtering the input signal, at the cost of somewhat lower high-frequency response.

Negative feedback is used to reduce nonlinear distortion and noise in the push-pull circuit. Negative feedback, however, usually increases high frequency transient distortion.

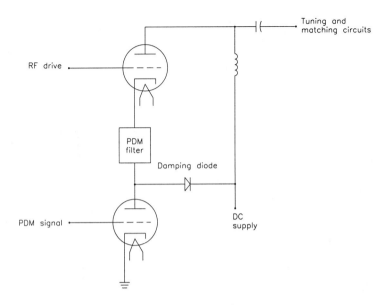

Figure 5.7 The pulse duration modulation (PDM) method of pulse width modulation.

5.2.5 Pulse Width Modulation

Pulse width modulation (PWM), also known as *pulse duration modulation* (PDM), is a common transmission system for modern vacuum tube transmitters. Figure 5.7 shows the PDM scheme (Harris) applied to an amplitude-modulated PA. The PDM system works by utilizing a square wave switching system, illustrated in Figure 5.8.

The PDM process begins with a signal generator (Figure 5.9). A 75 kHz sine wave is produced by an oscillator and used to drive a square wave generator, resulting in a simple 75 kHz square wave. The square wave is then integrated, resulting in a triangular waveform that is mixed with the input audio in a summing circuit. The resulting signal is a triangular waveform that rides on the incoming signal, typically audio. This composite signal is then applied to a threshold amplifier, which functions as a switch that is turned on whenever the value of the input signal exceeds a certain limit. The result is a string of pulses in which the width of the pulse is proportional to the period of time the triangular waveform exceeds the threshold. The pulse output is applied to an amplifier to obtain the necessary power to drive subsequent stages. A filter eliminates whatever transients may exist after the switching process is complete.

The PDM scheme is, in effect, a digital modulation system with the input information being sampled at a 75 kHz rate. The width of the pulses contains all the intelligence. The pulse-width-modulated signal is applied to a *switch* or *modulator tube*. The tube is simply turned *on*, to a fully saturated state, or *off* in accordance with the instanta-

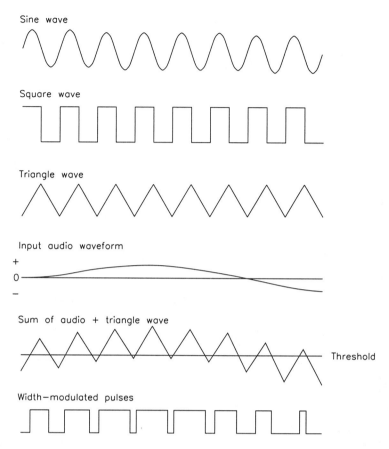

Figure 5.8 The principal waveforms of the PDM system.

neous value of the pulse. When the pulse goes positive, the modulator tube is turned on and the voltage across the tube drops to a minimum. When the pulse returns to its minimum value, the modulator tube turns off.

This PDM signal becomes the power supply to the final RF amplifier tube. When the modulator is switched on, the final amplifier will experience current flow and RF will be generated. When the switch or modulator tube goes off, the final amplifier current will cease. This system causes the final amplifier to operate in a highly efficient class D switching mode. A dc offset voltage to the summing amplifier is used to set the carrier (no modulation) level of the transmitter.

A high degree of third-harmonic energy will exist at the output of the final amplifier because of the switching-mode operation. This energy is eliminated by a third-harmonic trap. The result is a stable amplifier that normally operates in excess of 90 percent efficiency. The power consumed by the modulator and its driver is usually a fraction of a full class B amplifier stage.

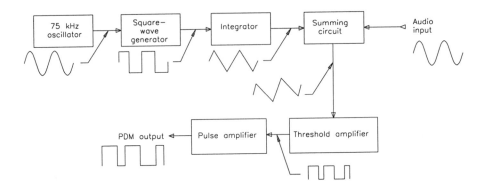

Figure 5.9 Block diagram of a PDM waveform generator.

The *damping diode* shown in Figure 5.9 is included to prevent potentially damaging transient overvoltages during the switching process. When the switching tube turns off the supply current during a period when the final amplifier is conducting, the high current through the inductors contained in the PDM filters could cause a large transient voltage to be generated. The energy in the PDM filter is returned to the power supply by the damping diode. If no alternative route were established, the energy would return by arcing through the modulator tube itself.

The PWM system makes it possible to eliminate audio frequency transformers in the transmitter. The result is wide frequency response and low distortion. Note that variations on this amplifier and modulation scheme have been used by other manufacturers for standard broadcast, shortwave service, and other applications.

5.3 Linear Amplification

The following features are desirable for vacuum tubes used in RF linear amplifier service [1]:

- High power gain
- Low plate-to-grid capacitance
- Good efficiency
- Linear characteristics that are maintained without degradation across the desired operating band

For linear service, RF amplifiers may be operated in class A, AB_1, AB_2, or B modes. Triode, tetrode, or pentode devices may be used, either grid- or cathode-driven. The choice of mode, tube, and driving method depends upon the operational specifications of each application.

5.3.1 Device Selection

The triode tube, having a large plate-to-grid interelectrode capacitance, requires neutralization in grid-driven linear service to prevent oscillation [1]. A triode having a low amplification factor is suitable for class AB and AB_2 grid-driven operation. The RF grid excitation voltage for this type of service will be high; grid excursions into the positive region are normal for class AB_2. A swamping resistor should be used across the input tuned circuit to maintain a constant input impedance to the stage and to provide for stability. With a low value of swamping resistance, the grid current drawn is only a small part of the total grid load, and the driver load impedance is relatively constant. The swamping resistor improves RF stability by providing a low impedance to ground for regenerative feedback through the plate-to-grid capacitance.

High-amplification-factor triodes perform exceptionally well in circuits where the grid is grounded and the cathode is driven. Under these conditions, the control grid acts as a shield between the input and output circuits. Neutralization, therefore, is not normally required. Zero-bias triodes operate in the class AB_2 mode and require only filament, plate, and drive power. For optimum linear operation, a tuned circuit is placed in the cathode RF return path to maintain a sinusoidal waveshape over the drive cycle. The tuned circuit will reduce the intermodulation distortion produced by the amplifier and also will reduce drive power requirements.

If the driver and PA stages are located close to each other, the tuned cathode circuit can be a part of the output circuit of the driver. If, however, the amplifiers are far removed and coupled by a length of coaxial cable, it is recommended that a tuned cathode circuit with a Q of between 2 and 4 be used.

Most tetrode and pentode amplifiers are designed to be grid-driven to take advantage of the high power gain of these devices.

In all linear amplifier systems, the driver output impedance should be kept low because of the nonlinear input loading characteristics of the amplifier tube as it approaches maximum power. The lower the driver amplifier impedance, the smaller the effect of the nonlinear input loading.

5.3.2 Grid-Driven Linear Amplifier

A linear amplifier utilizing a tetrode or pentode is usually grid-driven to take advantage of the inherent high gain of the tube [1]. Such an amplifier can be driven into the grid-current region under the proper circumstances. In any case, the input circuit will be loaded by the tube grid. For the no-grid-current case, the driver will work into the input conductance loading; for the grid-current case, the driver will work into the input conductance loading plus grid-current loading. It is therefore desirable (and necessary in the grid-current case) to swamp the input circuit with an appropriate noninductive resistor. The resistor will maintain an almost constant load on the driver and minimize the effects of any nonlinearity in grid loading.

5.3.3 Cathode-Driven Linear Amplifier

The cathode-driven amplifier may use triode, tetrode, or pentode tubes [1]. The drive signal is applied to the cathode in this class of operation. The cathode-driven amplifier is particularly suitable for high-power stages using high-μ triodes in the HF and VHF region. This class of operation normally eliminates the need for neutralization because the control grid screens the plate from the input circuit. The power gain for suitable triode class AB cathode-driven amplifiers is on the order of 7 to 20. The actual tube power gain is approximately equal to the ratio of RF plate voltage to RF cathode voltage. This relationship is true because the fundamental component of the plate current is common to the input and output circuits.

Tetrode tubes also can be used in cathode-driven operation. Power gain is considerably higher than that of the triode, on the order of 20 to 50. It is important to note that screen grid current loads the input circuit, just as control-grid current does.

For an amplifier located some distance from the driver, an improvement in intermodulation distortion can be realized by tuning the cathode circuit. When the driver is located close to the amplifier (0.1 wavelength or so) other means may be used to minimize the nonlinear loading of the cathode-driven stage.

5.3.4 Intermodulation Distortion

When an RF signal with varying amplitude is passed through a nonlinear device, many new products are produced [1]. (Intermodulation distortion is discussed in detail in Section 10.3.6.) The frequency and amplitude of each component can be determined mathematically because the nonlinear device can be represented by a power series expanded about the zero-signal operating point. The example of a typical two-tone signal serves to summarize this mathematical relationship. Assume that two equal-amplitude test signals ($f_1 = 2.001$ MHz, and $f_2 = 2.003$ MHz) are applied to a linear amplifier. The output spectrum of the amplifier is shown in Figure 5.10. Many of the distortion products fall outside the passband of the amplifier tuned circuits. If no impedance exists at the frequencies of the distortion components, then no voltage can be developed. Further study of this spectrum discloses that no even-order products fall near the two desired signals. Some odd-order products, however, fall near the desired frequencies and possibly within the passband of the tuned circuits.

The distortion products usually given in tube data sheets are the third- and fifth-order intermodulation distortion products that can fall within the amplifier passband. Using the same f_1 and f_2 frequencies of the previous example, the frequencies of the third-order products are:

$$2f_1 - f_2 = 1.999 \; MHz$$

$$2f_2 - f_1 = 2.005 \; MHz$$

The frequencies of the fifth-order products are:

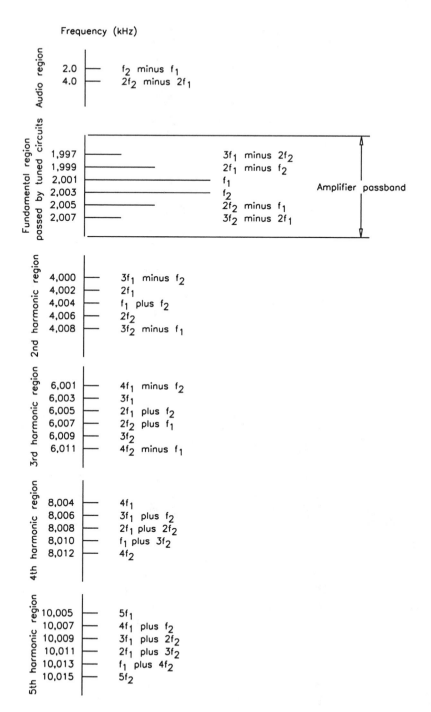

Figure 5.10 Spectrum at the output of a nonlinear device with an input of two equal-amplitude sine waves of $f_1 = 2.001$ MHz and $f_2 = 2.003$ MHz.

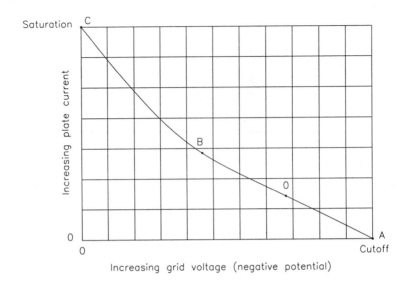

Figure 5.11 Ideal grid-plate transfer curve for class AB operation.

$$3f_1 - 2f_2 = 1.997 \ MHz$$

$$3f_2 - 2f_1 = 2.007 \ MHz$$

These frequencies are well within the passband of a tuned circuit intended to pass voice frequencies and, therefore, power will be delivered to the antenna at these frequencies. All intermodulation distortion power is wasted and serves no purpose other than to cause interference to adjacent channels.

Intermodulation distortion in a power amplifier tube is the result of variations of the transfer characteristics of the device from the ideal case. An ideal transfer characteristic curve is shown in Figure 5.11. Even-order products do not contribute to the intermodulation distortion problem because they fall outside the amplifier passband. Therefore, if the transfer characteristic produces an even-order curvature at the small-signal end of the curve (from point A to point B) and the remaining portion of the curve (point B to point C) is linear, the tube is considered to have an ideal transfer characteristic. If the operating point of the amplifier is set at point 0 (located midway horizontally between point A and point B), there will be no distortion in a class AB amplifier. However, no tube has this idealized transfer characteristic. It is possible, by manipulation of the electron ballistics within a given tube structure, to alter the transfer characteristic and minimize the distortion products.

5.4 High-Efficiency Linear Amplification

The class B RF amplifier is the classic means of achieving linear amplification of an AM signal. A significant efficiency penalty, however, is paid for linear operation. The plate efficiency at carrier level in a practical linear amplifier is about 33 percent, while a high-level (plate)-modulated stage can deliver a basic efficiency of 65 to 80 percent. In a real-world transmitter, power consumed by the modulator stage must also be taken into account. However, because the output power of the modulator is, at most, one-third of the total power output, the net efficiency of the plate-modulated amplifier is still higher than a conventional linear amplifier of comparable power.

Still, linear amplification is attractive to the users of high-power transmitters because low-level modulation can be employed. This eliminates the need for modulation transformers and reactors, which—at certain power levels and for particular applications—may be difficult to design and expensive to construct. To overcome the efficiency penalty of linear amplification, special systems have been designed that take advantage of the benefits of a linear mode, while not imposing an excessive efficiency burden.

5.4.1 Chireix Outphasing Modulated Amplifier

The *Chireix* amplifier employs two output stage tubes. The grids of the tubes are driven with signals whose relative phase varies with the applied modulating signal. The output waveforms of the tubes are then applied to a summing network, and finally to the load. Figure 5.12 shows a simplified schematic diagram.

Operation of the circuit is straightforward. When the grids of the tubes are driven with signals that are 180° out of phase, power output will be zero. This corresponds to a negative AM modulation peak of 100 percent. Full (100 percent) positive modulation occurs when both grids are driven with signals that are in phase. The AM modulating signal is produced by varying the relative phase of the drive signals.

The phase-modulated carrier-frequency grid waveforms are generated in a low-level balanced modulator and amplified to the level required by the PA tubes. Stability of the exciter in the Chireix system is critical. Any shift in the relative phase of the exciter outputs will translate to amplitude modulation of the transmitted waveform.

Because of the outphasing method employed to produce the AM signal, the power factor associated with the output stage is of special significance. The basic modulating scheme would produce a unity power factor at zero output. The power factor would then decrease with increasing RF output. To avoid this undesirable characteristic, the output plate circuits are detuned, one above resonance and one below, to produce a specified offset. This *dephasing*, coupled with adjustment of the relative phase of the driving signals, maintains the power factor within a reasonable range.

There are, however, two major disadvantages to this system. First, the efficiency of the amplifier is compromised because at all instantaneous levels of modulation except one, the anode circuits must work into a reactive load. Typical overall efficiency is on the order of 60 percent. Second, output carrier power adjustment is sensitive to tuning of any stage in the chain. This problem can be reduced by employing broadband amplifying circuits at lower-level stages.

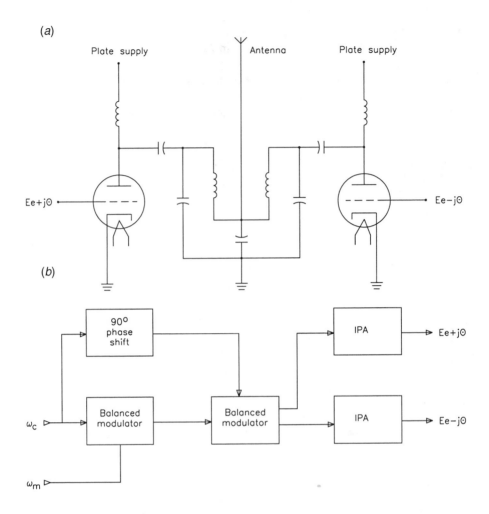

Figure 5.12 Simplified schematic diagram of the Chireix outphasing modulated amplifier: (*a*) output circuit, (*b*) drive signal generation circuit.

5.4.2 Doherty Amplifier

The *Doherty* modulated amplifier has seen considerable use in high-efficiency linear AM transmitters. The system is a true high-efficiency linear amplifier, rather than a hybrid amplifier/modulator. The Doherty amplifier employs two tubes and a 90° network as an *impedance inverter* to achieve load-line modification as a function of power level. The basic circuit is shown in Figure 5.13.

Two tubes are used in the Doherty amplifier, a *carrier tube* (V1) and a *peak tube* (V2). The carrier tube is biased class B. Loading and drive for the tube are set to provide

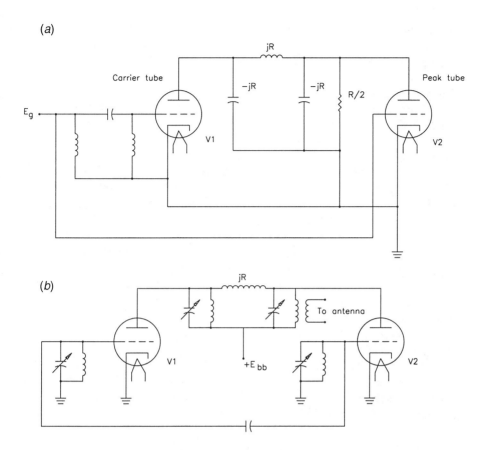

Figure 5.13 Doherty high-efficiency AM amplifier: (*a*) operating theory, (*b*) schematic diagram of a typical system.

maximum linear voltage swing at carrier level. The peak tube is biased class C; at carrier level it just begins to conduct plate current. Each tube delivers an output power equal to twice the carrier power when working into a load impedance of R Ω. At carrier, the reflected load impedance at the plate of V1 is equal to 2R Ω. This is the correct value of load impedance for carrier-level output at full plate voltage swing. (The impedance-inverting property of the 90° network is similar to a ¼-wave transmission line.)

A 90° phase lead network is included in the grid circuit of V1 to compensate for the 90° phase lag produced by the impedance-inverting network in the plate circuit. For negative modulation swings, the carrier tube performs as a linear amplifier with a load impedance of 2R Ω; the peak tube contributes no power to the output.

On positive modulation swings, the peak tube conducts and contributes power to the R/2 load resistance. This is equivalent to connecting a negative resistance in shunt with the R/2 load resistance, so that the value seen at the load end of the 90° network increases. This increase is reflected through the network as a decrease in load resistance at the plate of V1, causing an increase in output current from the tube, and an increase in output power. The drive levels on the tubes are adjusted so that each contributes the same power (equal to twice the carrier power) at a positive modulation peak. For this condition, a load of R Ω is presented to each tube.

The key aspect of the Doherty circuit is the change in load impedance on tube V1 with modulation. This property enables the device to deliver increased output power at a constant plate voltage swing. The result is high efficiency and good linearity. Overall efficiency of the Doherty amplifier ranges from 60 to 65 percent.

A practical application of the Doherty circuit is shown in Figure 5.13*b*. The shunt reactances of the phase-shift networks are achieved by detuning the related tuned circuits. The tuned circuits in the grid are tuned above the operating frequency; those in the plate are tuned below the operating frequency.

The Doherty amplifier has two important advantages in high-power applications. First, the peak anode voltage at either tube is (approximately) only one-half that required for an equivalent carrier power high-level pulse width modulation (PWM) or class B anode-modulated transmitter, thus significantly increasing reliability and usable tube life. Second, no large modulation transformer or special filtering components are used in the final amplifier stages that might cause transient overshoot distortion.

Disadvantages of the Doherty amplifier include nonlinear distortion performance and increased complexity of tuning. The major sources of nonlinear distortion include the nonlinearity of the carrier tube at or near the 100 percent negative modulation crest, and the nonlinearity of the peak tube at or near carrier level when it is just beginning to conduct. Both sources of distortion may be effectively reduced by using moderate amounts of overall envelope feedback.

5.4.3 Screen-Modulated Doherty-Type Amplifier

A variation on the basic Doherty scheme can be found in the screen-modulated power amplifier. The design is unique in that the system does not have to function as a linear amplifier. Instead, modulation is applied to the screen grids of both the carrier and peak tubes. The peak tube is modulated upward during the positive half of the modulation cycle, and the carrier tube is modulated downward during the negative half of the cycle. This technique results in performance improvements over the classic Doherty design in that RF excitation voltages and modulating voltages are isolated from each other, thereby eliminating a troublesome source of tuning-vs.-modulation interaction.

The basic screen modulation system is shown in Figure 5.14. The peak and carrier tubes are biased and driven in quadrature as class C amplifiers from the continuous wave RF drive source. At carrier level, the screen voltage of the carrier tube is adjusted so that the carrier device is near anode saturation and delivering approximately 96 percent of the carrier power. The screen voltage of the peak tube is adjusted so that it is just into conduction, and is supplying the remaining 4 percent (approximately) of carrier power. The combined anode efficiency at carrier level is greater than 77 percent.

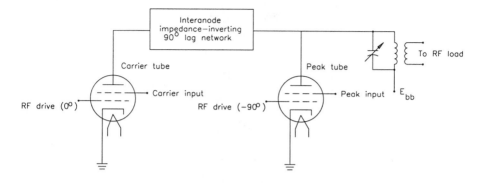

Figure 5.14 Screen-modulated Doherty-type circuit.

Modulation of the carrier occurs when the screen voltage of the peak tube begins to rise during the positive modulation half-cycle, thus causing the peak tube to supply more RF current to the output load. This increase of current into the output network causes the resistance seen by the interanode network to increase and, because of the impedance-inverting characteristics of the 90° interanode network, causes a proportional decrease in the load impedance presented to the carrier tube.

The carrier-tube resonant anode voltage drop is fully saturated over the entire positive modulation half-cycle, and therefore is effectively a constant voltage source. The power output of the carrier tube thus increases during the positive modulation half-cycle, caused by the modulated decreasing impedance at its anode, until both peak and carrier tubes deliver twice carrier power at the 100 percent positive modulation crest. During the negative modulation half-cycle, the peak tube is held out of conduction, while the carrier-tube output voltage decreases linearly to zero output at the 100 percent modulation trough.

A screen-modulated amplifier designed for MF operation at 1 MW is shown in Figure 5.15. The carrier tube delivers all of the 1 MW carrier power to the load through the 90° impedance-inverting network. When modulation is applied to the screen of the carrier tube (during the negative half-cycle), the carrier tube output follows the modulation linearly. Because the carrier tube is driven to full swing, its voltage excursion does not increase during the positive half-cycle. Positive modulation is needed on the screen only to maintain the full plate swing as the impedance load changes on the carrier tube.

The peak tube normally is operated with a negative screen voltage that maintains plate current at near cutoff. During the positive peak of modulation, the screen of the peak tube is modulated upward. As the peak tube is modulated positive, it delivers power into the output circuit. As the delivered power increases, the load seen by the interplate network changes as described for the general case.

The advantages of screen modulation are the same as those identified for the Doherty linear amplifier, plus higher efficiency at all depths of modulation. Furthermore, the screen modulation system is less sensitive to misadjustment of RF amplifier tuning.

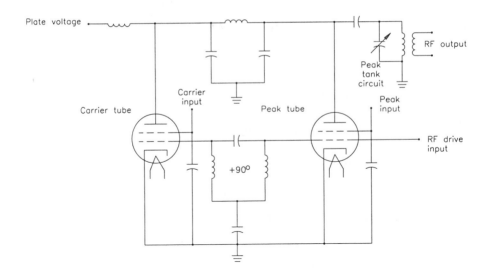

Figure 5.15 Practical implementation of a screen-modulated Doherty-type 1 MW power amplifier (Continental Electronics).

5.4.4 Terman-Woodyard Modulated Amplifier

The *Terman-Woodyard* configuration uses the basic scheme of Doherty for achieving high efficiency (the impedance-inverting property of a 90° phase-shift network). However, the Terman-Woodyard design also employs grid modulation of both the carrier tube and the peak tube, allowing both tubes to be operated class C. The result is an increase in efficiency over the Doherty configuration.

The basic circuit is shown in Figure 5.16. With no modulation, V1 operates as a class C amplifier, supplying the carrier power, and V2 is biased so that it is just beginning to conduct. The efficiency at carrier is, therefore, essentially that of a class C amplifier.

During positive modulation swings, V2 conducts. At 100 percent modulation, both tubes supply equal amounts of power to the load, as in the Doherty design. During negative modulation swings, V2 is cut off, and V1 performs as a standard grid-modulated amplifier.

The Terman-Woodyard system rates efficiency as a function of modulation percentage for sinusoidal waveforms. Typical efficiency at carrier is 80 percent, falling to a minimum of 68 percent at 50 percent modulation and 73 percent at 100 percent modulation.

5.4.5 Dome Modulated Amplifier

The *Dome* high-efficiency amplifier employs three power output tubes driven by different input signals. Modulation is achieved by load-line modification during positive

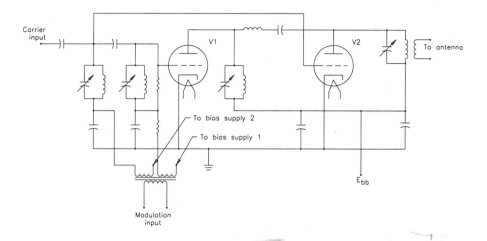

Figure 5.16 Terman-Woodyard high-efficiency modulated amplifier.

modulation excursions, and by linear amplification during negative modulation swings. Load-line modification is achieved by absorption of a portion of the generated RF power. However, most of the absorbed power is returned to the plate power supply, rather than being dissipated as heat. High efficiency is the result.

A basic Dome modulated amplifier is shown in Figure 5.17. Tube V1 is used in a plate-modulated configuration, supplying power to the grid of V2 (the power amplifier tube). The load impedance in the plate circuit of V2 is equivalent to the load impedance reflected into the primary of transformer T1 (from the antenna) in series with the impedance appearing across the C8 terminals of the 90° phase-shift network consisting of C8, C9, and L4. The impedance appearing across the C8 terminals of the 90° network is inversely related to the impedance across the other terminals of the network (in other words, the effective ac impedance of V3). Thus, with V3 cut off, a short circuit is reflected across the C8 terminals of the network. Tube V3 performs as a modulated rectifier with intelligence supplied to its grid. This tube is referred to as the *modifier.*

With no modulation (the carrier condition), tube V3 is biased to cutoff, and the drive to V2 is adjusted until power output is equal to four times the carrier power (corresponding to a positive modulation peak). Note that all the plate signal voltage of V2 appears across the primary of transformer T1 for this condition. The bias on V3 is then reduced, lowering the ac impedance of the tube and, therefore, reflecting an increasing impedance at the C8 terminals of the 90° network. The bias on V3 is reduced until the operating carrier power is delivered to the antenna.

An amount of power equal to the carrier is thus rectified by V3 and returned to the plate power supply, except for that portion dissipated on the V3 plate. The drive level to V2 is finally adjusted so that the tube is just out of saturation.

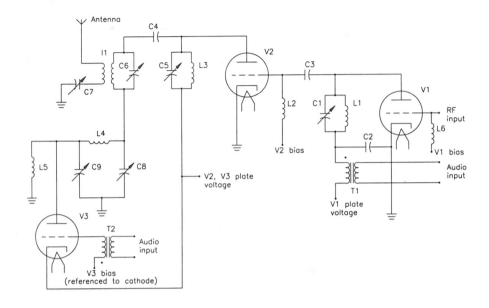

Figure 5.17 Dome high-efficiency modulated amplifier.

For positive modulation excursions, V3 grid voltage is driven negative, reaching cutoff for a positive modulation peak. For negative modulation swings, V2 acts as a linear amplifier. V3 does not conduct during the negative modulation swing because the peak RF voltage on the plate is less than the dc supply voltage on the cathode.

Because of the serial loss incurred in the power amplifier tube and in the modifier tube for energy returned to the power supply, the Dome circuit is not as efficient as the Doherty configuration. Typical efficiency at carrier ranges from 55 to 60 percent.

5.5 Television Power Amplifier Systems

Although VSBAM is used for a number of varied applications, TV transmission is the most common. A TV transmitter is divided into two basic subsystems:

- The *visual* section, which accepts the video input, amplitude-modulates an RF carrier, and amplifies the signal to feed the antenna system

- The *aural* section, which accepts the audio input, frequency-modulates a separate RF carrier, and amplifies the signal to feed the antenna system

The visual and aural signals usually are combined to feed a single radiating antenna. Different transmitter manufacturers have different philosophies with regard to the design and construction of a transmitter. Some generalizations are possible, however, with respect to basic system configurations. Transmitters can be divided into categories based on the following criteria:

- Output power

- Final-stage design

- Modulation system

5.5.1 System Considerations

When the power output of a TV transmitter is discussed, the visual section is the primary consideration. Output power refers to the peak power of the visual stage of the transmitter (*peak of sync*). The FCC-licensed ERP is equal to the transmitter power output times feedline efficiency times the power gain of the antenna.

A VHF station can achieve its maximum power output through a wide range of transmitter and antenna combinations. Reasonable pairings for a high-band VHF station range from a transmitter with a power output of 50 kW feeding an antenna with a gain of 8, to a 30 kW transmitter connected to a gain-of-12 antenna. These combinations assume reasonably low feedline losses. To reach the exact power level, minor adjustments are made to the power output of the transmitter, usually by a front-panel power control.

UHF stations that want to achieve their maximum licensed power output are faced with installing a very high power transmitter. Typical pairings include a transmitter rated for 220 kW and an antenna with a gain of 25, or a 110 kW transmitter and a gain-of-50 antenna. In the latter case, the antenna could pose a significant problem. UHF antennas with gains in the region of 50 are possible, but not advisable for most installations because of coverage problems that can result.

The amount of output power required of the transmitter will have a fundamental effect on system design. Power levels dictate the following parameters:

- Whether the unit will be of solid-state or vacuum tube design

- Whether air, water, or vapor cooling must be used

- The type of power supply required

- The sophistication of the high-voltage control and supervisory circuitry

- Whether *common amplification* of the visual and aural signals (rather than separate visual and aural amplifiers) is practical

Tetrodes generally are used for VHF transmitters above 25 kW, and specialized tetrodes can be found in UHF transmitters at the 15 kW power level and higher. As solid-state technology advances, the power levels possible in a reasonable transmitter design steadily increase, making solid-state systems more attractive options.

In the realm of UHF transmitters, the klystron (and its related devices) reigns supreme. Klystrons use an *electron-bunching* technique to generate high power—55 kW from a single tube is not uncommon—at ultrahigh frequencies. They are currently the first choice for high-power, high-frequency service. Klystrons, however, are not particularly efficient. A stock klystron with no special circuitry might be only 40 percent efficient. Various schemes have been devised to improve klystron efficiency, the best known of which is *beam pulsing*. Two types of pulsing are in common use:

- *Mod-anode pulsing*, a technique designed to reduce power consumption of the device during the color burst and video portion of the signal (and thereby improve overall system efficiency)

- *Annular control electrode* (ACE) pulsing, which accomplishes basically the same goal by incorporating the pulsing signal into a low-voltage stage of the transmitter, rather than a high-voltage stage (as with mod-anode pulsing)

Experience has shown the ACE approach—and other similar designs—to provide greater improvement in operating efficiency than mod-anode pulsing, and better reliability as well.

Several newer technologies offer additional ways to improve UHF transmitter efficiency, including:

- The *inductive output tube* (IOT), also known as the *Klystrode*. This device essentially combines the cathode/grid structure of the tetrode with the drift tube/collector structure of the klystron. (The Klystrode tube is a registered trademark of Varian Associates.)

- The *multistage depressed collector* (MSDC) klystron, a device that achieves greater efficiency through a redesign of the collector assembly. A multistage collector is used to recover energy from the electron stream inside the klystron and return it to the beam power supply.

Improved tetrode devices, featuring higher operating power at UHF and greater efficiency, have also been developed.

A number of approaches may be taken to amplitude modulation of the visual carrier. Current technology systems utilize low-level intermediate-frequency (IF) modulation. This approach allows superior distortion correction, more accurate vestigial sideband shaping, and significant economic advantages to the transmitter manufacturer.

A TV transmitter can be divided into four major subsystems:

- The exciter

- Intermediate power amplifier (IPA)

- Power amplifier

- High-voltage power supply

Figure 5.18 shows the audio, video, and RF paths for a typical design.

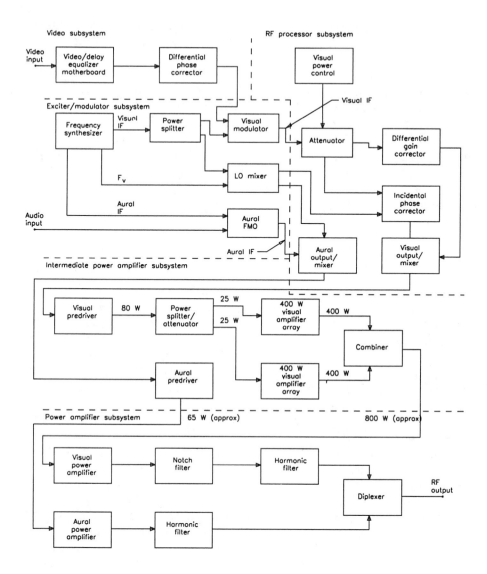

Figure 5.18 Basic block diagram of a TV transmitter. The three major subassemblies are the exciter, IPA, and PA. The power supply provides operating voltages to all sections, and high voltage to the PA stage.

5.5.2 Power Amplifier

The power amplifier raises the output energy of the transmitter to the required RF operating level. Tetrodes in TV service are usually operated in the class B mode to obtain reasonable efficiency while maintaining a linear transfer characteristic. Class B amplifiers, when operated in tuned circuits, provide linear performance because of the flywheel effect of the resonance circuit. This allows a single tube to be used instead of two in push-pull fashion. The bias point of the linear amplifier must be chosen so that the transfer characteristic at low modulation levels matches that at higher modulation levels. Even so, some nonlinearity is generated in the final stage, requiring differential gain correction. The plate (anode) circuit of a tetrode PA usually is built around a coaxial resonant cavity, providing a stable and reliable tank.

UHF transmitters using a klystron in the final output stage must operate class A, the most linear but also most inefficient operating mode for a vacuum tube. The basic efficiency of a nonpulsed klystron is approximately 40 percent. Pulsing, which provides full available beam current only when it is needed (during peak of sync), can improve device efficiency by as much as 25 percent, depending on the type of pulsing used.

Two types of klystrons are presently in service:

- Integral-cavity klystron

- External-cavity klystron

The basic theory of operation is identical for each tube, but the mechanical approach is radically different. In the integral-cavity klystron, the cavities are built into the klystron to form a single unit. In the external-cavity klystron, the cavities are outside the vacuum envelope and bolted around the tube when the klystron is installed in the transmitter.

A number of factors come into play in a discussion of the relative merits of integral- vs. external-cavity designs. Primary considerations include operating efficiency, purchase price, and life expectancy.

The PA stage includes a number of sensors that provide input to supervisory and control circuits. Because of the power levels present in the PA stage, sophisticated fault-detection circuits are required to prevent damage to components in the event of a problem inside or outside the transmitter. An RF sample, obtained from a directional coupler installed at the output of the transmitter, is used to provide automatic power-level control.

The transmitter system discussed in this section assumes separate visual and aural PA stages. This configuration is normally used for high-power transmitters. A combined mode also may be used, however, in which the aural and visual signals are added prior to the PA. This approach offers a simplified system, but at the cost of additional precorrection of the input video signal.

PA stages often are configured so that the circuitry of the visual and aural amplifiers is identical, providing backup protection in the event of a visual PA failure. The aural PA can then be reconfigured to amplify both the aural and the visual signals, at reduced power.

The aural output stage of a TV transmitter is similar in basic design to an FM broadcast transmitter. Tetrode output devices generally operate class C, providing good efficiency. Klystron-based aural PAs are used in UHF transmitters.

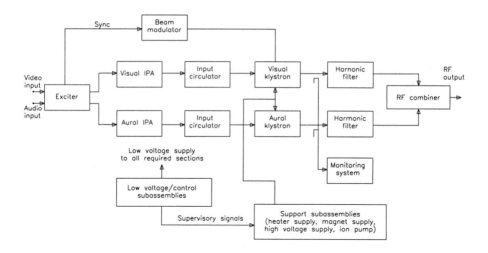

Figure 5.19 Schematic diagram of a 60 kW klystron-based TV transmitter.

Application Example

A 60 kW transmitter is shown in block diagram form in Figure 5.19. A single high-power klystron is used in the visual amplifier, and another is used in the aural amplifier. The tubes are driven from solid-state intermediate power amplifier modules. The transmitter utilizes ACE-type beam control, requiring additional predistortion to compensate for the nonlinearities of the final visual stage. Predistortion is achieved by correction circuitry at an intermediate frequency in the modulator. Both klystrons are driven from the output of a circulator, which ensures a minimum of driver-to-load mismatch problems.

A block diagram of the beam modulator circuit is shown in Figure 5.20. The system receives input signals from the modulator, which synchronizes ACE pulses to the visual tube with the video information. The pulse waveform is developed through a pulse amplifier, rather than a switch. This permits more accurate adjustments of operating conditions of the visual amplifier.

Although the current demand from the beam modulator is low, the bias is near cathode potential, which is at a high voltage relative to ground. The modulator, therefore, must be insulated from the chassis. This is accomplished with optical transmitters and receivers connected via fiber optic cables. The fiber optic lines carry supervisory, gain control, and modulating signals.

The four-cavity external klystrons will tune to any channel in the UHF-TV band. An adjustable *beam perveance* feature enables the effective electrical length of the device to be varied by altering the beam voltage as a function of operating frequency (see Section 6.3). Electromagnetic focusing is used on both tubes. The cavities, body, and gun

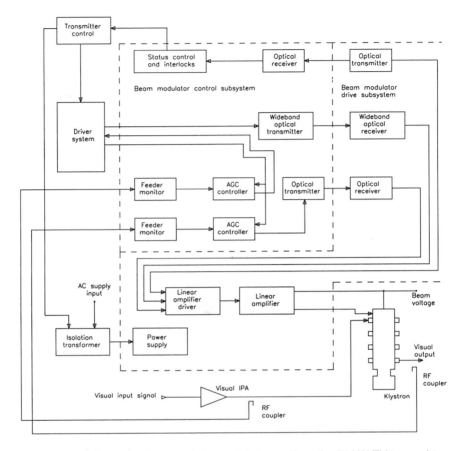

Figure 5.20 Schematic diagram of the modulator section of a 60 kW TV transmitter.

areas of the klystrons are air-cooled. The collectors are vapor-phase-cooled using an external heat exchanger system.

The outputs of the visual and aural klystrons are passed through harmonic filters to an RF combiner before being applied to the antenna system.

5.6 FM Power Amplifier Systems

The amplitude of an FM signal remains constant with modulation so that efficient class B and C amplifiers can be used.[1] Two basic circuit types have evolved:

- Amplifiers based on a tetrode or pentode in a grid-driven configuration

---

1 This section was contributed by Geoffrey N. Mendenhall, P. E., Cincinnati, OH, and Warren B. Bruene, P.E., Dallas, TX.

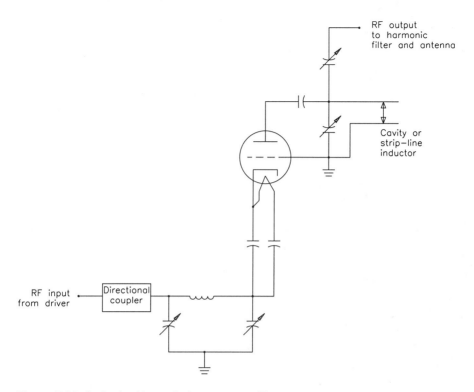

RF output
to harmonic
filter and antenna

Cavity or
strip−line
inductor

RF input
from driver

Directional
coupler

Figure 5.21 Cathode-driven triode power amplifier.

- Amplifiers based on a high-μ triode or tetrode device in a cathode-driven configuration (grounded grid)

High-power FM amplifiers operating in the VHF range typically employ resonant cavity systems. The basic configurations of FM amplifiers are discussed in this section, and amplifiers utilizing cavity resonators are discussed in Section 4.3.

5.6.1 Cathode-Driven Triode Amplifier

The characteristics of high-μ triodes are well adapted to FM power amplifier service. The grounded-grid circuit is simple, and no screen or grid bias power supplies are required. Figure 5.21 shows the basic circuit configuration. In this case, the grid is connected directly to chassis ground. The difference between dc cathode current and dc plate current is dc grid current. The output tank is a shorted coaxial cavity that is capacitively loaded by the tube output and stray circuit capacitance. A small capacitor is used for trimming the tuning, and another small variable capacitor is used for adjusting the loading. A pi network is used to match the 50 Ω input to the tube cathode input impedance.

The triode usually is operated in the class B mode to achieve maximum power gain, which is on the order of 20 (13 dB). The device can be driven into class C operation by providing negative grid bias or positive cathode bias. This increases the plate efficiency, but also requires increased drive power.

Most of the drive into a grounded-grid amplifier is fed through the tube and appears at the output. This increases the apparent efficiency so that the efficiency factor given by the transmitter manufacturer may be higher than the actual plate efficiency of the tube. The true plate efficiency is determined from the following:

$$Eff_p = \frac{P_o}{(I_p \times E_p) + (P_{in} \times \alpha)} \tag{5.2}$$

Where:
Eff_p = actual plate efficiency
Po = measured output power
I_p = dc plate input current to final stage
E_p = dc plate input voltage to final stage
P_{in} = RF input drive power
α = fraction of grid-to-plate transfer (0.9 typical)

Because most of the drive power is fed through the tube, any changes in loading of the output circuit also will affect the input impedance and the driver stage.

Because there is RF drive voltage on the cathode (filament), some means of decoupling must be used to block the signal from the filament transformer. One method employs high-current RF chokes; another feeds the filament power through the input tank circuit inductor.

5.6.2 Grounded-Grid vs. Grid-Driven Tetrode

Tetrodes also may be operated in the grounded-grid configuration by placing both the control grid and the screen grid at RF ground. Higher efficiency and gain can be achieved by placing negative bias on the control grid while placing a positive voltage on the screen grid of a cathode-driven tetrode.

The input capacitance of a tetrode in a grounded-grid configuration is much less than in a grid-driven configuration, and the input impedance is lower, providing wider bandwidth. The approximate input capacitances of some common tube types are listed in Table 5.1.

The typical drive power requirements of a high-gain tetrode, as a function of plate voltage, for a 5 kW power amplifier operating at 100 MHz are as follows:

- Grounded-grid configuration: 340 W input at 4.5 kV plate voltage, 280 W at 5.2 kV

- Grid-driven configuration: 190 W at 4.5 kV plate voltage, 140 W at 5.2 kV

Table 5.1 Approximate Input Capacitance of Common Power Grid Tubes Used in Grounded-Grid and Grounded-Cathode Service

Tube Type	Gounded Grid	Grounded Cathode
4CX3000A	67	140
4CX3500A	59	111
4CX5000A	53	115
4CX15,000A	67	161
4CX20,000A/8990	83	190

There are several tradeoffs in performance between the grounded-grid and the grid-driven modes of a tetrode PA with respect to gain, efficiency, amplitude bandwidth, phase bandwidth, and incidental AM under equivalent operating conditions:

- When the PA is driven into saturation, its bandwidth is limited by the output cavity bandwidth in the grounded-grid amplifier. The PA bandwidth in the grid-driven amplifier is limited by the input circuit Q.

- Ouput bandwidth under saturation can be improved in either configuration by reducing the plate voltage. This involves a tradeoff in efficiency with a smaller voltage swing. The bandwidth improvement can be obtained with a loss of PA gain and efficiency.

- A grounded-grid saturated PA improves bandwidth over a grid-driven saturated PA at the expense of amplifier gain. Best performance for FM operation is obtained when the amplifier is driven into saturation where little change in output power occurs with increasing drive power. Maximum efficiency also occurs at this point.

- The phase linearity in the 0.5 dB bandwidth of the amplifier is better in a grid-driven configuration. The class C grounded-grid PA exhibits a more nonlinear phase slope within the passband, yet has a wider amplitude bandwidth. This phenomenon is the result of interaction of the input and output circuits because they are effectively connected in series in the grounded-grid configuration. The neutralized grid-driven PA provides more isolation between these networks, so they behave more like independent filters.

5.6.3 Grid-Driven Tetrode/Pentode Amplifiers

RF generators using tetrode amplifiers throughout usually have one less stage than those using triodes. Because tetrodes have higher power gain, they are driven into class C operation for high plate efficiency. Countering these advantages is the requirement for neutralization, along with screen and bias power supplies. Figure 5.22

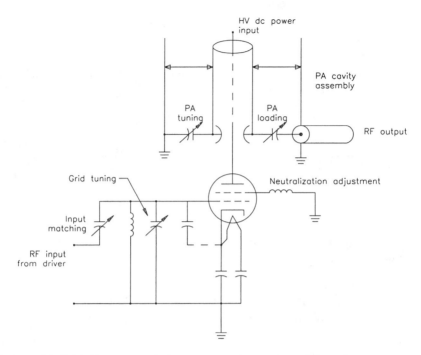

Figure 5.22 Grid-driven grounded-screen tetrode power amplifier.

shows a schematic of a grid-driven tetrode amplifier. In this example, the screen is operated at dc ground potential, and the cathode (filament) is operated below ground by the amount of screen voltage required (grounded-screen operation). The advantage of this configuration is that stability problems resulting from undesired resonances in the screen bypass capacitors are eliminated. With this arrangement, however, the screen supply must be capable of handling the combined plate and screen currents.

With directly heated tubes, it is necessary to use filament bypass capacitors. During grounded-screen operation, these bypass capacitors must have a higher breakdown voltage rating (relative to grounded-grid operation) because they will have the dc screen voltage across them. The filament transformer must have additional insulation to withstand the dc screen voltage. The screen power supply provides a negative voltage in series with the cathode-to-ground, and must have the additional capacity to handle the sum of the plate and screen currents. A coaxial cavity is used in the output circuit so that the circulating current is spread over large surfaces, keeping the losses low. This cavity is a shorted ¼-wavelength transmission line section that resonates the tube output capacitance. The length is preset to the desired carrier frequency, and a small-value variable capacitor is used to trim the system to resonance. (Amplifiers utilizing cavity resonator systems are discussed in Section 4.3.)

Pentode amplifiers have even higher gain than their tetrode counterparts. The circuit configuration and bias supply requirements for the pentode are similar to the tetrode's because the third (suppressor) grid is tied directly to ground. The additional isolating effect of the suppressor grid eliminates the need for neutralization in the pentode amplifier.

5.6.4 Impedance Matching into the Grid

The grid circuit usually is loaded (swamped) with added resistance. The purpose of this resistance is to broaden the bandwidth of the circuit by lowering the circuit Q and to provide a more constant load to the driver. The resistance also makes neutralization less critical so that the amplifier is less likely to become unstable with varying output circuit loading.

Cathode and filament lead inductance from inside the tube, through the socket and filament bypass capacitors to ground, can increase grid driver requirements. This effect is the result of RF current flowing from grid to filament through the tube capacitance and then through the filament lead inductance to ground. An RF voltage is developed on the filament that, in effect, causes the tube to be partly cathode-driven. This undesirable extra drive power requirement can be minimized by series resonating the cathode return path with the filament bypass capacitors or by minimizing the cathode-to-ground inductance by using a tube socket incorporating thin-film dielectric *sandwich* capacitors for coupling and bypassing.

High-power grid-driven class C amplifiers require a swing of several hundred RF volts on the grid. To develop this high voltage swing, the input impedance of the grid must be increased by the grid input matching circuit. Because the capacitance between the grid and the other tube elements may be 100 pF or more, the capacitive reactance at 100 MHz, for example, will be very low unless the input capacitance is parallel resonated with an inductor. Bandwidth can be maximized by minimizing any additional circuit capacitance and using a portion of the tube input capacitance as part of the impedance-transformation network. Figure 5.23 illustrates two popular methods of resonating and matching into the grid of a high-power tube. Both schemes can be analyzed by recognizing that the desired impedance transformation is produced by an equivalent L network.

In Figure 5.23a, a variable inductor (L_{in}) is used to raise the input reactance of the tube by bringing the tube input capacitance (C_{in}) almost to parallel resonance. Parallel resonance is not reached because a small amount of parallel capacitance (C_p) is required by the equivalent L network to transform the high impedance (Z_{in}) of the tube down to a lower value through the series matching inductor (L_s). This configuration has the advantage of providing a low-pass filter by using part of the tube's input capacitance to form C_p.

Figure 5.23b uses variable inductor L_{in} to take the input capacitance (C_{in}) past parallel resonance so that the tube's input impedance becomes slightly inductive. The variable series matching capacitor (C_s) forms the rest of the equivalent L network. This configuration constitutes a high-pass filter.

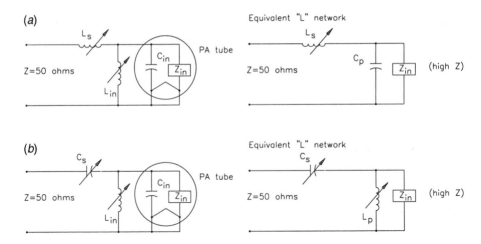

Figure 5.23 Input matching circuits: (a) inductive input matching, (b) capacitive input matching.

Interstage Coupling

The IPA output circuit and the final amplifier input circuit often are coupled by a co-axial transmission line. Impedance matching usually is accomplished at either end by one of the configurations shown in Figure 5.24. The circuits shown, except for 5.24d, require some interactive adjustment of the tuning and loading elements to provide a satisfactory impedance match for each operating frequency and RF drive level. The circuit in Figure 5.24d utilizes multiple LC sections, with each section providing a small step in the total impedance transformation. This technique provides a broad-band impedance match without adjustment, thereby improving transmitter stability, ease of operation, and maintainability. A single grid resonating control is sufficient to tune and match the 50 Ω driver impedance to the high input impedance of the grid over a relatively wide band of frequencies and RF power levels.

The transmission line matching problem is eliminated in some transmitter designs by integrating an IPA stage, using one or more tubes, into the grid circuit of the final amplifier by having the plate of the IPA and the grid of the final tube share a common tuned circuit. This technique has the advantage of simplicity by not transforming the impedance down to 50 Ω and then back up to the grid impedance level.

Solid-state RF power devices possess a low load impedance at the device output terminal, so an impedance transformation that goes through the 50 Ω intermediate impedance of a coaxial cable is necessary to couple these devices into the relatively high impedance of the final amplifier grid.

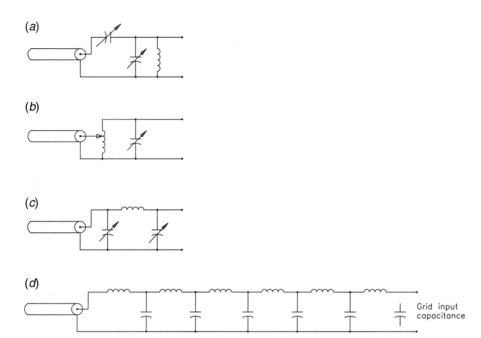

Figure 5.24 Interstage RF coupling circuits: (*a*) variable capacitance matching, (*b*) variable tank matching, (*c*) variable pi matching, (*d*) fixed broadband imped-ance-transformation network.

5.6.5 Neutralization

Most cathode-driven amplifiers using grounded-grid triodes do not require neutralization. It is necessary that the grid-to-ground inductance, both internal and external to the tube, be kept low to maintain this advantage. Omission of neutralization will allow a small amount of interaction between the output circuit and the input circuit through the plate-to-filament capacitance. This effect is not particularly noticeable because of the large coupling between the input and output circuits through the electron beam of the tube. Cathode-driven tetrodes have higher gain than triodes and, therefore, often require some form of neutralization.

Grid-driven high-gain tetrodes need accurate neutralization for best stability and performance. Self-neutralization can be accomplished simply by placing a small amount of inductance between the tube screen grid and ground. This inductance is usually in the form of several short adjustable-length straps. The RF current flowing from plate to screen in the tube also flows through this screen lead inductance. This develops a small RF voltage on the screen, of the opposite phase, which cancels the voltage fed back through the plate-to-grid capacitance. This method of lowering the self-neutralizing frequency of the tube works only if the self-neutralizing frequency of the tube/socket combination is above the desired operating frequency before the inductance is added. Special attention must be given to minimizing the inductances in the tube socket by integrating distributed bypass capacitors into the socket and cavity deck assembly. Pentodes normally do not require neutralization because the suppressor grid effectively isolates the plate from the grid.

5.7 Special-Application Amplifiers

The operating environment of tubes used in commercial, industrial, and military applications usually is characterized by widely varying supply voltages, heavy and variable vibration, and significant changes in loading. Scientific applications are no less demanding. Research projects span a wide range of powers and frequencies. Typical uses for power vacuum tubes as of this writing include:

- Super proton synchrotron delivering 2.4 MW at 200 MHz

- Fusion reactor using 33 MW of high-frequency heating in the 25 to 60 MHz band

- Plasma research using 12 MW in the 60 to 120 MHz range

- Fusion research using a 500 kW ion cyclotron resonance heating generator operating at 30 to 80 MHz

It can be seen from these examples that heat dissipation is a major consideration in the design of an amplifier or oscillator for scientific uses.

5.7.1 Distributed Amplification

Specialized research and industrial applications often require an amplifier covering several octaves of bandwidth. At microwave frequencies, a traveling wave tube (TWT) may be used. At MF and below, multiple tetrodes usually are combined to achieve the required bandwidth. A *distributed amplifier* consists of multiple tetrodes and lumped-constant transmission lines. The lines are terminated by load resistances with magnitudes equal to their characteristic impedances. A cyclical wave of current is present on the output transmission line circuit, with each tube contributing its part in proper phase. A typical configuration for a distributed amplifier includes 8 to 16 tubes. The efficiency of such a design is low.

5.7.2 Radar

The exciter stage of a radar set comprises oscillators, frequency multipliers, and mixers. The signals produced depend on whether the transmitter output device is operated as a power amplifier or an oscillator.

Transmitters using power oscillators such as *magnetrons* determine the radio frequency by tuning of the device itself. In a conventional (*noncoherent*) radar system, the only frequency required is the *local oscillator* (LO). The LO differs from the magnetron frequency by an intermediate frequency, and this difference usually is maintained

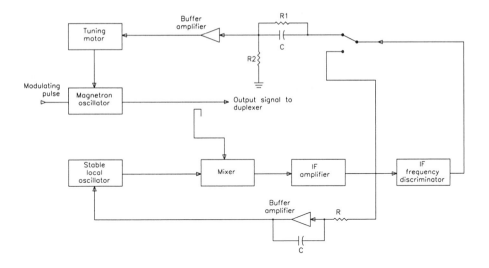

Figure 5.25 Two common approaches to automatic frequency control of a magnetron oscillator.

with an *automatic frequency control* (AFC) loop. Figure 5.25 shows a simple magnetron-based radar system with two methods of tuning:

- Slaving the magnetron to follow the *stable local oscillator* (STALO)

- Slaving the STALO to follow the magnetron

If the radar must use *coherent detection* (such as in Doppler applications), a second oscillator, called a *coherent oscillator* (COHO), is required. The COHO operates at the intermediate frequency and provides a reference output for signal processing circuits.

The synchronizer circuits in the exciter supply timing pulses to various radar subsystems. In a simple marine radar, this may consist of a single multivibrator that triggers the transmitter. In a larger system, 20 to 30 timing pulses may be required. These may turn the beam current on and off in various transmitter stages, start and stop RF pulse time attenuators, start display sweeps, and perform numerous other functions. Newer radar systems generate the required timing signals digitally. A digital synchronizer is illustrated in Figure 5.26.

Modulator

A radar system RF amplifier usually centers on one of two microwave devices: a *crossed-field tube* or *linear-beam tube*. Both are capable of high peak output power at microwave frequencies. To obtain high efficiency from a pulsed radar transmitter, it is necessary to cut off the current in the output tube between pulses. The modulator per-

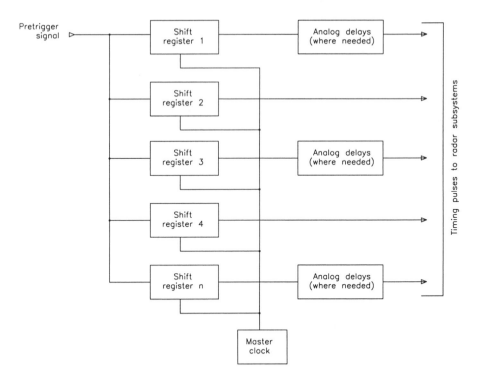

Figure 5.26 Digital synchronizer system for radar applications.

forms this function. Some RF tubes include control electrodes or grids to achieve the same result. Three common types of modulators are used in radar equipment:

- *Line-type modulator* (Figure 5.27). This common radar modulator is used most often to pulse a magnetron. Between pulses, a charge is stored in a *pulse-forming network* (PFN). A trigger signal fires a *thyratron* tube, short-circuiting the input to the PFN, which causes a voltage pulse to appear at the primary of transformer T1. The PFN components are chosen to produce a rectangular pulse at the magnetron cathode, with the proper voltage and current to excite the magnetron to oscillation. An advantage of this design is its simplicity. A drawback is the inability to electronically change the width of the transmitted pulse.

- *Active-switch modulator* (Figure 5.28). This system permits pulse width variation, within the limitations of the energy stored in the high-voltage power supply. A switch tube controls the generation of RF by completing the circuit path from the output tube to the power supply, or by causing stored energy to be dumped to the output device. The figure shows the basic design of an active-switch modulator and three variations on the scheme. The circuits differ in the method of coupling power supply energy to the output tube (capacitor-coupled, transformer-coupled, or a combination of the two methods).

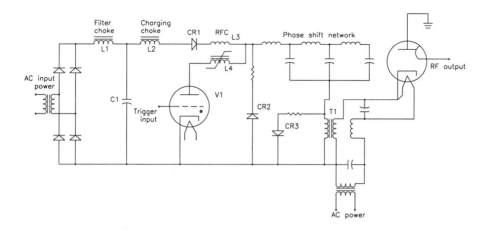

Figure 5.27 A line-type modulator for radar.

- *Magnetic modulator* (Figure 5.29). This design is the simplest of the three modulators discussed. No thyratron or switching device is used. Operation of the modulator is based on the saturation characteristics of inductors L1, L2, and L3. A
  long-duration low-amplitude pulse is applied to L1, which charges C1. As C1 approaches its fully charged state, L2 saturates and the energy in C1 is transferred in
  a resonant fashion to C2. This process continues to the next stage (L3 and C3).
  The transfer time is set by selection of the components to be about one-tenth that
  of the previous stage. At the end of the chain, a short-duration high-amplitude
  pulse is generated, exciting the RF output tube.

Because the applications for radar vary widely, so do antenna designs. Sizes range
from less than 1 ft to hundreds of feet in diameter. An antenna intended for radar applications must direct radiated power from the transmitter to the azimuth and elevation coordinates of the target. It also must serve as a receive antenna for the echo.

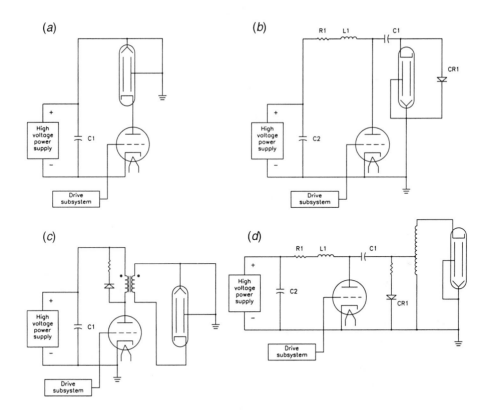

Figure 5.28 Active-switch modulator circuits: (a) direct-coupled system, (b) capacitor-coupled, (c) transformer-coupled, (d) capacitor- and transformer-coupled.

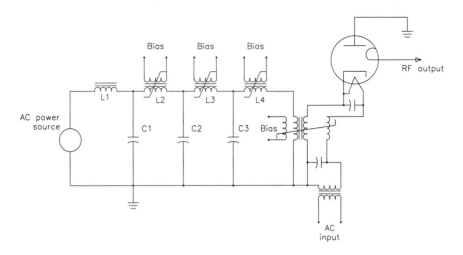

Figure 5.29 A magnetic modulator circuit.

5.8　References

1. Laboratory Staff, *The Care and Feeding of Power Grid Tubes*, Varian Eimac, San Carlos, CA, 1984.
2. Woodard, G. W., "AM Transmitters," in *NAB Engineering Handbook*, 8th Ed., E. B. Crutchfield (ed.), National Association of Broadcasters, Washington, DC, pg. 356, 1992.)

5.9　Bibliography

Chireix, Henry, "High Power Outphasing Modulation," *Proceedings of the IRE*, Vol. 23, no. 11, pg. 1370, November 1935.
Crutchfield, E. B., (ed.), *NAB Engineering Handbook*, 8th Ed., National Association of Broadcasters, Washington, DC, 1992.
Doherty, W. H., "A New High Efficiency Power Amplifier for Modulated Waves," *Proceedings of the IRE*, Vol. 24, no. 9, pg. 1163, September 1936.
Fink, D., and D. Christiansen, *Electronic Engineers' Handbook*, 3rd Ed., McGraw-Hill, New York, 1989.
Gray, T. S., *Applied Electronics*, Massachusetts Institute of Technology, 1954.
Heising, R. A., "Modulation in Radio Telephony," *Proceedings of the IRE*, Vol. 9, no. 3, pg. 305, June 1921.
Heising, R. A., "Transmission System," U.S. Patent no. 1,655,543, January 1928.
High Power Transmitting Tubes for Broadcasting and Research, Philips Technical Publication, Eindhoven, the Netherlands, 1988.
Honey, J. F., "Performance of AM and SSB Communications," *Tele-Tech.*, September 1953.
Jordan, Edward C., *Reference Data for Engineers: Radio, Electronics, Computers, and Communications*, 7th Ed., Howard W. Sams, Indianapolis, IN, 1985.
Martin, T. L., Jr., *Electronic Circuits*, Prentice Hall, Englewood Cliffs, NJ, 1955.
Mina and Parry, "Broadcasting with Megawatts of Power: The Modern Era of Efficient Powerful Transmitters in the Middle East," *IEEE Transactions on Broadcasting*, Vol. 35, no. 2, IEEE, Washington, D.C., June 1989.
Pappenfus, E. W., W. B. Bruene, and E. O. Schoenike, *Single Sideband Principles and Circuits*, McGraw-Hill, New York, 1964.
Ridgwell, J. F., "A New Range of Beam-Modulated High-Power UHF Television Transmitters," *Communications & Broadcasting*, no. 28, Marconi Communications, Clemsford Essex, England, 1988.
Skolnik, M. I. (ed.), *Radar Handbook*, McGraw-Hill, New York, 1980.
Terman, F. E., and J. R. Woodyard, "A High Efficiency Grid-Modulated Amplifier, *Proceedings of the IRE*, Vol. 26, no. 8, pg. 929, August 1938.
Terman, F. E., *Radio Engineering*, 3rd Ed., McGraw-Hill, New York, 1947.
Weldon, J. O., "Amplifiers," U. S. Patent no. 2,836,665, May 1958.
Whitaker, Jerry C., *AC Power Systems*, 2nd Ed., CRC Press, Boca Raton, FL, 1998.
Whitaker, Jerry C., *Radio Frequency Transmission Systems: Design and Operation*, McGraw-Hill, New York, 1991.
Woodard, George W., "AM Transmitters," in *NAB Engineering Handbook*, 9th Ed., Jerry C. Whitaker (ed.), National Association of Broadcasters, Washington, D.C., pp. 353-381, 1998.

Microwave Power Tubes

6.1 Introduction

Microwave power tubes span a wide range of applications, operating at frequencies from 300 MHz to 300 GHz with output powers from a few hundred watts to more than 10 MW. Applications range from the familiar to the exotic. The following devices are included under the general description of microwave power tubes:

- Klystron, including the *reflex* and *multicavity* klystron
- *Multistage depressed collector* (MSDC) klystron
- *Inductive output tube* (IOT)
- *Traveling wave tube* (TWT)
- *Crossed-field tube*
- *Coaxial magnetron*
- *Gyrotron*
- *Planar triode*
- High-frequency tetrode
- *Diacrode*

This wide variety of microwave devices has been developed to meet a broad range of applications. Some common uses include:

- UHF-TV transmission
- Shipboard and ground-based radar
- Weapons guidance systems
- *Electronic countermeasure* (ECM) systems
- Satellite communications
- Tropospheric scatter communications

- Fusion research

As new applications are identified, improved devices are designed to meet the needs. Microwave power tube manufacturers continue to push the limits of frequency, operating power, and efficiency. Microwave technology, therefore, is an evolving science. Figure 6.1 charts device type as a function of operating frequency and power output.

Two principal classes of microwave vacuum devices are in common use today:

- *Linear-beam* tubes

- *Crossed-field* tubes

Each class serves a specific range of applications. In addition to these primary classes, some power grid tubes also are used at microwave frequencies.

6.1.1 Linear-Beam Tubes

In a linear-beam tube, as the name implies, the electron beam and the circuit elements with which it interacts are arranged linearly. The major classifications of linear-beam tubes are shown in Figure 6.2. In such a device, a voltage applied to an anode accelerates electrons drawn from a cathode, creating a beam of kinetic energy. Power supply potential energy is converted to kinetic energy in the electron beam as it travels toward the microwave circuit. A portion of this kinetic energy is transferred to microwave energy as RF waves slow down the electrons. The remaining beam energy is either dissipated as heat or returned to the power supply at the collector. Because electrons will repel one another, there usually is an applied magnetic focusing field to maintain the beam during the interaction process. The magnetic field is supplied either by a solenoid or permanent magnets. Figure 6.3 shows a simplified schematic of a linear-beam tube.

6.1.2 Crossed-Field Tubes

The magnetron is the pioneering device of the family of crossed-field tubes. The family tree of this class of devices is shown in Figure 6.4. Although the physical appearance differs from that of linear-beam tubes, which are usually circular in format, the major difference is in the interaction physics that requires a magnetic field at right angles to the applied electric field. Whereas the linear-beam tube sometimes requires a magnetic field to maintain the beam, the crossed-field tube always requires a magnetic focusing field.

Figure 6.5 shows a cross section of the magnetron, including the magnetic field applied perpendicular to the cathode-anode plane. The device is basically a diode with the anode composed of a plurality of resonant cavities. The interaction between the electrons emitted from the cathode and the crossed electric and magnetic fields produces a series of space-charge spokes that travel around the anode-cathode space in a manner that transfers energy to the RF signal supported by the multicavity circuit. The mechanism is highly efficient.

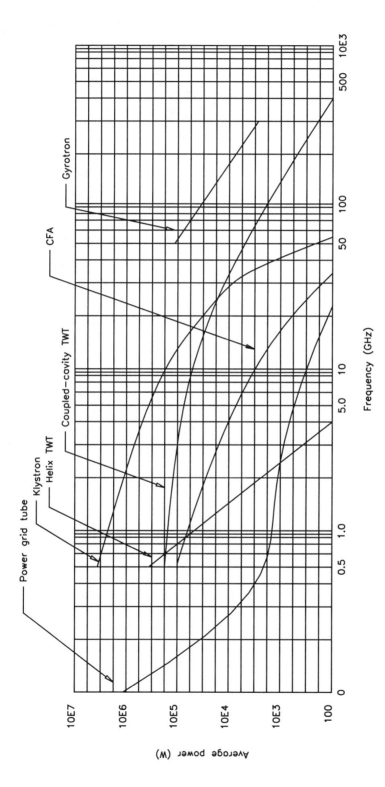

Figure 6.1 Microwave power tube type as a function of frequency and output power.

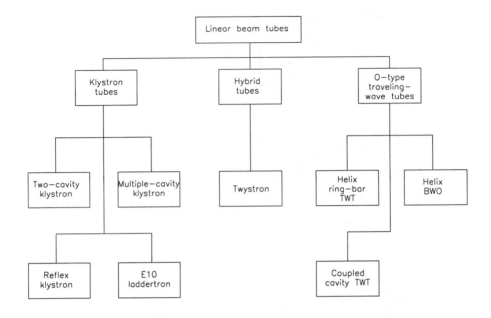

Figure 6.2 Types of linear-beam microwave tubes.

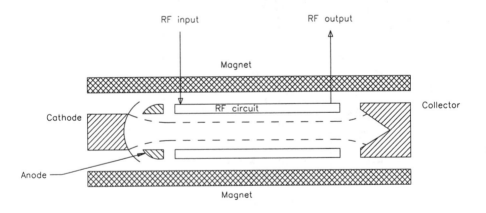

Figure 6.3 Schematic diagram of a linear-beam tube.

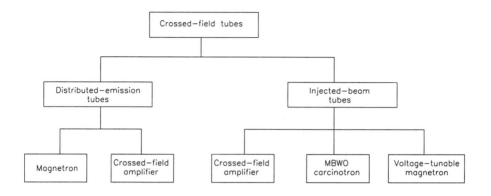

Figure 6.4 Types of crossed-field microwave tubes.

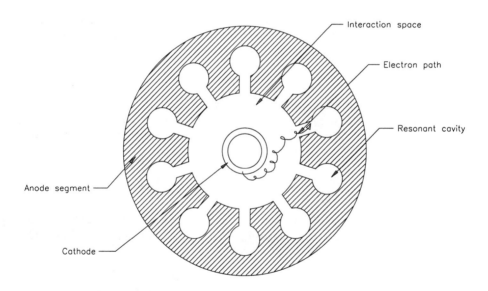

Figure 6.5 Magnetron electron path looking down into the cavity with the magnetic field applied.

Crossed-Field Amplifiers

Figure 6.6 shows the family tree of the crossed-field amplifier (CFA). The configuration of a typical present-day distributed emission amplifier is similar to that of the magnetron except that the device has an input for the introduction of RF energy into the circuit. Current is obtained primarily by secondary emission from the negative

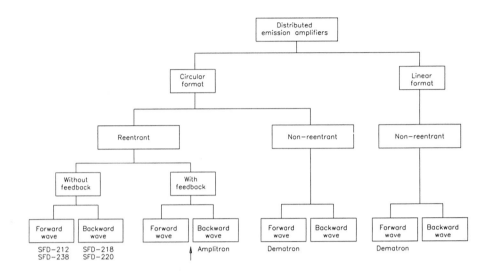

Figure 6.6 Family tree of the distributed emission crossed-field amplifier (CFA).

electrode that serves as a cathode throughout all or most of the interaction space. The earliest versions of this tube type were called *amplitrons*.

The CFA is deployed in radar systems operating from UHF to the Ku-band, and at power levels up to several megawatts. In general, bandwidth ranges from a few percent to as much as 25 percent of the center frequency.

6.2 Grid Vacuum Tubes

The physical construction of a vacuum tube causes the output power and available gain to decrease with increasing frequency. The principal limitations faced by grid-based devices include the following:

- Physical size. Ideally, the RF voltages between electrodes should be uniform, but this condition cannot be realized unless the major electrode dimensions are significantly less than $\frac{1}{4}$ wavelength at the operating frequency. This restriction presents no problems at VHF, but as the operating frequency increases into the microwave range, severe restrictions are placed on the physical size of individual tube elements.

- Electron transit time. Interelectrode spacing, principally between the grid and the cathode, must be scaled inversely with frequency to avoid problems associated with electron transit time. Possible adverse conditions include: 1) excessive loading of the drive source, 2) reduction in power gain, 3) back-heating of the cathode as a result of electron bombardment, and 4) reduced conversion efficiency.

- Voltage standoff. High-power tubes operate at high voltages. This presents significant problems for microwave vacuum tubes. For example, at 1 GHz the grid-cathode spacing must not exceed a few mils. This places restrictions on the operating voltages that may be applied to the individual elements.

- Circulating currents. Substantial RF currents may develop as a result of the inherent interelectrode capacitances and stray inductances/capacitances of the device. Significant heating of the grid, connecting leads, and vacuum seals may result.

- Heat dissipation. Because the elements of a microwave grid tube must be kept small, power dissipation is limited.

Still, a number of grid-based vacuum tubes find applications at high frequencies. For example, planar triodes are available that operate at several gigahertz, with output powers of 1 to 2 kW in pulsed service. Efficiency (again for pulsed applications) ranges from 30 to 60 percent, depending on the frequency.

6.2.1 Planar Triode

A cross-sectional diagram of a planar triode is shown in Figure 6.7. The envelope is made of ceramic, with metal members penetrating the ceramic to provide for connection points. The metal members are shaped either as disks or as disks with cylindrical projections.

The cathode is typically oxide-coated and indirectly heated. The key design objective for a cathode is high emission density and long tube life. Low-temperature emitters are preferred because high cathode temperatures typically result in more evaporation and shorter life.

The grid of the planar triode is perhaps the greatest design challenge for tube manufacturers. Close spacing of small-sized elements is needed, at tight tolerances. Good thermal stability also is required, because the grid is subjected to heating from currents in the element itself, plus heating from the cathode and bombardment of electrons from the cathode.

The anode, usually made of copper, conducts the heat of electron bombardment to an external heat sink. Most planar triodes are air-cooled.

Planar triodes designed for operation at 1 GHz and above are used in a variety of circuits. The grounded-grid configuration is most common. The plate resonant circuit is cavity-based, using waveguide, coaxial line, or stripline. Electrically, the operation of the planar triode is much more complicated at microwave frequencies than at low frequencies. Figure 6.8a compares the elements at work for a grounded-grid amplifier operating at low frequencies and Figure 6.8b compares the situation at microwave frequencies. The equivalent circuit is made more complex by:

- Stray inductance and capacitance of the tube elements

- Effects of the tube contact rings and socket elements

- Distributed reactance of cavity resonators and the device itself

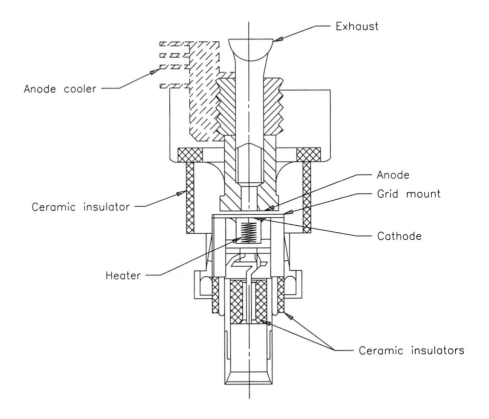

Figure 6.7 Cross section of a 7289 planar triode.

• Electron transit-time effects, which result in resistive loading and phase shifts

Reasonable gains of 5 to 10 dB may be achieved with a planar triode. Increased gain is available by cascading stages. Interstage coupling may consist of waveguide or coaxial-line elements. Tuning is accomplished by varying the cavity inductance or capacitance. Additional bandwidth is possible by stagger tuning of cascaded stages.

6.2.2 High-Power UHF Tetrode

New advancements in vacuum tube technology have permitted the construction of high-power UHF transmitters based on tetrodes. Such devices are attractive because they inherently operate in a relatively efficient class AB mode. UHF tetrodes operating at high power levels provide essentially the same specifications, gain, and efficiency as tubes operating at lower powers. The anode power supply is much lower in voltage than the collector potential of a klystron- or IOT-based system (8 kV is common). Also, the tetrode does not require a focusing magnet system.

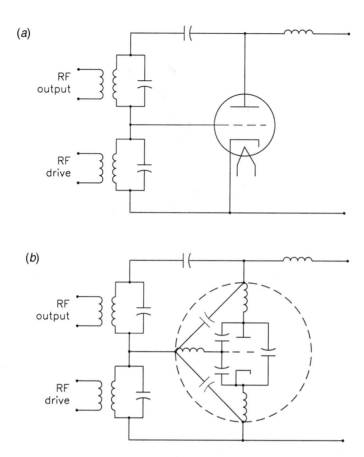

Figure 6.8 Grounded-grid equivalent circuits: (*a*) low-frequency operation, (*b*) micro-wave-frequency operation. The cathode-heating and grid-bias circuits are not shown.

Efficient removal of heat is the key to making a tetrode practical at high power levels. Such devices typically use water or vapor-phase cooling. Air cooling at such levels is impractical because of the fin size that would be required. Also, the blower for the tube would have to be quite large, reducing the overall transmitter ac-to-RF efficiency.

Another drawback inherent in tetrode operation is that the output circuit of the device appears electrically in series with the input circuit and the load [1]. The parasitic reactance of the tube elements, therefore, is a part of the input and output tuned circuits. It follows, then, that any change in the operating parameters of the tube as it ages can affect tuning. More importantly, the series nature of the tetrode places stringent limitations on internal element spacings and the physical size of those elements in order to minimize the electron transit time through the tube vacuum space. It is also fair to point out, however, the tetrode's input-to-output circuit characteristic has at least one advantage: power delivered to the input passes through the tube and contributes to the total power output of the transmitter. Because tetrodes typically exhibit low gain compared

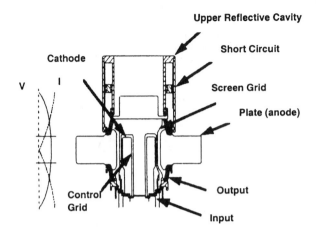

Figure 6.9 Cutaway view of the tetrode (*left*) and the Diacrode (*right*). Note that the RF current peaks above and below the Diacrode center while on the tetrode there is only one peak at the bottom. (*After* [2].)

to a klystron-based device, significant power can be required at the input circuit. The pass-through effect, therefore, contributes to the overall operating efficiency of the transmitter.

The expected lifetime of a tetrode in UHF service is usually shorter than a klystron of the same power level. Typical lifetimes of 8,000 to 15,000 hours have been reported. Intensive work, however, has led to products that offer higher output powers and extended operating lifetime, while retaining the benefits inherent in tetrode devices.

6.2.3 Diacrode

The *Diacrode* (Thomson) is a promising adaptation of the high-power UHF tetrode. The operating principle of the Diacrode is basically the same as that of the tetrode. The anode current is modulated by an RF drive voltage applied between the cathode and the power grid. The main difference is in the position of the active zones of the tube in the resonant coaxial circuits, resulting in improved reactive current distribution in the electrodes of the device.

Figure 6.9 compares the conventional tetrode with the Diacrode. The Diacrode includes an electrical extension of the output circuit structure to an external cavity [2]. The small dc-blocked cavity rests on top of the tube, as illustrated in Figure 6.10.

The cavity is a quarter-wave transmission line, as measured from the top of the cavity to the vertical center of the tube. The cavity is short-circuited at the top, reflecting an open circuit (current minimum) at the vertical center of the tube and a current maximum at the base of the tube, like the conventional tetrode, and a second current maximum above the tube at the cavity short-circuit.

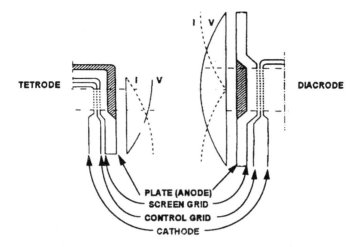

Figure 6.10 The elements of the Diacrode, including the upper cavity. Double current, and consequently, double power, is achieved with the device because of the current peaks at the top and bottom of the tube, as shown. (*After* [2].)

With two current maximums, the RF power capability of the Diacrode is double that of the equivalent tetrode, while the element voltages remain the same. All other properties and aspects of the Diacrode are basically identical to the TH563 high-power UHF tetrode (Thomson), upon which the Diacrode is patterned.

Some of the benefits of such a device, in addition to the robust power output available, is the low high-voltage requirements (low relative to a klystron/IOT-based system, that is), small size, and simple replacement procedures. On the downside, there is little installed service lifetime data at this writing because the Diacrode is relatively new to the market.

6.3 Klystron

The klystron is a *linear-beam* device that overcomes the transit-time limitations of a grid-controlled tube by accelerating an electron stream to a high velocity before it is modulated. Modulation is accomplished by varying the velocity of the beam, which causes the drifting of electrons into *bunches* to produce RF *space current*. One or more cavities reinforce this action at the operating frequency. The output cavity acts as a transformer to couple the high-impedance beam to a low-impedance transmission line. The frequency response of a klystron is limited by the impedance-bandwidth product of the cavities, but may be extended through stagger tuning or the use of multiple-resonance filter-type cavities.

The klystron is one of the primary means of generating high power at UHF and above. Output powers for multicavity devices range from a few thousand watts to 10

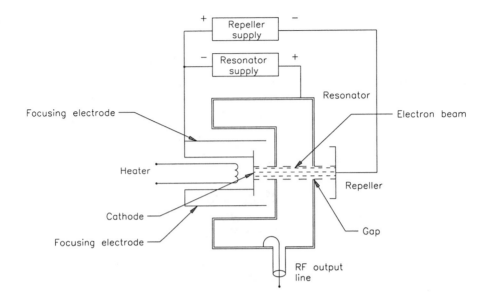

Figure 6.11 Schematic representation of a reflex klystron.

MW or more. The klystron provides high gain and requires little external support circuitry. Mechanically, the klystron is relatively simple. It offers long life and requires minimal routine maintenance.

6.3.1 Reflex Klystron

The reflex klystron uses a single-cavity resonator to modulate the RF beam and extract energy from it. The construction of a reflex klystron is shown in Figure 6.11. In its basic form, the tube consists of the following elements:

- A cathode
- Focusing electrode at cathode potential
- Coaxial line or reentrant-type cavity resonator, which also serves as an anode
- *Repeller* or *reflector* electrode, which is operated at a moderately negative potential with respect to the cathode

The cathode is so shaped that, in relation to the focusing electrode and anode, an electron beam is formed that passes through a gap in the resonator, as shown in the figure, and travels toward the repeller. Because the repeller has a negative potential with respect to the cathode, it turns the electrons back toward the anode, where they pass through the anode gap a second time. By varying the applied voltage on the reflector

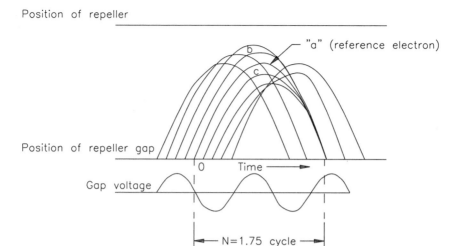

Position of repeller

"a" (reference electron)

b

c

Position of repeller gap

0   Time

Gap voltage

N=1.75 cycle

Figure 6.12 Position-time curves of electrons in the anode-repeller space, showing the tendency of the electrons to bunch around the electron passing through the anode at the time when the alternating gap voltage is zero and becoming negative.

electrode, phasing of the beam can be varied to produce the desired oscillating mode and to control the frequency of oscillation.

The variation of position with time for electrons in the anode-repeller space is illustrated in Figure 6.12. Path *a* corresponds to an electron that emerges from the anode with a velocity corresponding to the anode voltage. This electron follows a parabolic path, as shown, determined by the electric field in the anode-repeller space.

Operation of the reflex klystron can best be understood by examining the movement of electrons inside the device. Assume that oscillations exist in the resonator so that an alternating voltage develops across the gap. Assume further that the electron corresponding to path *a* passed through the gap at the instant that this alternating voltage across the gap was zero and becoming negative. An electron passing through the gap just before electron *a* will encounter an accelerating voltage across the gap and, therefore, will emerge from the anode with greater velocity than the first or *reference* electron. This second electron, accordingly, penetrates farther toward the repeller against the retarding field and, as a result, takes longer to return to the anode. Consequently, this electron follows path *b*, as shown in Figure 6.12, and tends to arrive at the anode on its return path at the same time as the reference electron because its earlier start is more or less compensated for by increased transit time. In a similar manner, an electron passing through the anode gap slightly later than the reference electron will encounter a negative or retarding field across the gap, and so will emerge from the anode with less velocity than the electron that follows path *a*. This third electron will then follow trajectory *c* and return to the anode more quickly than electron *a*. Electron *c*, therefore, tends to return to the anode at about the same time as electron *a*, because the later start of electron *c* is compensated for by the reduced transit time.

This variation with time of the velocity of electrons emerging from the anode is termed *velocity modulation*. The effect of this phenomenon can be seen in Figure 6.12 to cause a *bunching* of electrons about the electron that passed through the gap when the resonator voltage was zero and becoming negative. This bunching causes the electrons that are returned toward the anode by the repeller to pass through the anode gap in bursts or pulses, one each cycle. When these pulses pass through the gap at such a time that the electrons in the pulse are slowed as a result of the alternating voltage existing across the gap at the instant of their return passage, energy will be delivered to the oscillations in the resonator, thereby assisting in maintaining the oscillations. This condition corresponds to a transit time N from the resonator toward the repeller and back to the resonator of approximately:

$$N = n + \tfrac{3}{4} \tag{6.1}$$

Where:
n = an integer (including zero)

The transit time in the anode-repeller space in any particular case depends upon the following:

- The anode voltage
- Repeller voltage
- Geometry of the anode-repeller space

The extent of the bunching action that takes place when the transit time of the reference electron has the correct value for sustaining oscillations is determined by the following:

- The amplitude of the alternating voltage across the resonator gap in relation to the anode and repeller voltage
- The geometry of the repeller space

The reflex klystron typically includes a grid to concentrate the electric field so that it may efficiently couple to the electron beam. Such a device is illustrated in Figure 6.13.

The reflex klystron may be used as a local oscillator, low-power FM transmitter, or test signal source. Reflex tubes are used primarily from 4 to 40 GHz. Power outputs of 1 W or less are common.

The reflex tube is the only klystron in which *beam feedback* is used to produce output energy. In klystrons with more than one cavity, the electron beam passes through each cavity in succession.

6.3.2 The Two-Cavity Klystron

The two-cavity klystron operates on the same bunching principle as the reflex klystron, but incorporates two cavities connected by a *drift tube*. Figure 6.14 shows a cross section of a classic device. The heater/cathode element (shown as A in the fig-

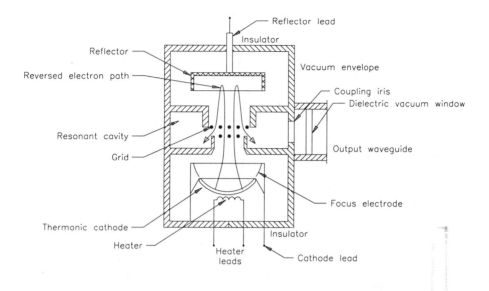

Figure 6.13 Schematic cross section of a reflex klystron oscillator.

ure) produces an electron beam in conjunction with a focusing electrode. The route taken by the electrons is as follows:

- The beam passes through grid elements D in the side of a reentrant cavity resonator (the buncher).

- The beam then passes through the drift tube, which is at the same electrical potential as the buncher.

- Finally, the beam enters a second resonator termed the *collector*, which is provided with a grid E.

The cathode and its associated focusing electrode are maintained at a high negative potential with respect to the remaining part of the structure, all of which is at the same dc potential. The entire arrangement illustrated in the figure is enclosed in a vacuum.

The operational principles of the two-cavity klystron are similar in nature to those of the reflex klystron. Assume, first, that oscillations exist in the buncher so that an alternating voltage is present across the gap D. When this voltage is zero but just becoming positive, an electron passing through the buncher travels through the grids D, down the drift tube, and into the collector resonator with unchanged velocity. However, an electron that passes through the buncher slightly later receives acceleration while passing through, because of the positive alternating field that it encounters between grids D, and enters the drift tube with increased velocity. This later electron, therefore, tends to overtake the earlier electron. Similarly, an electron that arrives at the buncher slightly

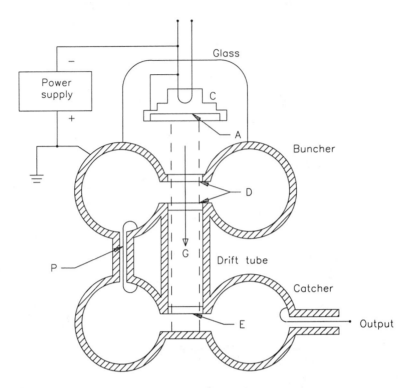

Figure 6.14 Cross section of a classic two-cavity klystron oscillator.

earlier than the first (reference electron) encounters a field between grids D that opposes its motion. Hence, this early electron enters the drift tube with reduced velocity and tends to drop back and be overtaken by the reference electron.

As a result of these actions, the electrons bunch together as they travel down the drift tube. This effect is more pronounced at certain distances from the buncher. If the collector is located at a distance where the bunching is pronounced, the electrons enter the element in pulses, one pulse per cycle.

With proper adjustment, the amount of power required to produce the bunching effect is relatively small compared with the amount of energy delivered by the electron beam to the collector. As a result, the klystron operates as an amplifying device.

The Two-Cavity Klystron Oscillator

The two-cavity klystron oscillator is designed for applications requiring moderate power (up to 100 W), stable frequency output, and low sideband noise. The device has a coupling iris on the wall between the two cavities. The tube can be frequency-modulated by varying the cathode voltage about the center of the oscillating mode. Although it is more efficient and powerful than the reflex klystron, the two-cavity klystron requires more modulator power. The two-cavity klystron typically is used in Doppler radar systems.

The Two-Cavity Klystron Amplifier

Similar in design to the two-cavity oscillator, the two-cavity klystron amplifier provides limited power output (10 W or less) and moderate gain (about 10 dB). A driving signal is coupled into the input cavity, which produces velocity modulation of the beam. After the drift space, the density-modulated beam induces current in the output resonator. Electrostatic focusing of the beam is common.

The two-cavity klystron finds only limited applications because of its restrictions on output power and gain. For many applications, solid-state amplifiers are a better choice.

6.3.3 The Multicavity Klystron

The multicavity klystron is an important device for amplifying signals to high power levels at microwave frequencies [3]. Each cavity tuned to the operating frequency adds about 20 dB gain to the 10 dB gain offered by the basic two-cavity klystron amplifier. Overall gains of 60 dB are practical. Cavities may be tuned to either side of resonance to broaden the operating bandwidth of the device. Klystrons with up to eight cavities have been produced. Operating power for continuous wave klystrons ranges up to 1 MW per device, and as much as 50 MW per device for pulsed applications.

The primary physical advantage of the klystron over a grid-based power tube is that the cathode-to-collector structure is virtually independent of transit-time effects. Therefore, the cathode can be made large and the electron beam density kept low.

The operating frequency of a klystron may be fixed—determined by the mechanical characteristics of the tube and its cavities—or tunable. Cavities are tuned mechanically using one of several methods, depending on the operating power and frequency. Tuning is accomplished by changing the physical dimensions of the cavities using one or more of the following techniques:

- *Cavity wall deformation*, in which one wall of the cavity consists of a thin diaphragm that is moved in and out by a tuning mechanism. About 3 percent frequency shift may be accomplished using this method, which varies the inductance of the cavity.

- *Movable cavity wall*, in which one wall of the cavity is moved in or out by a tuning mechanism. About 10 percent frequency shift is possible with this approach, which varies the inductance of the cavity.

- *Paddle element*, in which an element inside the cavity moves perpendicularly to the beam and adds capacitance across the interaction gap. A tuning range of about 25 percent is provided by this approach.

- *Combined inductive-capacitive tuning*, which uses a combination of the previous methods. Tuning variations of 35 percent are possible.

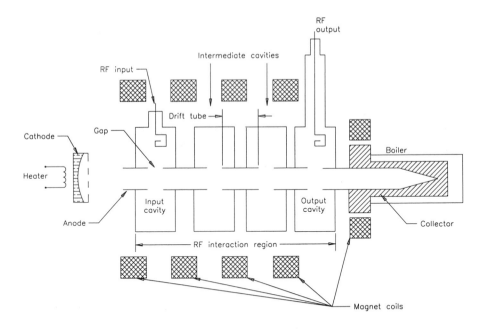

Figure 6.15 Principal elements of a klystron.

Each of these tuning methods may be used whether the cavity is inside or outside the vacuum envelope of the tube. Generally speaking, however, tubes that use external cavities provide more adjustment range, usually on the order of 35 percent. Bandwidth may be increased by stagger tuning of the cavities, at the expense of gain.

High conversion efficiency requires the formation of electron bunches, which occupy a small region in velocity space, and the formation of *interbunch regions* with low electron density. The latter is particularly important because these electrons are phased to be accelerated into the collector at the expense of the RF field. Studies show that the energy loss as a result of an electron accelerated into the collector may exceed the energy delivered to the field by an equal but properly phased electron. Therein lies a key in improving the efficiency of the klystron: Recover a portion of this wasted energy.

Klystrons are cooled by air or liquid for powers up to 5 kW. Tubes operating in excess of 5 kW are usually water- or vapor-cooled.

Operating Principles

A high-velocity electron beam emitted from the cathode passes through the anode and into the RF interaction region, as shown in Figure 6.15 [3]. An external magnetic field is employed to prevent the beam from spreading as it passes through the tube. At the other end of the device, the electron beam impinges on the collector electrode, which dissipates the beam energy and returns the electron current to the beam power supply.

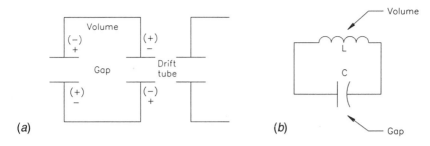

Figure 6.16 Klystron cavity: (*a*) physical arrangement, (*b*) equivalent circuit.

The RF interaction region, where amplification occurs, contains resonant cavities and field-free drift spaces. The first resonant cavity encountered by an electron in the beam (the input cavity) is excited by the microwave signal to be amplified, and an alternating voltage of the signal frequency is developed across the gap. This action can be best explained by drawing an analogy between a resonant cavity and a conventional parallel resonant LC circuit (see Figure 6.16). The cavity gap corresponds to the capacitor, and the volume of the cavity to the inductor. If the cavity is of the correct physical dimensions, it will resonate at the desired microwave frequency. At resonance, opposite sides of the gap become alternately positive and negative at a frequency equal to the microwave input signal frequency.

An electron passing through the gap when the voltage across the gap is zero continues with unchanged velocity along the drift tube toward the next cavity gap; this electron is the *reference electron*. An electron passing through the same gap slightly later is accelerated by the positive field at the gap. This electron speeds up and tends to overtake the reference electron ahead of it in the drift tube. However, an electron that passes through the gap slightly ahead of the reference electron encounters a negative field and is slowed down. This electron tends to fall back toward the following reference electron. As a result of passing the alternating field of the input-cavity gap, the electrons gradually bunch together as they travel down the drift tube, as illustrated in Figure 6.17. Because electrons approach the input-cavity gap with equal velocities and emerge with different velocities, which are a function of the microwave signal, the electron beam is said to be velocity-modulated. As the electrons travel down the drift tube, bunching develops, and the density of electrons passing a given point varies cyclically with time. This bunching is identical in nature to the action in a two-cavity klystron, discussed in Section 6.3.2.

The modulation component of the beam current induces current in each of the following cavities. Because each cavity is tuned near resonance, the resulting increase in field at each gap produces successively better-defined electron bunches and, consequently, amplification of the input signal. The RF energy produced in this interaction with the beam is extracted from the beam and fed into a coaxial or waveguide transmission line by means of a coupling loop in the output cavity. The dc beam input power not converted to RF energy is dissipated in the collector.

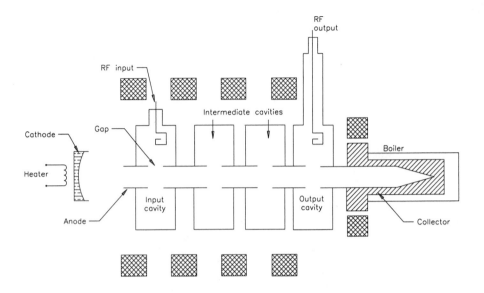

Figure 6.17 Bunching effect of a multicavity klystron.

Emission

The electron gun section of a klystron, shown in Figure 6.18, consists of the following elements [3]:

- A heater
- An emitter
- A beam-forming focusing electrode
- A modulating anode

When the emitter temperature is raised to the proper value by the heater, electrons are released from the emitter surface. The electrons are accelerated toward the modulating anode, which is at a positive potential with respect to the emitter. As the electrons travel between the emitter and the modulating anode, they are formed into a beam by the lens action of the focusing electrode and modulating anode. Figure 6.19 shows how this lens is formed.

All cathodes have a specific optimum range of operating temperature. The temperature of the cathode must be sufficiently high to prevent variations in heater power from affecting the electron emission current (beam current) in the klystron. However, the temperature of the emitting surface must not be higher than necessary because excessive temperature can shorten emission life. Figure 6.20 shows beam current (emission current) as a function of the emitter temperature, which varies directly with heater power. When the heater voltage (E_n) is too low, the emitter will not be hot enough to

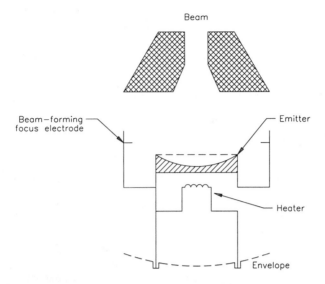

Figure 6.18 Diode section of a klystron electron gun.

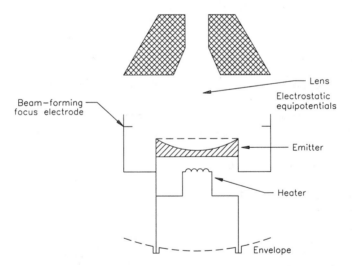

Figure 6.19 Beam forming in the diode section of a klystron electron gun.

produce the desired beam current. In addition, even small variations in heater voltage will change the beam current significantly. With the proper heater voltage (E_{f1}), constant beam current will be maintained even with minor variations in heater voltage. The same is true for a higher heater voltage value (E_{f2}), but in this case the emitter temperature is greater than that needed for the desired beam current. Reduced tube life will re-

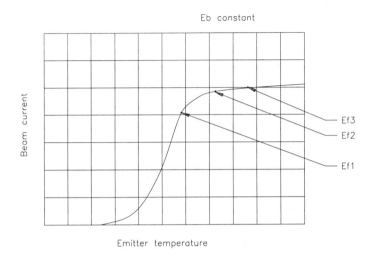

Figure 6.20 Klystron beam-current variation as a function of emitter temperature.

sult. The correct value of heater voltage and/or heater current is included in the data shipped with each klystron.

Modulating Anode

Because the modulating anode is electrically isolated from both the cathode and klystron body (the RF structure, between polepieces) the voltage applied to it provides a convenient means for controlling beam current independently of the beam voltage applied between the cathode and body, as shown in Figure 6.21. When the cathode is operated in the space-charge-limited region, E_{j2} and E_{j3} of Figure 6.20, the emission current will be a specific function of the applied voltage:

$$I_b = k E^{3/2} \tag{6.2}$$

Where:
I_b = beam current in amps
E = beam potential in volts

The constant k is a function of the geometry of the cathode-anode structure, and is termed *perveance*. Because the modulating anode is physically positioned between the RF structure (body) and the cathode, even if the full beam voltage is maintained between cathode and body, the actual beam current into the tube may be reduced at will by biasing the modulating anode to any voltage between cathode and body. Figure 6.22 shows the relationship between beam current and voltage described in the previous equation.

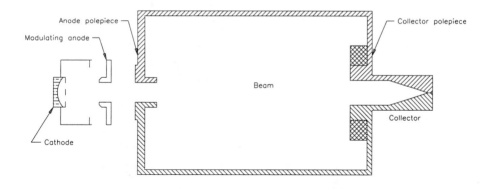

Figure 6.21 Modulating-anode electrode in a multicavity klystron.

Magnetic Field

Beam focusing is used in multielement klystrons to keep the electron beam uniformly small. Focusing may be accomplished by one or more electrostatic *lenses* or external magnetic fields placed parallel to the beam. Magnetic focusing is the most common method. Permanent magnets are used at operating powers of about 5 kW or less. Electromagnets are used at higher powers.

Electromagnetic coils typically are placed around the klystron to develop the magnetic field along the axis of the RF circuit. This field controls the size of the electron beam and keeps it aligned with the drift tubes. Figure 6.23 illustrates the beam-forming portion at the cathode end of the klystron and RF section, where the magnetic field is developed between two cylindrical disks (polepieces). The electron beam in this illustration is shown traveling two paths. One path shows the beam spreading out to points A; the other path shows the beam confined by the magnetic field to a constant size throughout the distance between polepieces. The beam spreads toward points A when the magnetic field is inadequate.

Figure 6.24 shows the magnetic field pattern of a typical solenoid used for klystrons. When direct current passes through the magnetic coils, a magnetic field is generated along the axis of the tube. The strength of this field can be controlled by changing the current flow through the magnetic coils. The shape of the field is determined by polepiece geometry and winding distribution inside the solenoid. Figure 6.25 illustrates the field pattern and the shape of the beam for a properly adjusted field.

Figure 6.26 demonstrates how the beam of a klystron is distorted when an external magnetic material is placed near the RF circuit of the tube. Electrons in the beam will follow the bent magnetic field lines and may strike the walls of the drift tube. Klystron damage can result. Magnetic materials such as screwdrivers, wrenches, bolts, and nuts must not be left near the magnetic circuit or near the cathode or collector. Magnetic tools must not be used to tune a klystron.

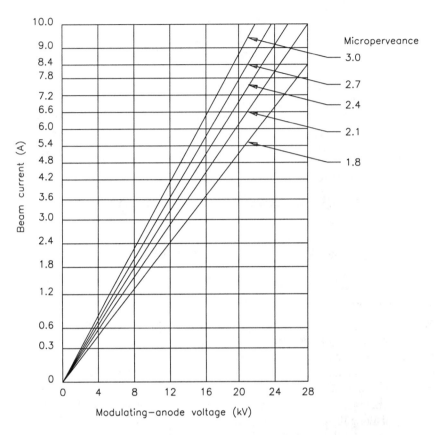

Figure 6.22 Beam-current variation as a function of modulating-anode voltage.

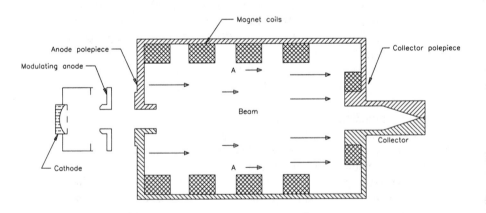

Figure 6.23 The effect of a magnetic field on the electron beam of a klystron.

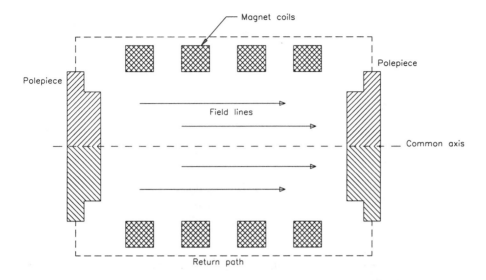

Figure 6.24 Field pattern of a klystron electromagnet.

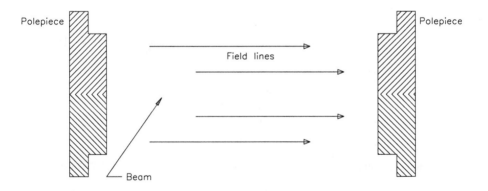

Figure 6.25 Field pattern and beam shape in a properly adjusted magnetic field.

RF Structure

The RF structure of a klystron amplifier consists of tunable resonant circuits (cavities) positioned along the axis of the electron beam. The electron beam traveling through the cavities provides the necessary coupling between each of the RF circuits.

The cavities of a klystron are high-frequency parallel resonant circuits constructed so that they provide an RF voltage across the capacitive component (gap), which interacts with the dc beam. Figure 6.27a illustrates the polarity near the drift-tube tips within a cavity excited by an alternating voltage of microwave signal frequency. Figure 6.27b

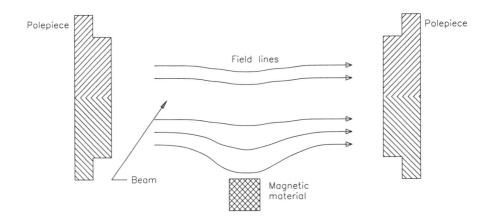

Figure 6.26 Distortion of the field pattern and beam shape due to magnetic material in the magnetic field.

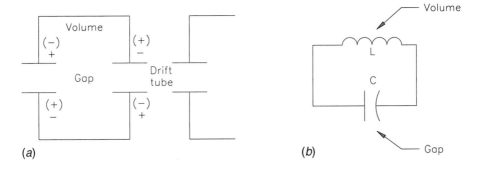

Figure 6.27 Klystron cavity: (a) physical element, (b) equivalent electric circuit.

is the equivalent circuit of a simple cavity. To achieve circuit resonance, the inductive and capacitive reactances of each of the components must be equal. The reactance of each of the components shown in the figure can be measured separately. However, the reactances of the components within a klystron cavity are difficult to determine, because they cannot be measured individually. Therefore, the regions of voltage maxima or minima are used to define each component of a klystron cavity in the following way:

- The capacitance of a cavity is developed across the gap at the drift tubes where the voltage is at maximum.

- The inductance of a cavity is located in the outer volume of the cavity where the voltage is at minimum.

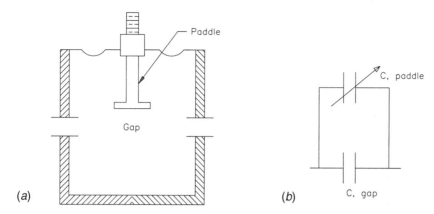

Figure 6.28 Capacitance-tuned klystron cavity: (*a*) physical element, (*b*) equivalent circuit.

By defining each component of a cavity in these terms, it is easy to visualize changes in the volume as changes in inductance, and changes affecting the gap as changes in capacitance.

The resonant frequency of each cavity can be adjusted to the operating frequency of the RF generator. This can be accomplished in one of two ways:

- Change the inductance by changing the volume of the cavity.

- Change the capacitance of the drift-tube gaps.

Figure 6.28 illustrates how the capacitance of the cavity gap can be modified by attaching a post to a thin-wall diaphragm with a paddle close to the drift-tube gap. In (*a*), the mechanical configuration of a cavity with this type of tuning is illustrated; (*b*) shows the equivalent circuit capacitance formed between the paddle and the drift tubes at the gap. Moving the paddle away from the drift tubes decreases the gap capacitance and increases the resonant frequency of the cavity.

Figure 6.29 shows a schematic diagram of the equivalent circuits of a four-cavity klystron. Circuit 1 is the input, and circuit 4 is the output. Figure 6.30*a* illustrates magnetic-loop coupling, where the RF energy is fed through a coaxial line with its center conductor inserted into the klystron cavity. The end of the center conductor is formed into a loop. This forms a simple one-turn transformer that couples RF energy into or out of the cavity through a coaxial transmission line. Figure 6.30*b* shows the equivalent circuit. The transformer formed by the loop and cavity is an impedance-matching device between the transmission line and the cavity.

Klystron cavities may be externally loaded to improve their instantaneous bandwidth characteristics. These loads lower the Q of the cavities slightly and thereby increase the bandwidth of the klystron.

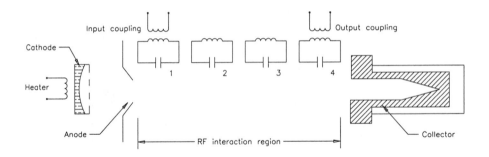

Figure 6.29 Schematic equivalent circuit of a four-cavity klystron.

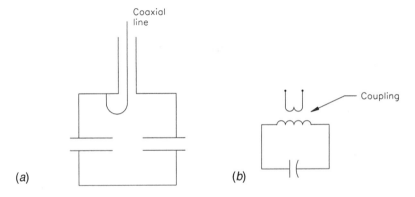

Figure 6.30 Klystron loop coupling: (a) mechanical arrangement, (b) equivalent electric circuit.

Phased Electron Operation

By properly phasing the second-harmonic fields of a klystron, a favorable electron density distribution pattern can be established at the output gap. The result is the generation of additional RF energy.

A phase-space diagram for a high-power klystron is shown in Figure 6.31. The curves represent a plot of the relative phase of the reference electrons as a function of axial distance along the tube. Electrons having negative slope have been decelerated. Electrons having positive slope have been accelerated with respect to a nonaccelerated electron parallel to the axis. The diagram shows how the electrons are nicely grouped at the output cavity gap while the interbunch regions are relatively free of electrons.

This interaction can be viewed another way, as shown in Figure 6.32, which plots the normalized RF beam currents as a function of distance along the tube. The curves show that the fundamental component of the plasma wave has a negative slope at the third gap. This normally would be a poor condition, but because of the drift of the interbunch

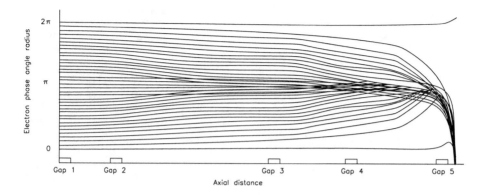

Figure 6.31 Plot of the relative phase of the reference electrons as a function of axial distance in a high-efficiency klystron.

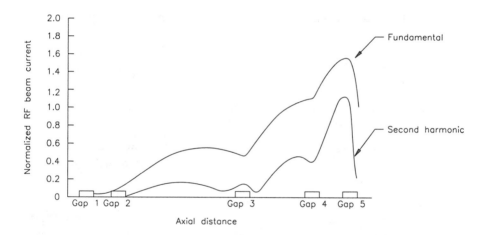

Figure 6.32 Plot of the normalized RF beam currents as a function of distance along the length of a high-efficiency klystron.

electrons, the fundamental current peaks at nearly 1.8 times the dc beam current. The theoretical limit for perfect bunching in a delta function is 2. The second harmonic of the plasma wave also peaks at the output gap, which adds to the conversion efficiency.

Types of Devices

Klystrons can be classified according to the following basic parameters:

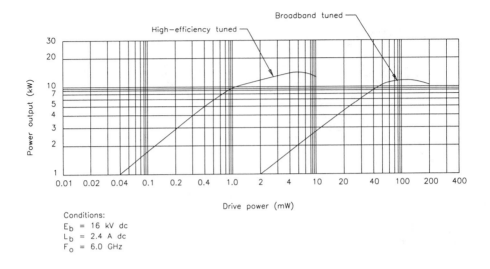

Conditions:
E_b = 16 kV dc
L_b = 2.4 A dc
F_o = 6.0 GHz

Figure 6.33 Typical gain, output power, and drive requirements for a klystron.

- Power operating level. Klystrons are available ranging from a few hundred watts to more than 10 MW.

- Operating frequency. Klystrons typically are used over the frequency range of 300 MHz to 40 GHz.

- Number of cavities. The number of resonant cavities may range from one to five or more. Furthermore, the cavities may be *integral* or *external* to the vacuum envelope of the device.

The klystron is a true linear amplifier from zero signal level up to 2 to 3 dB below saturated output. Figure 6.33 shows a typical transfer characteristic for a klystron. RF modulation is applied to the input drive signal. Amplitude modulation typically is limited to the linear portion of the gain transfer characteristic (class A operation). The result is low efficiency, because the beam power is always on. For applications requiring frequency modulation, the drive power is set for saturated output.

Pulse modulation of the klystron may be obtained by applying a negative rectangular voltage, instead of a dc voltage, to the cathode. The RF drive, set to a saturation value, usually is pulsed on for a slightly shorter time than the beam pulse.

Because of the high power levels typically used at UHF frequencies, device efficiency is a critical parameter. Klystrons usually are rated in terms of *saturated efficiency*, determined by dividing the saturated RF output power by the dc input power.

Saturated efficiency governs the maximum *peak-of-sync efficiency* available when beam-pulsing techniques are employed for UHF-TV service. Peak-of-sync efficiency is the commonly used *figure of merit* (FOM) expression, defined as the peak-of-sync output power divided by the dc input power.

Table 6.1 Typical Operating Parameters for an Integral-Cavity Klystron for UHF-TV Service (*Courtesy of Varian.*)

Parameter	Typical Value
Operating frequency (visual)	519 MHz
Output power, peak-of-sync	64 kW
Drive power, peak-of-sync	15 W
Gain, peak-of-sync	36 dB
Efficiency, saturated	55%
Bandwidth, −1 dB	6 MHz
Beam voltage	24.5 kV
Beam current	4.8 A
Body current	10 MA dc
Modulating-anode voltage	17 kV dc
Modulating-anode current	0.5 mA dc
Focusing current	30 A dc
Load VSWR	> 1.1:1
Collector temperature	130°C

Table 6.1 lists typical operating parameters for an integral-cavity klystron.

6.3.4 Beam Pulsing

Beam pulsing is a common method of improving the efficiency of a broadband linear klystron amplifier. Depending on the transmitted waveform, efficiency may be boosted by 25 percent or more. This technique typically is used in UHF-TV transmitters to reduce visual klystron beam dissipation during video portions of the transmission. *Sync pulsing*, as the technique is commonly known, is accomplished by changing the operating point of the tube during the synchronizing interval, when peak power is required, and returning it to a linear transfer characteristic during the video portion of the transmission.

This control is accomplished through the application of a voltage to an electrode placed near the cathode of the klystron. Biasing toward cathode potential increases the beam current, and biasing toward ground (collector potential) decreases beam current.

In the composite TV waveform, video information occupies 75 percent of the amplitude, and sync occupies the remaining 25 percent. The *tip of sync* represents the peak power of the transmitted waveform. Black (the *blanking level*) represents 56 percent of the peak power. If the blanking level could be made to represent 100 percent modulation and the sync pulsed in, as in a radar system, efficiency would be increased significantly. Unfortunately, the color-burst signal extends 50 percent into the sync region, and any attempt to completely pulse the sync component would distort the color-burst reference waveform. Sync pulsing is, therefore, limited to 12.5 percent above black to protect color-burst. Two common implementations of beam pulsing can be found:

- Modulating-anode (mod-anode) pulsing

- Annular beam control electrode (ACE) pulsing

The mod-anode system (first discussed in Section 6.3.3) utilizes an additional electrode after the cathode to control beam power. The ACE-type tube operates on a similar principle, but the annular ring is placed close to the cathode so that the ring encloses the electron beam. Because of the physical design, pulsing with the ACE-type gun is accomplished at a much lower voltage than with a mod-anode device. The ACE element, in effect, grid-modulates the beam.

In theory, the amount of beam-current reduction achievable and the resulting efficiency improvement are independent of whether mod-anode or ACE-type pulsing are used. In practice, however, differences are noted. With existing mod-anode pulsers, an efficiency improvement of about 19 percent in beam current over nonpulsed operation may be achieved. A beam reduction of 30 to 35 percent may be achieved through use of an ACE-equipped tube. The effect of ACE voltage on beam current is shown in Figure 6.34 for an external-cavity klystron.

The ACE electrode, positioned in the gun assembly, is driven by a negative-directed narrowband video signal of a few hundred volts peak. The annular ring varies the beam density through a *pinching action* that effectively reduces the cross-sectional area of the stream of electrons emitted by the cathode. The klystron thus operates in a quasi-class-AB condition rather than the normal class A (for linear TV service).

A peak-of-sync FOM for an integral-cavity UHF-TV klystron without ACE or equivalent control is 0.67 to 0.68. Through the use of an ACE-type tube, the FOM may be increased to 0.80. Similar improvements in efficiency can be realized for external-cavity klystrons. A typical switching-type mod-anode pulser is shown in Figure 6.35.

Pulsing is not without its drawbacks, however. The greater the pulsing, the greater the precorrection required from the modulator. Precorrection is needed to compensate for nonlinearities of the klystron transfer characteristic during the video modulation period. Level-dependent RF phase precorrection also may be required. Switching between different klystron characteristics produces phase modulation of the visual carrier. If not corrected, this phase modulation may result in intercarrier "buzz" in the received audio. These and other considerations limit the degree of pulsing that may be achieved on a reliable basis.

6.3.5 Integral vs. External Cavity

In an integral-cavity klystron, the resonant cavities are located within the vacuum envelope of the tube. In an external-cavity klystron, the cavities are located outside the envelope in a mechanical assembly that wraps around the drift tube.

Fundamentally, klystron theory applies equally to integral- and external-cavity tubes. In both cases, a velocity-modulated electron beam interacts with multiple resonant cavities to provide an amplified output signal. The resonant cavity *interaction gap* and drift length requirements for optimum performance—including conversion efficiency—are independent of whether the tuning mechanism is inside or outside the vacuum envelope. High-efficiency integral- and external-cavity klystrons have been de-

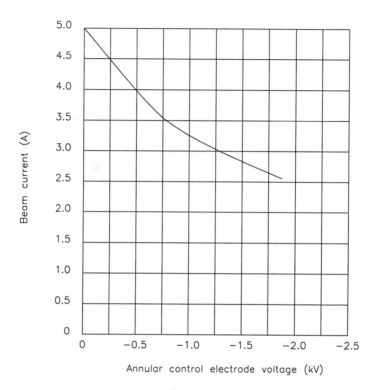

Figure 6.34 Klystron beam current as a function of annular control electrode voltage.

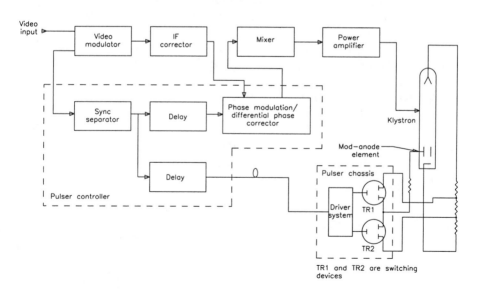

Figure 6.35 Block diagram of a switching mod-anode pulser.

signed to provide maximum conversion performance consistent with signal bandwidth requirements. The saturated conversion efficiency is essentially identical for integral- or external-cavity klystrons with equal numbers of resonant cavities.

Number of Cavities

The relative benefits of integral- vs. external-cavity klystrons can be debated at length. Discussion typically centers on operating efficiency, life expectancy, and replacement device cost. In a comparison of the efficiency of integral- vs. external-cavity klystrons for UHF-TV applications, however, the question really boils down to the number of cavities (four or five) used in the device. Four-cavity external tubes are standard. It is usually not practical to produce high-power five-cavity external tubes for mechanical and electrical (voltage standoff) reasons.

Integral devices permit the addition of a fifth cavity to the design for two primary reasons. First, the device itself is mechanically more robust. The addition of a fifth cavity to an external klystron increases the length and weight of the device. Because of the size of the drift tube in an external device designed for UHF-TV frequencies, the additional weight is difficult to support. Second, because the cavities are enclosed in a vacuum in the integral design, voltage standoff problems are greatly reduced.

The tuning mechanism of an integral-cavity klystron is enclosed in a rigid steel shell. In the external-cavity unit, the tuning mechanism is in air. Within the region of the tuning mechanism, RF fields can reach high levels, especially at the high end of the UHF-TV band. The highest energy field occurs in the area of the last cavity. Because of the high dielectric properties of a vacuum, dielectric breakdown is less of a problem in the integral design, which places the tuning mechanism within a vacuum envelope. This situation is of particular concern at the high end of the UHF band, where the spacing of tuning elements is closer.

Four-cavity integral-type klystrons are manufactured, but for power outputs of 30 kW or less. The 60 kW integral klystron, typically used for UHF-TV broadcasting, is produced as a five-cavity design.

The physics of the integral- and external-cavity klystrons are essentially the same. In a comparison of integral and external units of like design—that is, with the same number of cavities—performance should be identical. The two units follow the same laws of physics and use basically the same components up to the beam stick. It is when manufacturers take advantage of the relative merits of each design that differences in performance are realized.

Efficiency

For the sake of comparison, the data presented in this section will assume:

- The power level is 60 kW
- All integral-cavity devices utilize five cavities
- All external-cavity devices utilize four cavities

The five-cavity integral *S-tuned* klystron is inherently at least 20 percent more efficient than a four-cavity tube. (*S-tuning* refers to the method of stagger tuning the cavities.) The five-cavity klystron generally is specified by the manufacturer for a minimum efficiency of 52 percent (saturated efficiency). Typical efficiency is 55 percent. Four-cavity devices are characteristically specified at 42 percent minimum and 45 percent typical.

This efficiency advantage is possible because the fifth cavity of the integral design permits tuning patterns that allow maximum transfer of RF energy while maintaining adequate bandpass response. Tighter bunching of electrons in the beam stick, a function of the number of cavities, also contributes to the higher efficiency operation. In actuality, the fifth cavity allows design engineers to trade gain for efficiency. Still, the five-cavity tube has significantly more gain than the four-cavity device. The five-cavity unit, therefore, requires less drive, which simplifies the driving circuit. A five-cavity klystron requires approximately 25 W of drive power, while a four-cavity tube needs as much as 90 W for the same power output.

Under pulsed operation, approximately the same reduction in beam current is realized with both integral- and external-cavity klystrons. In a comparison of peak-of-sync FOM, the efficiency differences will track. There is, fundamentally, no reason that one type of klystron should pulse differently than the other.

Performance Tradeoffs

It is the designer's choice whether to build a transmitter with the klystron cavities located inside or outside the vacuum envelope. There are benefits and drawbacks to each approach.

When the cavity resonators are a part of the tube, the device becomes more complicated and more expensive. However, the power generating system is all together in one package, which simplifies installation significantly.

When the resonator is separate from the tube, as in an external-cavity device, it can be made with more *compliance* (greater room for adjustment). Consequently, a single device may be used over a wider range of operating frequencies. In terms of UHF-TV, a single external-cavity device may be tuned for operation over the entire UHF-TV band. This feature is not possible if the resonant cavities are built into the device. To cover the entire UHF-TV band, three integral-cavity tubes are required. The operational divisions are:

- Channels 14 to 29 (470 to 566 MHz)
- Channels 30 to 51 (566 to 698 MHz)
- Channels 52 to 69 (698 to 746 MHz)

This practical limitation to integral-cavity klystron construction may be a drawback for some facilities. For example, it is not uncommon for group operations to share one or more spare klystrons. If the facilities have operating frequencies outside the limits of a single integral device, it may be necessary to purchase more than one spare. Also, when the cavities are external, the resonators are in air and can be accessed to permit fine adjustments of the tuning stages for peak efficiency.

The advantages of tube changing are significant with an integral device. Typical tube change time for an integral klystron is 1 hour, as opposed to 4 to 6 hours for an external device. The level of experience of the technician is also more critical when an external-type device is being changed. Tuning procedures must be carefully followed by maintenance personnel to avoid premature device failure.

6.3.6 MSDC Klystron

Developmental work on the multistage depressed collector (MSDC) klystron began in the mid-1980s. The project[1] produced a working tube capable of efficiency in UHF service that had been impossible with previous klystron-based technology.

The MSDC device may be used in a number of varied applications. NASA originally became involved in the project as a way to improve the efficiency of satellite transmitters. With limited power available onboard a space vehicle, efficient operation is critically important. Such transmitters traditionally operate in a linear, inefficient mode. UHF-TV broadcasters were interested in the MSDC because it promised to reduce the huge operating costs associated with high-power operation.

Experimentation with depressed collector klystrons dates back to at least the early 1960s. Early products offered a moderate improvement in efficiency, but at the price of greater mechanical and electrical complexity. The MSDC design, although mechanically complex, offers a significant gain in efficiency.

Theory of Operation

MSDC tubes have been built around both integral-cavity and external-cavity klystrons. The devices are essentially identical to a standard klystron, except for the collector assembly. Mathematical models provided researchers with detailed information on the interactions of electrons in the collector region. Computer modeling also provided the basis for optimization of a *beam-reconditioning* scheme incorporated into the device. Beam reconditioning is achieved by including a *transition region* between the RF interaction circuit and the collector under the influence of a magnetic field. It is interesting to note that the mathematical models made for the MSDC project translated well into practice when the actual device was constructed.

From the electrical standpoint, the more stages of a multistage depressed collector klystron, the better. The tradeoff, predictably, is increased complexity and, therefore, increased cost for the product. There is also a point of diminishing returns that is reached as additional stages are added to the depressed collector system. A four-stage device was chosen for TV service because of these factors. As more stages are added (beyond four), the resulting improvement in efficiency is proportionally smaller.

---

1 A joint effort of the National Aeronautics and Space Administration (NASA), several UHF-TV transmitter manufacturers, Varian Associates, and other concerns.

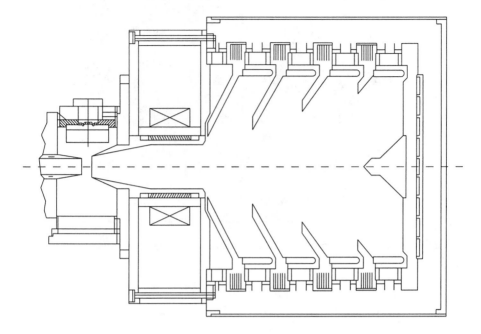

Figure 6.36 Mechanical design of the multistage depressed collector assembly. Note the "V" shape of the 4-element system.

Figure 6.36 shows the mechanical configuration of the four-stage MSDC klystron. Note the "V" shape that was found, through computer modeling, to provide the best *capture* performance, minimizing electron feedback. A partially assembled collector assembly is shown in Figure 6.37.

Because the MSDC device is identical to a conventional klystron except for the collector, efficiency improvement techniques used for klystrons can be incorporated into the MSDC. ACE-type pulsing commonly is used to improve efficiency in TV applications. Figure 6.38 illustrates the effects of ACE voltage on beam current.

Electron Trajectories

The dispersion of electrons in the multistage collector is the key element in recovering power from the beam and returning it to the power supply. This is the mechanism that permits greater operating efficiency from the MSDC device.

Figure 6.39 illustrates the dispersion of electrons in the collector region during a carrier-only operating mode. Note that there is little dispersion of electrons between stages of the MSDC. Most are attracted to electrode 4, the element at the lowest potential (6.125 kV), referenced to the cathode.

Figure 6.37 A partially assembled MSDC collector. (*Courtesy of Varian.*)

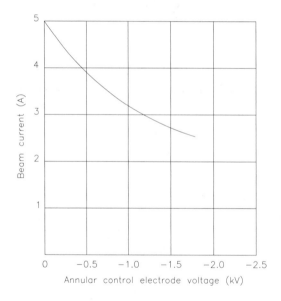

Conditions:
Beam voltage = 24.5 kV
Mod−anode voltage = 18 kV

Figure 6.38 Beam current as a function of annular control electrode voltage for an MSDC klystron.

Figure 6.40 shows collector electron trajectories at 25 percent saturation. The electrons exhibit predictable dispersion characteristics during the application of modulation, which varies the velocity of the electrons. This waveform is a reasonable approximation of *average modulation* for a typical video image.

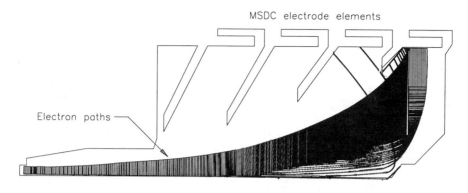

Figure 6.39 Collector electron trajectories for the carrier-only condition. Note that nearly all electrons travel to the last electrode (4), producing electrode current I_4.

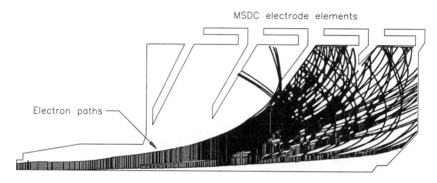

Figure 6.40 Collector trajectories at 25 percent saturation. With the application of modulation, the electrons begin to sort themselves out.

Figure 6.41 shows electron trajectories at 50 percent saturation, approximately the *blanking* level. Note the increased number of electrons attracted to electrodes 2 and 3, the higher-potential electrodes (referenced to the cathode).

Figure 6.42 illustrates electron dispersion at 90 percent saturation, approximately the level of sync in a video waveform. As the modulation level is increased, more electrons are attracted to the higher-voltage electrodes. The dramatic increase in electron capture by electrode 1, the highest-potential element of the device (at a voltage of 24.5 kV, referenced to the cathode), can be observed.

The electrons, thus, sort themselves out in a predictable manner. Notice the arc that is present on many electron traces. The electrons penetrate the electrostatic field of the collector, then are pulled back to their respective potentials.

A savings in power is realized because the electrostatic forces set up in the MSDC device slow down the electrons before they contact the copper collector electrode. The heat that would be produced in the collector is, instead, returned to the power supply in

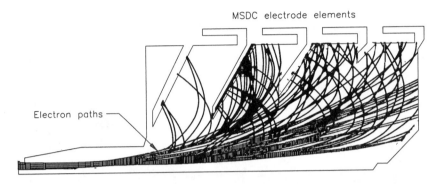

Figure 6.41 Collector electron trajectories at 50 percent saturation, approximately the blanking level. The last three electrodes (2, 3, and 4) share electrons in a predictable manner, producing currents I_2, I_3, and I_4.

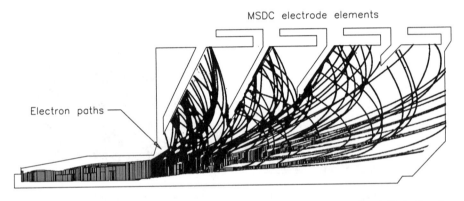

Figure 6.42 Collector trajectories at 90 percent saturation, the sync level. Note the significant increase in the number of electrons attracted to the first electrode, producing I_1.

the form of electric energy. In theory, peak efficiency would occur if the electrons were slowed down to zero velocity. In practice, however, that is not possible.

Figure 6.43 shows the distribution of collector current as a function of drive power. With no RF drive, essentially all current goes to electrode 4, but as drive is increased, I_4 drops rapidly as collector current is distributed among the other elements. Note that the current to electrode 5 (cathode potential) peaks at about 10 percent of beam current. This suggests that the *secondary yield* of the collector surfaces is within acceptable limits.

Inserted between the klystron and the collector assembly is a refocusing electromagnet that controls the electron beam as it enters the collector region.

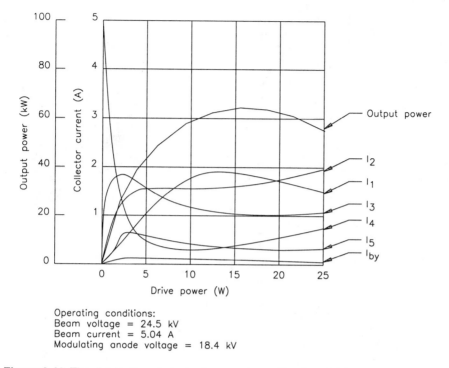

Operating conditions:
Beam voltage = 24.5 kV
Beam current = 5.04 A
Modulating anode voltage = 18.4 kV

Figure 6.43 The distribution of electrode current as a function of drive power. Note the significant drop in I_4 as drive power is increased. I_5 is the electrode at cathode potential.

Mechanical Construction

The completed MSDC assembly is shown in Figure 6.44 with the collector shield partially removed to allow visibility of the collector elements. The collector of the four-stage MSDC design actually is composed of five elements mounted between ceramic rings for electrical insulation. The fifth electrode is at cathode potential.

Cooling for the MSDC is, not surprisingly, more complicated than for a conventional klystron. The tradeoff, however, is that there is less heat to remove because of the higher efficiency of the device. Water cooling is provided on each electrode of the MSDC tube.

Figure 6.45 illustrates the overall mechanical design of an external-cavity version of the MSDC tube, including placement of the device in its cavity bay.

Although the MSDC has not been in service for a sufficient length of time to completely characterize the product life expectancy, researchers believe that the MSDC design will have little, if any, effect on the lifetime of the klystron. The electron beam is essentially unchanged. The tube is identical to a conventional integral- or external-cavity klystron except in collector assembly.

Figure 6.44 The MSDC collector assembly with the protective shield partially removed. (*Courtesy of Varian.*)

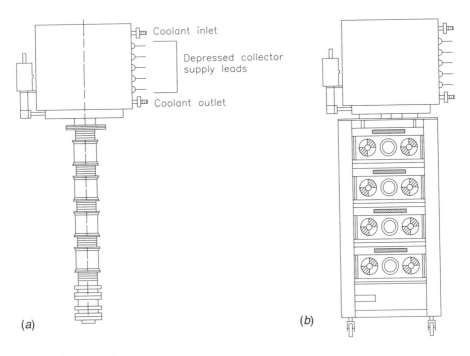

Coolant inlet

Depressed collector supply leads

Coolant outlet

(a)

(b)

Figure 6.45 Mechanical construction of an external-cavity MSDC tube: (a) device, (b) device in cavity assembly.

MSDC Power Supply

Design criteria for the collector power supply system provide a mixed bag of requirements. The critical parameter is the degree of regulation between the cathode and anode. The relative differences between the elements of the collector are not, in most applications, significant. Consequently, the bulk of the power supplied to the tube does not need to be well regulated. This is in contrast with conventional klystron operation,

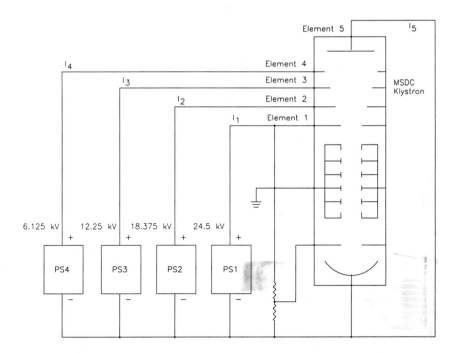

Figure 6.46 Parallel configuration of the MSDC power supply. Each supply section has an output voltage that is an integral multiple of 6.125 kV.

in which the entire beam power supply must be regulated. This factor effectively decreases the amount of power that must be regulated to 1 to 2 percent of the dc input, offsetting to some extent the additional cost involved in constructing multiple supplies to facilitate the 4-stage MSDC design. Two approaches can be taken to collector power supply design:

- Parallel arrangement, shown in Figure 6.46
- Series arrangement, shown in Figure 6.47

Note that, in both cases, the collector electrodes are stepped at a 6.125 kV potential difference for each element.

Device Performance

The efficiency improvement of the MSDC klystron over the conventional klystron is impressive. For TV applications, the MSDC—in pulsed operation—is more than twice as efficient as a conventional klystron.

The bandpass performance of the device is another critical parameter. Figure 6.48 charts power output as a function of frequency and RF drive at full power. (Drive power is charted from 0.5 to 16 W.) Figure 6.49 charts power output as a function of frequency

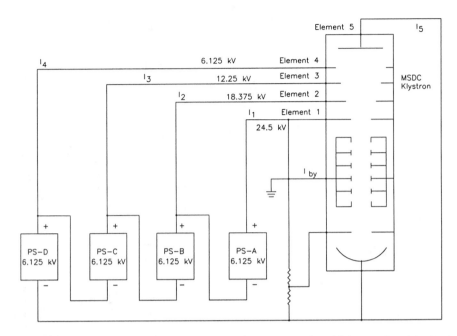

Figure 6.47 Series configuration of the MSDC power supply, which uses four identical 6.125 kV power supplies connected in series to achieve the needed voltages.

and RF drive with the tube in a beam-pulsing mode. (Drive power is charted from 0.5 to 32 W.) Note that the traces provide good linearity over a 6 MHz bandwidth. Gain, as a function of frequency and power, is essentially constant and undisturbed.

Table 6.2 lists typical operating parameters for an external-cavity MSDC device.

Applying the MSDC Klystron

A block diagram of a typical system is shown in Figure 6.50. The 60 kW TV transmitter incorporates two external-cavity MSDC klystrons, one for the visual and another for the aural. Design of the transmitter is basically identical to a non-MSDC system, with the exception of the power supply and cooling system. The efficiency available from the MSDC makes further device improvements subject to the law of diminishing returns, as mentioned previously. Support equipment begins to consume an increasingly large share of the power budget as the output device efficiency is improved. Tuning of the MSDC klystron is the same as for a conventional klystron, and the same magnetic circuit is typically used.

The power supply arrangement for the example MSDC system is shown in Figure 6.51. A series beam supply was chosen for technical and economic reasons. Although the current to each collector varies widely with instantaneous output level, the total current stays within narrow limits. Using a single transformer for all supplies, therefore, minimizes the size of the iron core required. A 12-pulse rectifier bank provides low rip-

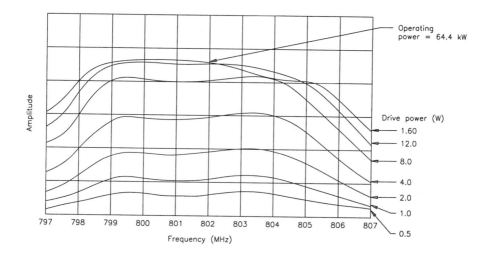

Figure 6.48 Device bandwidth as a function of frequency and drive power. Beam voltage is 24.5 kV, and beam current is 5.04 A for an output power of 64 kW. These traces represent the full-power test of the MSDC device.

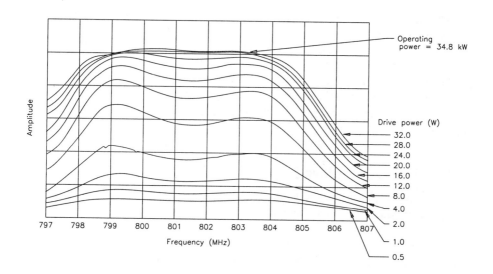

Figure 6.49 Device bandwidth in the beam-pulsing mode as a function of frequency and drive power. Beam voltage is 24.5 kV, and beam current is 3.56 A with an output power of 34.8 kW.

Table 6.2 Typical Operating Parameters for an External-Cavity MSDC Klystron for UHF-TV service (*Courtesy of Varian.*)

Parameter	Typical Value
Operating frequency (visual)	519 MHz
Output power, peak-of-sync	64 kW
Drive power, peak-of-sync	20 W
Gain, peak-of-sync	35 dB
Figure of merit	1.3[1]
Bandwidth, −1 dB	6 MHz
Beam voltage	24.5 kV
Beam current	5.3 A
Body current	50 mA dc
Modulating-anode voltage	19.5 kV dc
Modulating-anode current	0.5 mA dc
Focusing current	11 A dc
Ion pump voltage	3.2 kV dc
Load VSWR	> 1.1:1
Refocusing coil current	7 A dc
Refocusing coil voltage	10 V dc

[1] *Figure of merit* is equal to the quotient of peak-of-sync output power and the average dc beam input power.

ple and reduces the need for additional filtering. The size and complexity of the rectifier stack are increased little beyond a normal beam supply because the total potential of the four supplies is similar to that of a normal klystron transmitter (24.5 to 27.5 kV).

The collector stages of the MSDC device are water-cooled by a single water path that loops through each electrode element. Because high voltage is present on the individual elements, purity of the water is critical to proper operation. A two-stage system is used with a water-to-water plate heat exchanger separating the primary and secondary systems.

6.4   Klystrode/Inductive Output Tube (IOT)

The Klystrode tube is the result of a development program started in 1980[2] with UHF-TV in mind as a primary application. The basic concept of the Klystrode dates back to the late 1930s,[3] but it was not until the early 1980s that serious engineering effort was put into the tube to make it a viable product for high-power UHF service. The fundamental advantage of the Klystrode, also known as the *inductive output tube*

---

2   Varian/Eimac
3   Andrew Haeff, 1938

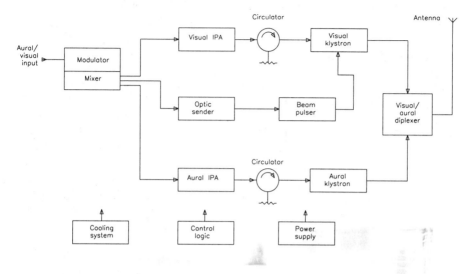

Figure 6.50 Simplified block diagram of a 60 kW (TVT) MSDC transmitter. The aural klystron may utilize a conventional or MSDC tube at the discretion of the user.

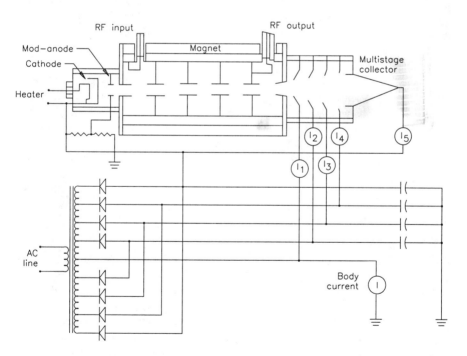

Figure 6.51 Simplified schematic diagram of the power supply for an MSDC klystron transmitter. A single high-voltage transformer with multiple taps is used to provide the needed collector voltage potentials.

(IOT), is its ability to operate class B. The result is higher efficiency when compared with a conventional klystron.

6.4.1 Theory of Operation

As its name implies, the Klystrode tube is a hybrid between a klystron and a tetrode. The high reliability and power-handling capability of the klystron is due, in part, to the fact that electron beam dissipation takes place in the collector electrode, quite separate from the RF circuitry. The electron dissipation in a tetrode is at the anode and the screen grid, both of which are an inherent part of the RF circuit and, therefore, must be physically small at UHF frequencies. The tetrode, however, has the advantage that modulation is produced directly at the cathode by a grid so that a long drift space is not required to produce density modulation. The Klystrode/IOT has a similar advantage over the klystron—high efficiency in a small package [4]. The Klystrode tube is a registered trademark of Varian. The IOT is the generic description for the device, and it will be used in this discussion.

The IOT is shown schematically in Figure 6.52. The electron beam is formed at the cathode, density-modulated with the input RF signals by a grid, then accelerated through the anode aperture. In its bunched form, the beam drifts through a field-free region, then interacts with the RF field in the output cavity. Power is extracted from the beam in the same way as in a klystron. The input circuit resembles that of a typical UHF power grid tube. The output circuit and collector resemble those in a klystron.

A production version of a 60 kW device is shown in Figure 6.53. Double-tuned cavities are used to obtain the required operating bandwidth. The load is coupled at the second cavity, as shown in Figure 6.54. This arrangement has proved to be an attractive way to couple power out of the device because no coupling loop or probe is required in the primary cavity, which can be a problem at the high end of the UHF band.

Because the IOT provides both beam power variation during sync pulses (as in a pulsed klystron) and variation of beam power over the active modulating waveform, it is capable of high efficiency. The device provides full-time beam modulation as a result of its inherent structure and class B operation.

6.4.2 Electron Gun

Many design elements from the klystron have been borrowed for use in the IOT [4]. A barium aluminate cathode is used for the gun, taken from a conventional klystron together with its heater structure. The methods used to support the cathode, heater, and necessary heat shields in the klystron gun have been retained for use in the IOT. Figure 6.55 shows the construction of the IOT gun and output cavity.

In the IOT, the RF input voltage is applied between the cathode and a grid that allows extra electrons to be drawn from the cathode into a low-quiescent-current electron beam according to the instantaneous RF voltage appearing between the grid and cathode. The resulting density-modulated beam is then passed into the klystron-like RF output interaction region of the tube.

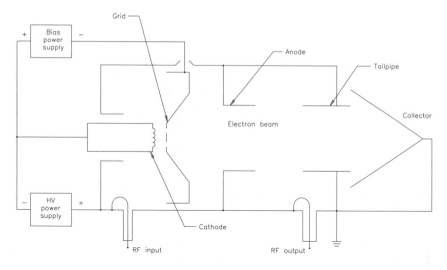

Figure 6.52 Simplified schematic diagram of the Klystrode tube.

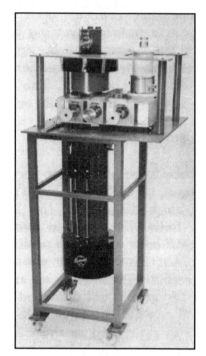

Figure 6.53 A 60 kW Klystrode tube mounted in its support stand with the output cavity attached. (*Courtesy of Varian.*)

The grid is clamped in place in front of the cathode, supported on a metal cylinder, and isolated from the cathode by a ceramic insulator. A second ceramic insulator supports the complete gridded electron gun at the correct distance from the grounded an-

Figure 6.54 A close-up view of the double-tuned output cavity of a 60 kW Klystrode tube. (*Courtesy of Varian.*)

ode. This ceramic insulator completes the vacuum envelope and holds off the full beam voltage of approximately 30 kV. The IOT gun operates at cathode-to-anode voltages approximately 50 percent higher than those of conventional klystron guns. Stray capacitance in the grid support structure has been minimized to reduce losses.

The grid-to-cathode space of the electron gun forms the end of a long and complex RF transmission line from the RF input connector of the input cavity system. This design has a significant effect upon the final frequency range of the IOT input cavity. Because the grid-to-cathode distance is crucial to the physics of the device, the grid requires a rigid support structure to maintain proper spacing at high operating temperatures.

6.4.3 Grid Structure

A pyrolytic graphite grid is used in the IOT because of its strength and its ability to maintain desired specifications at elevated temperatures [4]. Pyrolytic graphite has a unique advantage over other common grid materials in that its strength increases as the temperature climbs to 2500°C and above, whereas the strength of pure metals universally decreases as the temperature increases. This allows the designer to produce a thin grid, with fine grid wires that may be accurately positioned and will retain their position and shape when raised to operating temperatures of approximately 1000°C.

During production of the grid, a hydrocarbon, typically methane, is fed into a low-pressure chamber containing a graphite rod of the correct form to produce the required graphite shell. A shell typically consists of a graphite cylinder with a closed, shaped end. The cold gas is passed into a hot zone in the reactor, which is heated by an RF eddy current system from outside the vessel. An ordered carbon structure, as opposed to an amorphous structure, is required. To provide the needed processing, the temperature of the reactor is set to approximately 2000°C at a pressure of 10 torr. The graphite shells produced in this way have a layered structure with anisotropic properties and are physically durable.

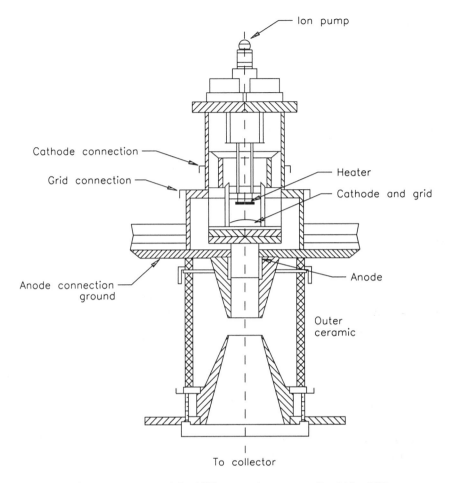

Figure 6.55 Overall structure of the IOT gun and output cavity. (*After* [4].)

The shells are then machined to the desired shape and size, and holes are cut into the resulting grid back using a laser under computer control. The process of producing a pyrolytic graphite grid is illustrated in Figure 6.56.

6.4.4 Input Cavity

Although similar, the Klystrode and the IOT are not identical. There are both subtle and significant differences in each design. The greatest departure can be seen in the configuration of the input cavity [4]. The design of the IOT is shown in Figure 6.57. A cylindrical resonant cavity containing an annular sliding tuning door is used. The cavity is folded at the IOT electron gun end to make contact with the tube via an RF choke structure. Because both the cathode and grid operate at beam potential (30 kV), it is necessary to maintain the body of the cavity and its tuning mechanism at ground

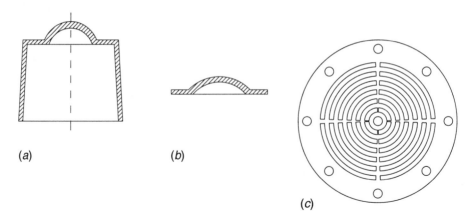

Figure 6.56 Pyrolytic graphite grid: (*a*) shell, (*b*) grid blank, (*c*) grid cut.

potential by using RF chokes to prevent leakage of RF energy, while holding off the full beam voltage.

At a given point on the grid connection within the RF choke, an insulated high-voltage cable is attached and exits so that the –30 kV dc can be supplied to the grid. This cable is screened and fitted with RF chokes to prevent UHF energy from passing into the beam power supply. RF input energy is fed via a coaxial cable to the input cavity, which is excited by means of a loop antenna carried by the annular tuning door.

Cooling air for the IOT electron gun and grid connection is fed through an insulated air pipe down the inside of the input cavity.

The tuning door of the cavity is driven by three two-stage tuning screws coupled to the outside via a rubber-toothed belt and bevel-gear drive. A mechanical turns-counter is provided to relate cavity door position with operating frequency.

6.4.5 Output Cavity

The primary output cavity is clamped around the output ceramic of the IOT, as on a conventional klystron [4]. The output cavity, shown in Figure 6.58, contains an RF coupling loop that may be rotated about a horizontal axis to adjust the degree of coupling through a short transmission line section into a secondary cavity via a doorknob-type antenna. The secondary cavity contains a dome structure adjusted in size so that the cavity can be made to cover the required frequency band. An output coupler, of standard klystron design incorporating a loop antenna, connects the secondary cavity of the RF output feeder system via a standard output interface. This design provides for instantaneous bandwidths of 8 MHz or greater.

Because RF voltages on the order of the IOT beam voltage can be expected in the output system, RF arc detectors are fitted into both cavities.

Cooling of the output cavities is accomplished by means of filtered forced air. Air enters the primary cavity, passes over the output ceramic and coupling loop, then exits

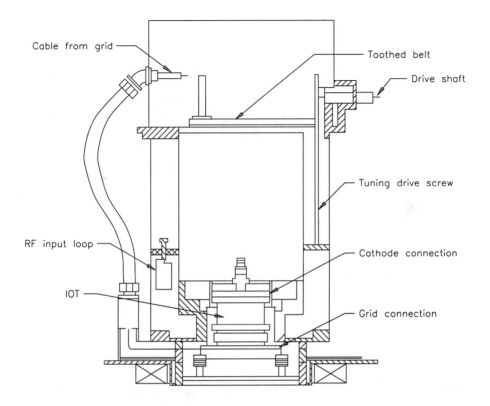

Figure 6.57 IOT input cavity. (*After* [4].)

via holes in the coupling hub assembly into the secondary cavity. After cooling the secondary cavity, some of the air exits via the contact fingers on the tuning doors, while a small portion traverses the output loop and coupler, exiting at a stub-pipe fitting for that purpose.

Typical operating parameters for the IOT are shown in Table 6.3.

6.4.6 Application Considerations

An active crowbar circuit is included in most applications of the Klystrode/IOT to protect the pyrolytic graphite grid in the event of an arc condition inside the tube [4]. The crowbar provides the added benefit of reducing the amount of gas generated in the tube during an arc. A block diagram of a typical crowbar is shown in Figure 6.59. The response time of the crowbar is typically less than 10 μs. The peak current permissible through the discharge tube may be 3000 A or greater.

Because of the IOT's class B operation, the response of the power supply to a varying load is an important design parameter. If a standard klystron beam supply were used in an IOT-equipped RF generator, performance under varying modulation levels could be

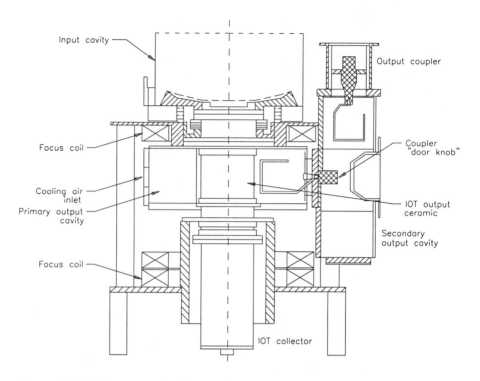

Figure 6.58 IOT output cavity. (*After* [4].)

Table 6.3 Typical Operating Parameters of the IOT in TV Service

Parameter	Typical Value
Beam voltage	28 kV
Mean beam current	1.15 A
Peak beam current	2.50 A
Grid bias	–72 V
Grid current	1 mA
Body current	6 mA
Input power	399 W
Output power	40.3 kW

unacceptable. Consider the application of a *bounce* (black-to-white) signal to a TV transmitter. Beam current for a typical IOT 60 kW transmitter would change from approximately 400 mA to 2 A. The effect with a conventional klystron supply would be ringing of about 20 percent on the beam voltage. The IOT power supply, therefore, must be designed for tight transient regulation.

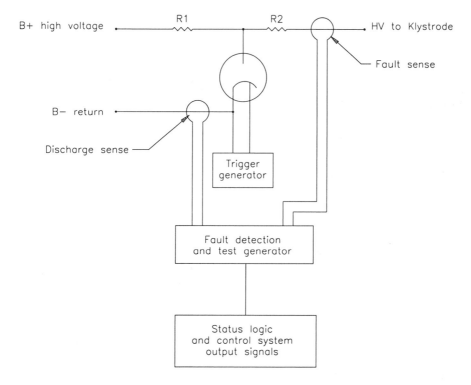

Figure 6.59 Simplified block diagram of the crowbar circuit designed into the Klystrode SK-series (Comark) transmitters. The circuit is intended to protect the tube from potentially damaging fault currents.

The IOT does not have the same hard saturation characteristic as a klystron. The IOT's high-power transfer curve flattens more slowly than a klystron's and continues to increase with increasing drive power, requiring a different approach to linearity correction.

6.4.7 Continuing Research Efforts

Several variations on the basic Klystrode/IOT theme have been developed, including air-cooled devices operating at 30 kW and above. Air cooling is practical at these power levels because of the improved efficiency that class B operation provides.

Research also is being conducted to extend the operating power of the IOT to 500 kW or more. Designed for scientific research applications, such devices offer numerous benefits over conventional klystron technology. The IOT is much smaller than a klystron of similar power, and it requires less support circuitry. Because of the improved efficiency, power supply requirements are reduced, and device cooling is simplified.

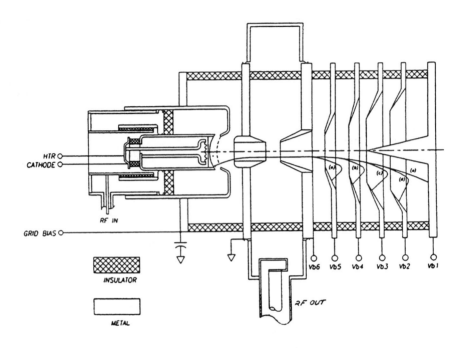

Figure 6.60 Schematic overview of the MSDC IOT or *constant efficiency amplifier*. (*After* [5].)

6.5 Constant Efficiency Amplifier

Because of the similarity between the spent electron beam in an IOT and that of a klystron or traveling-wave tube, it is possible to consider the use of a multistage depressed collector on an IOT to improve the operating efficiency [5]. This had been considered by Priest and Shrader [6] and by Gilmore [7], but the idea was rejected because of the complexity of the multistage depressed collector assembly and because the IOT already exhibited fairly high efficiency. Subsequent development by Symons [5, 8] has led to a prototype device (at this writing). An inductive output tube, modified by the addition of a multistage depressed collector, has the interesting property of providing linear amplification with (approximately) constant efficiency.

6.5.1 Theory of Operation

Figure 6.60 shows a schematic representation of the *constant efficiency amplifier* (CEA) [5]. The cathode, control grid, anode and output gap, and external circuitry are essentially identical with those of the IOT amplifier. Drive power introduced into the input cavity produces an electric field between the control grid and cathode, which draws current from the cathode during positive half-cycles of the input RF signal. For operation as a linear amplifier, the peak value of the current—or more accurately, the fundamental component of the current—is made (as nearly as possible) proportional to the square root of the drive power, so that the product of this current and the voltage it induces in the output cavity will be proportional to the drive power.

Following the output cavity is a multistage depressed collector in which several typical electron trajectories are shown. These are identified by the letters a through e. The collector electrodes are connected to progressively lower potentials between the anode potential and the cathode potential so that more energetic electrons penetrate more deeply into the collector structure and are gathered on electrodes of progressively lower potentials.

In considering the difference between an MSDC IOT and an MSDC klystron, it is important to recognize that in a class B device, no current flows during the portion of the RF cycle when the grid voltage is below cutoff and the output gap fields are accelerating. As a result, it is not necessary to have any collector electrode at a potential equal to or below cathode potential. At low output powers, when the RF output gap voltage is just equal to the difference in potential between the lowest-potential collector electrode and the cathode, all the current will flow to that electrode. Full class B efficiency is thus achieved under these conditions.

As the RF output gap voltage increases with increased drive power, some electrons will have lost enough energy to the gap fields so they cannot reach the lowest potential collector, and so current to the next-to-the-lowest potential electrode will start increasing. The efficiency will drop slightly and then start increasing again until all the current is just barely collected by the two lowest-potential collectors, and so forth.

Maximum output power is reached when the current delivered to the output gap is sufficient to build up an electric field or voltage that will just stop a few electrons. At this output power, the current is divided between all of the collector electrodes and the efficiency will be somewhat higher than the efficiency of a single collector, class B amplifier. Computer simulations have demonstrated that it is possible to select the collector voltages so as to achieve very nearly constant efficiency from the MSDC IOT device over a wide range of output powers [5].

The challenge of developing a multistage depressed collector for an IOT is not quite the same as that of developing a collector for a conventional klystron [8]. It is different because the dc component of beam current rises and falls in proportion to the square root of the output power of the tube. The dc beam current is not constant as it is in a klystron (or a traveling-wave tube for that matter). As a result, the energy spread is low because the output cavity RF voltage is low at the same time that the RF and dc beam currents are low. Thus, there will be small space-charge forces, and the beam will not spread as much as it travels deep into the collector toward electrodes having the lowest potential. For this reason, the collector is likely to be rather long and thin when compared to the multistage depressed collector for a conventional klystron, as described previously.

Figure 6.61 charts the calculated efficiency for an IOT with six depressed collector stages, set at 0.1, 0.2, 0.3, 0.45, 0.7, and 1.0 times the beam voltage (compared to that of a conventional IOT).

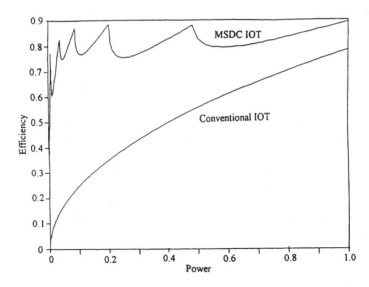

Figure 6.61 Calculated efficiency of an MSDC IOT compared to an unmodified IOT. (*After* [5].)

6.6 Traveling Wave Tube

The traveling wave tube (TWT) is a linear-beam device finding extensive applications in communications and research. Power levels range from a few watts to 10 MW. Gain ranges from 40 to 70 dB for small drive signals. The TWT consists of four basic elements:

- *Electron gun.* The gun forms a high-current-density beam of electrons that interact with a wave traveling along the RF circuit to increase the amplitude of the RF signal. In a typical application, electrons are emitted from a cathode and converged to the proper beam size by focusing electrodes.

- *RF interaction circuit.* The RF wave is increased in amplitude as a result of interaction with the electron beam from the gun. The fundamental principle on which the TWT operates is that an electron beam, moving at approximately the same velocity as an RF wave traveling along a circuit, gives up energy to the RF wave.

- *Magnetic electron beam focusing system.* The beam size is maintained at the proper dimensions through the interaction structure by the focusing system. This may be accomplished by using either a permanent magnet or an electromagnetic focusing element.

- *Collector.* The electron beam is received at the collector after it has passed through the interaction structure. The remaining beam energy is dissipated in the collector.

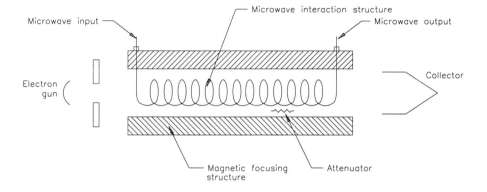

Figure 6.62 Basic elements of a traveling wave tube.

Figure 6.62 shows the basic elements of a TWT.

The primary differences between types of TWT devices involve the RF interaction structure employed. In Figure 6.62, the interaction structure is a helix. A variety of other structures may be employed, depending on the operating power and frequency. Three common approaches are used to provide the needed magnetic beam focusing. Illustrated in Figure 6.63, they are:

- Electromagnetic focusing, used primarily on high-power tubes, where tight beam focusing is required.

- Permanent-magnet focusing, used where the interaction structure is short.

- Periodic permanent-magnet focusing, used on most helix TWT and coupled-cavity tubes. The magnets are arranged with alternate polarity in successive cells along the interaction region.

6.6.1 Theory of Operation

The interaction structure acts to slow the RF signal so that it travels at approximately the same speed as the electron beam. Electrons enter the structure during both positive and negative portions of the RF cycle. Electrons entering during the positive portion are accelerated; those entering during the negative portion are decelerated. The result is the creation of *electron bunches* that produce an alternating current superimposed on the dc beam current. This alternating current induces the growth of an RF *circuit wave* that encourages even tighter electron bunching.

One or more *severs* are included to absorb reflected power that travels in a backward direction on the interaction circuit. This reflected power is the result of a mismatch between the output port and the load. Without the sever, regenerative oscillations could occur.

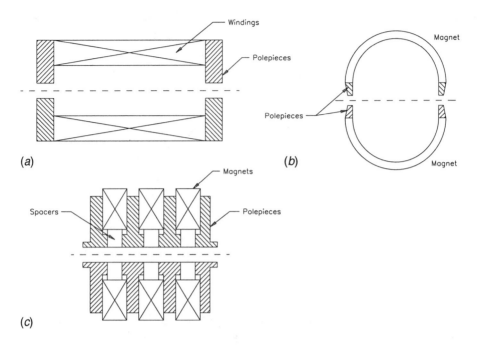

Figure 6.63 Magnetic focusing for a TWT: (*a*) solenoid-type, (*b*) permanent-magnet-type, (*c*) periodic permanent-magnet structure.

At a given frequency, a particular level of drive power will result in maximum bunching and power output. This operating point is referred to as *saturation*.

Interaction Circuit

The key to TWT operation lies in the interaction element. Because RF waves travel at the speed of light, a method must be provided to slow down the forward progress of the wave to approximately the same velocity as the electron beam from the cathode. The beam speed of a TWT is typically 10 to 50 percent the speed of light, corresponding to cathode voltages of 4 to 120 kV. Two mechanical structures commonly are used to slow the RF wave:

- Helix circuit. The helix is used where bandwidths of an octave or more are required. Over this range the velocity of the signal carried by the helix is basically constant with frequency. Typical operating frequencies range from 500 MHz to 40 GHz. Operating power, however, is limited to a few hundred watts. TWTs intended for higher-frequency operation may use a variation of the helix, shown in Figure 6.64. The *ring-loop* and *ring-bar* designs permit peak powers of hundreds of kilowatts. The average power, however, is about the same as that of a conventional helix because the structure used to support the interaction circuit is the same.

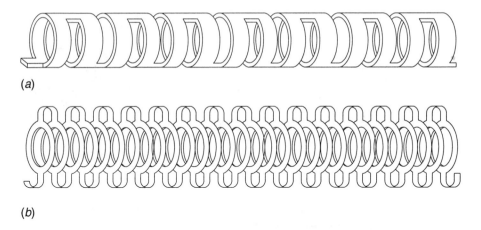

(a)

(b)

Figure 6.64 Helix structures for a TWT: (a) ring-loop circuit, (b) ring-bar circuit.

- *Coupled-cavity circuit.* The coupled-cavity interaction structure permits operation at high peak and average power levels, and moderate bandwidth (10 percent
  being typical). TWTs using coupled-cavity structures are available at operating
  frequencies from 2 to 100 GHz. The basic design of a coupled-cavity interaction
  circuit is shown in Figure 6.65. Resonant cavities, coupled through slots cut in the
  cavity end walls, resemble a folded waveguide. Two basic schemes are used: the
  cloverleaf and the *single-slot space harmonic* circuit.

The cloverleaf, also known as the *forward fundamental* circuit, is used primarily on
high-power tubes. The cloverleaf provides operation at up to 3 MW peak power and 5
kW average at S-band frequencies. The single-slot space harmonic interaction circuit is
more common than the cloverleaf. The mechanical design is simple, as shown in the
figure. The single-slot space harmonic structure typically provides peak power of up to
50 kW and average power of 5 kW at X-band frequencies.

Pulse Modulation

The electron beam from the gun may be pulse-modulated using one of four methods:

- *Cathode pulsing.* The cathode is pulsed in a negative direction with respect to the
  grounded anode. This approach requires the full beam voltage and current to be
  switched.

- *Anode pulsing.* This approach is similar to cathode pulsing, except that the full
  beam voltage is switched between cathode potential and ground. The current
  switched, however, is only that value intercepted on the anode. Typically, the intercepted current is a few percent of the full beam potential.

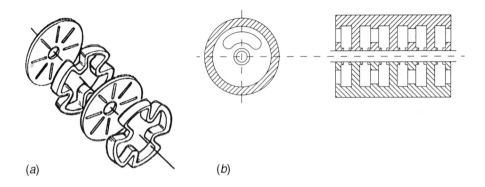

(a)   (b)

Figure 6.65 Coupled-cavity interaction structures: (a) forward fundamental circuit or "cloverleaf," (b) single-slot space harmonic circuit.

- *Focus electrode pulsing.* If the focus electrode, which normally operates at or near cathode potential, is biased negatively with respect to the cathode, the beam will be turned off. The voltage swing required is typically one-third of the full cathode voltage. This approach is attractive because the focus electrode draws essentially no current, making implementation of a switching modulator relatively easy.

- *Grid pulsing.* The addition of a grid to the cathode region permits control of beam intensity. The voltage swing required for the grid, placed directly in front of the cathode, is typically 5 percent of the full beam potential.

Electron Gun

The electron gun of a TWT is a device that supplies the electron beam to the tube [9]. A schematic diagram of a generic electron gun is given in Figure 6.66. The device consists of a hot cathode heated by an electric heater, a negatively biased focusing electrode or *focuser,* and a positively biased *accelerating anode.* The cross-sectional view given in the figure can be a two-dimensional or three-dimensional coaxial structure [10, 11].

An axially symmetrical solid cylindrical electron beam is produced by the gun structure shown in Figure 6.66 if the structure is axially cylindrically symmetrical. If the middle of the hot cathode is made nonemitting and only the edge of the cathode is emitting, the cathode becomes an *annular cathode* [9]. The annular cathode produces a hollow beam. The annular electron beam can be used to reduce beam current for a given microwave output power.

If the gun structure shown in Figure 6.66 is two dimensional, then a ribbon-shaped electron beam is produced. A ribbon-shaped beam is used for a TWT of a two-dimensional structure.

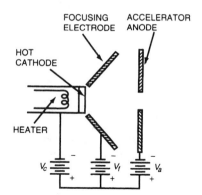

Figure 6.66 Generic TWT electron gun structure. (*From* [9]. *Used with permission.*)

If the angle of the focusing electrode against the axis of the electron beam is 67.5° and the anode is also tilted forward to produce a *rectilinear flow* (electron flow parallel to the z axis in Figure 6.66), then such an electron gun is termed the *Pierce gun*.

In practice, the hot cathode surface is curved as shown in Figure 6.67 to increase the electron emitting surface and to obtain a high-density electron beam.

Beam Focusing

Electrons in an electron beam mutually repel each other by the electron's own coulomb force because of their negative charge [9]. In addition, the electron beam usually exists in proximity to the positively biased slow-wave structure, as shown in Figure 6.68. Therefore, the electron beam tends to diverge. The process of confining the electron beam within the desired trajectory against the mutual repulsion and diverging force from the slow-wave structure is termed *electron beam focusing*.

The electron beam in a TWT is usually focused by a dc magnetic flux applied parallel to the direction of the electron beam, which is coaxial to the slow-wave transmission line. Variations on this basic technique include:

- *Brillouin flow*, where the output of the electron gun is not exposed to the focusing magnetic flux [12].

- *Immersed flow*, where the electron gun itself is exposed to and unshielded from the focusing para-axial longitudinal magnetic flux [12].

- *Generic flow*, where the electron gun is not shielded from the focusing magnetic flux, and focusing flux is not para-axia. (In other words, neither Brillouin nor immersed flow.)

Collector Assembly

Various configurations are used for the collector assembly of a TWT. Figure 6.69 shows a selection of the more common, including [9]:

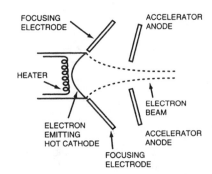

Figure 6.67 Cross-sectional view of a TWT electron gun with a curved hot cathode. (*From* [9]. *Used with permission.*)

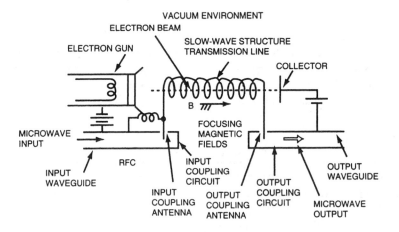

Figure 6.68 Generic configuration of a traveling wave tube. (*From* [9]. *Used with permission.*)

- Plate collector
- Cone collector
- Curved cone
- Cylinder collector
- Depressed potential cylinder
- Two-stage collector
- Three-stage collector

Cooling options include conduction, air, and water.

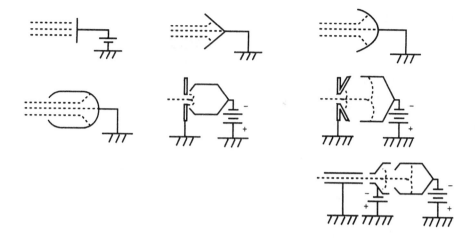

Figure 6.69 Cross-sectional view of various collector configurations for a TWT. (*From* [9]. *Used with permission.*)

Cooling of a low-power TWT is accomplished by clamping the tube to a metal base-plate, mounted in turn on an air- or liquid-cooled heat sink. Coupled-cavity tubes below 1 kW average power are convection-cooled by circulating air over the entire length of the device. Higher-power coupled-cavity tubes are cooled by circulating liquid over the tube body and collector.

6.6.2 Operating Efficiency

Efficiency is not one of the TWT's strong points. Early traveling wave tubes offered only about 10 percent dc-to-RF efficiency. Wide bandwidth and power output are where the TWT shines. TWT efficiency may be increased in two basic ways: (1) collector depression for a single-stage collector, or (2) use of a multistage collector.

Collector depression refers to the practice of operating the collector at a voltage lower than the full beam voltage. This introduces a potential difference between the interaction structure and the collector, through which electrons pass. The amount by which a single-stage collector can be depressed is limited by the remaining energy of the slowest electrons. In other words, the potential reduction can be no greater than the amount of energy of the slowest electrons, or they will turn around and reenter the interaction structure, causing oscillations.

By introducing multiple depressed collector stages, still greater efficiency can be realized. This method provides for the collection of the slowest electrons at one collector potential, while allowing those with more energy to be collected on other stages that are depressed still further. This approach, similar to the MSDC design discussed previously, is illustrated in Figure 6.70.

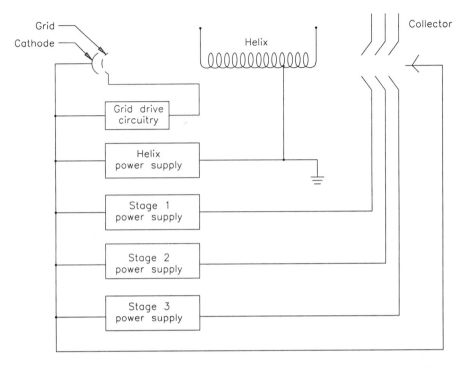

Figure 6.70 Power supply configuration for a multistage depressed collector TWT.

6.6.3 Operational Considerations

Although traveling wave tubes offer numerous benefits to the end user, they are not without their drawbacks.

Intermodulation Distortion

TWTs are susceptible to intermodulation (IM) distortion when multiple carriers are introduced, as in double illuminated or multiple SCPC (*single channel per carrier*) satellite transponders. These IM products may be found at frequencies that are displaced from the fundamental carriers by the difference in frequency between them. When multiple carriers are present, the potential for IM exists. This potential is reduced by operating the TWT below saturation. Power must be reduced (backed off) in proportion to the number of carriers and their relative power.

Second-Harmonic Content

Because of the wide bandwidth and nonlinear operating characteristics under saturation conditions, a TWT may generate significant second-harmonic energy. It is not uncommon to measure second-harmonic energy at the output of a TWT that is down only 10 dB from the operating carrier.

Reduction of harmonic content usually involves injecting a coherent harmonic signal with controlled phase and amplitude along with the fundamental carrier so that they interact, minimizing harmonic energy at the output of the device.

AM/PM Conversion

AM/PM conversion is the change in phase angle between the input and output signals as the input varies in amplitude. The root cause of this distortion in a TWT centers on the reduction of electron beam velocity as the input signal level increases. This causes a greater energy exchange between the electron beam and the RF wave. At a level 20 dB below the input power required for saturation, AM/PM conversion is negligible. At higher levels, AM/PM distortion may increase.

Phase Variation

When the velocity of the electron beam in the TWT is changed, the phase of the output signal also will vary. The primary causes of beam velocity variations include changes in one or more of the following:

- Cathode temperature
- Grid voltage
- Anode voltage
- Cathode voltage

The TWT power supply must be well regulated (to less than 0.2 percent) to prevent beam velocity changes that may result in output signal phase variations.

6.7 Crossed-Field Tubes

A crossed-field microwave tube is a device that converts dc into microwave energy using an electronic energy-conversion process. These devices differ from *beam tubes* in that they are *potential-energy converters*, rather than kinetic-energy converters. The term *crossed field* is derived from the orthogonal characteristics of the dc electric field supplied by the power source and the magnetic field required for beam focusing in the interaction region. This magnetic field typically is supplied by a permanent-magnet structure. Such devices also are referred to as *M-tubes*.

Practical devices based on the crossed-field principles fall into two broad categories:

- *Injected-beam crossed-field tubes*. The electron stream is produced by an electron gun located external to the interaction region, similar to a TWT. The concept is illustrated in Figure 6.71.

- *Emitting-sole tubes*. The electron current for interaction is produced directly within the interaction region by secondary electron emissions, which result when some electrons are driven to the negative electrode and allowed to strike it. The

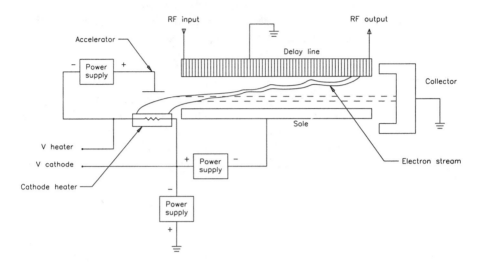

Figure 6.71 Linear injected-beam microwave tube.

negative electrode is formed using a material capable of producing significant numbers of secondary-emission electrons. The concept is illustrated in Figure 6.72.

6.7.1 Magnetron

The magnetron encompasses a class of devices finding a wide variety of applications. Pulsed magnetrons have been developed that cover frequency ranges from the low UHF band to 100 GHz. Peak power from a few kilowatts to several megawatts has been obtained. Typical overall efficiencies of 30 to 40 percent may be realized, depending on the power level and operating frequency. CW magnetrons also have been developed, with power levels of a few hundred watts in a tunable tube, and up to 25 kW or more in a fixed-frequency device. Efficiencies range from 30 percent to as much as 70 percent.

The magnetron operates electrically as a simple diode. Pulsed modulation is obtained by applying a negative rectangular voltage waveform to the cathode with the anode at ground potential. Operating voltages are less critical than for beam tubes; line-type modulators often are used to supply pulsed electric power. The physical structure of a conventional magnetron is shown in Figure 6.73.

High-power pulsed magnetrons are used primarily in radar systems. Low-power pulsed devices find applications as beacons. Tunable CW magnetrons are used in ECM (electronic countermeasures) applications. Fixed-frequency devices are used as microwave heating sources.

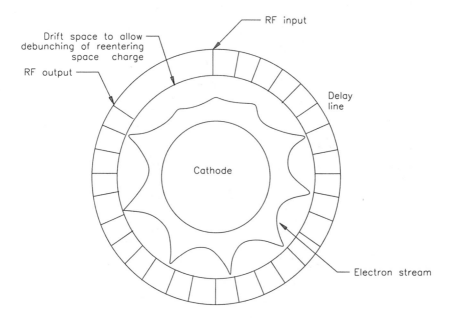

Figure 6.72 Reentrant emitting-sole crossed-field amplifier tube.

Tuning of conventional magnetrons is accomplished by moving capacitive tuners or by inserting symmetrical arrays of plungers into the inductive portions of the device. Tuner motion is produced by a mechanical connection through flexible bellows in the vacuum wall. Tuning ranges of 10 to 12 percent of bandwidth are possible for pulsed tubes, and as much as 20 percent for CW tubes.

Operating Principles

Most magnetrons are built around a cavity structure of the type shown in Figure 6.74. The device consists of a cylindrical cathode and anode, with cavities in the anode that open into the cathode-anode space—the *interaction space*—as shown. Power can be coupled out of the cavities by means of a loop or a tapered waveguide.

Cavities, together with the spaces at the ends of the anode block, form the resonant system that determines the frequency of the generated oscillations. The actual shape of the cavity is not particularly important, and various types are used, as illustrated in Figure 6.75. The oscillations associated with the cavities are of such a nature that alternating magnetic flux lines pass through the cavities parallel to the cathode axis, while the alternating electric fields are confined largely to the region where the cavities open into the interaction space. The most important factors determining the resonant frequency of the system are the dimensions and shape of the cavities in a plane perpendicular to the axis of the cathode. Frequency also is affected by other factors such as the end space and the axial length of the anode block, but to a lesser degree.

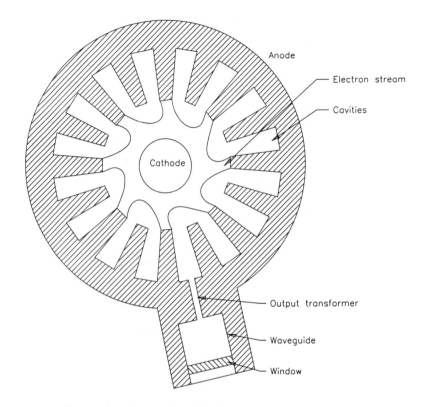

Figure 6.73 Conventional magnetron structure.

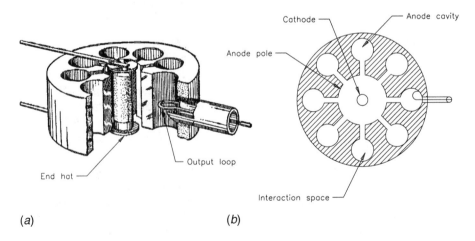

Figure 6.74 Cavity magnetron oscillator: (a) cutaway view, (b) cross section view perpendicular to the axis of the cathode.

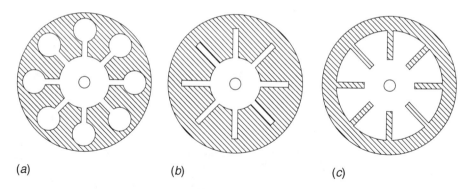

Figure 6.75 Cavity magnetron oscillator anode: (*a*) hole-and-slot type, (*b*) slot type, (*c*) vane type.

The magnetron requires an external magnetic field with flux lines parallel to the axis of the cathode. This field usually is provided by a permanent-magnet or electromagnet.

The cathode is commonly constructed as a cylindrical disk.

Coaxial Magnetron

The frequency stability of a conventional magnetron is affected by variations in the load impedance and by cathode-current fluctuations. Depending on the extent of these two influences, the magnetron occasionally may fail to produce a pulse. The co-axial magnetron minimizes these effects by using the anode geometry shown in Figure 6.76. Alternate cavities are slotted to provide coupling to a surrounding coaxial cavity.

The oscillating frequency is controlled by the combined vane system and the resonant cavity. Tuning may be accomplished through the addition of a movable end plate in the cavity, as shown in Figure 6.77.

Frequency-Agile Magnetron

Tubes developed for specialized radar and ECM applications permit rapid tuning of the magnetron. A conventional device may be tuned using one of the following methods:

- A rapidly rotating capacitive element. Tubes of this type are referred to as *spin-tuned magnetrons*.

- A hydraulic-driven tuning mechanism. Tubes of this type are referred to as *mechanically tuned magnetrons*.

Electronic tuning of magnetrons is also possible, with tuning rates as high as several megahertz per microsecond.

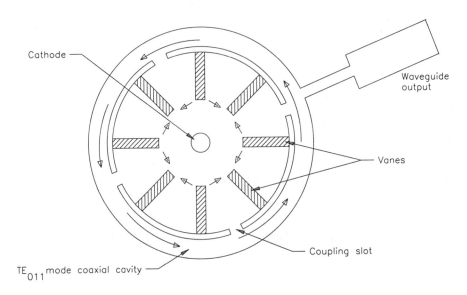

Figure 6.76 Structure of a coaxial magnetron.

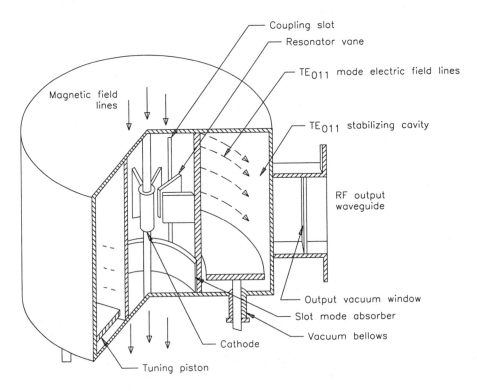

Figure 6.77 Structure of a tunable coaxial magnetron.

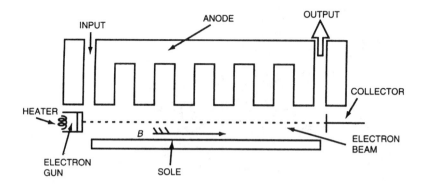

Figure 6.78 Cross-sectional view of a linear magnetron. (*From* [13]. *Used with permission.*)

Linear Magnetron

Although the most common types of magnetrons are radial in nature, the *linear magnetron* and *inverted magnetron* may also be used, depending upon the application [13]. A cross-sectional view of a linear magnetron is given in Figure 6.78. Shown in the figure is the *O-type* linear magnetron, in which the electron beam emitted from the electron gun is focused by a longitudinally applied dc magnetic flux density (B), as in the case of the traveling wave tube.

As shown in the figure, a number of slots are included in the basic structure. These slots are cut 1/4-wavelength deep, functioning as quarter-wave cavity resonators. This structure forms a series of microwave cavity resonators coupling to an electron beam, in a similar manner to the multicavity klystron. The velocity modulated electrons are bunched, and the tightly bunched electrons produce amplified microwave energy at the output cavity, which is coupled to an external circuit. The linear magnetron typically offers high gain, but narrow frequency bandwidth.

6.7.2 Backward Wave Oscillator

In a traveling wave tube, if the microwave signal to be amplified is propagating in the slow-wave structure backwardly to the direction of the electron beam, the device is termed a *backward wave oscillator* (BWO) [13]. Microwaves traveling in a backward direction carry positive feedback energy toward the electron gun and yield stronger velocity modulation and bunching. Thus, the system is inherently an oscillator rather than a stable amplifier. The input is typically terminated by an impedance-matched reflectionless termination device. The oscillation frequency is a function of the speed of the electrons and the time constant of the feedback mechanism. The speed of electron motion is controlled by the anode voltage.

An *M-type* radial BWO is shown in Figure 6.79. The direction of electron pole motion and the direction of microwave propagation along the annular reentrant type

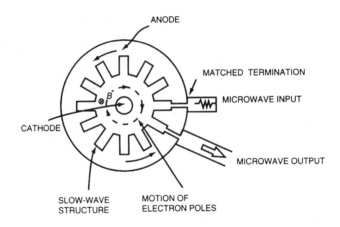

Figure 6.79 Functional schematic of the M-type radial BWO. (*From* [13]. *Used with permission.*)

slow-wave structure are opposite each other. It should be noted that the depths of the slits cut in the inner surface of the anode is very shallow—much less than 1/4-wavelength deep. In other words, the slits are not in resonance; they are not cavity resonators, as is the case of a magnetron. Rather, the slits are nonresonating, as in the case of a TWT. In the M-type radial BWO, the electron beam is focused by a magnetic flux density applied perpendicular to the beam, as seen from Figure 6.79.

An M-type radial BWO is sometimes termed the *Carcinotron*, a trade name. A key feature of the Carcinotron is its wide voltage tunability over a broad frequency range.

6.7.3 Strap-Fed Devices

A radial magnetron can be configured so that every other pole of the anode resonators are conductively tied for microwave potential equalization, as shown in Figure 6.80*a* [13]. These conducting tie rings are termed *straps*; the technique of using strap rings is termed *strapping*. Strapping ensures good synchronization of microwaves in the magnetron resonators with the rotation of electron poles.

The technique of strapping is extended and modified for an M-type radial BWO, as shown in Figure 6.80*b*. Strapping rings tie every other pole of the radial slow-wave structure, as in the case of a strapped radial magnetron, but the strapping rings are no longer reentrant. Microwave energy to be amplified is fed to the strap at one end, and the amplified output is extracted from the other end. This type of electron tube is termed a *strap-fed device*.

If an M-type radial BWO is strapped, usually it does not start oscillation by itself. But, if microwave energy is fed through the strap from the outside using an external microwave power source to the microwave input, then the oscillation starts—and even if

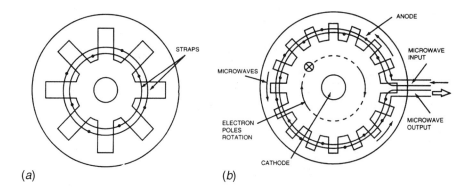

Figure 6.80 Strap-fed devices: (*a*) strapped radial magnetron, (*b*) nonreentrant strapping of a BWO. (*From* [13]. *Used with permission.*)

the exciter source is turned off, the oscillation continues. This type of M-type radial BWO is termed the *platinotron* [14].

In a platinotron, if the output of the tube is fed back to the input through a high-Q cavity resonator, it becomes a self-starting oscillator. The oscillation frequency is stabilized by the high-Q cavity resonator. This type of high-power frequency stabilized strapped radial BWO is termed the *stabilotron* [14]. The operating powers are at kilowatt and megawatt levels.

Performance of the platinotron depends on, among other things, the design of the slow-wave structure. For example, the interdigital slow-wave structure as shown in Figure 6.80 has a limited power handling capability and frequency bandwidth. Design of a slow-wave structure with greater power handling capacity and stability, with broader frequency bandwidth, is possible. For example, instead of an anode with an interdigital slow-wave structure, the anode could be made of an annular open conducting duct, loaded with a number of pairs of conducting posts across the open duct. Strapping is done at every other tip of the pairs of conducting posts. This type of strapping loads the slow-wave structure, stabilizing it and preventing oscillation. The structure of the anode with an annular duct and pairs of posts increases the power handling capability. This type of loaded radial BWO is termed the *amplitron* [14, 10]. The amplitron is capable of amplifying high-power microwave signals with pulses and continuous waves. It is used for long-range pulsed radar transmitter amplifiers and industrial microwave heating generators. The operating power levels range from kilowatt to megawatt levels.

6.7.4 Gyrotron

The *gyrotron* is a *cyclotron resonance maser.*[4] The device includes a cathode, collector, and circular waveguide of gradually varying diameter. Electrons are emitted at the cathode with small variations in speed. The electrons then are accelerated by an electric field and guided by a static magnetic field through the device. The nonuniform induction field causes the rotational speed of the electrons to increase. The linear velocity of the electrons, as a result, decreases. The interaction of the microwave field within the waveguide and the rotating (helical) electrons causes bunching similar to the bunching within a klystron. A decompression zone at the end of the device permits decompression and collection of the electrons.

The power available from a gyrotron is 100 times greater than that possible from a classic microwave tube at the same frequency. Power output of 100 kW is possible at 100 GHz, with 30 percent efficiency. At 300 GHz, up to 1.5 kW may be realized, but with only 6 percent efficiency.

Theory of Operation

The trajectory of an electron in an electron beam focused by a longitudinally applied magnetic field is a helix [13]. If the electron velocity, electron injection angle, and applied longitudinal magnetic flux density are varied, then an electron beam of helical form with different size and pitch will be formed. A coil-shaped electron beam will be produced by adjusting the acceleration voltage, applied magnetic flux density, and the electron injection angle to the focusing magnetic field. The coil of the electron beam can be a simple single coil, or—depending on the adjustment of the aforementioned three parameters—it can be an electron beam of a double coil, or a large coil made of thin small coils. In the case of the double-coil trajectory, the large coil-shaped trajectory is termed the *major orbit* and smaller coil trajectory is termed the *minor orbit*.

If a single coil-shaped electron beam is launched in a waveguide, as shown in Figure 6.81, then microwaves in the waveguide will interact with the helical beam. This type of vacuum tube is termed the *helical beam tube* [14]. In this class of device, a single-coil helical beam is launched into a TE_{10} mode rectangular waveguide. Inside the waveguide, microwaves travel from right to left and the helical beam travels in an opposite direction. Therefore, the microwave-electron beam interaction is of the *backward wave* type. If the microwave frequency, the focusing magnetic flux density B, and the acceleration voltage V_a are properly adjusted, this device will function as a *backward wave amplifier*. Electrons in the helical beam interact with the transverse microwave electric fields and are velocity modulated at the left-hand side of the waveguide as the beam enters into the waveguide. The velocity modulated electrons in the helical beam are bunched as they travel toward the right. If the alternating microwave electric field syn-

---

4   Maser is an acronym for *microwave amplification by simulated emission of radiation*. Maser is a general class of microwave amplifiers based on molecular interaction with electromagnetic radiation.

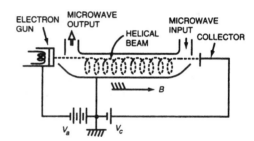

Figure 6.81 Basic structure of a helical beam tube. (*From* [13]. *Used with permission.*)

chronizes its period and phase with the helical motion of bunched electrons so that the electrons always receive retardation from microwave transverse electric fields, then the electrons lose their kinetic energy and the microwave signal gains in electric field energy according to the principle of kinetic energy conservation. Thus, the amplified microwave power emerges at the waveguide output at the left (because the microwaves travel backward).

In Figure 6.81, if the microwave input port and the output port are interchanged with each other, then the system becomes a *forward wave amplifier*. Such a forward wave amplifier is termed a *peniotron* [14].

If the electron gun is modified to incorporate a side emitting cathode and the waveguide is changed to TE_{11} mode oversized circular waveguide, as shown in Figure 6.82a, the gyrotron is formed. In this device, both ends of the waveguide are open and there are sufficient reflections in the waveguide for positive feedback. The gyrotron is, thus, a forward wave oscillator.

A *double-coil helical beam gyrotron* is shown in Figure 6.82b. The device is formed by readjusting the anode voltage and the focusing flux density so that the electron beam is made into a double helical coil (as shown in the figure), and operating the oversized circular waveguide in the TE_{01} mode. In the TE_{01} mode, the microwave transverse electric fields exist as concentric circles. Therefore, the tangential electric fields interact with electrons in the small coil trajectory. The alternating tangential microwave electric fields are made to synchronize with the tangential motion of electrons in the minor coil-shaped trajectory. Thus, electron velocity modulation takes place near the cathode and bunching takes place in the middle of the tube. Microwave kinetic energy transfer takes place as the beam approaches the right. The focusing magnetic flux density B is applied only in the interaction region. Therefore, if the electron beam comes out of the interaction region, it is defocused and collected by the anode waveguide (as depicted in Figure 6.82b). If the circular waveguide is operated in an oversized TE_{11} mode, with the double-coil helical beam, then the device is referred to as a *tornadotron* [14]. Microwave-electron interaction occurs between the parallel component of tangential motion of the small helical trajectory and the TE_{11} mode microwave electric field. If the phase of microwave electric field decelerates bunched electrons, then the lost kinetic energy of the bunched electrons is transferred to the microwave signal and oscillation begins.

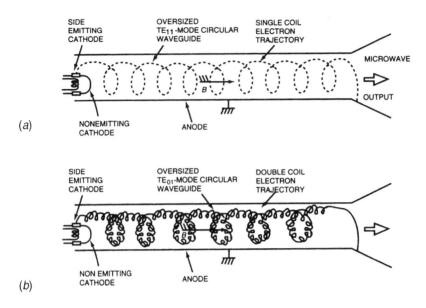

Figure 6.82 Functional schematic diagram of the gyrotron: (*a*) single coiled helical beam gyrotron, (*b*) electron trajectory of double-coil helical beam gyrotron. (*From* [13]. *Used with permission.*)

Gyrotron Design Variations

The gyrotron exists in a number of design variations, each optimized for a particular feature or application [13, 15, 16].

When the gyrotron circular waveguide is split as shown in Figure 6.83, the tube is termed the *gyroklystron amplifier* [13]. Both waveguides resonate to the input frequency, and there are strong standing waves in both waveguide resonators. The input microwave signal to be amplified is fed through a side opening to the input waveguide resonator. This is the *buncher resonator*, which functions in a manner similar to the klystron. The buncher resonator imparts velocity modulation to gyrating electrons in the double helical coil-shaped electron beam. There is a drift space between the buncher resonator and the catcher resonator at the output. While drifting electrons bunch and bunched electrons enter into the output waveguide catcher resonator, electron speed is adjusted in such a manner that electrons are decelerated by the resonating microwave electric field. This lost kinetic energy in bunched electrons is transformed into microwave energy and microwaves in the catcher resonator are, thus, amplified. The amplified power appears at the output of the tube.

If the gyrotron waveguide is an unsplit one-piece waveguide that is impedance-matched and not resonating, as shown in Figure 6.84, the tube is termed the *gyrotron traveling wave tube amplifier* [13]. In this tube, the input microwaves are fed through an opening in the waveguide near the electron gun. Microwaves in the wave-

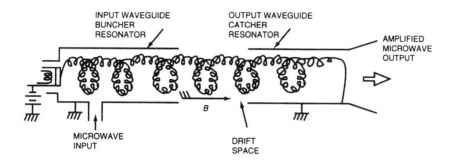

Figure 6.83 Basic structure of the gyroklystron amplifier. (*From* [13]. *Used with permission.*)

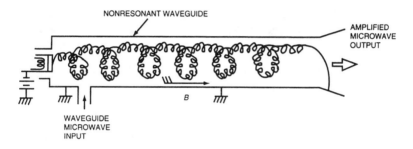

Figure 6.84 Basic structure of the gyroklystron traveling wave tube amplifier. (*From* [13]. *Used with permission.*)

guide are amplified gradually as they travel toward the output port by interacting with the double-coiled helical electron beam, which is velocity-modulated and bunched. There are no significant standing waves in the waveguide. Microwaves grow gradually in the waveguide as they travel toward the output port as a result of interaction with electrons.

If the electron gun of the gyrotron is moved to the side of the waveguide and microwave power is extracted from the waveguide opening in proximity to the electron gun, as shown in Figure 6.85, then the device is termed a *gyrotron backward oscillator* [13]. The principle involved is similar to the backward wave oscillator, and the process of velocity modulation, drifting, bunching, and catching is similar to that of the klystron. Microwave energy induced in the waveguide travels in both directions, but the circuit is adjusted to emphasize the waves traveling in a backward direction. The backward waves become the output of the tube and, at the same time, carry the positive feedback energy to the electrons just emitted and to be velocity-modulated. The system, thus, goes into oscillation.

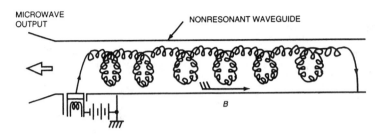

Figure 6.85 Basic structure of the gyrotron backward oscillator. (*From* [13]. *Used with permission.*)

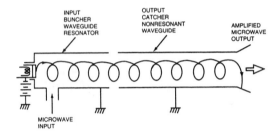

Figure 6.86 Basic structure of the gyrotwystron amplifier. (*From* [13]. *Used with permission.*)

If the gyrotron waveguide is split into two again, but this time the input side waveguide is short and the output side waveguide is long, as shown in Figure 6.86, then the tube is termed a *gyrotwystron amplifier* [13]. This device is a combination of the gyroklystron and gyrotron traveling wave tube amplifier, thus the name gyrotwystron amplifier. The input side waveguide resonator is the same as the input resonator waveguide of a gyroklystron. There are strong standing waves in the input bunched-waveguide resonator. There is no drift space between the two waveguides. The output side waveguide is a long impedance-matched waveguide and there is no microwave standing wave in the waveguide (a traveling-wave waveguide). As microwaves travel in the waveguide, they interact with bunched electrons and the microwaves grow as they move toward the output port.

6.8   Other Microwave Devices

There are a number of variations on the basic microwave devices outlined previously in this chapter. Some find widespread usage, others have little commercial interest.

Among the important specialty device classifications are *quasiquantum devices* and modified klystrons.

6.8.1 Quasiquantum Devices

In a gyrotron, if electrons are accelerated by extremely high voltage, they become relativistic [13]. That is, the mass of an electron becomes a function of the velocity as,

$$m = \frac{m_0}{\sqrt{1 - \left(\dfrac{c}{v}\right)^2}} \qquad (6.3)$$

Where:
m = the relativistic mass of an electron
m_0 = the static mass of an electron
c = the speed of light in vacuum
v = the speed of the electron in question

The energy states of electrons are specified by a set of quantum numbers, and the transition of energy states occurs only between those numbers so described. A set of quantum numbers specifies the high-energy orbit and low-energy orbit of an electron.

One device based on this principle is the *free electron laser*, or the *ubitron* [13]. A schematic diagram of the device is given in Figure 6.87. A high-speed relativistic electron beam is emitted from an electron gun focused by a longitudinally applied dc magnetic flux density B and is periodically deflected by magnetic means. The repetitious deflections create among relativistic electrons high-energy states and low-energy states. If the high-energy electrons in the deflection waveguides are stimulated by the resonance frequency of the waveguide resonator, then the *downward transition* (the transition of electrons in a high-energy state to a low-energy state) occurs. Microwave emissions result because of the energy transitions at the stimulation frequency.

The ubitron operates at millimeter or submillimeter wave frequencies, and typically at high power levels.

6.8.2 Variations on the Klystron

Because the multiple cavities of a conventional klystron are difficult to fabricate at millimeter wavelengths, variations on the basis design have been produced [13]. Instead of using multiple cavity resonators, the cavities can instead be combined into a single cavity. Such a design can be accomplished through the use of ladder-shaped grids, as shown in Figure 6.88a by the dashed lines with the single cavity. This type of klystron is referred to as the *laddertron* [14]. Because of the ladder structure, a standing wave will exist on on the ladder line. Velocity modulation results near the electron gun and bunching begins near the middle of the electron stream. Catching is done in the ladder structure near the collector. Because this device is a one-cavity resonator,

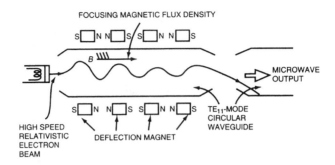

Figure 6.87 Basic structure of the ubitron. (*From* [13]. *Used with permission.*)

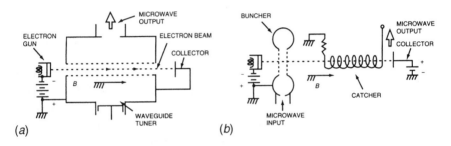

Figure 6.88 Functional schematic diagrams of modified klystrons: (*a*) laddertron, (*b*) twystron. (*From* [13]. *Used with permission.*)

there is built-in positive feedback. Oscillation, therefore, begins in a manner similar to a reflex klystron, also a single cavity tube.

At millimeter wavelengths, a dominant mode cavity resonator becomes very small and loses its power-handling capability. In a laddertron, however, the cavity resonator is a multiple higher-mode resonator. Therefore, the Q of the cavity is high and the size of the cavity resonator is reasonably large. This being the case, the laddertron can handle tens of watts of power at frequency as high as 50 GHz.

The catcher of a two-cavity klystron amplifier can be replaced by the slow-wave structure of a traveling wave tube, as shown in Figure 6.88*b*. This type of tube is termed the *twystron* [14]. The slow-wave structure provides a broader frequency bandwidth than a regular two-cavity klystron. A microwave input signal fed into the buncher cavity results in velocity modulation of the electron beam. Bunching occurs while the beam drifts, and the bunched electrons induce a microwave voltage in the slow-wave structure. The electron beam is focused by the use of longitudinally applied magnetic flux density B.

Commercial twystrons typically operate at 2–6 GHz and are used in pulsed radar transmitter power amplifiers. Pulsed peak output powers range from 1 to 7 MW (pulse width of 10–50 μs) [10].

6.9 Microwave Tube Life

Any analysis of microwave tube life must first identify the parameters that define *life*. The primary *wear-out* mechanism in a microwave power tube is the electron gun at the cathode. In principle, the cathode eventually will evaporate the activating material and cease to produce the required output power. Tubes, however, rarely fail because of low emission, but for a variety of other reasons that are usually external to the device.

Power tubes designed for microwave applications provide long life when operated within their designed parameters. The point at which the device fails to produce the required output power can be predicted with some accuracy, based on design data and in-service experience. Most power tubes, however, fail because of mechanisms other than predictable chemical reactions inside the device itself. External forces, such as transient overvoltages caused by lightning, cooling system faults, and improper tuning, more often than not lead to the failure of a microwave tube.

6.9.1 Life-Support System

Transmitter control logic usually is configured for two states of operation:

- An *operational level*, which requires all of the "life-support" systems to be present before the high-voltage (HV) command is enabled.

- An *overload level*, which removes HV when one or more fault conditions occur.

The cooling system is the primary life-support element in most RF generators. The cooling system should be fully operational before the application of voltages to the tube. Likewise, a cool-down period usually is recommended between the removal of beam and filament voltages and shutdown of the cooling system.

Most microwave power tubes require a high-voltage removal time of less than 100 ms after the occurrence of an overload. If the trip time is longer, damage to the device may result. *Arc detectors* often are installed in the cavities of high-power tubes to sense fault conditions and shut down the high-voltage power supply before damage can be done to the tube. Depending on the circuit parameters, arcs can be sustaining, requiring removal of high voltage to squelch the arc. A number of factors can cause RF arcing, including:

- Overdrive condition

- Mistuning of one or more cavities

- Poor cavity fit (applies to external types only)

- Undercoupling of the output to the load

- Lightning strike at the antenna

- High VSWR

Regardless of the cause, arcing can destroy internal elements or the vacuum seal if drive and/or high voltage are not removed quickly. A lamp usually is included with each arc detector photocell for test purposes.

6.9.2 Protection Measures

A microwave power tube must be protected by control devices in the amplifier system. Such devices offer either visual indications or aural alarm warnings, or they actuate interlocks within the system. Figure 6.89 shows a klystron amplifier and the basic components associated with its operation, including metering for each of the power supplies. Other types of microwave power devices use similar protection schemes. Sections of coaxial transmission line, representing essential components, are shown in the figure attached to the RF input and RF output ports of the tube. A single magnetic coil is shown to represent any coil configuration that may exist; its position in the drawing is for convenience only and does not represent its true position in the system.

Heater Supply

The heater power supply can be either ac or dc. If it is dc, the positive terminal must be connected to the common heater-cathode terminal and the negative terminal to the second heater terminal. The amount of power supplied to the heater is important because it establishes the cathode operating temperature. The temperature must be high enough to provide ample electron emission, but not so high that emission life is jeopardized.

Because the cathode and heater are connected to the negative side of the beam supply, they must be insulated to withstand the full beam potential. This dictates the use of a filament transformer designed specifically for the specific application.

Beam Supply

The high-voltage beam supply furnishes the dc input power to the microwave device. In a klystron, the positive side of the beam supply is connected to the body and collector. The negative terminal is connected to the common heater-cathode terminal. Never connect the negative terminal of the beam supply to the heater-only terminal because the beam current will then flow through the heater to the cathode and cause premature heater failure. The voltmeter, E_b in Figure 6.89, measures the beam voltage applied between the cathode and the body of the klystron.

Current meter I_c measures collector current, typically 95 percent or more of the total device current. Current meter I_{by} measures body current. An interlock should interrupt the beam supply if the body current exceeds a specified maximum value.

The sum of the body current (I_{by}) and collector current (I_c) is equal to the beam current (I_b), which should stay constant as long as the beam voltage and modulating-anode voltage are held constant.

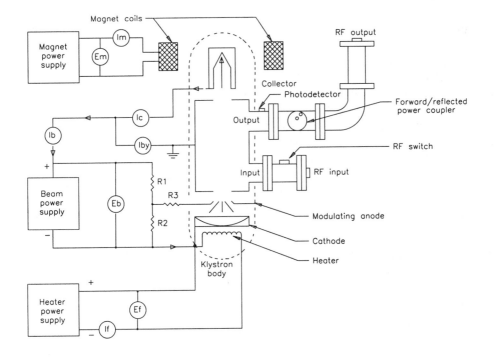

Figure 6.89 Protection and metering system for a klystron amplifier.

Magnet Supply

Electrical connections to the dc magnet supply typically include two meters, one for measuring current through the circuit (I_m in Figure 6.89), and one for measuring voltage (E_m). When a microwave device is installed in its magnet assembly, both parameters should be measured and recorded for future reference. If excessive body current or other unusual symptoms occur, this data will be valuable for system analysis.

Undercurrent protection should be provided to remove beam voltage if the magnetic circuit current falls below a preset value. The interlock also should prevent the beam voltage from being applied if the magnetic circuit is not energized. This scheme, however, will not provide protection if windings of the magnet are short-circuited. Short-circuited conditions can be determined by measuring the normal values of voltage and current and recording them for future reference.

The body-current overload protection should actuate if the magnetic field is reduced for any reason.

RF Circuits

In Figure 6.89, monitoring devices are shown on the RF input and output of the klystron. These monitors protect the device in case of a failure in the RF output circuit. Two directional couplers and a photodetector are attached to the output of the device. These components and an RF switching device on the input form a protective network against output transmission line mismatch. The RF switch is activated by the photodetector or the reflected power monitor and must be capable of removing RF drive power from the tube in less than 10 ms (typically).

In the RF output circuit, the forward power coupler is used to measure the relative power output of the microwave device. The reflected power coupler measures the RF energy reflected by the output circuit components or antenna. Damaged components or foreign material in the RF line will increase the RF reflected power. The amount of reflected power should be no more than 5 percent of the actual forward RF output power of the tube in most applications. An interlock monitors the reflected power and removes RF drive to the device if the reflected energy reaches an unsafe level. To protect against arcs occurring between the monitor and the output window, a photodetector is placed between the monitor and the window. Light from an arc will trigger the photodetector, which actuates the interlock system to remove RF drive before the window is damaged.

6.9.3 Filament Voltage Control

Extending the life of a microwave tube begins with accurate adjustment of filament voltage, as noted in the previous section. The filament should not be operated at a reduced voltage in an effort to extend tube life, as in the case of the thoriated tungsten grid tube. In a microwave tube, reduced filament voltage may cause uneven emission from the surface of the cathode with little or no improvement in cathode life.

Voltage should be applied to the filament for a specified warm-up period before the application of beam current to minimize thermal stress on the cathode/gun structure. However, voltage should not be applied to the filaments for extended periods (typically 2 hours or more) if no beam voltage is present. The net rate of evaporation of emissive material from the cathode surface is greater without beam voltage. Subsequent condensation of material on gun components may lead to voltage standoff problems.

6.9.4 Cooling System

The cooling system is vital to any RF generator. In a high-power microwave transmitter, the cooling system may need to dissipate as much as 70 percent of the input ac power in the form of waste heat. For vapor-phase-cooled devices, pure (distilled or demineralized) water must be used. Because the collector is usually only several volts above ground potential, it is generally not necessary to use deionized water. (Cooling considerations for microwave power tubes are discussed in Chapter 8. Cooling system maintenance is examined in Chapter 9.)

6.9.5 Reliability Statistics

Determination of the *mean time between failure* (MTBF) of a microwave tube provides valuable information on the operation of a given device in a given installation. The following formulas can be used to predict tube life in operating years, and to estimate the number of replacement tubes that will be needed during the life of an RF generator.

$$Y = \frac{MTBF}{H} \qquad (6.4)$$

$$N = \frac{L \times S}{Y} \qquad (6.5)$$

$$R = N - S \qquad (6.6)$$

Where:
Y = Tube life in operating years
$MTBF$ = Tube mean time between failure (gathered from manufacturer literature or on-site experience)
H = Hours of operation per year (= $365 \times$ hours per operating day)
N = Number of tubes needed over the life of the transmitter
L = Anticipated life of the transmitter in years
S = Number of tubes per transmitter
R = Number of replacement tubes needed over the life of the transmitter

MTTR

The *mean time to repair* (MTTR) of a given piece of RF equipment is an important consideration for any type of facility. MTTR defines the *maintainability* of a system. In the case of a microwave generator, the time required to change a tube is an important factor, especially if standby equipment is not available. The example time-change estimates given previously in this chapter (1 hour for an integral-cavity klystron and 4 to 6 hours for a 60 kW UHF external-cavity tube) assume that no preparation work has been performed on the spare device. Pretuning a tube is one way to shorten the MTTR for an external-cavity device. The spare tube is installed during a maintenance period, and the system is tuned for proper operation. After the positions of all tuning controls are documented, the tube is removed and returned to its storage container, along with the list of tuning control settings. In this way, the external-cavity tube can be placed into service much faster during an emergency. This procedure probably will not result in a tube tuned for optimum performance, but it may provide a level of performance that is acceptable on a temporary basis.

Consideration of MTTR is important for a facility because most microwave tube failure is caused by mechanisms other than reduced cathode emission, which is a *soft failure* that can be anticipated with some degree of accuracy. Catastrophic failures, on the other hand, offer little, if any, warning.

Reliability engineering is discussed in greater detail in Section 9.3.

6.10 References

1. Ostroff, Nat S., "A Unique Solution to the Design of an ATV Transmitter," *Proceedings of the 1996 NAB Broadcast Engineering Conference*, National Association of Broadcasters, Washington, DC, pg. 144, 1996.

2. Hulick, Timothy P., "60 kW Diacrode UHF TV Transmitter Design, Performance and Field Report," *Proceedings of the 1996 NAB Broadcast Engineering Conference*, National Association of Broadcasters, Washington, DC, pg. 442, 1996.

3. "Integral Cavity Klystrons for UHF-TV Transmitters," Varian Associates, Palo Alto, CA.

4. Clayworth, G. T., H. P. Bohlen, and R. Heppinstall, "Some Exciting Adventures in the IOT Business," *NAB 1992 Broadcast Engineering Conference Proceedings*, National Association of Broadcasters, Washington, DC, pp. 200 - 208, 1992.

5. Symons, Robert S., "The Constant Efficiency Amplifier," *Proceedings of the NAB Broadcast Engineering Conference*, National Association of Broadcasters, Washington, D.C., pp. 523–530, 1997.

6. Priest, D. H., and M.B. Shrader, "The Klystrode—An Unusual Transmitting Tube with Potential for UHF-TV," *Proc. IEEE*, Vol. 70, No. 11, pp. 1318–1325, November 1982.

7. Gilmore, A. S., *Microwave Tubes*, Artech House, Dedham, Mass., pp. 196–200, 1986.

8. Symons, R., M. Boyle, J. Cipolla, H. Schult, and R. True, "The Constant Efficiency Amplifier—A Progress Report," *Proceedings of the NAB Broadcast Engineering Conference*, National Association of Broadcasters, Washington, D.C., pp. 77–84, 1998.

9. Ishii, T. K., "Traveling Wave Tubes," in *The Electronics Handbook*, Jerry C. Whitaker (ed.), CRC Press, Boca Raton, Fla., pp. 428–443, 1996.

10. Liao, S. Y., *Microwave Electron Tube Devices*, Prentice-Hall, Englewood Cliffs, N.J., 1988.

11. Sims, G. D., and I. M Stephenson, *Microwave Tubes and Semiconductor Devices*, Interscience, London, 1963.

12. Hutter, R. G. E., *Beam and Wave Electronics in Microwave Tubes*, Interscience, London, 1960.

13. Ishiik, T. K., "Other Microwave Vacuum Devices," in *The Electronics Handbook*, Jerry C. Whitaker (ed.), CRC Press, Boca Raton, Fla., pp. 444–457, 1996.

14. Ishii, T. K., *Microwave Engineering*, Harcourt-Brace-Jovanovich, San Diego, CA, 1989.

15. Coleman, J. T., *Microwave Devices*, Reston Publishing, Reston, VA, 1982.

16. McCune, E. W., "Fision Plasma Heating with High-Power Microwave and Millimeter Wave Tubes," *Journal of Microwave Power*, Vol. 20, pp. 131–136, April, 1985.

6.11 Bibliography

Badger, George, "The Klystrode: A New High-Efficiency UHF-TV Power Amplifier," *Proceedings of the NAB Engineering Conference*, National Association of Broadcasters, Washington, DC, 1986.

Collins, G. B., *Radar System Engineering*, McGraw-Hill, New York, 1947.

Crutchfield, E. B., (ed.), *NAB Engineering Handbook*, 8th Ed., National Association of Broadcasters, Washington, DC, 1992.

Dick, Bradley, "New Developments in RF Technology," *Broadcast Engineering*, Intertec Publishing, Overland Park, KS, May 1986.

Fink, D., and D. Christiansen (eds.), *Electronics Engineers' Handbook*, 3rd Ed., McGraw-Hill, New York, 1989.

Fisk, J. B., H. D. Hagstrum, and P. L. Hartman, "The Magnetron As a Generator of Centimeter Waves," *Bell System Tech J.*, Vol. 25, pg. 167, April 1946.

Ginzton, E. L., and A. E. Harrison, "Reflex Klystron Oscillators," *Proc. IRE*, Vol. 34, pg. 97, March 1946.

IEEE Standard Dictionary of Electrical and Electronics Terms, Institute of Electrical and Electronics Engineers, Inc., New York, 1984.

McCune, Earl, "Final Report: The Multi-Stage Depressed Collector Project," *Proceedings of the NAB Engineering Conference*, National Association of Broadcasters, Washington, DC, 1988.

Ostroff, N., A. Kiesel, A. Whiteside, and A. See, "Klystrode-Equipped UHF-TV Transmitters: Report on the Initial Full Service Station Installations," *Proceedings of the NAB Engineering Conference*, National Association of Broadcasters, Washington, DC, 1989.

Ostroff, N., A. Whiteside, A. See, and A. Kiesel, "A 120 kW Klystrode Transmitter for Full Broadcast Service," *Proceedings of the NAB Engineering Conference*, National Association of Broadcasters, Washington, DC, 1988.

Ostroff, N., A. Whiteside, and L. Howard, "An Integrated Exciter/Pulser System for Ultra High-Efficiency Klystron Operation," *Proceedings of the NAB Engineering Conference*, National Association of Broadcasters, Washington, DC, 1985.

Pierce, J. R., "Reflex Oscillators," *Proc. IRE*, Vol. 33, pg. 112, February 1945.

Pierce, J. R., "Theory of the Beam-Type Traveling Wave Tube," *Proc. IRE*, Vol. 35, pg. 111, February 1947.

Pierce, J. R., and L. M. Field, "Traveling-Wave Tubes," *Proc. IRE*, Vol. 35, pg. 108, February 1947.

Pond, N. H., and C. G. Lob, "Fifty Years Ago Today or On Choosing a Microwave Tube," *Microwave Journal*, pp. 226 - 238, September 1988.

Priest, D., and M. Shrader, "The Klystrode—An Unusual Transmitting Tube with Potential for UHF-TV," *Proceedings of the IEEE*, Vol. 70, no. 11, IEEE, New York, November 1982.

Shrader, Merrald B., "Klystrode Technology Update," *Proceedings of the NAB Engineering Conference*, National Association of Broadcasters, Washington, DC, 1988.

Spangenberg, Karl, *Vacuum Tubes*, McGraw-Hill, New York, 1947.

Terman, F. E., *Radio Engineering*, 3rd Ed., McGraw-Hill, New York, 1947.

Varian, R., and S. Varian, "A High-Frequency Oscillator and Amplifier," *J. Applied Phys.*, Vol. 10, pg. 321, May 1939.

Webster, D. L., "Cathode Bunching," *J. Applied Physics*, Vol. 10, pg. 501, July 1939.

Whitaker, Jerry C., and T. Blankenship, "Comparing Integral and External Cavity Klystrons," *Broadcast Engineering*, Intertec Publishing, Overland Park, KS, November 1988.

Whitaker, Jerry C., *Radio Frequency Transmission Systems: Design and Operation*, McGraw-Hill, New York, 1991.

Whitaker, Jerry C., (ed.), *NAB Engineering Handbook*, 9th Ed., National Association of Broadcasters, Washington, DC, 1998.

RF Interconnection and Switching

7.1   Introduction

The components that connect, interface, transfer, and filter RF energy within a given system—or between systems—are critical elements in the operation of vacuum tube devices. Such hardware, usually passive, determines to a large extent the overall performance of the RF generator. To optimize the performance of power vacuum devices, first it is necessary to understand and optimize the components upon which the tube depends.

The mechanical and electrical characteristics of the transmission line, waveguide, and associated hardware that carry power from a power source (usually a transmitter) to the load (usually an antenna) are critical to proper operation of any RF system. Mechanical considerations determine the ability of the components to withstand temperature extremes, lightning, rain, and wind. In other words, they determine the overall reliability of the system.

7.1.1   Skin Effect

The effective resistance offered by a given conductor to radio frequencies is considerably higher than the ohmic resistance measured with direct current. This is because of an action known as the *skin effect*, which causes the currents to be concentrated in certain parts of the conductor and leaves the remainder of the cross section to contribute little or nothing toward carrying the applied current.

When a conductor carries an alternating current, a magnetic field is produced that surrounds the wire. This field continually expands and contracts as the ac wave increases from zero to its maximum positive value and back to zero, then through its negative half-cycle. The changing magnetic lines of force cutting the conductor induce a voltage in the conductor in a direction that tends to retard the normal flow of current in the wire. This effect is more pronounced at the center of the conductor. Thus, current within the conductor tends to flow more easily toward the surface of the wire. The higher the frequency, the greater the tendency for current to flow at the surface. The depth of current flow is a function of frequency, and it is determined from the following equation:

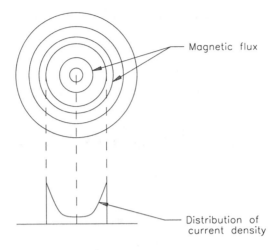

Figure 7.1 Skin effect on an isolated round conductor carrying a moderately high frequency signal.

$$d = \frac{2.6}{\sqrt{\mu f}} \qquad (7.1)$$

Where:
d = depth of current in mils
μ = permeability (copper = 1, steel = 300)
f = frequency of signal in mHz

It can be calculated that at a frequency of 100 kHz, current flow penetrates a conductor by 8 mils. At 1 MHz, the skin effect causes current to travel in only the top 2.6 mils in copper, and even less in almost all other conductors. Therefore, the series impedance of conductors at high frequencies is significantly higher than at low frequencies. Figure 7.1 shows the distribution of current in a radial conductor.

When a circuit is operating at high frequencies, the skin effect causes the current to be redistributed over the conductor cross section in such a way as to make most of the current flow where it is encircled by the smallest number of flux lines. This general principle controls the distribution of current regardless of the shape of the conductor involved. With a flat-strip conductor, the current flows primarily along the edges, where it is surrounded by the smallest amount of flux.

7.2 Coaxial Transmission Line

Two types of coaxial transmission line are in common use today: *rigid line* and corrugated (*semiflexible*) line. Rigid coaxial cable is constructed of heavy-wall copper tubes with Teflon or ceramic spacers. (Teflon is a registered trademark of DuPont.)

Rigid line provides electrical performance approaching an ideal transmission line, including:

- High power-handling capability

- Low loss

- Low VSWR (*voltage standing wave ratio*)

Rigid transmission line is, however, expensive to purchase and install.

The primary alternative to rigid coax is semiflexible transmission line made of corrugated outer and inner conductor tubes with a spiral polyethylene (or Teflon) insulator. The internal construction of a semiflexible line is shown in Figure 7.2. Semiflexible line has four primary benefits:

- It is manufactured in a continuous length, rather than the 20-ft sections typically used for rigid line.

- Because of the corrugated construction, the line may be shaped as required for routing from the transmitter to the antenna.

- The corrugated construction permits differential expansion of the outer and inner conductors.

- Each size of line has a minimum bending radius. For most installations, the flexible nature of corrugated line permits the use of a single piece of cable from the transmitter to the antenna, with no elbows or other transition elements. This speeds installation and provides for a more reliable system.

7.2.1 Electrical Parameters

A signal traveling in free space is unimpeded; it has a free-space velocity equal to the speed of light. In a transmission line, capacitance and inductance slow the signal as it propagates along the line. The degree to which the signal is slowed is represented as a percentage of the free-space velocity. This quantity is called the *relative velocity of propagation* and is described by the equation:

$$V_p = \frac{1}{\sqrt{L \times C}} \tag{7.2}$$

Where:
L = inductance in henrys per foot
C = capacitance in farads per foot

and

$$V_r = \frac{V_p}{c} \times 100\% \tag{7.3}$$

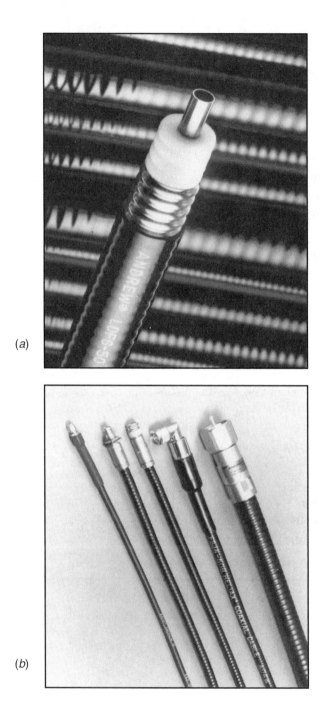

(a)

(b)

Figure 7.2 Semiflexible coaxial cable: (a) a section of cable showing the basic construction, (b) cable with various terminations. (*Courtesy of Andrew.*)

Where:
V_p = velocity of propagation
$c = 9.842 \times 10^8$ feet per second (free-space velocity)
V_r = velocity of propagation as a percentage of free-space velocity

Transverse Electromagnetic Mode

The principal mode of propagation in a coaxial line is the *transverse electromagnetic mode* (TEM). This mode will not propagate in a waveguide, and that is why coaxial lines can propagate a broad band of frequencies efficiently. The cutoff frequency for a coaxial transmission line is determined by the line dimensions. Above cutoff, modes other than TEM can exist and the transmission properties are no longer defined. The cutoff frequency is equivalent to:

$$F_c = \frac{7.50 \times V_r}{D_i + D_o} \tag{7.4}$$

Where:
F_c = cutoff frequency in gigahertz
V_r = velocity (percent)
D_i = inner diameter of outer conductor in inches
D_o = outer diameter of inner conductor in inches

At dc, current in a conductor flows with uniform density over the cross section of the conductor. At high frequencies, the current is displaced to the conductor surface. The effective cross section of the conductor decreases, and the conductor resistance increases because of the skin effect.

Center conductors are made from copper-clad aluminum or high-purity copper and can be solid, hollow tubular, or corrugated tubular. Solid center conductors are found on semiflexible cable with $\frac{1}{2}$ in or smaller diameter. Tubular conductors are found in $\frac{7}{8}$ in or larger-diameter cables. Although the tubular center conductor is used primarily to maintain flexibility, it also can be used to pressurize an antenna through the feeder.

Dielectric

Coaxial lines use two types of dielectric construction to isolate the inner conductor from the outer conductor. The first is an air dielectric, with the inner conductor supported by a dielectric spacer and the remaining volume filled with air or nitrogen gas. The spacer, which may be constructed of spiral or discrete rings, typically is made of Teflon or polyethylene. Air-dielectric cable offers lower attenuation and higher average power ratings than foam-filled cable but requires pressurization to prevent moisture entry.

Foam-dielectric cables are ideal for use as feeders with antennas that do not require pressurization. The center conductor is surrounded completely by foam-dielectric material, resulting in a high dielectric breakdown level. The dielectric materials are polyethylene-based formulations, which contain antioxidants to reduce dielectric deterioration at high temperatures.

Impedance

The expression *transmission line impedance* applied to a point on a transmission line signifies the vector ratio of line voltage to line current at that particular point. This is the impedance that would be obtained if the transmission line were cut at the point in question, and the impedance looking toward the receiver were measured.

Because the voltage and current distribution on a line are such that the current tends to be small when the voltage is large (and vice versa), as shown in Figure 7.3, the impedance will, in general, be oscillatory in the same manner as the voltage (large when the voltage is high and small when the voltage is low). Thus, in the case of a short-circuited receiver, the impedance will be high at distances from the receiving end that are odd multiples of $\frac{1}{4}$ wavelength, and it will be low at distances that are even multiples of $\frac{1}{4}$ wavelength.

The extent to which the impedance fluctuates with distance depends on the standing wave ratio (ratio of reflected to incident waves), being less as the reflected wave is proportionally smaller than the incident wave. In the particular case where the load impedance equals the characteristic impedance, the impedance of the transmission line is equal to the characteristic impedance at all points along the line.

The power factor of the impedance of a transmission line varies according to the standing waves present. When the load impedance equals the characteristic impedance, there is no reflected wave, and the power factor of the impedance is equal to the power factor of the characteristic impedance. At radio frequencies, the power factor under these conditions is accordingly resistive. However, when a reflected wave is present, the power factor is unity (resistive) only at the points on the line where the voltage passes through a maximum or a minimum. At other points the power factor will be reactive, alternating from leading to lagging at intervals of $\frac{1}{4}$ wavelength. When the line is short-circuited at the receiver, or when it has a resistive load less than the characteristic impedance so that the voltage distribution is of the short-circuit type, the power factor is inductive for lengths corresponding to less than the distance to the first voltage maximum. Thereafter, it alternates between capacitive and inductive at intervals of $\frac{1}{4}$ wavelength. Similarly, with an open-circuited receiver or with a resistive load greater than the characteristic impedance so that the voltage distribution is of the open-circuit type, the power factor is capacitive for lengths corresponding to less than the distance to the first voltage minimum. Thereafter, the power factor alternates between capacitive and inductive at intervals of $\frac{1}{4}$ wavelength, as in the short-circuited case.

Resonant Characteristics

A transmission line can be used to perform the functions of a resonant circuit. If the line, for example, is short-circuited at the receiver, at frequencies in the vicinity of a frequency at which the line is an odd number of $\frac{1}{4}$ wavelengths long, the impedance will be high and will vary with frequency in the vicinity of resonance. This character-

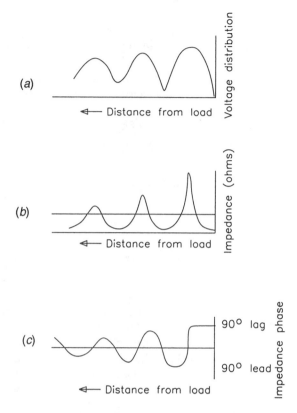

Figure 7.3 Magnitude and power factor of line impedance with increasing distance from the load for the case of a short-circuited receiver and a line with moderate attenuation: (*a*) voltage distribution, (*b*) impedance magnitude, (*c*) impedance phase.

istic is similar in nature to a conventional parallel resonant circuit. The difference is that with the transmission line, there are a number of resonant frequencies, one for each of the infinite number of frequencies that make the line an odd number of ¼ wavelengths long. At VHF, the parallel impedance at resonance and the circuit Q obtainable are far higher than can be realized with lumped circuits. Behavior corresponding to that of a series resonant circuit can be obtained from a transmission line that is an odd number of ¼ wavelengths long and open-circuited at the receiver.

Transmission lines also can be used to provide low-loss inductances or capacitances if the proper combination of length, frequency, and termination is employed. Thus, a line short-circuited at the receiver will offer an inductive reactance when less than ¼ wavelength, and a capacitive reactance when between ¼ and ½ wavelength. With an open-circuited receiver, the conditions for inductive and capacitive reactances are reversed.

7.2.2 Electrical Considerations

VSWR, attenuation, and power-handling capability are key electrical factors in the application of coaxial cable. High VSWR can cause power loss, voltage breakdown, and thermal degradation of the line. High attenuation means less power delivered to the antenna, higher power consumption at the transmitter, and increased heating of the transmission line itself.

VSWR is a common measure of the quality of a coaxial cable. High VSWR indicates nonuniformities in the cable that can be caused by one or more of the following conditions:

- Variations in the dielectric core diameter

- Variations in the outer conductor

- Poor concentricity of the inner conductor

- Nonhomogeneous or periodic dielectric core

Although each of these conditions may contribute only a small reflection, they can add up to a measurable VSWR at a particular frequency.

Rigid transmission line typically is available in a standard length of 20 ft, and in alternative lengths of 19.5 ft and 19.75 ft. The shorter lines are used to avoid VSWR buildup caused by discontinuities resulting from the physical spacing between line section joints. If the section length selected and the operating frequency have a $\frac{1}{2}$-wave correlation, the connector junction discontinuities will add. This effect is known as *flange buildup*. The result can be excessive VSWR. The critical frequency at which a $\frac{1}{2}$-wave relationship exists is given by:

$$F_{cr} = \frac{490.4 \times n}{L} \qquad (7.5)$$

Where:
F_{cr} = the critical frequency
n = any integer
L = transmission line length in feet

For most applications, the critical frequency for a chosen line length should not fall closer than ±2 MHz of the passband at the operating frequency.

Attenuation is related to the construction of the cable itself and varies with frequency, product dimensions, and dielectric constant. Larger-diameter cable exhibits lower attenuation than smaller-diameter cable of similar construction when operated at the same frequency. It follows, therefore, that larger-diameter cables should be used for long runs.

Air-dielectric coax exhibits less attenuation than comparable-size foam-dielectric cable. The attenuation characteristic of a given cable also is affected by standing waves present on the line resulting from an impedance mismatch. Table 7.1 shows a representative sampling of semiflexible coaxial cable specifications for a variety of line sizes.

Table 7.1 Representative Specifications for Various Types of Flexible Air-Dielectric Co-axial Cable

Cable size (in.)	Maximum frequency (MHz)	Velocity (percent)	Peak power 1 MHz (kW)	Average power		Attenuation[1]	
				100 MHz (kW)	1 MHz (kW)	100 MHz (dB)	1 Mhz (dB)
1⅝	2.7	92.1	145	145	14.4	0.020	0.207
3	1.64	93.3	320	320	37	0.013	0.14
4	1.22	92	490	490	56	0.010	0.113
5	0.96	93.1	765	765	73	0.007	0.079
[1] Attenuation specified in dB/100 ft.							

7.2.3 Coaxial Cable Ratings

Selection of a type and size of transmission line is determined by a number of parameters, including power-handling capability, attenuation, and phase stability.

Power Rating

Both *peak* and *average* power ratings are required to fully describe the capabilities of a transmission line. In most applications, the peak power rating limits the low frequency or pulse energy, and the average power rating limits high-frequency applications, as shown in Figure 7.4. Peak power ratings usually are stated for the following conditions:

- VSWR = 1.0

- Zero modulation

- One atmosphere of absolute dry air pressure at sea level

The peak power rating of a selected cable must be greater than the following expression, in addition to satisfying the average-power-handling criteria:

$$E_{pk} > P_t \times (1 + M)^2 \times VSWR \qquad (7.6)$$

Where:
P_{pk} = cable peak power rating in kilowatts
P_t = transmitter power in kilowatts
M = amplitude modulation percentage expressed decimally (100 percent = 1.0)
VSWR = voltage standing wave ratio

From this equation, it can be seen that 100 percent amplitude modulation will increase the peak power in the transmission line by a factor of 4. Furthermore, the peak power in the transmission line increases directly with VSWR.

The peak power rating is limited by the voltage breakdown potential between the inner and outer conductors of the line. The breakdown point is independent of frequency.

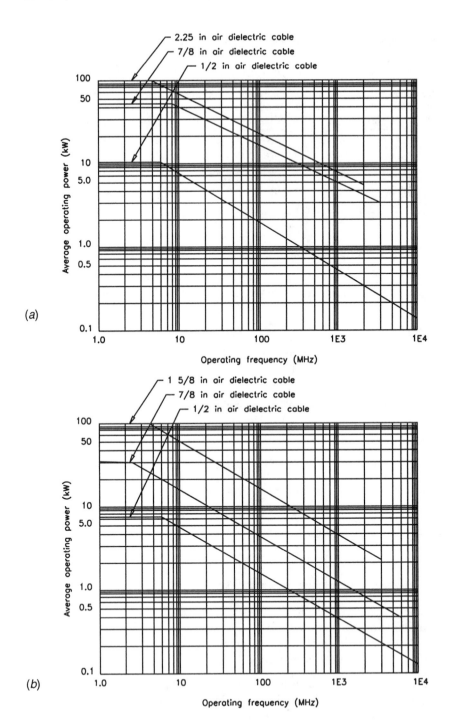

Figure 7.4 Power rating data for a variety of coaxial transmission lines: (*a*) 50 Ω line, (*b*) 75 Ω line.

It varies, however, with the line pressure (for an air-dielectric cable) and the type of pressurizing gas.

The average power rating of a transmission line is limited by the safe long-term operating temperature of the inner conductor and the dielectric. Excessive temperatures on the inner conductor will cause the dielectric material to soften, leading to mechanical instability inside the line.

The primary purpose of pressurization of an air-dielectric cable is to prevent the ingress of moisture. Moisture, if allowed to accumulate in the line, can increase attenuation and reduce the breakdown voltage between the inner and outer conductors. Pressurization with high-density gases can be used to increase both the average power and the peak power ratings of a transmission line. For a given line pressure, the increased power rating is more significant for peak power than for average power. High-density gases used for such applications include Freon 116 and sulfur hexafluoride. Figure 7.5 illustrates the effects of pressurization on cable power rating.

An adequate safety factor is necessary for peak and average power ratings. Most transmission lines are tested at two or more times their rated peak power before shipment to the customer. This safety factor is intended as a provision for transmitter transients, lightning-induced effects, and high-voltage excursions resulting from unforeseen operating conditions.

Connector Effects

Foam-dielectric cables typically have a greater dielectric strength than air-dielectric cables of similar size. For this reason, foam cables would be expected to exhibit higher peak power ratings than air lines. Higher values, however, usually cannot be realized in practice because the connectors commonly used for foam cables have air spaces at the cable/connector interface that limit the allowable RF voltage to "air cable" values.

The peak-power-handling capability of a transmission line is the smaller of the values for the cable and the connectors attached to it. Table 7.2 lists the peak power ratings of several common RF connectors at standard conditions (defined in the previous section).

Attenuation

The attenuation characteristics of a transmission line vary as a function of the operating frequency and the size of the line itself. The relationships are shown in Figure 7.6.

The *efficiency* of a transmission line dictates how much power output by the transmitter actually reaches the antenna. Efficiency is determined by the length of the line and the attenuation per unit length.

The attenuation of a coaxial transmission line is defined by the equation:

$$\alpha = 10 \times \log \left\{ \frac{P_1}{P_2} \right\} \qquad (7.7)$$

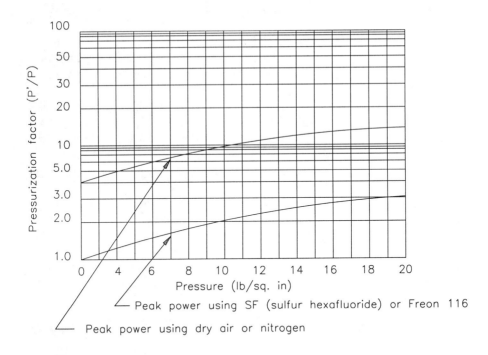

Figure 7.5 Effects of transmission line pressurization on peak power rating. Note that P′ = the rating of the line at the increased pressure, and P = the rating of the line at atmospheric pressure.

Table 7.2 Electrical Characteristics of Common RF Connectors

Connector Type	DC Test Voltage (kW)	Peak Power (kW)
SMA	1.0	1.2
BNC, TNC	1.5	2.8
N, UHF	2.0	4.9
GR	3.0	11
HN, $\frac{7}{16}$	4.0	20
LC	5.0	31
$\frac{7}{8}$ EIA, F Flange	6.0	44
$1\frac{5}{8}$ EIA	11.0	150
$3\frac{1}{8}$ EIA	19.0	4

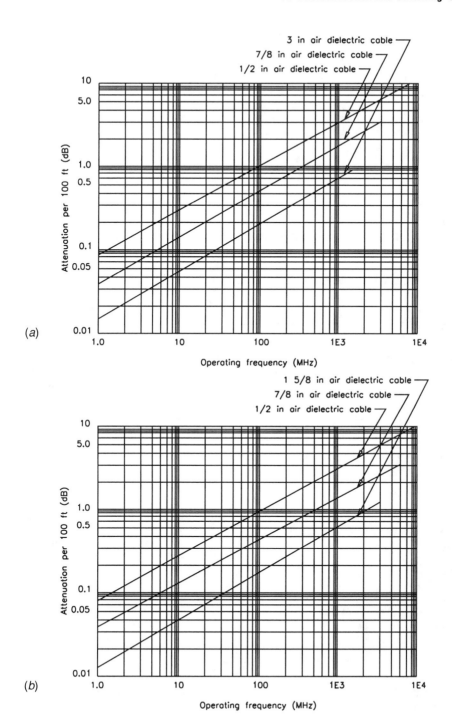

Figure 7.6 Attenuation characteristics for a selection of coaxial cables: (*a*) 50 Ω line, (*b*) 75 Ω line.

Where:

α = attenuation in decibels per 100 meters

P_1 = input power into a 100-meter line terminated with the nominal value of its characteristic impedance

P_2 = power measured at the end of the line

Stated in terms of efficiency (E, percent):

$$E = 100 \times \left\{ \frac{P_o}{P_i} \right\} \tag{7.8}$$

Where:

P_i = power delivered to the input of the transmission line

P_o = power delivered to the antenna

The relationship between efficiency and loss in decibels (*insertion loss*) is illustrated in Figure 7.7.

Manufacturer-supplied attenuation curves typically are guaranteed to within approximately ±5 percent. The values given usually are rated at 24°C (75°F) ambient temperature. Attenuation increases slightly with higher temperature or applied power. The effects of ambient temperature on attenuation are illustrated in Figure 7.8.

Loss in connectors is negligible, except for small (SMA and TNC) connectors at frequencies of several gigahertz and higher. Small connectors used at high frequencies typically add 0.1 dB of loss per connector.

When a transmission line is attached to a load, such as an antenna, the VSWR of the load increases the total transmission loss of the system. This effect is small under conditions of low VSWR. Figure 7.9 illustrates the interdependence of these two elements.

Phase Stability

A coaxial cable expands as the temperature of the cable increases, causing the electrical length of the line to increase as well. This factor results in phase changes that are a function of operating temperature. The phase relationship can be described by the equation:

$$\theta = 3.66 \times 10^{-7} \times P \times L \times T \times F \tag{7.9}$$

Where:

θ = phase change in degrees

P = phase-temperature coefficient of the cable

L = length of coax in feet

T = temperature range (minimum-to-maximum operating temperature)

F = frequency in Mhz

Phase changes that are a function of temperature are important in systems utilizing multiple transmission lines, such as a directional array fed from a single phasing

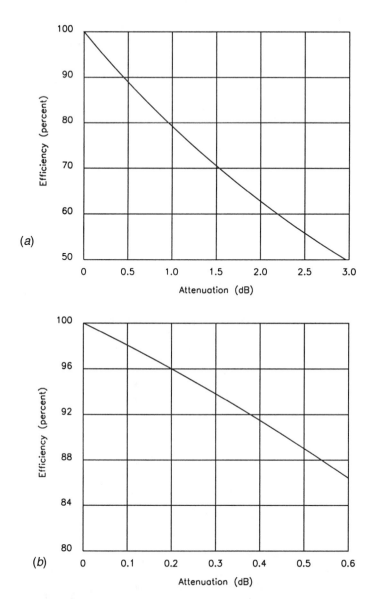

Figure 7.7 Conversion chart showing the relationship between decibel loss and efficiency of a transmission line: (a) high-loss line, (b) low-loss line.

source. To maintain proper operating parameters, the phase changes of the cables must be minimized. Specially designed coaxial cables that offer low-phase-temperature characteristics are available. Two types of coax are commonly used for this purpose:

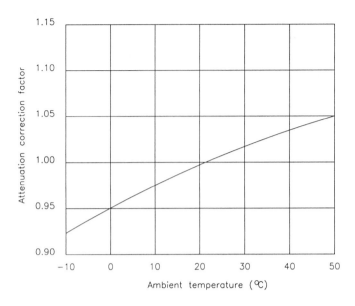

Figure 7.8 The variation of coaxial cable attenuation as a function of ambient temperature.

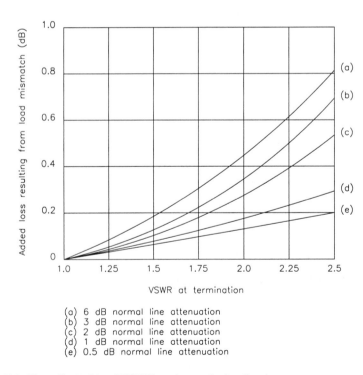

(a) 6 dB normal line attenuation
(b) 3 dB normal line attenuation
(c) 2 dB normal line attenuation
(d) 1 dB normal line attenuation
(e) 0.5 dB normal line attenuation

Figure 7.9 The effect of load VSWR on transmission line loss.

- *Phase-stabilized* cables, which have undergone extensive temperature cycling until such time as they exhibit their minimum phase-temperature coefficient

- *Phase-compensated* cables, in which changes in the electrical length have been minimized through adjustment of the mechanical properties of the dielectric and inner/outer conductors

7.2.4 Mechanical Parameters

Corrugated copper cables are designed to withstand bending with no change in properties. Low-density foam- and air-dielectric cables generally have a minimum bending radius of 10 times the cable diameter. Super flexible versions provide a smaller allowable bending radius.

Rigid transmission line will not tolerate bending. Instead, transition elements (elbows) of various sizes are used. Individual sections of rigid line are secured by multiple bolts around the circumference of a coupling flange.

When a large cable must be used to meet attenuation requirements, short lengths of a smaller cable (jumpers or *pigtails*) may be used on either end for ease of installation in low-power systems. The tradeoff is slightly higher attenuation and some additional cost.

The *tensile strength* of a cable is defined as the axial load that may be applied to the line with no more than 0.2 percent permanent deformation after the load is released. When divided by the weight per foot of cable, this gives an indication of the maximum length of cable that is self-supporting and therefore can be installed readily on a tower with a single hoisting grip. This consideration usually applies only to long runs of corrugated line; rigid line is installed one section at a time.

The *crush strength* of a cable is defined as the maximum force per linear inch that may be applied by a flat plate without causing more than a 5 percent deformation of the cable diameter. Crush strength is a good indicator of the ruggedness of a cable and its ability to withstand rough handling during installation.

Cable jacketing affords mechanical protection during installation and service. Semiflexible cables typically are supplied with a jacket consisting of low-density polyethylene blended with 3 percent carbon black for protection from the sun's ultraviolet rays, which can degrade plastics over time. This approach has proved to be effective, yielding a life expectancy of more than 20 years. Rigid transmission line has no covering over the outer conductor.

For indoor applications, where fire-retardant properties are required, cables can be supplied with a fire-retardant jacket, usually listed by Underwriters Laboratories. Note that under the provisions of the National Electrical Code, outside plant cables such as standard black polyethylene-jacketed coaxial line may be run as far as 50 ft inside a building with no additional protection. The line also may be placed in conduit for longer runs.

Low-density foam cables are designed to prevent water from traveling along their length, should it enter through damage to the connector or the cable sheath. This is accomplished by mechanically locking the outer conductor to the foam dielectric by annular corrugations. Annular or ring corrugations, unlike helical or screw-thread-type corrugations, provide a water block at each corrugation. Closed-cell polyethylene dielectric foam is bonded to the inner conductor, completing the moisture seal.

A coaxial cable line is only as good as the connectors used to tie it together. The connector interface must provide a weatherproof bond with the cable to prevent water from penetrating the connection. This is ensured by using O-ring seals. The cable-connector interface also must provide a good electrical bond that does not introduce a mismatch and increase VSWR. Good electrical contact between the connector and the cable ensures that proper RF shielding is maintained.

7.3 Waveguide

As the operating frequency of a system reaches into the UHF band, waveguide-based transmission line systems become practical. From the mechanical standpoint, waveguide is simplicity itself. There is no inner conductor; RF energy is *launched* into the structure and propagates to the load. Several types of waveguide are available, including rectangular, square, circular, and elliptical. Waveguide offers several advantages over coax. First, unlike coax, waveguide can carry more power as the operating frequency increases. Second, efficiency is significantly better with waveguide at higher frequencies.

Rectangular waveguide commonly is used in high-power transmission systems. Circular waveguide also may be used, especially for applications requiring a cylindrical member, such as a rotating joint for an antenna feed. The physical dimensions of the guide are selected to provide for propagation in the *dominant* (lowest-order) mode.

Waveguide is not without its drawbacks, however. Rectangular or square guide constitutes a large windload surface, which places significant structural demands on a tower. Because of the physical configuration of rectangular and square guide, pressurization is limited, depending on the type of waveguide used (0.5 psi is typical). Excessive pressure can deform the guide shape and result in increased VSWR. Wind also may cause deformation and ensuing VSWR problems. These considerations have led to the development of circular and elliptical waveguide.

7.3.1 Propagation Modes

Propagation modes for waveguide fall into two broad categories:

- *Transverse-electric* (TE) waves
- *Transverse-magnetic* (TM) waves

With TE waves, the electric vector (*E vector*) is perpendicular to the direction of propagation. With TM waves, the magnetic vector (*H vector*) is perpendicular to the direction of propagation. These propagation modes take on integers (from 0 or 1 to infinity) that define field configurations. Only a limited number of these modes can be propagated, depending on the dimensions of the guide and the operating frequency.

Energy cannot propagate in waveguide unless the operating frequency is above the *cutoff frequency*. The cutoff frequency for rectangular guide is:

$$F_c = \frac{C}{2 \times a} \tag{7.10}$$

Where:
F_c = waveguide cutoff frequency
$c = 1.179 \times 10^{10}$ inches per second (the velocity of light)
a = the wide dimension of the guide

The cutoff frequency for circular waveguide is defined by the following equation:

$$F_c = \frac{c}{3.41 \times a'} \tag{7.11}$$

Where:
a' = the radius of the guide

There are four common propagation modes in waveguide:

- $TE_{0,1}$, the principal mode in rectangular waveguide.

- $TE_{1,0}$, also used in rectangular waveguide.

- $TE_{1,1}$, the principal mode in circular waveguide. $TE_{1,1}$ develops a complex propagation pattern with electric vectors curving inside the guide. This mode exhibits the lowest cutoff frequency of all modes, which allows a smaller guide diameter for a specified operating frequency.

- $TM_{0,1}$, which has a slightly higher cutoff frequency than $TE_{1,1}$ for the same size guide. Developed as a result of discontinuities in the waveguide, such as flanges and transitions, $TM_{0,1}$ energy is not coupled out by either dominant or cross-polar transitions. The parasitic energy must be filtered out, or the waveguide diameter chosen carefully to reduce the unwanted mode.

The field configuration for the dominant mode in rectangular waveguide is illustrated in Figure 7.10. Note that the electric field is vertical, with intensity maximum at the center of the guide and dropping off sinusoidally to zero intensity at the edges. The magnetic field is in the form of loops that lie in planes that are at right angles to the electric field (parallel to the top and bottom of the guide). The magnetic field distribution is the same for all planes perpendicular to the Y-axis. In the X direction, the intensity of the component of magnetic field that is transverse to the axis of the waveguide (the component in the direction of X) is at any point in the waveguide directly proportional to the intensity of the electric field at that point. This entire configuration of fields travels in the direction of the waveguide axis (the Z direction in Figure 7.10).

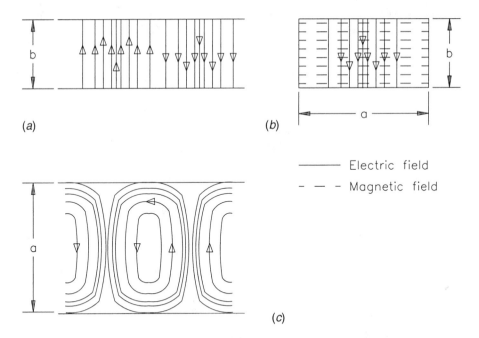

Electric field
Magnetic field

Figure 7.10 Field configuration of the dominant or TE$_{1,0}$ mode in a rectangular waveguide: (a) side view, (b) end view, (c) top view.

The field configuration for the TE$_{1,1}$ mode in circular waveguide is illustrated in Figure 7.11. The TE$_{1,1}$ mode has the longest cutoff wavelength and is, accordingly, the dominant mode. The next higher mode is the TM$_{0,1}$, followed by TE$_{2,1}$.

Dual-Polarity Waveguide

Waveguide will support dual-polarity transmission within a single run of line. A combining element (*dual-polarized transition*) is used at the beginning of the run, and a splitter (polarized transition) is used at the end of the line. Square waveguide has found numerous applications in such systems. Theoretically, the TE$_{1,0}$ and TE$_{0,1}$ modes are capable of propagation without cross coupling, at the same frequency, in lossless waveguide of square cross section. In practice, surface irregularities, manufacturing tolerances, and wall losses give rise to TE$_{1,0}$- and TE$_{0,1}$-mode cross conversion. Because this conversion occurs continuously along the waveguide, long guide runs usually are avoided in dual-polarity systems.

Efficiency

Waveguide losses result from the following:

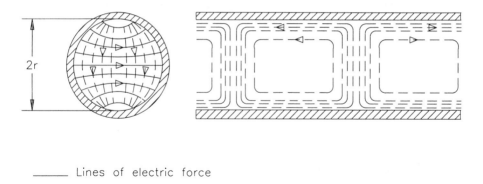

_____ Lines of electric force
— — — Lines of magnetic force

Figure 7.11 Field configuration of the dominant mode in circular waveguide.

- Power dissipation in the waveguide walls and the dielectric material filling the enclosed space
- Leakage through the walls and transition connections of the guide
- Localized power absorption and heating at the connection points

The operating power of waveguide may be increased through pressurization. Sulfur hexafluoride commonly is used as the pressurizing gas.

7.3.2 Ridged Waveguide

Rectangular waveguide may be ridged to provide a lower cutoff frequency, thereby permitting use over a wider frequency band. As illustrated in Figure 7.12, one- and two-ridged guides are used. Increased bandwidth comes at the expense of increased attenuation, relative to an equivalent section of rectangular guide.

7.3.3 Circular Waveguide

Circular waveguide offers several mechanical benefits over rectangular or square guide. The windload of circular guide is two-thirds that of an equivalent run of rectangular waveguide. It also presents lower and more uniform windloading than rectangular waveguide, reducing tower structural requirements.

The same physical properties of circular waveguide that give it good power handling and low attenuation also result in electrical complexities. Circular waveguide has two potentially unwanted modes of propagation: the cross-polarized $TE_{1,1}$ and $TM_{0,1}$ modes.

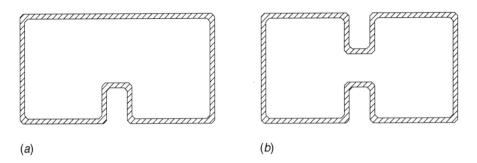

Figure 7.12 Ridged waveguide: (a) single-ridged, (b) double-ridged.

Circular waveguide, by definition, has no short or long dimension and, consequently, no method to prevent the development of cross-polar or *orthogonal* energy. Cross-polar energy is formed by small ellipticities in the waveguide. If the cross-polar energy is not trapped out, the parasitic energy can recombine with the dominant-mode energy.

Parasitic Energy

Hollow circular waveguide works as a high-Q resonant cavity for some energy and as a transmission medium for the rest. The parasitic energy present in the cavity formed by the guide will appear as increased VSWR if not disposed of. The polarization in the guide meanders and rotates as it propagates from the source to the load. The end pieces of the guide, typically circular-to-rectangular transitions, are polarization-sensitive. See Figure 7.13(a). If the polarization of the incidental energy is not matched to the transition, energy will be reflected.

Several factors can cause this undesirable polarization. One cause is out-of-round guides that result from nonstandard manufacturing tolerances. In Figure 7.13, the solid lines depict the situation at launching: perfectly circular guide with perpendicular polarization. The dashed lines show how certain ellipticities cause polarization rotation into unwanted states, while others have no effect. A 0.2 percent change in diameter can produce a –40 dB cross-polarization component per wavelength. This is roughly 0.03 in for 18 in of guide length.

Other sources of cross polarization include twisted and bent guides, out-of-roundness, offset flanges, and transitions. Various methods are used to dispose of this energy trapped in the cavity, including absorbing loads placed at the ground and/or antenna level.

7.3.4 Doubly Truncated Waveguide

The design of *doubly truncated waveguide* (DTW) is intended to overcome the problems that may result from parasitic energy in a circular waveguide. As shown in Fig-

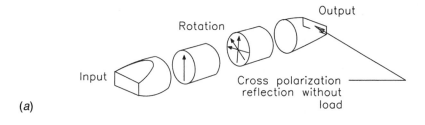

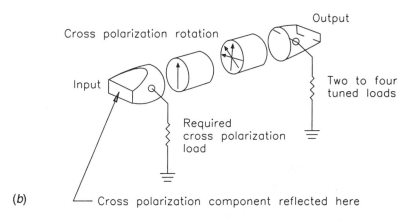

(a)

(b)

Figure 7.13 The effects of parasitic energy in circular waveguide: (a) trapped cross-polarization energy, (b) delayed transmission of the trapped energy.

ure 7.14, DTW consists of an almost elliptical guide inside a circular shell. This guide does not support cross polarization; tuners and absorbing loads are not required. The low windload of hollow circular guide is maintained, except for the flange area.

Each length of waveguide is actually two separate pieces: a doubly truncated center section and a circular outer skin, joined at the flanges on each end. A large hole in the broadwall serves to pressurize the circular outer skin. Equal pressure inside the DTW and inside the circular skin ensures that the guide will not "breathe" or buckle as a result of rapid temperature changes.

DTW exhibits about 3 percent higher windloading than an equivalent run of circular waveguide (because of the transition section at the flange joints), and 32 percent lower loading than comparable rectangular waveguide.

7.3.5 Impedance Matching

The efficient flow of power from one type of transmission medium to another requires matching of the field patterns across the boundary to launch the wave into the second medium with a minimum of reflections. Coaxial line typically is matched into rectangular waveguide by extending the center conductor of the coax through the broadwall

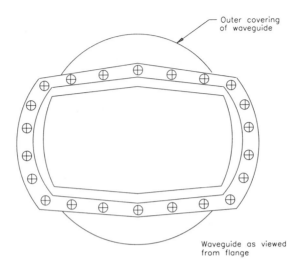

Outer covering
of waveguide

Waveguide as viewed
from flange

Figure 7.14 Physical construction of doubly truncated waveguide.

of the guide, parallel to the electric field lines across the guide. Alternatively, the center conductor can be formed into a loop and oriented to couple the magnetic field to the guide mode.

Standing waves are generally to be avoided in waveguide for the same reasons that they are to be avoided in transmission lines. Accordingly, it is usually necessary to provide impedance-matching systems in waveguides to eliminate standing waves. One approach involves the introduction of a compensating reflection in the vicinity of a load that neutralizes the standing waves that would exist in the system because of an imperfect match. A probe or tuning screw commonly is used to accomplish this, as illustrated in Figure 7.15. The tuning screw projects into the waveguide in a direction parallel to the electric field. This is equivalent to shunting a capacitive load across the guide. The susceptance of the load increases with extension into the guide up to $\frac{1}{4}$ wavelength. When the probe is exactly $\frac{1}{4}$ wavelength long, it becomes resonant and causes the guide to behave as though there were an open circuit at the point of the resonant probe. Probes longer than $\frac{1}{4}$ wavelength but shorter than $\frac{3}{4}$ wavelength introduce inductive loading. The extent to which such a probe projects into the waveguide determines the compensating reflection, and the position of the probe with respect to the standing wave pattern to be eliminated determines the phasing of the reflected wave.

Dielectric slugs produce an effect similar to that of a probe. The magnitude of the effect depends upon the following considerations:

- The dielectric constant of the slug

- Thickness of the slug in an axial direction

- ˙Whether the slug extends entirely across the waveguide

The phase of the reflected wave is controlled by varying the axial position of the slug.

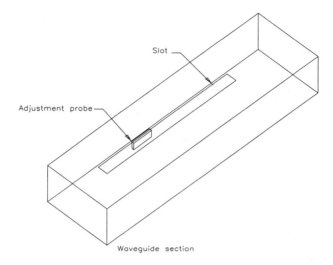

Figure 7.15 A probe configured to introduce a reflection in a waveguide that is adjustable in magnitude and phase.

There are several alternatives to the probe and slug for introducing controllable irregularity for impedance matching, including a metallic barrier or *window* placed at right angles to the axis of the guide, as illustrated in Figure 7.16. Three configurations are shown:

- The arrangement illustrated in Figure 7.16a produces an effect equivalent to shunting the waveguide with an inductive reactance.

- The arrangement shown in Figure 7.16b produces the effect of a shunt capacitive susceptance.

- The arrangement shown in Figure 7.16c produces an inductive shunt susceptance.

The waveguide equivalent of the coaxial cable tuning stub is a *tee* section, illustrated in Figure 7.17. The magnitude of the compensating effect is controlled by the position of the short-circuiting plug in the branch. The phase of the compensating reflected wave produced by the branch is determined by the position of the branch in the guide.

Waveguide Filters

A section of waveguide beyond cutoff constitutes a simple high-pass reflective filter. Loading elements in the form of posts or stubs may be employed to supply the reactances required for conventional lumped-constant filter designs.

Absorption filters prevent the reflection of unwanted energy by incorporating lossy material in secondary guides that are coupled through *leaky walls* (small sections of

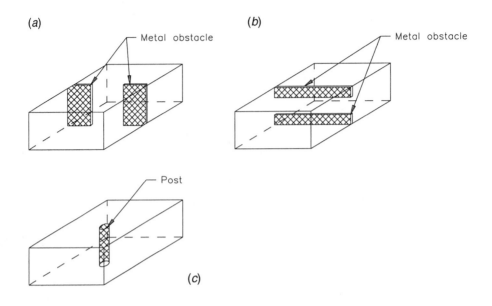

Figure 7.16 Waveguide obstructions used to introduce compensating reflections: (*a*) inductive window, (*b*) capacitive window, (*c*) post (inductive) element.

guide beyond cutoff in the passband). Such filters typically are used to suppress harmonic energy.

7.3.6 Installation Considerations

Waveguide system installation is both easier and more difficult than traditional transmission line installation. There is no inner conductor to align, but alignment pins must be set and more bolts are required per flange. Transition hardware to accommodate loads and coax-to-waveguide interfacing also is required.

Flange reflections can add up in phase at certain frequencies, resulting in high VSWR. The length of the guide must be chosen so that flange reflection buildup does not occur within the operating bandwidth.

Flexible sections of waveguide are used to join rigid sections or components that cannot be aligned otherwise. Flexible sections also permit controlled physical movement resulting from thermal expansion of the line. Such hardware is available in a variety of forms. Corrugated guide commonly is produced by shaping thin-wall seamless rectangular tubing. Flexible waveguide can accommodate only a limited amount of mechanical movement. Depending on the type of link, the manufacturer may specify a maximum number of bends.

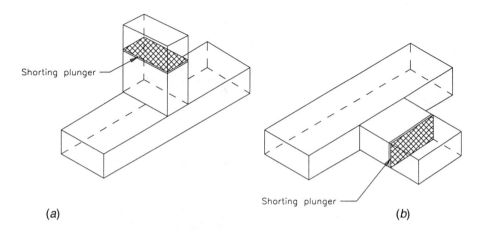

Figure 7.17 Waveguide stub elements used to introduce compensating reflections: (a) series *tee* element, (b) shunt *tee* element.

Tuning

Circular waveguide must be tuned. This requires a two-step procedure. First, the cross-polar $TE_{1,1}$ component is reduced, primarily through *axial ratio compensators* or *mode optimizers*. These devices counteract the net system ellipticity and indirectly minimize cross-polar energy. The cross-polar filters also may be rotated to achieve maximum isolation between the dominant and cross-polar modes. Cross-polar energy manifests itself as a net signal rotation at the end of the waveguide run. A perfect system would have a net rotation of zero.

In the second step, tuning slugs at both the top and bottom of the waveguide run are adjusted to reduce the overall system VSWR. Tuning waveguide can be a complicated and time-consuming procedure. Once set, however, tuning normally does not drift and must be repeated only if major component changes are made.

Waveguide Hardware

Increased use of waveguide has led to the development of waveguide-based hardware for all elements from the output of the RF generator to the load. Waveguide-based filters, elbows, directional couplers, switches, combiners, and diplexers are currently available. The RF performance of a waveguide component is usually better than the same item in coax. This is especially true in the case of diplexers and filters. Waveguide-based hardware provides lower attenuation and greater power-handling capability for a given physical size.

7.3.7 Cavity Resonators

Any space completely enclosed by conducting walls can contain oscillating electro-magnetic fields. Such a cavity possesses certain frequencies at which it will resonate when excited by electrical oscillations. These *cavity resonators* find extensive use as resonant circuits at VHF and above. (The use of cavities in power grid tube amplifiers is discussed in Chapter 4.) Advantages of cavity resonators over conventional *LC* circuits include:

- Simplicity in design.

- Relatively large physical size compared with alternative methods of obtaining resonance. This attribute is important in high-power, high-frequency applications.

- High *Q*.

- The capability to configure the cavity to develop an extremely high shunt impedance.

Cavity resonators commonly are used at wavelengths on the order of 10 cm or less.

The simplest cavity resonator is a section of waveguide shorted at each end with a length *l* equal to:

$$l = \frac{\lambda_g}{2} \qquad (7.12)$$

Where:
λ_g = the guide wavelength

This configuration results in a resonance similar to that of a $\frac{1}{2}$-wavelength transmission line short-circuited at the receiving end.

A sphere or any other enclosed surface (irrespective of how irregular the outline) also may be used to form a cavity resonator.

Any given cavity is resonant at a number of frequencies, corresponding to the different possible field conditions that can exist within the space. The resonance having the longest wavelength (lowest frequency) is termed the *dominant* or *fundamental resonance*. The resonant wavelength is proportional to the size of the resonator. If all dimensions are doubled, the wavelength corresponding to resonance will likewise be doubled. The resonant frequency of a cavity can be changed by incorporating one or more of the following mechanisms:

- Altering the mechanical dimensions of the cavity. Small changes can be achieved by flexing walls, but large changes require some form of sliding member.

- Coupling reactance into the resonator through a coupling loop.

- Introducing a movable copper paddle into the cavity. A paddle placed inside the resonator will affect the normal distribution of flux and tend to raise the resonant frequency by an amount determined by the orientation of the paddle.

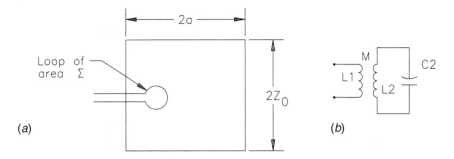

Figure 7.18 Cavity resonator coupling: (*a*) coupling loop, (*b*) equivalent circuit.

The Q of a cavity resonator has the same significance as for a conventional resonant circuit. Q can be defined for a cavity by the following relationship:

$$Q = 2\pi \frac{E_s}{E_l} \tag{7.13}$$

Where:
E_s = energy stored
E_l = energy lost per cycle

The energy stored is proportional to the square of the magnetic flux density integrated throughout the volume of the resonator. The energy lost in the walls is proportional to the square of the magnetic flux density integrated over the surface of the cavity. To obtain high Q, the resonator should have a large ratio of volume to surface area, because it is the volume that stores energy and the surface area that dissipates energy.

Coupling can be obtained from a resonator by means of a coupling loop or coupling electrode. Magnetic coupling is accomplished through the use of a loop oriented so as to enclose magnetic flux lines existing in the desired mode of operation. This technique is illustrated in Figure 7.18. A current passed through the loop will excite oscillations of this mode. Conversely, oscillations existing in the resonator will induce a voltage in the coupling loop. The magnitude of the coupling can be controlled by rotating the loop; the coupling reduces to zero when the plane of the loop is parallel to the magnetic flux.

Coupling of a resonator also may be accomplished through the use of a probe or opening in one wall of the cavity.

7.4 RF Combiner and Diplexer Systems

The basic purpose of an RF combiner is to add two or more signals to produce an output signal that is a composite of the inputs. The combiner performs this signal addition while providing isolation between inputs. Combiners perform other functions as well, and can be found in a wide variety of RF equipment utilizing power vacuum tubes. Combiners are valuable devices because they permit multiple amplifiers to drive a single load. The isolation provided by the combiner permits tuning adjustments to be made on one amplifier—including turning it on or off—without significantly affecting the operation of the other amplifier. In a typical application, two amplifiers drive the hybrid and provide two output signals:

- A combined output representing the sum of the two input signals, typically directed toward the antenna.

- A difference output representing the difference in amplitude and phase between the two input signals. The difference output typically is directed toward a dummy (reject) load.

For systems in which more than two amplifiers must be combined, two or more combiners are cascaded.

Diplexers are similar in nature to combiners but permit the summing of output signals from two or more amplifiers operating at different frequencies. This allows, for example, the outputs of several transmitters operating on different frequencies to utilize a single broadband antenna.

7.4.1 Passive Filters

A *filter* is a multiport-network designed specifically to respond differently to signals of different frequency [1]. This definition excludes *networks*, which incidentally behave as filters, sometimes to the detriment of their main purpose. Passive filters are constructed exclusively with passive elements (i.e., resistors, inductors, and capacitors). Filters are generally categorized by the following general parameters:

- Type
- Alignment (or class)
- Order

Filter Type

Filters are categorized by type, according to the magnitude of the frequency response, as one of the following [1]:

- *Low-pass* (LP)
- *High-pass* (HP)
- *Band-pass* (BP)
- *Band-stop* (BS).

The terms *band-reject* or *notch* are also used as descriptive of the BS filter. The term *all-pass* is sometimes applied to a filter whose purpose is to alter the phase angle without affecting the magnitude of the frequency response. Ideal and practical interpretations of the types of filters and the associated terminology are illustrated in Figure 7.19.

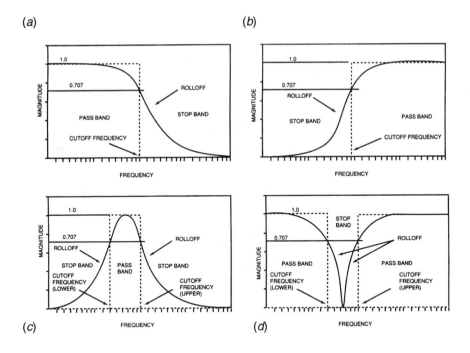

Figure 7.19 Filter characteristics by type: (a) low-pass, (b) high-pass, (c) bandpass, (d) bandstop. (*From* [1]. *Used with permission.*)

In general, the voltage gain of a filter in the *stop band* (or *attenuation band*) is less than $\sqrt{2}/2$ (≈ 0.707) times the maximum voltage gain in the pass band. In logarithmic terms, the gain in the stop band is at least 3.01 dB less than the maximum gain in the pass band. The cutoff (*break* or *corner*) frequency separates the pass band from the stop band. In BP and BS filters, there are two cutoff frequencies, sometimes referred to as the *lower* and *upper* cutoff frequencies. Another expression for the cutoff frequency is *half-power frequency*, because the power delivered to a resistive load at cutoff frequency is one-half the maximum power delivered to the same load in the pass band. For BP and BS filters, the center frequency is the frequency of maximum or minimum response magnitude, respectively, and bandwidth is the difference between the upper and lower cutoff frequencies. *Rolloff* is the transition from pass band to stop band and is specified in gain unit per frequency unit (e.g., gain unit/Hz, dB/decade, dB/octave, etc.)

Filter Alignment

The *alignment* (or class) of a filter refers to the shape of the frequency response [1]. Fundamentally, filter alignment is determined by the coefficients of the filter network

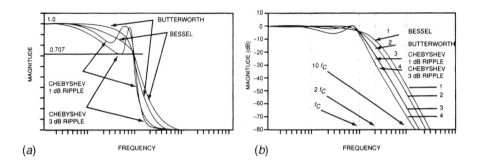

Figure 7.20 Filter characteristics by alignment, third-order, all-pole filters: (*a*) magnitude, (*b*) magnitude in decibels. (*From* [1]. *Used with permission.*)

transfer function, so there are an indefinite number of filter alignments, some of which may not be realizable. The more common alignments are:

- Butterworth
- Chebyshev
- Bessel
- Inverse Chebyshev
- Elliptic (or Cauer)

Each filter alignment has a frequency response with a characteristic shape, which provides some particular advantage. (See Figure 7.20.) Filters with Butterworth, Chebyshev, or Bessel alignment are called *all-pole filters* because their low-pass transfer functions have no zeros. Table 7.3 summarizes the characteristics of the standard filter alignments.

Filter Order

The *order* of a filter is equal to the number of poles in the filter network transfer function [1]. For a lossless *LC* filter with resistive (nonreactive) termination, the number of reactive elements (inductors or capacitors) required to realize a LP or HP filter is equal to the order of the filter. Twice the number of reactive elements are required to realize a BP or a BS filter of the same order. In general, the order of a filter determines the slope of the rolloff—the higher order, the steeper the rolloff. At frequencies greater than approximately one octave above cutoff (i.e., $f > 2f_c$), the rolloff for all-pole filters is $20n$ dB/decade (or approximately $6n$ dB/octave), where n is the

Table 7.3 Summary of Standard Filter Alignments (*After* [1].)

Alignment	Pass Band Description	Stop Band Description	Comments
Butterworth	Monotonic	Monotonic	All-pole; maximally flat
Chebyshev	Rippled	Monotonic	All-pole
Bessel	Monotonic	Monotonic	All-pole; constant phase shift
Inverse Chebyshev	Monotonic	Rippled	
Elliptic (or Cauer)	Rippled	Rippled	

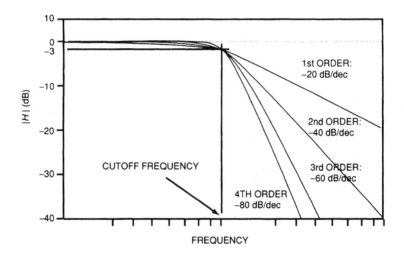

Figure 7.21 The effects of filter order on rolloff (Butterworth alignment). (*From* [1]. *Used with permission.*)

order of the filter (Figure 7.21). In the vicinity of f_c, both filter alignment and filter order determine rolloff.

7.4.2 Four-Port Hybrid Combiner

A hybrid combiner (coupler) is a reciprocal four-port device that can be used for either splitting or combining RF energy over a wide range of frequencies. An exploded view of a typical 3 dB 90° hybrid is illustrated in Figure 7.22. The device consists of two identical parallel transmission lines coupled over a distance of approximately one-quarter wavelength and enclosed within a single outer conductor. Ports at the

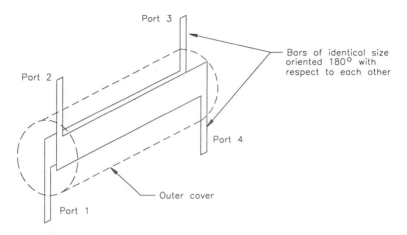

Port 3

Bars of identical size
oriented 180° with
respect to each other

Port 2

Port 4

Outer cover

Port 1

Figure 7.22 Physical model of a 90° hybrid combiner.

same end of the coupler are in phase, and ports at the opposite end of the coupler are in quadrature (90° phase shift) with respect to each other.

The phase shift between the two inputs or outputs is always 90° and is virtually independent of frequency. If the coupler is being used to combine two signals into one output, these two signals must be fed to the hybrid in phase quadrature. When the coupler is used as a power splitter, the division is equal (half-power between the two ports). The hybrid presents a constant impedance to match each source.

Operation of the combiner can best be understood through observation of the device in a practical application. Figure 7.23 shows a four-port hybrid combiner used to add the outputs of two transmitters to feed a single load. The combiner accepts one RF source and splits it equally into two parts. One part arrives at output port *C* with 0° phase (no phase delay; it is the *reference phase*). The other part is delayed by 90° at port *D*. A second RF source connected to input port *B*, but with a phase delay of 90°, also will split in two, but the signal arriving at port *C* now will be in phase with source 1, and the signal arriving at port *D* will cancel, as shown in the figure.

Output port *C*, the summing point of the hybrid, is connected to the load. Output port *D* is connected to a resistive load to absorb any residual power resulting from slight differences in amplitude and/or phase between the two input sources. If one of the RF inputs fails, half of the remaining transmitter output will be absorbed by the resistive load at port *D*.

The four-port hybrid works only when the two signals being mixed are identical in frequency and amplitude, and when their relative phase is 90°.

Operation of the hybrid can best be described by a *scattering matrix* in which vectors are used to show how the device operates. Such a matrix is shown in Table 7.4. In a 3 dB hybrid, two signals are fed to the inputs. An input signal at port 1 with 0° phase will arrive in phase at port 3, and at port 4 with a 90° lag (–90°) referenced to port 1. If the signal at port 2 already contains a 90° lag (–90° referenced to port 1), both input signals will combine in phase at port 4. The signal from port 2 also experiences another 90°

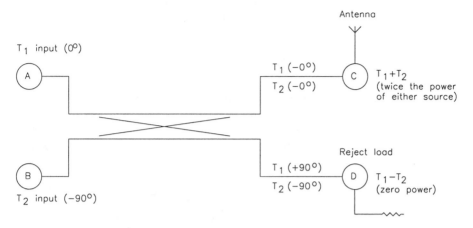

Figure 7.23 Operating principles of a hybrid combiner. This circuit is used to add two identical signals at inputs A and B.

change in the hybrid as it reaches port 3. Therefore, the signals from ports 1 and 2 cancel each other at port 3.

If the signal arriving at port 2 leads by 90° (mode 1 in the table), the combined power from ports 1 and 2 appears at port 4. If the two input signals are matched in phase (mode 4), the output ports (3 and 4) contain one-half of the power from each of the inputs.

If one of the inputs is removed, which would occur in a transmitter failure, only one hybrid input receives power (mode 5). Each output port then would receive one-half the input power of the remaining transmitter, as shown.

The input ports present a predictable load to each amplifier with a VSWR that is lower than the VSWR at the output port of the combiner. This characteristic results from the action of the difference port, typically connected to a dummy load. Reflected power coming into the output port will be directed to the reject load, and only a portion will be fed back to the amplifiers. Figure 7.24 illustrates the effect of output port VSWR on input port VSWR, and on the isolation between ports.

As noted previously, if the two inputs from the separate amplifiers are not equal in amplitude and not exactly in phase quadrature, some power will be dissipated in the difference port reject load. Figure 7.25 plots the effect of power imbalance, and Figure 7.26 plots the effects of phase imbalance. The power lost in the reject load can be reduced to a negligible value by trimming the amplitude and/or phase of one (or both) amplifiers.

7.4.3 Non-Constant-Impedance Diplexer

Diplexers are used to combine amplifiers operating on different frequencies (and at different power levels) into a single output. Such systems typically are utilized to sum different transmitter outputs to feed a single broadband antenna.

Table 7.4 Single 90° Hybrid System Operating Modes

MODE	INPUT		SCHEMATIC	OUTPUT	
	1	2		3	4
1	$P_1 / 0°$	$P_2 \angle -90°$	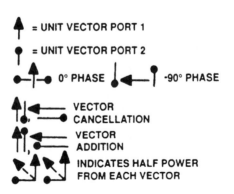	0	$P_1 + P_2$
2	$P_1 / 0°$	$P_2 / 90°$		$P_1 + P_2$	0
3	$P_1 / 0°$	$P_2 / 0°$		$P_{1/2} + P_{2/2}$	$P_{1/2} + P_{2/2}$
4	$P_1 / 0°$	$P_2 = 0$		$P_{1/2}$	$P_{1/2}$
5	$P_1 = 0$	$P_2 / 0°$		$P_{2/2}$	$P_{2/2}$

↑ = UNIT VECTOR PORT 1

● = UNIT VECTOR PORT 2

●—↑—● 0° PHASE   ↓←● -90° PHASE

VECTOR CANCELLATION

VECTOR ADDITION

INDICATES HALF POWER FROM EACH VECTOR

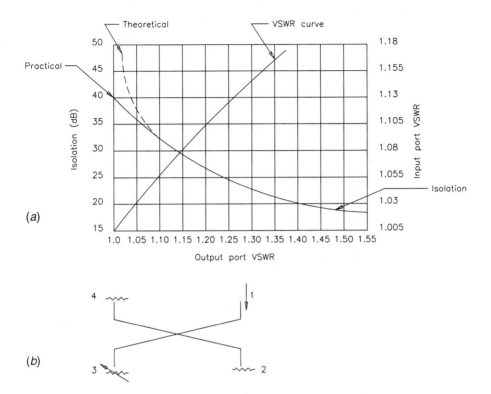

Figure 7.24 The effects of load VSWR on input VSWR and isolation: (*a*) respective curves, (*b*) coupler schematic.

The *branch diplexer* is the typical configuration for a diplexer that does not exhibit constant-impedance inputs. As shown in Figure 7.27, the branch diplexer consists of two banks of filters each feeding into a coaxial tee. The electrical length between each filter output and the centerline of the tee is frequency-sensitive, but this fact is more of a tuning nuisance than a genuine user concern.

For this type of diplexer, all of the electrical parameters are a function of the filter characteristics. The VSWR, insertion loss, group delay, and rejection/isolation will be the same for the overall system as they are for the individual banks of cavities. The major limitation of this type of combiner is the degree of isolation that can be obtained for closely spaced channels.

7.4.4 Constant-Impedance Diplexer

The constant-impedance diplexer employs 3 dB hybrids and filters with a terminating load on the isolated port. The filters in this type of combiner can be either notch-type or bandpass-type. The performance characteristics are noticeably different for each design.

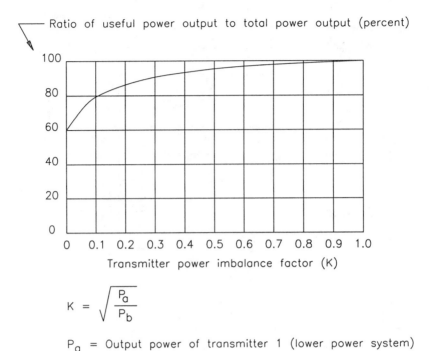

$$K = \sqrt{\frac{P_a}{P_b}}$$

P_a = Output power of transmitter 1 (lower power system)
P_b = Output power of transmitter 2

Figure 7.25 The effects of power imbalance at the inputs of a hybrid coupler.

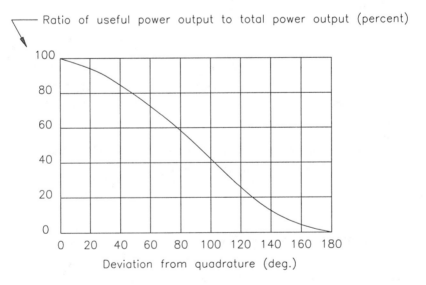

Figure 7.26 Phase sensitivity of a hybrid coupler.

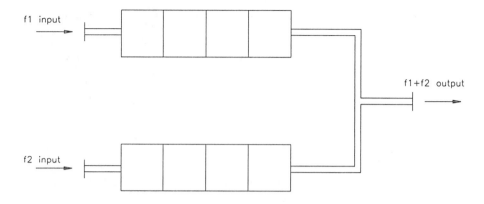

Figure 7.27 Non-constant-impedance branch diplexer. In this configuration, two banks of filters feed into a coaxial tee.

Band-Stop Diplexer

The band-stop (notch) constant-impedance diplexer is configured as shown in Figure 7.28. For this design, the notch filters must have a high Q response to keep insertion loss low in the passband skirts. The high Q characteristic results in a sharp notch. Depending on the bandwidth required of the diplexer, two or more cavities may be located in each leg of the diplexer. They are typically stagger tuned, one high and one low for the two-cavity case. With this dual-cavity reject response in each leg of the band-stop diplexer system, the following analysis explains the key performance specifications.

If frequency f_1 is fed into the top left port of Figure 7.28, it will be split equally into the upper and lower legs of the diplexer. Both of these signals will reach the filters in their respective leg and be rejected/reflected back toward the input hybrid, recombine, and emerge through the lower left port, also known as the *wideband output*.

The VSWR looking into the f_1 input is near 1:1 at all frequencies in the band. Within the bandwidth of the reject skirts, the observed VSWR is equal to the termination of the wideband output. Outside of the passband, the signals will pass by the cavities, enter the rightmost hybrid, recombine, and emerge into the dummy load. Consequently, the out-of-band VSWR is, in fact, the VSWR of the load.

The insertion loss from the f_1 input to the wideband output is low, typically on the order of 0.1 dB at carrier. This insertion loss depends on perfect reflection from the cavities. As the rejection diminishes on the skirts of the filters, the insertion loss from f_1 to the wideband output increases.

The limitation in reject bandwidth of the cavities causes the insertion loss to rise at the edges of the passband. The isolation from f_1 to the wideband input consists of a combination of the reject value of the cavities plus the isolation of the rightmost hybrid.

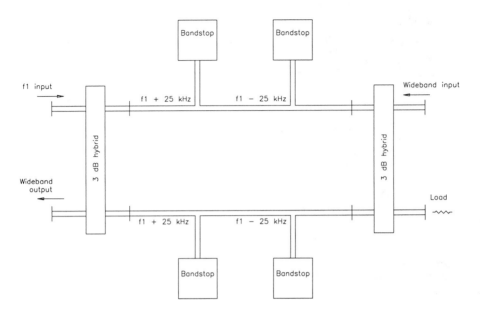

Figure 7.28 Band-stop (notch) constant-impedance diplexer module. This design incorporates two 3 dB hybrids and filters, with a terminating load on the isolated port.

A signal entering at f_1 splits and proceeds in equal halves rightward through both the upper and lower legs of the diplexer. It is rejected by the filters at carrier and, to a lesser extent, on both sides of the carrier. Any residual signal that gets by the cavities reaches the rightmost hybrid. There it recombines and emerges from the *load port*. The hybrid provides a specified isolation from the load port to the wideband input port. This hybrid isolation must be added to the filter rejection to obtain the total isolation from the f_1 input to the wideband input.

If a signal is fed into the wideband input of the combiner shown in Figure 7.28, the energy will split equally and proceed leftward along the upper and lower legs of the diplexer. Normally, f_1 is not fed into the wideband input. If it were, f_1 would be rejected by the notch filters and recombine into the load. All other signals sufficiently removed from f_1 will pass by the cavities with minimal insertion loss and recombine into the wideband output.

The VSWR looking into the wideband input is equal to the VSWR at the output for frequencies other than f_1. If f_1 were fed into the wideband input, the VSWR would be equal to the VSWR of the load.

Isolation from the wideband input to the f_1 input is simply the isolation available in the leftmost 3 dB hybrid. This isolation is usually inadequate for high-power applications. To increase the isolation from the wideband input to the f_1 input, it is necessary to use additional cavities between the f_1 transmitter and the f_1 input that will reject the frequencies fed into the wideband input. Unfortunately, adding these cavities to the input

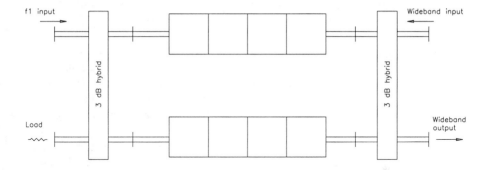

Figure 7.29 Bandpass constant-impedance diplexer.

line also cancels the constant-impedance input. Although the notch diplexer as shown in the figure is truly a constant-impedance type of diplexer, the constant impedance is presented to the two inputs by virtue of using the hybrids at the respective inputs.

The hybrids essentially cause the diplexer to act as an absorptive type of filter to all out-of-band signals. The out-of-band signals generated by the transmitter are absorbed by the load rather than reflected to the transmitter.

If a filter is added to the input to supplement isolation, the filter will reflect some out-of-band signals back at the transmitter. The transmitter then will be seeing the passband impedance of this supplemental filter rather than the constant impedance of the notch diplexer. This is a serious deficiency for applications that require a true constant-impedance input.

Bandpass Constant-Impedance Diplexer

The bandpass constant-impedance diplexer is shown in Figure 7.29. This system takes all of the best features of diplexers and combines them into one unit. It also provides a constant-impedance input that need not be supplemented with input cavities that rob the diplexer of its constant-impedance input.

The bandpass filters exhibit good bandwidth, providing near 1:1 VSWR across the operating bandpass. Insertion is low at carrier (0.28 dB is typical), rising slightly at bandpass extremes. Diplexer rejection, when supplemented by isolation of the hybrids, provides ample transmitter-to-transmitter isolation. Group delay is typically exceptional, providing performance specifications similar to those of a branch-style bandpass system. This configuration has the additional capability of providing high port-to-port isolation between closely spaced operating channels, as well as a true constant-impedance input.

The hybrids shown in Figure 7.29 work in a manner identical to those described for a band-stop diplexer. However, the bandpass filters cause the system to exhibit performance specifications that exceed the band-stop system in every way. Consider a signal entering at the f_1 input.

Within the passbands of the filters, which are tuned to f_1, the VSWR will be near 1:1 at carrier, rising slightly at the bandpass extremes. Because of the characteristic of the leftmost hybrid, the VSWR is, in fact, a measure of the similarity of response of the top and bottom bands of filters. The insertion loss looking from the f_1 input to the wideband output will be similar to the insertion loss of the top and bottom filters individually. A value of approximately 0.28 dB at carrier is typical.

Both the insertion loss and group delay can be determined by the design bandwidth of the filters. Increasing bandwidth causes the insertion loss and group delay deviation to decrease. Unfortunately, as the bandwidth increases with a given number of cavities, the isolation suffers for closely spaced channels because the reject skirt of the filter decreases with increasing bandwidth.

Isolation of f_1 to the wideband input is determined as follows: A signal enters at the f_1 input, splits equally into the upper and lower banks of filters, passes with minimal loss through the filters, and recombines into the wideband output of the rightmost hybrid. Both the load and the wideband input ports are isolated by their respective hybrids to some given value below the f_1 input level. Isolation of the f_1 input to the wideband input is supplemented by the reject skirt of the next module.

A signal fed into the wideband input could be any frequency removed from f_1 by some minimal amount. As the wideband signal enters, it will be split into equal halves by the hybrid, then proceed to the left until the two components reach the reject skirts of the filters. The filters will shunt all frequencies removed from f_1 by some minimal amount. If the shunt energy is in phase for the given frequency when the signal is reflected back to the right hybrid, it will recombine into the wideband output. The VSWR under these conditions will be equal to the termination at the wideband output. If the reject skirts of the filter are sufficient, the insertion loss from wideband input to wideband output will be minimal (0.03 dB is typical).

The isolation from the wideband input to f_1 can be determined as follows: A signal enters at the wideband input, splits equally into upper and lower filters, and is rejected by the filters. Any residual signal that passes through the filters in spite of the rejection still will be in the proper phase to recombine into the load, producing additional isolation to the f_1 input port. Thus, the isolation of the wideband input to the f_1 input is the sum of the rejection from the filters and from the left hybrid.

Isolation of f_1 to f_2

Extending the use of the diplexer module into a multiplexer application supplements the deficient isolation described previously (narrowband input performance), while maintaining the constant-impedance input. In a multiplexer system, the wideband input of one module is connected to the wideband output of the next module, as illustrated in Figure 7.30.

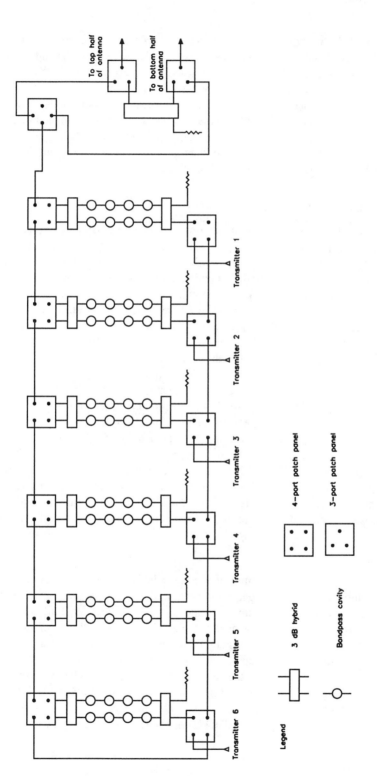

Figure 7.30 Schematic diagram of a six-module bandpass multiplexer. This configuration accommodates a split antenna design and incorporates patch panels for bypass purposes.

It has already been stated that the isolation from the f_1 input to the wideband input is deficient, but additional isolation is provided by the isolation of the wideband output to the f_2 input of the next module. Consider that f_1 has already experienced 30 dB isolation to the wideband input of the same module. When this signal continues to the next module through the wideband output of module 2, it will be split into equal halves and proceed to the left of module 2 until it reaches the reject skirts of the filters in module 2. Assume that these filters are tuned to f_2 and reject f_1 by at least 25 dB. The combined total isolation of f_1 to f_2 is the sum of the 30 dB of the right hybrid in module 1, plus the 25 dB of the reject skirts of module 2, for a total of 55 dB.

Intermodulation Products

The isolation just described is equal in magnitude to that for a band-stop module, but it provides further protection against the generation of intermodulation products. The most troublesome intermodulation (*intermod*) products usually occur when an incoming (secondary) signal mixes with the second harmonic of a primary transmitter. (The properties of intermodulation are discussed in more detail in Chapter 10.)

When the primary transmitter is operating on frequency A, the intermod will occur at the frequency $(2 \times A) - B$. This formula invariably places the intermod from the primary transmitter symmetrically about the operating frequency. By an interesting coincidence, the bandpass filters in the bandpass module also provide symmetrical reject response on both sides of the primary operating frequency.

Assume again that the incoming signal is attenuated by 30 dB in the respective hybrid and by 25 dB in the filter, for a total of 55 dB. If an intermod still is generated in spite of this isolation, it will emerge on the other skirt of the filter attenuated by 25 dB. In the bandpass system, the incoming signal is attenuated by 55 dB and the resulting outgoing spur by 25 dB, for a total of 80 dB suppression.

Interestingly, the entire 80 dB of attenuation is supplied by the diplexer regardless of the *turnaround loss* of the transmitter. The tendency toward wideband final stage amplifiers in transmitters requires constant-impedance inputs. The transmitters also require increased isolation because they offer limited turnaround loss.

Group Delay

Group delay in the bandpass multiplexer module is equal to the sum of the narrowband input group delay and the wideband input group delay of all modules between the input and the load (antenna). (Group delay is discussed in detail in Chapter 10.) The narrowband input group delay is a U-shaped response, with minimum at center and rising to a maximum on both sides at the frequency where the reject rises to 3 dB. Group delay then decreases rapidly at first, then more slowly.

If the bandwidth of the pass band is made such that the group delay is ±25 ns over ±150 kHz (in an example system operating near 100 MHz), the 3 dB points will be at ±400 kHz and the out-of-band group delay will fall rapidly at ±800 kHz and possibly ±1.0 MHz. If there are no frequencies 800 kHz or 1.0 MHz removed upstream in other modules, this poses no problem.

If modules upstream are tuned to 800 kHz or 1.0 MHz on either side, then the group delay (when viewed at the upstream module) will consist of its own narrowband input group delay plus the rapidly falling group delay of the wideband input of the closely spaced downstream module. Under these circumstances, if good group delay is desired, it is possible to utilize a *group delay compensation* module.

A group delay compensation module consists of a hybrid and two cavities used as notch cavities. It typically provides a group delay response that is inverted, compared with a narrowband input group delay. Because group delay is additive, the inverted response subtracts from the standard response, effectively reducing the group delay deviation.

It should be noted that the improvement in group delay is obtained at a cost of insertion loss. In large systems (8 to 10 modules), the insertion loss can be high because of the cumulative total of all wideband losses. Under these conditions, it may be more prudent to accept higher group delay and retain minimal insertion loss.

7.4.5 Microwave Combiners

Hybrid combiners typically are used in microwave amplifiers to add the output energy of individual power modules to provide the necessary output from an RF generator. Quadrature hybrids effect a VSWR-canceling phenomenon that results in well-matched power amplifier inputs and outputs that can be broadbanded with proper selection of hybrid tees. Several hybrid configurations are possible, including the following:

- *Split-tee*

- *Branch-line*

- *Magic-tee*

- *Backward-wave*

Key design parameters include coupling bandwidth, isolation, and ease of fabrication. The equal-amplitude quadrature-phase reverse-coupled TEM $\frac{1}{4}$-wave hybrid is particularly attractive because of its bandwidth and amenability to various physical implementations. Such a device is illustrated in Figure 7.31.

7.4.6 Hot Switching Combiners

Switching RF is nothing new. Typically, the process involves coaxial switches, coupled with the necessary logic to ensure that the "switch" takes place with no RF energy on the contacts. This process usually takes the system off-line for a few seconds while the switch is completed. Through the use of hybrid combiners, however, it is possible to redirect RF signals without turning the carrier off. This process is referred to as *hot switching*. Figure 7.32 illustrates two of the most common switching functions (SPST and DPDT) available from hot switchers.

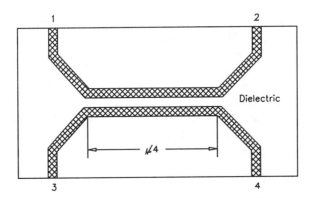

Figure 7.31 Reverse-coupled ¼-wave hybrid combiner.

The unique phase-related properties of an RF hybrid make it possible to use the device as a switch. The input signals to the hybrid in Figure 7.33*a* are equally powered but differ in phase by 90°. This phase difference results in the combined signals being routed to the output terminal at port 4. If the relative phase between the two input signals is changed by 180°, the summed output then appears on port 3, as shown in Figure 7.33*b*. The 3 dB hybrid combiner, thus, functions as a switch.

This configuration permits the switching of two RF generators to either of two loads. Remember, however, that the switch takes place when the phase difference between the two inputs is 90°. To perform the switch in a useful way requires adding a high-power phase shifter to one input leg of the hybrid. The addition of the phase shifter permits the full power to be combined and switched to either output. This configuration of hybrid and phase shifter, however, will not permit switching a main or standby generator to a main or auxiliary load (DPDT function). To accomplish this additional switch, a second hybrid and phase shifter must be added, as shown in Figure 7.34. This configuration then can perform the following switching functions:

- RF source 1 routed to output *B*
- RF source 2 routed to output *A*
- RF source 1 routed to output *A*
- RF source 2 routed to output *B*

The key element in developing such a switch is a high-power phase shifter that does not exhibit reflection characteristics. In this application, the phase shifter allows the line between the hybrids to be electrically lengthened or shortened. The ability to adjust the relative phase between the two input signals to the second hybrid provides the needed control to switch the input signal between the two output ports.

If a continuous analog phase shifter is used, the transfer switch shown in Figure 7.34 also can act as a hot switchless combiner where RF generators 1 and 2 can be combined

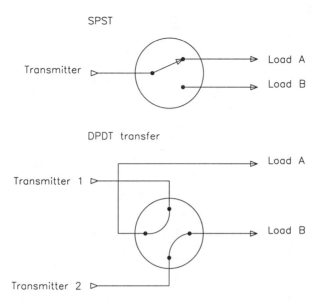

Figure 7.32 Common RF switching configurations.

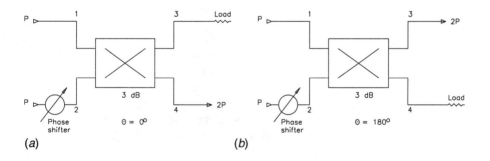

Figure 7.33 Hybrid switching configurations: (*a*) phase set so that the combined energy is delivered to port 4, (*b*) phase set so that the combined energy is delivered to port 3.

and fed to either output *A* or *B*. The switching or combining functions are accomplished by changing the physical position of the phase shifter.

Note that it does not matter whether the phase shifter is in one or both legs of the system. It is the phase difference ($\theta 1 - \theta 2$) between the two input legs of the second hybrid that is important. With 2-phase shifters, dual drives are required. However, the phase shifter needs only two positions. In a 1-phase shifter design, only a single drive is required, but the phase shifter must have four fixed operating positions.

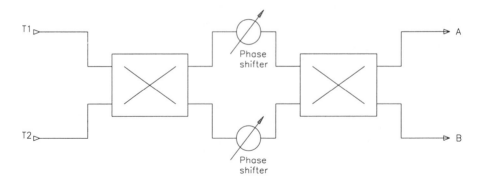

Figure 7.34 Additional switching and combining functions enabled by adding a second hybrid and another phase shifter to a hot switching combiner.

Phase Relationships

To better understand the dual-hybrid switching and combining process, it is necessary to examine the primary switching combinations. Table 7.5 lists the various combinations of inputs, relative phase, and output configurations that are possible with the single-phase shifter design.

Using vector analysis, note that when two input signals arrive in phase (mode 1) at ports 1 and 2 with the phase shifter set to 0°, the circuit acts like a crossover network with the power from input port 1 routed to output port 4. Power from input port 2 is routed to output port 3. If the phase shifter is set to 180°, the routing changes, with port 1 being routed to port 3 and port 2 being routed to port 4.

Mode 2 represents the case where one of the dual-input RF generators has failed. The output signal from the first hybrid arrives at the input to the second hybrid with a 90° phase difference. Because the second hybrid introduces a 90° phase shift, the vectors add at port 4 and cancel at port 3. This effectively switches the working transmitter connected to port 1 to output port 4, the load.

By introducing a 180° phase shift between the hybrids, as shown in modes 4 and 5, it is possible to reverse the circuit. This allows the outputs to be on the same side of the circuit as the inputs. This configuration might be useful if generator 1 fails, and all power from generator 2 is directed to a diplexer connected to output 4.

Normal operating configurations are shown in modes 6 and 7. When both generators are running, it is possible to have the combined power routed to either output port. The switching is accomplished by introducing a ±90° phase shift between the hybrids.

As shown in the table, it is possible to operate in all the listed modes through the use of a single-phase shifter. The phase shifter must provide four different phase positions. A similar analysis would show that a 2-phase shifter design, with two positions for each shifter, is capable of providing the same operational modes.

Table 7.5 Operating Modes of the Dual 90° Hybrid/Single-Phase Shifter Combiner System

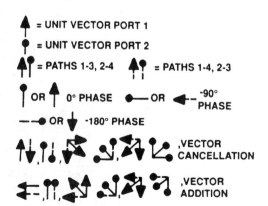

MODE	INPUT		ϕ	INPUT VECTOR	SCHEMATIC	OUTPUT VECTOR	OUTPUT	
	1	2					3	4
1	$P_1 / 0°$	$P_2 / 0°$	0°				P_2	P_1
	$P_1 / 0°$	$P_2 / 0°$	180°				P_1	P_2
2	$P_1 / 0°$	$P_2 = 0$	0°				0	P_1
3	$P_1 = 0$	$P_2 / 0°$	0°				P_2	0
4	$P_1 / 0°$	$P_2 = 0$	180°				P_1	0
5	$P_1 = 0$	$P_2 / 0°$	180°				0	P_2
6	$P_1 / 0°$	$P_2 / 0°$	90°				0	$P_1 + P_2$
	$P_1 / 0°$	$P_2 / 0°$	-90°				$P_1 + P_2$	0

= UNIT VECTOR PORT 1

= UNIT VECTOR PORT 2

= PATHS 1-3, 2-4   = PATHS 1-4, 2-3

OR   0° PHASE   OR   -90° PHASE

OR   -180° PHASE

, VECTOR CANCELLATION

, VECTOR ADDITION

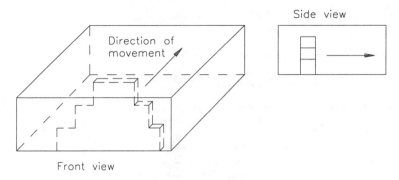

Figure 7.35 The dielectric vane switcher, which consists of a long dielectric sheet mounted within a section of rectangular waveguide.

The key to making hybrid switches work in the real world lies in the phase shifter. The dual 90° hybrid combiner just discussed requires a phase shifter capable of introducing a fixed phase offset of −90°, 0°, +90°, and +180°. This can be accomplished easily at low power levels through the use of a sliding short-circuit (trombone-type) line stretcher. However, when high-frequency and high-power signals are being used, the sliding short circuit is not an appropriate design choice. In a typical case, the phase shifter must be able to handle 100 kW or more at UHF. Under these conditions, sliding short-circuit designs are often unreliable. Therefore, three other methods have been developed:

- *Variable-dielectric vane*
- *Dielectric post*
- *Variable-phase hybrid*

Variable-Dielectric Vane

The variable-dielectric vane consists of a long dielectric sheet mounted in a section of rectangular waveguide, as illustrated in Figure 7.35. The dielectric sheet is long enough to introduce a 270° phase shift when located in the center of the waveguide. As the dielectric sheet is moved toward the wall, into the lower field, the phase shift decreases. A single-sided phase shifter easily can provide the needed four positions. A 2-stage ¼-wave transformer is used on each end of the sheet to maintain a proper match for any position over the desired operating band. The performance of a typical switchless combiner, using the dielectric vane, is given in Table 7.6.

Dielectric Posts

Dielectric posts, shown in Figure 7.36, operate on the same principle as the dielectric vane. The dielectric posts are positioned ¼-wavelength apart from each other to cancel any mismatch, and to maintain minimal VSWR.

Table 7.6 Typical Performance of a Dielectric Vane Phase Shifter

Type	Input	Phase Change (deg)	VSWR	Input Attenuation		Output Attenuation	
				1 (dB)	2 (dB)	3 (dB)	4 (dB)
Single input	T1	180	1.06	-	39	0.1	39
	T1	0	1.05	-	39	39	0.1
	T2	180	1.05	39	-	39	0.1
	T2	0	1.06	39	-	0.1	39
Dual input	T1+T2	270	1.06	-	-	0.1	36
	T1+T2	90	1.06	-	-	36	0.1

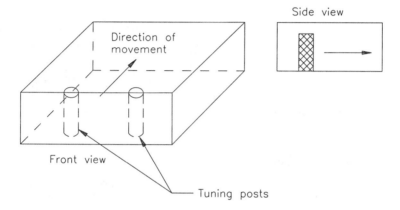

Figure 7.36 Dielectric post waveguide phase shifter.

Variable-Phase Hybrid

The variable-phase hybrid, shown in Figure 7.37, relies on a 90° hybrid, similar to those used in a combiner. With a unit vector incident on port 1, the power is split by the 90° hybrid. The signal at ports 3 and 4 is reflected by the short circuit. These reflected signals are out of phase at port 1 and in phase at port 2. The relative phase of the hybrid can be changed by moving the short circuit.

The variable-phase hybrid is linear with respect to position. Noncontacting choke-type short-circuits, with high front-to-back ratios, are typically used in the device. The performance available from a typical high-power variable-phase switchless combiner is given in Table 7.7.

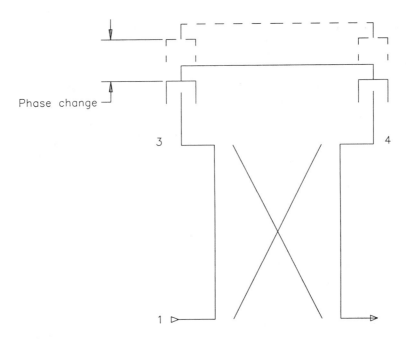

Figure 7.37 Variable-phase hybrid phase shifter.

7.4.7 Phased-Array Antenna Systems

The phased-array antenna, commonly used in radar applications, is a variation on the hot switching combiner systems discussed in the previous section. The hot switching combiner is used to control the flow of RF energy from one or more generators to two or more loads, while the phased-array antenna is used to modify—in a controlled fashion—the radiating characteristics of an antenna.

Phased-array antennas are *steered* by tilting the phase front independently in two orthogonal directions called the *array coordinates*. Scanning in either array coordinate causes the beam to move along a cone whose center is at the center of the array. As the beam is steered away from the array normal, the projected aperture in the beam's direction varies, causing the beamwidth to vary proportionally.

Arrays can be classified as either active or passive. Active arrays contain duplexers and amplifiers behind every element or group of elements of the array. Passive arrays are driven from a single feed point. Active arrays are capable of high-power operation. Both passive and active arrays must divide the signal from a single transmission line among all the elements of the system. This can be accomplished through one of the following methods:

- *Optical feed*: a single feed, usually a monopulse horn, is used to illuminate the array with a spherical phase front, illustrated in Figure 7.38. Power collected by the rear elements of the array is transmitted through the phase shifters to produce a planar front and steer the array. The energy then may be radiated from the other

Table 7.7 Typical Performance of a Variable-Phase Hybrid Phase Shifter

Type	Input	Phase Change (deg)	VSWR	Input Attenuation		Output Attenuation	
				1 (dB)	2 (dB)	3 (dB)	4 (dB)
Single input	T1	180	1.06	-	36	0.1	52
	T1	0	1.04	-	36	50	0.1
	T2	180	1.06	36	-	52	0.1
	T2	0	1.07	36	-	0.1	50
Dual input	T1+T2	270	1.06	-	-	0.1	36
	T1+T2	90	1.06	-	-	36	0.1

side of the array or reflected and reradiated through the collecting elements. In the latter case, the array acts as a *steerable reflector*.

- *Corporate feed*: a system utilizing a series-feed network (Figure 7.39) or parallel-feed network (Figure 7.40). Both designs use transmission-line components to divide the signal among the elements. Phase shifters can be located at the elements or within the dividing network. Both the series- and parallel-feed systems have several variations, as shown in the figures.

- *Multiple-beam network*: a system capable of forming simultaneous beams with a given array. The *Butler matrix*, shown in Figure 7.41, is one such technique. It connects the N elements of a linear array to N feed points corresponding to N beam outputs. The phase shifter is one of the most critical components of the system. It produces controllable phase shift over the operating band of the array. Digital and analog phase shifters have been developed using both ferrites and *pin* diodes.

Frequency scan is another type of multiple-beam network, but one that does not require phase shifters, dividers, or beam-steering computers. Element signals are coupled from points along a transmission line. The electrical path length between elements is longer than the physical separation, so a small frequency change will cause a phase change between elements that is large enough to steer the beam. This technique can be applied only to one array coordinate. If a two-dimensional array is required, phase shifters normally are used to scan the other coordinate.

Phase-Shift Devices

The design of a phase shifter must meet two primary criteria:

- Low transmission loss
- High power-handling capability

(a)

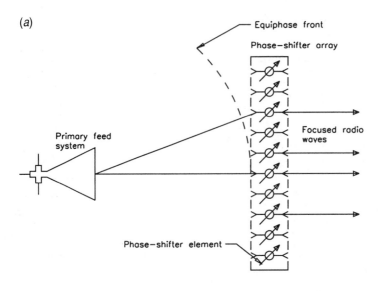

(b)

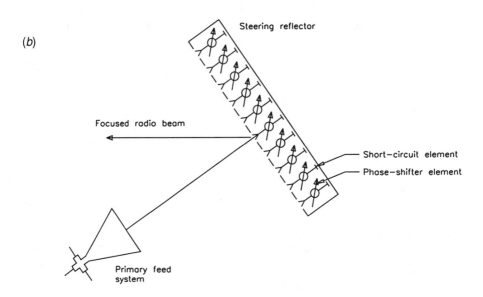

Figure 7.38 Optical antenna feed systems: (a) lens, (b) reflector.

The *Reggia-Spencer* phase shifter meets both requirements. The device, illustrated in Figure 7.42, consists of a ferrite rod mounted inside a waveguide that delays the RF signal passing through the waveguide, permitting the array to be steered. The amount of phase shift can be controlled by the current in the solenoid, because of the effect a

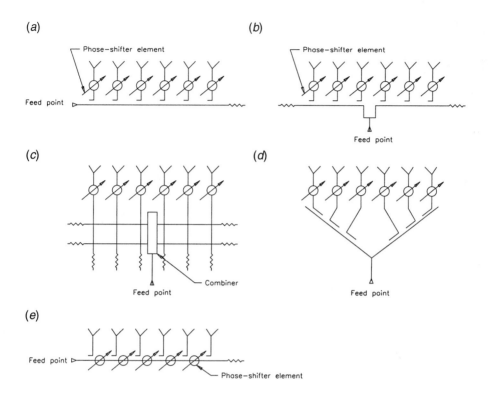

Figure 7.39 Series-feed networks: (*a*) end feed, (*b*) center feed, (*c*) separate optimization, (*d*) equal path length, (*e*) series phase shifters.

magnetic field has on the permeability of the ferrite. This design is a *reciprocal phase shifter*, meaning that the device exhibits the same phase shift for signals passing in either direction (forward or reverse). Nonreciprocal phase shifters also are available, where phase-shift polarity reverses with the direction of propagation.

Phase shifters also may be developed using pin diodes in transmission line networks. One configuration, shown in Figure 7.43, uses diodes as switches to change the signal path length of the network. A second type uses pin diodes as switches to connect reactive loads across a transmission line. When equal loads are connected with ¼-wave separation, a pure phase shift results.

Radar System Duplexer

The duplexer is an essential component of any radar system. The switching elements used in a duplexer include gas tubes, ferrite circulators, and pin diodes. Gas tubes are the simplest. A typical gas-filled *TR tube* is shown in Figure 7.44. Low-power RF sig-

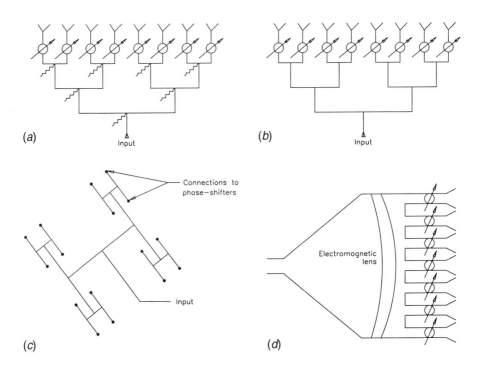

Figure 7.40 Types of parallel-feed networks: (a) matched corporate feed, (b) reactive corporate feed, (c) reactive stripline, (d) multiple reactive divider.

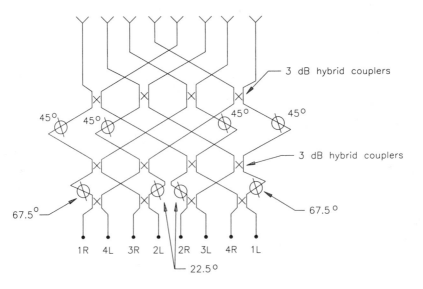

Figure 7.41 The Butler beam-forming network.

End view

Perspective cutaway view

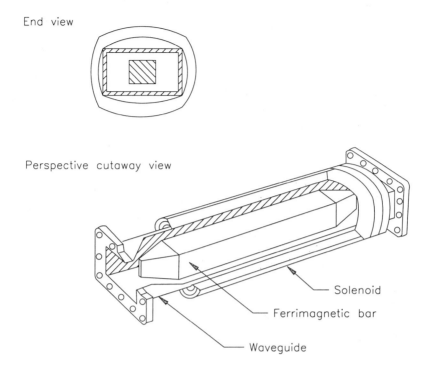

Solenoid

Ferrimagnetic bar

Waveguide

Figure 7.42 Basic concept of a Reggia-Spencer phase shifter.

nals pass through the tube with little attenuation. Higher-power signals, however, cause the gas to ionize and present a short circuit to the RF energy.

Figure 7.45 illustrates a *balanced duplexer* using hybrid junctions and TR tubes. When the transmitter is on, the TR tubes fire and reflect the RF power to the antenna port of the input hybrid. During the receive portion of the radar function, signals picked up by the antenna are passed through the TR tubes and on to the receiver port of the output hybrid.

Newer radar systems often use a ferrite circulator as the duplexer. A TR tube is required in the receiver line to protect input circuits from transmitter power reflected by the antenna because of an imperfect match. A four-port circulator generally is used with a load between the transmitter and receiver ports so that power reflected by the TR tube is properly terminated.

Pin diode switches also have been used in duplexers to perform the protective switching function of TR tubes. Pin diodes are more easily applied in coaxial circuitry, and at lower microwave frequencies. Multiple diodes are used when a single diode cannot withstand the expected power.

Microwave filters sometimes are used in the transmit path of a radar system to suppress spurious radiation, or in the receive signal path to suppress spurious interference.

Harmonic filters commonly are used in the transmission chain to absorb harmonic energy output by the system, preventing it from being radiated or reflected back from

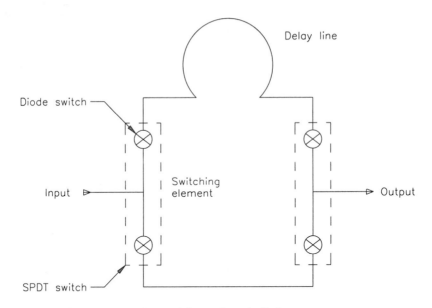

Figure 7.43 Switched-line phase shifter using *pin* diodes.

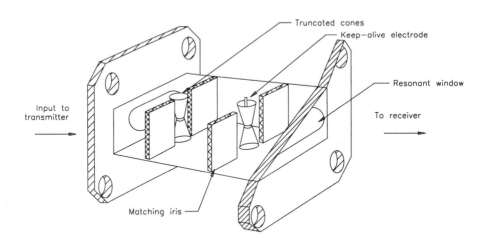

Figure 7.44 Typical construction of a TR tube.

the antenna. Figure 7.46 shows a filter in which harmonic energy is coupled out through holes in the walls of the waveguide to matched loads.

Narrowband filters in the receive path, often called *preselectors*, are built using mechanically tuned cavity resonators or electrically tuned *TIG resonators*. Preselectors

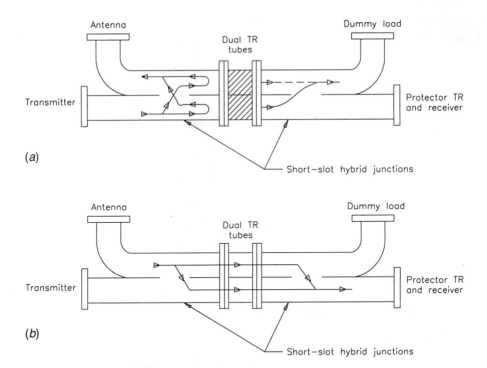

Figure 7.45 Balanced duplexer circuit using dual TR tubes and two short-slot hybrid junctions: (*a*) transmit mode, (*b*) receive mode.

can provide up to 80 dB suppression of signals from other radar transmitters in the same RF band, but at a different operating frequency.

7.5   High-Power Isolators

The high-power ferrite isolator offers the ability to stabilize impedance, isolate the RF generator from load discontinuities, eliminate reflections from the load, and absorb harmonic and intermodulation products. The isolator also can be used to switch between an antenna or load under full power, or to combine two or more generators into a common load.

Isolators commonly are used in microwave transmitters at low power to protect the output stage from reflections. Until recently, however, the insertion loss of the ferrite made use of isolators impractical at high-power levels (25 kW and above). Ferrite isolators are now available that can handle 500 kW or more of forward power with less than 0.1 dB of forward power loss.

7.5.1 Theory of Operation

High-power isolators are three-port versions of a family of devices known as *circulators*. The circulator derives its name from the fact that a signal applied to one

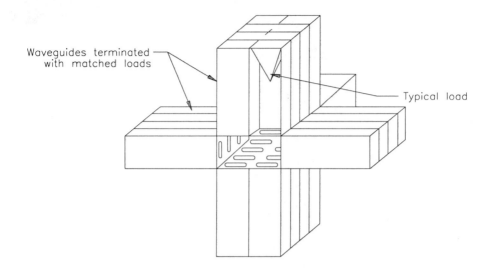

Waveguides terminated with matched loads

Typical load

Figure 7.46 Construction of a dissipative waveguide filter.

of the input ports can travel in only one direction, as shown in Figure 7.47. The input port is isolated from the output port. A signal entering port 1 appears only at port 2; it does not appear at port 3 unless reflected from port 2. An important benefit of this one-way power transfer is that the input VSWR at port 1 is dependent only on the VSWR of the load placed at port 3. In most applications, this load is a resistive (dummy) load that presents a perfect load to the RF generator.

The unidirectional property of the isolator results from magnetization of a ferrite alloy inside the device. Through correct polarization of the magnetic field of the ferrite, RF energy will travel through the element in only one direction (port 1 to 2, port 2 to 3, and port 3 to 1). Reversing the polarity of the magnetic field makes it possible for RF flow in the opposite direction. Recent developments in ferrite technology have resulted in high isolation with low insertion loss.

In the basic design, the ferrite is placed in the center of a Y-junction of three transmission lines, either waveguide or coax. Sections of the material are bonded together to form a thin cylinder perpendicular to the electric field. Even though the insertion loss is low, the resulting power dissipated in the cylinder can be as high as 2 percent of the forward power. Special provisions must be made for heat removal. It is efficient heat-removal capability that makes high-power operation possible.

The insertion loss of the ferrite must be kept low so that minimal heat is dissipated. Values of ferrite loss on the order of 0.05 dB have been produced. This equates to an efficiency of 98.9 percent. Additional losses from the transmission line and matching structure contribute slightly to loss. The overall loss is typically less than 0.1 dB, or 98 percent efficiency. The ferrite element in a high-power system is usually water-cooled in a closed-loop path that uses an external radiator.

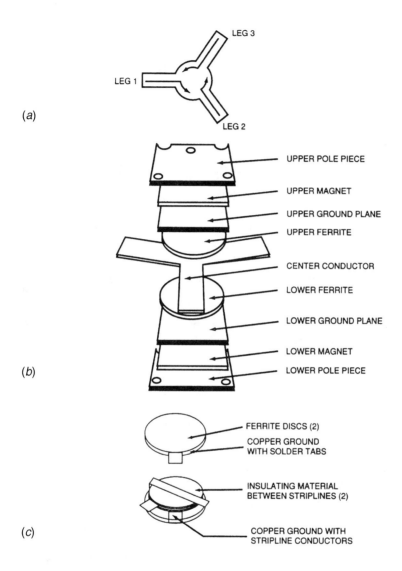

LEG 3

LEG 1

(a)

LEG 2

UPPER POLE PIECE

UPPER MAGNET

UPPER GROUND PLANE

UPPER FERRITE

CENTER CONDUCTOR

LOWER FERRITE

LOWER GROUND PLANE

LOWER MAGNET

(b)

LOWER POLE PIECE

FERRITE DISCS (2)

COPPER GROUND
WITH SOLDER TABS

INSULATING MATERIAL
BETWEEN STRIPLINES (2)

(c)

COPPER GROUND WITH
STRIPLINE CONDUCTORS

Figure 7.47 Basic characteristics of a circulator: (a) operational schematic, (b) distributed constant circulator, (c) lump constant circulator. (*From* [2]. *Used with permission.*)

The two basic circulator implementations are shown in Figures 7.47a and 7.47b. These designs consist of Y-shaped conductors sandwiched between magnetized ferrite discs [2]. The final shape, dimensions, and type of material varies according to frequency of operation, power handling requirements, and the method of coupling. The *distributed constant circulator* is the older design; it is a broad-band device, not quite as efficient in terms of insertion loss and leg-to-leg isolation, and considerably more expensive to produce. It is useful, however, in applications where broad-band isolation is required. More commonly today is the *lump constant circulator*, a less expensive and more efficient, but narrow-band, design.

At least one filter is always installed directly after an isolator, because the ferrite material of the isolator generates harmonic signals. If an ordinary band-pass or band-reject filter is not to be used, a harmonic filter will be needed.

7.5.2 Applications

The high-power isolator permits an RF generator to operate with high performance and reliability despite a load that is less than optimum. The problems presented by ice formations on a transmitting antenna provide a convenient example. Ice buildup will detune an antenna, resulting in reflections back to the transmitter and high VSWR. If the VSWR is severe enough, transmitter power will have to be reduced to keep the system on the air. An isolator, however, permits continued operation with no degradation in signal quality. Power output is affected only to the extent of the reflected energy, which is dissipated in the resistive load.

A high-power isolator also can be used to provide a stable impedance for devices that are sensitive to load variations, such as klystrons. This allows the device to be tuned for optimum performance regardless of the stability of the RF components located after the isolator. Figure 7.48 shows the output of a wideband (6 MHz) klystron operating into a resistive load, and into an antenna system. The power loss is the result of an impedance difference. The periodicity of the ripple shown in the trace is a function of the distance of the reflections from the source.

Hot Switch

The circulator can be made to perform a switching function if a short circuit is placed at the output port. Under this condition, all input power will be reflected back into the third port. The use of a high-power stub on port 2, therefore, permits redirecting the output of an RF generator to port 3.

At odd ¼-wave positions, the stub appears as a high impedance and has no effect on the output port. At even ¼-wave positions, the stub appears as a short circuit. Switching between the antenna and a test load, for example, can be accomplished by moving the shorting element ¼ wavelength.

Diplexer

An isolator may be configured to combine the aural and visual outputs of a TV transmitter into a single output for the antenna. The approach is shown in Figure 7.49. A single notch cavity at the aural frequency is placed on the visual transmitter output (circulator input), and the aural signal is added (as shown). The aural signal will be routed to the antenna in the same manner as it is reflected (because of the hybrid action) in a conventional diplexer.

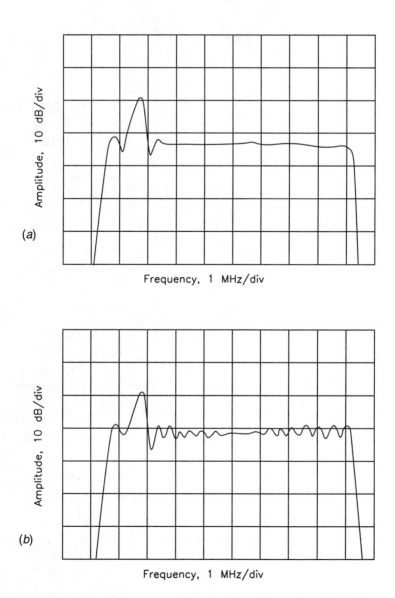

Figure 7.48 Output of a klystron operating into different loads through a high-power isolator: (*a*) resistive load, (*b*) an antenna system.

Multiplexer

A multiplexer can be formed by cascading multiple circulators, as illustrated in Figure 7.50. Filters must be added, as shown. The primary drawback of this approach is the increased power dissipation that occurs in circulators nearest the antenna.

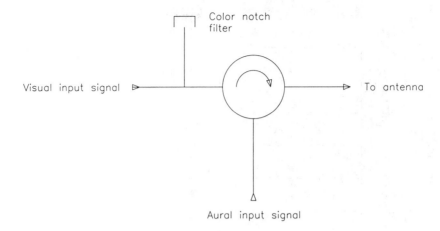

Figure 7.49 Use of a circulator as a diplexer in TV applications.

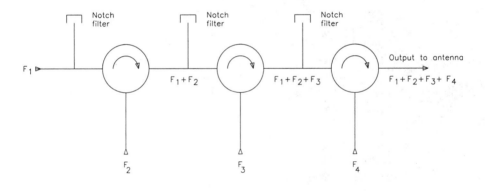

Figure 7.50 Using multiple circulators to form a multiplexer.

7.6   References

1. Harrison, Cecil, "Passive Filters," in *The Electronics Handbook*, Jerry C. Whitaker (ed.), CRC Press, Boca Raton, Fla., pp. 279–290, 1996.
2. Surette, Robert A., "Combiners and Combining Networks," in *The Electronics Handbook*, Jerry C. Whitaker (ed.), CRC Press, Boca Raton, Fla., pp. 1368–1381, 1996.

7.7   Bibliography

Andrew Corporation, "Broadcast Transmission Line Systems," Technical Bulletin 1063H, Orland Park, IL, 1982.

Andrew Corporation, "Circular Waveguide: System Planning, Installation and Tuning," Technical Bulletin 1061H, Orland Park, IL, 1980.

Ben-Dov, O., and C. Plummer, "Doubly Truncated Waveguide," *Broadcast Engineering*, Intertec Publishing, Overland Park, KS, January 1989.

Benson, K. B., and J. C. Whitaker, *Television and Audio Handbook for Technicians and Engineers*, McGraw-Hill, New York, 1989.

Cablewave Systems, "The Broadcaster's Guide to Transmission Line Systems," Technical Bulletin 21A, North Haven, CT, 1976.

Cablewave Systems, "Rigid Coaxial Transmission Lines," Cablewave Systems Catalog 700, North Haven, CT, 1989.

Crutchfield, E. B. (ed.), *NAB Engineering Handbook*, 8th Ed., National Association of Broadcasters, Washington, DC, 1992.

DeComier, Bill, "Inside FM Multiplexer Systems," *Broadcast Engineering*, Intertec Publishing, Overland Park, KS, May 1988.

Fink, D., and D. Christiansen (eds.), *Electronics Engineers' Handbook*, 3rd Ed., McGraw-Hill, New York, 1989.

Fink, D., and D. Christiansen (eds.), *Electronics Engineers' Handbook*, 2nd Ed., McGraw-Hill, New York, 1982.

Heymans, Dennis, "Hot Switches and Combiners," *Broadcast Engineering*, Overland Park, KS, December 1987.

Jordan, Edward C., *Reference Data for Engineers: Radio, Electronics, Computer and Communications*, 7th Ed., Howard W. Sams, Indianapolis, IN, 1985.

Krohe, Gary L., "Using Circular Waveguide," *Broadcast Engineering*, Intertec Publishing, Overland Park, KS, May 1986.

Perelman, R., and T. Sullivan, "Selecting Flexible Coaxial Cable," *Broadcast Engineering*, Intertec Publishing, Overland Park, KS, May 1988.

Stenberg, James T., "Using Super Power Isolators in the Broadcast Plant," *Proceedings of the Broadcast Engineering Conference*, Society of Broadcast Engineers, Indianapolis, IN, 1988.

Terman, F. E., *Radio Engineering*, 3rd Ed., McGraw-Hill, New York, 1947.

Vaughan, T., and E. Pivit, "High Power Isolator for UHF Television," *Proceedings of the NAB Engineering Conference*, National Association of Broadcasters, Washington, DC, 1989.

Whitaker, Jerry C., G. DeSantis, and C. Paulson, *Interconnecting Electronic Systems*, CRC Press, Boca Raton, FL, 1993.

Whitaker, Jerry C., *Radio Frequency Transmission Systems: Design and Operation*, McGraw-Hill, New York, 1990.

8.1 Introduction

Adequate cooling of the tube envelope and seals is one of the principal factors affecting vacuum tube life [1]. Deteriorating effects increase directly with the temperature of the tube envelope and ceramic/metal seals. Inadequate cooling is almost certain to invite premature failure of the device.

Tubes operated at VHF and above are inherently subjected to greater heating action than tubes operated at lower frequencies. This results directly from the following:

- The flow of larger RF charging currents into the tube capacitances at higher frequencies

- Increased dielectric losses at higher frequencies

- The tendency of electrons to bombard parts of the tube structure other than the normal elements as the operating frequency is increased

Greater cooling, therefore, is required at higher frequencies. The technical data sheet for a given tube type specifies the maximum allowable operating temperature. For forced-air- and water-cooled tubes, the recommended amount of air or water is also specified in the data sheet. Both the temperature and quantity of coolant should be measured to be certain that cooling is adequate.

8.1.1 Thermal Properties

In the commonly used model for materials, heat is a form of energy associated with the position and motion of the molecules, atoms, and ions of the material [2]. The position is analogous with the state of the material and is *potential energy*, while the motion of the molecules, atoms and ions is *kinetic energy*. Heat added to a material makes it hotter, and heat withdrawn from a material makes it cooler. Heat energy is measured in *calories* (cal), *British Thermal Units* (Btu), or *joules*. A calorie is the amount of energy required to raise the temperature of one gram of water one degree Celsius (14.5 to 15.5 °C). A Btu is the unit of energy necessary to raise the temperature of one pound of water by one degree Fahrenheit. A joule is an equivalent amount of energy equal to the work done when a force of one newton acts through a distance of one meter.

Temperature is a measure of the average kinetic energy of a substance. It can also be considered a relative measure of the difference of the heat content between bodies.

Heat capacity is defined as the amount of heat energy required to raise the temperature of one mole or atom of a material by one °C without changing the state of the material. Thus, it is the ratio of the change in heat energy of a unit mass of a substance to its change in temperature. The heat capacity, often referred to as *thermal capacity*, is a characteristic of a material and is measured in cal/gram per °C or Btu/lb per °F.

Specific heat is the ratio of the heat capacity of a material to the heat capacity of a reference material, usually water. Because the heat capacity of water is one Btu/lb and one cal/gram, the specific heat is numerically equal to the heat capacity.

Heat transfers through a material by conduction resulting when the energy of atomic and molecular vibrations is passed to atoms and molecules with lower energy. As heat is added to a substance, the kinetic energy of the lattice atoms and molecules increases. This, in turn, causes an expansion of the material that is proportional to the temperature change, over normal temperature ranges. If a material is restrained from expanding or contracting during heating and cooling, internal stress is established in the material.

8.1.2 Heal Transfer Mechanisms

The process of heat transfer from one point or medium to another is a result of temperature differences between the two. Thermal energy may be transferred by any of three basic modes:

- Conduction

- Convection

- Radiation

A related mode is the convection process associated with the change of phase of a fluid, such as condensation or boiling. A vacuum tube amplifier or oscillator may use several of these mechanisms to maintain the operating temperature of the device within specified limits.

Conduction

Heat transfer by conduction in solid materials occurs whenever a hotter region with more rapidly vibrating molecules transfers a portion of its energy to a cooler region with less rapidly vibrating molecules. Conductive heat transfer is the most common form of thermal exchange in electronic equipment. Thermal conductivity for solid materials used in electronic equipment spans a wide range of values, from excellent (high conductivity) to poor (low conductivity). Generally speaking, metals are the best conductors of heat, whereas insulators are the poorest. Table 8.1 lists the thermal conductivity of materials commonly used in the construction (and environment) of

Table 8.1 Thermal Conductivity of Common Materials

Material	Btu/(h ft ºF)	W/(m ºC)
Silver	242	419
Copper	228	395
Gold	172	298
Beryllia	140	242
Phosphor bronze	30	52
Glass (borosilicate)	0.67	1.67
Mylar	0.11	0.19
Air	0.015	0.026

Table 8.2 Relative Thermal Conductivity of Various Materials As a Percentage of the Thermal Conductivity of Copper

Material	Relative Conductivity
Silver	105
Copper	100
Berlox high-purity BeO	62
Aluminum	55
Beryllium	39
Molybdenum	39
Steel	9.1
High-purity alumina	7.7
Steatite	0.9
Mica	0.18
Phenolics, epoxies	0.13
Fluorocarbons	0.05

power vacuum tubes. Table 8.2 compares the thermal conductivity of various substances as a percentage of the thermal conductivity of copper.

With regard to vacuum tube devices, conduction is just one of the cooling mechanisms involved in maintaining element temperatures within specified limits. At high power levels, there is always a secondary mechanism that works to cool the device. For example, heat generated within the tube elements is carried to the cooling surfaces of the device by conduction. Convection or radiation then is used to remove heat from the tube itself.

Thermal conduction has been used successfully to cool vacuum tubes through direct connection of the anode to an external heat sink, but only at low power levels [1].

Convection

Heat transfer by natural convection occurs as a result of a change in the density of a fluid (including air), which causes fluid motion. Convective heat transfer between a heated surface and the surrounding fluid is always accompanied by a mixing of fluid adjacent to the surface. Power vacuum tube devices relying on convective cooling invariably utilize forced air or water passing through a heat-transfer element, typically the anode in a grid tube or the collector in a microwave tube [1]. This *forced convection* provides for a convenient and relatively simple cooling system. In such an arrangement, the temperature gradient is confined to a thin layer of fluid adjacent to the surface so that the heat flows through a relatively thin *boundary layer*. In the main stream outside this layer, isothermal conditions exist.

Radiation

Cooling by radiation is a function of the transfer of energy by electromagnetic wave propagation. The wavelengths between 0.1 and 100 m are referred to as *thermal radiation wavelengths*. The ability of a body to radiate thermal energy at any particular wavelength is a function of the body temperature and the characteristics of the surface of the radiating material. Figure 8.1 charts the ability to radiate energy for an ideal radiator, a *blackbody*, which, by definition, radiates the maximum amount of energy at any wavelength. Materials that act as perfect radiators are rare. Most materials radiate energy at a fraction of the maximum possible value. The ratio of the energy radiated by a given material to that emitted by a blackbody at the same temperature is termed *emissivity*. Table 8.3 lists the emissivity of various common materials.

8.1.3 The Physics of Boiling Water

The *Nukiyama curve* shown in Figure 8.2 charts the heat-transfer capability (measured in watts per square centimeter) of a heated surface, submerged in water at various temperatures [1]. The first portion of the curve—*zone A*—indicates that from 100 to about 108°C, heat transfer is a linear function of the temperature differential between the hot surface and the water, reaching a maximum of about 5 W/cm^2 at 108°C. This linear area is known as the *convection cooling zone*. Boiling takes place in the heated water at some point away from the surface.

From 108 to 125°C—*zone B*—heat transfer increases as the fourth power of ΔT until, at 125°C, it reaches 135 W/cm^2. This zone is characterized by *nucleate boiling*; that is, individual bubbles of vapor are formed at the hot surface, break away, and travel upward through the water to the atmosphere.

Above 125°C, an unstable portion of the Nukiyama curve is observed, where increasing the temperature of the heated surface actually reduces the unit thermal conductivity. At this area—*zone C*—the vapor partially insulates the heated surface from the water until a temperature of approximately 225°C is reached on the hot surface. At this point—called the *Leidenfrost point*—the surface becomes completely covered with a sheath of vapor, and all heat transfer is accomplished through this vapor cover. Thermal conductivity of only 30 W/cm^2 is realized at this region.

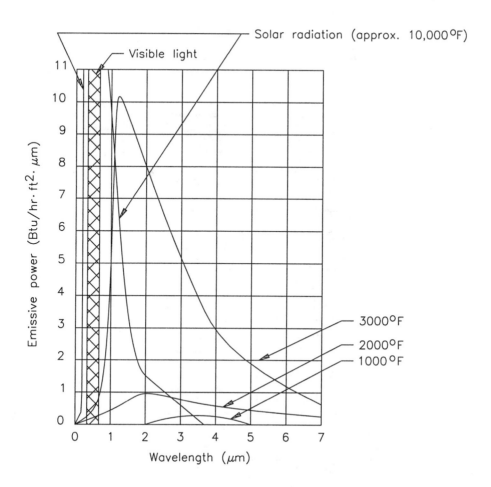

Figure 8.1 Blackbody energy distribution characteristics.

Table 8.3 Emissivity Characteristics of Common Materials at 80°F

Surface Type	Finish	Emissivity
Metal	Copper (polished)	0.018
Metal	Nickel	0.21
Metal	Silver	0.10
Metal	Gold	0.04–0.23
Glass	Smooth	0.9–0.95
Ceramic	Cermet[1]	0.58
[1] Ceramic containing sintered metal		

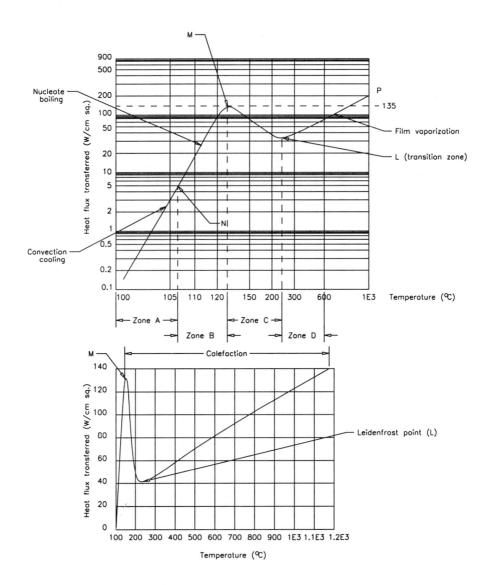

Figure 8.2 Nukiyama heat-transfer curves: (a) logarithmic, (b) linear.

From the Leidenfrost point on through *zone D*—the *film vaporization zone*—heat transfer increases with temperature until at about 1000°C the value of 135 W/cm² again is reached.

The linear plot of the Nukiyama curve indicates that zones *A* and *B* are relatively narrow areas and that a heated surface with unlimited heat capacity will tend to pass from zone *A* to zone *D* in a short time. This irreversible superheating is known as *calefaction*. For a cylindrical vacuum tube anode, the passing into total calefaction would not be tolerated, because any unit heat-transfer density above 135 W/cm^2 would result in temperatures above 1000°C, well above the safe limits of the tube.

These properties dictate many of the physical parameters of water- and vapor-cooled power vacuum tubes. For a given dissipation, the anode structure that comes in contact with the coolant must be sufficient to keep the tube operating within the desired cooling zone of the Nukiyama curve and, in any event, at less than 135 W/cm^2 dissipation on the anode structure.

8.2 Application of Cooling Principles

Excessive dissipation is perhaps the single greatest cause of catastrophic failure in a power tube [1]. PA tubes used in communications, industrial, and research applications can be cooled using one of three methods: forced-air, liquid, and vapor-phase cooling (discussed in Chapter 3). Forced-air cooling is the simplest and most common method used.

The critical points of almost every PA tube type are the metal-to-ceramic junctions or seals. At temperatures below 250°C these seals remain secure, but above that temperature, the bonding in the seal may begin to disintegrate. Warping of grid structures also may occur at temperatures above the maximum operating level of the device. The result of prolonged overheating is shortened tube life or catastrophic failure. Several precautions usually are taken to prevent damage to tube seals under normal operating conditions. Air directors or sections of tubing may be used to provide spot cooling at critical surface areas of the device. Airflow sensors typically prevent operation of the RF system in the event of a cooling system failure.

Temperature control is important for vacuum tube operation because the properties of many of the materials used to build a tube change with increasing temperature. In some applications, these changes are insignificant. In others, however, such changes can result in detrimental effects, leading to—in the worst case—catastrophic failure. Table 8.4 details the variation of electrical and thermal properties with temperature for various substances.

8.2.1 Forced-Air Cooling Systems

Air cooling is the simplest and most common method of removing waste heat from a vacuum tube device [1]. The basic configuration is illustrated in Figure 8.3. The normal flow of cooling air is upward, making it consistent with the normal flow of convection currents. In all cases, the socket is an open structure or has adequate vent holes to allow cooling of the base end of the tube. Cooling air enters at the grid circuit compartment below the socket through a screened opening, passes through the socket to cool the base end of the tube, sweeps upward to cool the envelope, and enters the output circuit compartment.

The output compartment is provided with a mesh-covered opening that permits the air to vent out readily. High-power tubes typically include a chimney to direct airflow, as shown in Figure 8.4. For a forced-air design, a suitable fan or blower is used to pressurize the compartment below the tube. No holes should be provided for the passage of

Table 8.4 Variation of Electrical and Thermal Properties of Common Insulators As a Function of Temperature

Parameters		20°C	120°C	260°C	400°C	538°C
Thermal conductivity[1]	99.5% BeO	140	120	65	50	40
	99.5% Al_2O_3	20	17	12	7.5	6
	95.0% Al_2O_3	13.5				
	Glass	0.3				
Power dissipation[2]	BeO	2.4	2.1	1.1	0.9	0.7
Electrical resistivity[3]	BeO	10^{16}	10^{14}	5×10^{12}	10^{12}	10^{11}
	Al_2O_3	10^{14}	10^{14}	10^{12}	10^{12}	10^{11}
	Glass	10^{12}	10^{10}	10^{8}	10^{6}	
Dielectric constant[4]	BeO	6.57	6.64	6.75	6.90	7.05
	Al_2O_3	9.4	9.5	9.6	9.7	9.8
Loss tangent[4]	BeO	0.00044	0.00040	0.00040	0.00049	0.00080

[1] Heat transfer in Btu/ft^2/hr/°F
[2] Dissipation in W/cm/°C
[3] Resistivity in Ω-cm
[4] At 8.5 GHz

air from the lower to the upper compartment other than the passages through the socket and tube base. A certain amount of pressure must be built up to force the proper amount of air through the socket to cool the anode.

Attention must be given to airflow efficiency and turbulence in the design of a cooling system. Consider the case shown in Figure 8.5. Improper layout has resulted in inefficient movement of air because of circulating thermal currents. Anode cooling will be insufficient in this case.

The cooling arrangements illustrated in Figures 8.3 and 8.4 provide for the uniform passage of cooling air over the tube base and anode. An arrangement that forces cooling air transversely across the tube base and/or anode will not provide the same cooling effectiveness and may result in hot spots on the device.

In many cases, packaged air system sockets and chimneys are designed specifically for a tube or family of tube types. The technical data sheet specifies the recommended socketing for adequate cooling. The tube data sheet also will specify the back pressure, in inches of water, and the cubic feet per minute required for adequate cooling. In an actual application, the back pressure may be measured by means of a simple manometer. This device consists of a U-shaped glass tube partially filled with water, as shown in Figure 8.6.

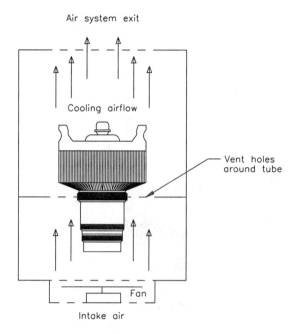

Figure 8.3 Cooling system design for a power grid tube.

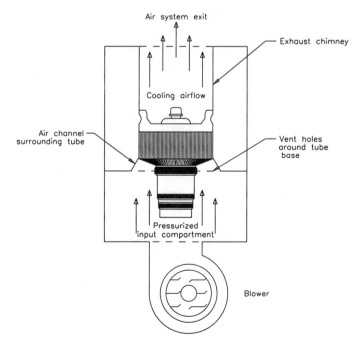

Figure 8.4 The use of a chimney to improve cooling of a power grid tube.

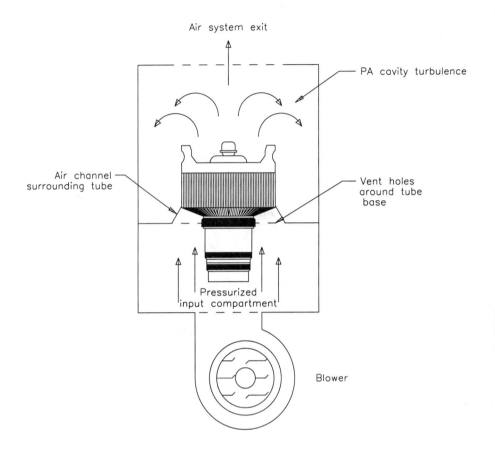

Figure 8.5 A poorly designed cooling system in which circulating air in the output compartment reduces the effectiveness of the heat-removal system.

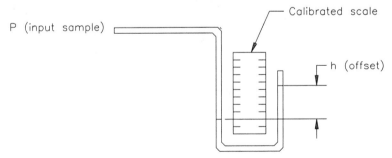

Figure 8.6 A manometer, used to measure air pressure.

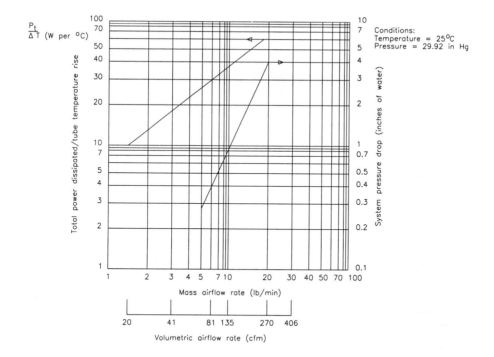

Figure 8.7 Cooling airflow requirements for a power grid tube. (*Data courtesy of Varian.*)

Cooling Airflow Data

Power tube manufacturers typically outline minimum cooling airflow requirements for large external-anode tubes in the form of one or more charts [1]. Figure 8.7 plots $P_t/\Delta T$ as a function of A_m, where P_t = total power dissipated in watts, ΔT = tube temperature rise in degrees Celsius, and A_m = mass airflow rate in pounds of air per minute. This type of graph is used in calculating the cooling requirements of a vacuum tube device. The graph applies to a specified tube and socket-chimney combination; furthermore, the direction of airflow is specified. When reverse airflow (from anode to base) is used, the cooling requirements are sharply increased because the air applied to the base seals already will have been heated by its passage through the anode cooler, losing much of its cooling effectiveness.

To use the type of graph shown in Figure 8.7, first determine the minimum cooling requirements using the following steps:

1. Calculate the total power dissipated (P_t) by adding all of the power dissipated by the tube during operation in a given installation. This value includes plate and fila-

ment dissipations, plus maximum anticipated grid and screen dissipations (as applicable).

2. Calculate the tube temperature rise (ΔT) by taking the difference between the maximum rated tube temperature specified in the appropriate data sheet and the maximum expected air-inlet temperature.

3. To convert the mass airflow rate M (pounds per minute) to volumetric airflow rate Q (cubic feet per minute) at 25°C and at sea level, divide the mass airflow rate by the density of air at 25°C and 29.9 in mercury.

4. The curve on the right side of the graph is the pressure drop (ΔP) in inches of water across the tube and its specified socket-chimney combination. This value is valid at 25°C at sea level only.

To adjust the 25°C sea-level laboratory test conditions to any other atmospheric (socket-inlet) condition, multiply both the Q and ΔP values by the ratio of the laboratory standard density (0.074 lb/ft³; 25°C at sea level) to the density at the new socket-inlet condition.

A shorter method may be used to approximately correct the 25°C sea level requirements to both a different temperature and/or barometric socket-inlet condition. These corrections are made by multiplying the Q and ΔP values by the appropriate correction factors listed in Table 8.5.

Figure 8.8 is a graph of the combined correction factors that can be applied to the 25°C sea-level information for land-based installations located at elevations up to 10,000 ft, and for socket-inlet air temperatures between 10 and 50°C. Figure 8.9 provides a method to convert the mass airflow rate in pounds per minute into volumetric airflow rate (cfm) at 25°C and sea level.

Good engineering practice must be used when applying altitude and temperature corrections to the 25°C sea-level cooling requirement for airborne applications. Although the air outside the aircraft may be very cold at high altitudes, the air actually entering the tube socket may be many degrees warmer. This inlet temperature (and pressure) is affected by the installation design (compressed, ram, static, or recirculating air in a pressurized heat exchanger).

Blower Selection

In the previous section, a method of determining minimum air cooling requirements for external-anode tubes was described, pertaining to any altitude and air temperature [1]. Because most blower manufacturers furnish catalog data on their products in the form of volumetric airflow (Q, cfm) vs. operating back pressure (ΔP, inches of water) for sea level conditions only, the information gained from the foregoing procedure cannot be compared directly with data furnished by manufacturers for the purpose of selecting the proper device. The following method is recommended for use in selecting a blower for applications above sea level from existing blower catalog data:

Table 8.5 Correction Factors for Q and ΔP

Socket-Inlet Air Temperature (°C)	Q and ΔP Correction Factor	
0	0.917	
5	0.933	
10	0.950	
15	0.967	
20	0.983	
25	1.000	
30	1.017	
35	1.034	
40	1.051	
45	1.067	
50	1.084	
Socket-Inlet Air Pressure[1]	Altitude (ft)	Q and ΔP Correction Factor
29.92	0	1.00
24.90	5000	1.20
20.58	10,000	1.46
16.89	15,000	1.77
13.75	20,000	2.17
11.10	25,000	2.69
8.89	30,000	3.37
7.04	35,000	4.25
[1] Pressure in inches of mercury		

1. Determine the Q and ΔP requirements for the tube socket-chimney combination for an ambient air temperature of 25°C at sea level. Include the estimated ΔP of ducting and filters in the system.

2. Determine the corrected Q and ΔP system requirements for the actual inlet temperature and altitude conditions by multiplying by the correction factor shown in Figure 8.7.

3. Multiply the ΔP—but not the Q—requirement by the correction factor cited in step 2.

4. Use the corrected Q factor and doubly-corrected ΔP value to select a blower from the manufacturer's published sea-level curves. Although this blower will overcool the tube at sea level when operated in an ambient temperature of 25°C, it will provide adequate cooling at the actual inlet temperature and at high-altitude conditions.

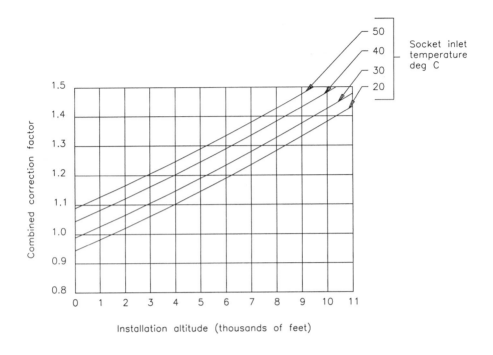

Figure 8.8 Combined correction factors for land-based tube applications.

8.2.2 Water Cooling

A water-cooled tube depends upon an adequate flow of fluid to remove heat from the device and transport it to an external heat sink [1]. The recommended flow as specified in the technical data sheet should be maintained at all times when the tube is in operation. Inadequate water flow at high temperature may cause the formation of steam bubbles at the anode surface where the water is in direct contact with the anode. This condition can contribute to premature tube failure.

Circulating water can remove about 1.0 kW/cm^2 of effective internal-anode area. In practice, the temperature of water leaving the tube is limited to 70°C to preclude the possibility of spot boiling. The water then is passed through a heat exchanger where it is cooled to 30 to 40°C before being pumped over the tube anode again.

Cooling System Design

A liquid cooling system consists of the following principal components:

- A source of coolant
- Circulation pump

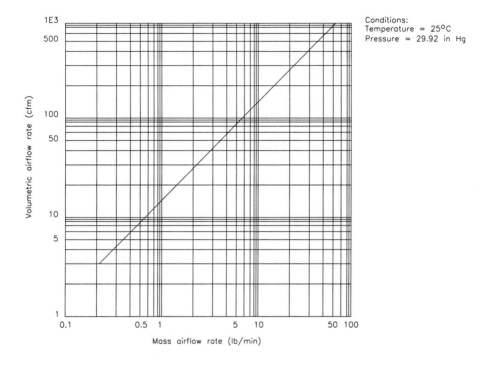

Figure 8.9 Conversion of mass airflow rate to volumetric airflow rate.

- Heat exchanger
- Coolant purification loop
- Various connection pipes, valves, and gauges
- Flow interlocking devices (required to ensure coolant flow anytime the equipment is energized)

Such a system is shown schematically in Figure 8.10. In most cases the liquid coolant will be water, but if there is a danger of freezing, it will be necessary to use an antifreeze solution such as ethylene glycol. In these cases, coolant flow must be increased or plate dissipation reduced to compensate for the poorer heat capacity of the ethylene glycol solution. A mixture of 60 percent ethylene glycol to 40 percent water by weight will be about 75 percent as efficient as pure water at 25°C. Regardless of the choice of liquid, the system volume must be maintained above the minimum required to ensure proper cooling of the vacuum tube(s).

The main circulation pump must be of sufficient size to ensure necessary flow and pressure as specified on the tube data sheet. Care must be taken when connecting the coolant lines to the tube to be certain that flow is in the direction specified. Improper direction of flow may result in inadequate cooling as well as excessive pressure and possi-

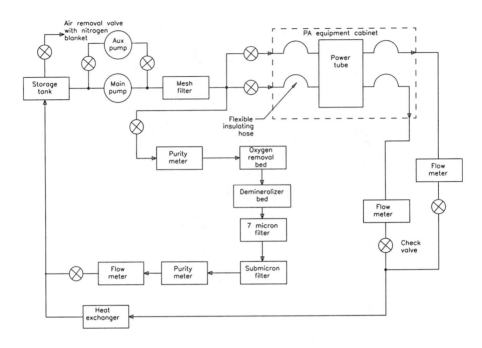

Figure 8.10 Functional schematic of a water cooling system for a high-power amplifier or oscillator.

ble mechanical deformation of the anode. Precautions should be taken during system design to protect against water *hammering* of the anode. A filter screen of at least 60 mesh usually is installed in the pump outlet line to trap any circulating debris that might clog coolant passages within the tube.

The heat exchanger system is sized to maintain the outlet temperature such that the outlet water from the tube at full plate dissipation does not exceed 70°C. Filament or grid coolant courses may be connected in parallel or series with the main supply as long as the maximum outlet temperature is not exceeded.

Valves and pressure meters are installed on the inlet lines to the tube to permit adjustment of flow and measurement of pressure drop, respectively. A pressure meter and check valve are employed in the outlet line. In addition, a flow meter—sized in accordance with the tube data sheet—and a thermometer are included in the outlet line of each coolant course. These flow meters are equipped with automatic interlock switches, wired into the system electrical controls, so that the tube will be completely deenergized in the event of a loss of coolant flow in any one of the coolant passages. In some tubes, filament power alone is sufficient to damage the tube structure in the absence of proper water flow.

The lines connecting the plumbing system to the inlet and outlet ports of the tube are made of a flexible insulating material configured so as to avoid excessive strain on the tube flanges. Polypropylene tubing is the best choice for this service, but chlorinated polyvinyl chloride (CPVC) pipe, which is stronger, also is acceptable. Reinforced poly-

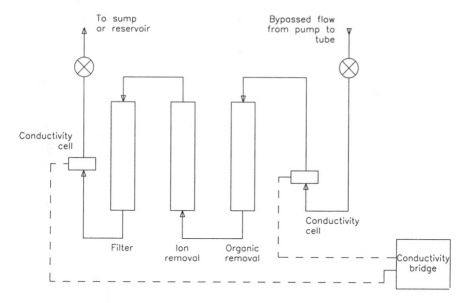

Figure 8.11 Typical configuration of a water purification loop.

propylene, such as *Nylobraid*, is expensive, but excellent in this application. The coolant lines must be of sufficient length to keep electrical leakage below 4 mA at full operating power.

The hoses are coiled or otherwise supported so that they do not contact each other or any conducting surface between the high-voltage end and ground. Conducting hose *barbs*, connected to ground, are provided at the low-potential end so that the insulating column of water is broken and grounded at the point where it exits the equipment cabinet.

All metallic components within the water system, including the pump, must be of copper, stainless steel, or unleaded brass or bronze; copper or stainless steel is preferred. If brass or bronze is used, some zinc may be leached out of the metal into the system. Any other material, such as iron or cold rolled steel, will grossly contaminate the water.

Even if the cooling system is constructed using the recommended materials, and is filled with distilled or deionized water, the solubility of the metals, carbon dioxide, and free oxygen in the water make the use of a coolant purification (*regeneration*) loop essential. The integration of the purification loop into the overall cooling system is shown in Figure 8.10. The regeneration loop typically taps 5 to 10 percent of the total cooling system capacity, circulating it through oxygen scavenging and deionization beds, and submicron filters before returning it into the main system. Theoretically, such a purification loop can process water to 18 MΩ-cm resistivity at 25°C. In practice, resistivity will be somewhat below this value.

Figure 8.11 shows a typical purification loop configuration. Packaged systems such as the one illustrated are available from a number of manufacturers. In general, such

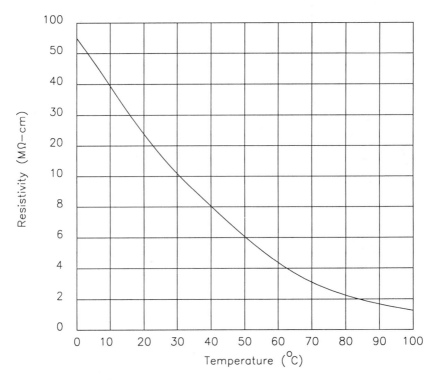

Figure 8.12 The effect of temperature on the resistivity of ultrapure water.

systems consist of replaceable cartridges that perform the filtering, ion exchange, and organic-solid-removal functions. The system usually will include flow and pressure gauges and valves, and conductivity cells for continuous evaluation of the condition of the water and filters.

Water Purity and Resistivity

The purity of the cooling water is an important operating parameter for any water-cooled amplifier or oscillator. The specific resistivity must be maintained at 1 MΩ-cm minimum at 25°C. Distilled or deionized water should be used and the purity and flow protection periodically checked to ensure against degradation. Oxygen and carbon dioxide in the coolant will form copper oxide, reducing cooling efficiency, and electrolysis may destroy the coolant passages. In addition, a filter screen should be installed in the tube inlet line to trap any circulating debris that might clog coolant passages within the tube.

After normal operation for an extended period, the cooling system should be capable of holding 3 to 4 MΩ-cm until the filter beds become contaminated and must be replaced. The need to replace the filter bed resins is indicated when the purification loop output water falls below 5 MΩ-cm. Although the resistivity measurement is not a test for free oxygen in the coolant, the oxygen filter bed always should be replaced when the deionizing bed is replaced.

The resistivity of the coolant also is affected by the temperature of the water. The temperature-dependence of the resistivity of pure water is charted in Figure 8.12.

It is recommended that the coolant water be circulated at all times. This procedure provides the following benefits:

- Maintains high resistivity

- Reduces bacteria growth

- Minimizes oxidation resulting from coolant stagnation

If it is undesirable to circulate the coolant at the regular rate when the tube is deenergized, a secondary circulating pump can be used to move the coolant at a lower rate to purge any air that might enter the system and to prevent stagnation. Recommended minimum circulation rates within the coolant lines are as follows:

- 2 m/s during normal operation

- 30 cm/s during standby mode

It is recommended that a circulation rate of 10 m/s be established within the tube itself. When it is necessary to turn the coolant flow completely off, the cooling system should be restarted, and allowed to return to a minimum of 1 MΩ-cm resistivity before energizing the tube.

The regeneration loop is typically capable of maintaining the cooling system such that the following maximum contaminant levels are not exceeded:

- Copper: 0.05 ppm by weight

- Oxygen: 0.5 ppm by weight

- CO_2: 0.5 ppm by weight

- Total solids: 3 ppm by weight

These parameters represent maximum levels; if the precautions outlined in this section are taken, actual levels will be considerably lower.

If the cooling system water temperature is allowed to reach 50°C, it will be necessary to use cartridges in the coolant regeneration loop that are designed to operate at elevated temperatures. Most ordinary cartridges will decompose in high-temperature service.

Condensation

The temperature of the input cooling water is an important consideration for proper operation of the cooling system [1]. If the air is humid and the cooling water is cold, condensation will accumulate on the surfaces of pipes, tube jackets, and other parts carrying water. This condensation may decrease the surface leakage resistance, or drops of water may fall on electric components, causing erratic operation or system failure. Some means is necessary, therefore, to control the temperature of the incoming water to keep it above the dew point. Such control is rather easy in a closed cooling system, but in a system that employs tap water and drains the exhaust water into a sewer, control is difficult.

Preventive Maintenance

Through electrolysis and scale formation, hard water may cause a gradual constriction of some parts of the water system [1]. Therefore, water flow and plumbing fittings must be inspected regularly. The fittings on the positive-potential end of an insulating section of hose or ceramic water coil or column are particularly subject to corrosion or electrolysis unless they have protective *targets*. The target elements should be checked periodically and replaced when they have disintegrated.

To extend the life of the resin beds in the purification loop, all the coolant lines should be flushed with a nonsudsing detergent and a citric acid solution,[1] then rinsed repeatedly with tap water before initially connecting the resin beds and filling with distilled or filtered deionized water. It is also good practice to sterilize the coolant lines with a chlorine solution[2] before filling to prevent algae and/or bacteria growth.

8.2.3 Vapor-Phase Cooling

Vapor-phase cooling offers several advantages over conventional water cooling systems by exploiting the latent heat of the evaporation of water [1]. Raising the temperature of 1 g of water from 40 to 70°C (as in a water system) requires 30 calories of energy. Transforming 1 g of water at 100°C to steam vapor requires 540 calories. In a vapor cooling system, then, a given quantity of water will remove nearly 20 times as much energy as in a water cooling system. Power densities as high as 135 W/cm^2 of effective internal-anode surface can be attained through vapor cooling.

A typical vapor-phase installation consists of a tube with a specially designed anode immersed in a *boiler* filled with distilled water. When power is applied to the tube, anode dissipation heats the water to 100°C; further applied energy causes the water to boil and be converted into steam vapor. The vapor is passed through a condenser where it gives up its energy and is converted back into the liquid state. This condensate is then returned to the boiler, completing the cycle.

To achieve the most efficient heat transfer, the anode must be structured to provide for optimum contact with the water in the boiler. Figure 8.13 shows several examples.

Because the boiler usually is at a high potential relative to ground, it must be insulated from the rest of the system. The boiler typically is mounted on insulators, and the steam and water connections are made through insulated tubing. High-voltage standoff

---

1 Cirtic acid solution mixtures vary; consult tube manufacturer.
2 Chlorine solution: sodium hypochlorite bleach added in an amount sufficient to give the odor of chlorine in the circulating water.

Figure 8.13 Vapor-phase-cooled tubes removed from their companion boilers. (*Courtesy of Varian/Eimac.*)

is more easily accomplished in a vapor-phase system than in a water-cooled system because of the following factors:

- There is a minimum of contamination because the water is constantly being redistilled.

- There is inherently higher resistance in the system because of the lower water-flow rate. The water inlet line is of a smaller diameter, resulting in greater resistance.

In a practical system, a 2-ft section of insulating tubing on the inlet and outlet ports is capable of 20 kV standoff.

Because of the effects of *thermosiphoning*, natural circulation of the water eliminates the need for a pump in most systems.

The dramatic increase in heat absorption that results from converting hot water to steam is repeated in reverse in the condenser. As a result, a condenser of much smaller thermal capacity is required for a vapor-phase system as opposed to a water-cooled system. For example, in a practical water-cooled system, water enters the heat exchanger at 70°C and exits at 40°C; the mean temperature is 55°C. For an ambient external temperature of 25°C, the mean differential between the water and the heat sink (air) is 30°C. The greater the differential, the more heat transferred. In a practical vapor-phase-cooled system, water enters the heat exchanger as steam at 100°C and exits as water at 100°C; the mean temperature is 100°C. For an ambient external temperature of 25°C, the mean differential between the vapor/water and the heat sink is 75°C. In this example, the vapor-phase condenser is nearly three times more efficient than its water-cooled counterpart.

Note that the condenser in a vapor-phase system may use either air or water as a heat sink. The water-cooled condenser provides for isolation of the PA tube cooling system

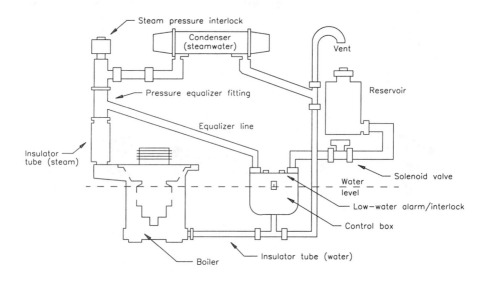

Figure 8.14 Vapor-phase cooling system for a power tube.

from the outside world. This approach may be necessary because of high voltage or safety reasons. In such a system, the water is considered a *secondary coolant*.

Where air-cooled condensers are preferred, the higher thermal gradient can be exploited in reducing the size of condenser equipment and in lowering the blower horsepower requirement. In some instances where sufficient area is available, natural convection alone is used to cool the steam condensers, resulting in complete elimination of the condenser blower.

A typical vapor-phase cooling system is shown in Figure 8.14. It consists of the following elements:

- Power tube
- Boiler
- Condenser
- Insulated tubing
- Control box
- Reservoir
- Associated plumbing

Anode Design

The success of vapor-phase cooling is dependent on anode and boiler designs that allow a tube to operate at a temperature that results in maximum heat dissipation [1].

The most common approach has been to incorporate thick vertical fins on the exterior of the anode to achieve a radial temperature gradient on the surfaces submerged in water. In this way, a hot spot does not cause instantaneous runaway of the tube; only the average fin temperature increases, and this merely shifts the temperature range to a somewhat higher level. The temperatures of the fins typically vary from approximately 110°C at the tip to about 180°C at the root. With an average overall fin temperature of approximately 115°C, for example, the average transferred heat flux is on the order of 60 W/cm^2.

When operating at low dissipation levels, boiling takes place at the root of the fin. Increasing power density causes this boiling area to move out toward the end of the fin until, at rated dissipation levels, boiling takes place on the outer half of the fins. Good design dictates that the anode fin outer edge always remain at less than 125°C.

Anode shape is also important in assuring good cooling by breaking up the sheath of vapor that might tend to form at the surfaces. To a point, the more complicated the shape, the more efficient the anode cooler. Horizontal slots often are milled into the vertical fins to provide more area and to break up bubbles. The more vigorous the boiling action, the lower the possibility that an insulating vapor sheath will form. One design—known as the "pineapple"—carries this idea to the extreme, and incorporates square "knobs" all around the anode to provide dozens of heat-radiating surfaces. The more conventional finned anode, however, is generally adequate and less costly to fabricate.

The *holed anode* is another popular design for lower power densities (up to 100 W/cm^2). In this design, a heavy cylindrical anode is made with vertical holes in the outside wall. Vapor, formed by the boiling of water within the holes, is siphoned upward through the holes to the top of the boiler. This type of cooler always operates below the calefaction point and is used in smaller tube types. The concept is illustrated in Figure 8.15.

Boiler

The boiler supports the power tube and contains the water used for cooling [1]. In addition, it acts as the high-voltage anode connector. The boiler should be mounted so that the axis of the tube is vertical. For effective cooling, tilt should be limited to less than 2° to ensure that the anode is covered with water and that the steam outlet is clear. Figure 8.16 shows a 4CV35,000A tetrode mounted in a boiler.

The anode flange of the tube must seal securely against the O-ring provided on the boiler. A locking flange presses the anode flange against the O-ring for a vapor-tight seal. The steam outlet at the top of the steam separation chamber on the boiler and the water inlet at the bottom of the boiler are equipped with fittings for attaching the Pyrex-insulated tubing. A *target* to inhibit electrolytic action is provided in the inlet water fitting.

In most cases the boiler is at a high potential relative to ground. Therefore, it must be electrically insulated from the rest of the system. The boiler is mounted on insulators, and the steam and water connections are made through insulating tubing. Boilers can be

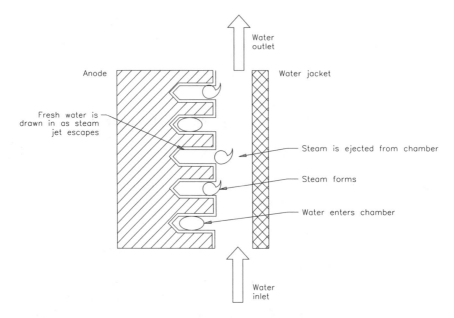

Figure 8.15 Cross section of a *holed anode* design for vapor-phase cooling.

Figure 8.16 A 4CV35,000A tetrode mounted in its boiler assembly. (*Courtesy of Varian.*)

constructed with provisions for mounting multiple tubes in parallel. In such an arrangement, a single water inlet and steam outlet fitting typically would be used.

Common anode/boiler hardware is well suited to RF amplifiers operating up into the HF band, which typically use conventional tuned circuits. Operation in systems at higher frequencies, however, becomes more complicated than water cooling, and cer-

Figure 8.17 Control box for a vapor-phase cooling system. (*Courtesy of Varian.*)

tainly more complicated than air cooling. The cooling system in an air-cooled RF generator does not interfere with the proper operation of the cavity; it is invisible to the cavity. Water cooling introduces some compromises and/or adjustments in the cavity, which can be readily overcome. Vapor-cooled tubes, however, require special design attention for successful use.

Insulating Tubing

The length of the steam and water insulating lines varies with individual installation requirements, but will always be shorter than would be needed in a circulating water system [1]. The length of the insulating tubing is determined by the following considerations:

- The voltage to be applied to the tube anode
- Purity of the water
- Volume of returned cooling water

Control Box

The function of the control box is to monitor and adjust the water level in the boiler [1]. A typical box is shown in Figure 8.17. The control box, an airtight vessel containing an overflow siphon and two float switches, also serves as a partial reservoir. When the water level drops approximately $\frac{1}{4}$ in below the recommended level, the first switch is closed. This signal may be used, at the same time, to activate a solenoid-controlled water valve to admit more distilled water from an external reservoir; it also may actuate a warning alarm.

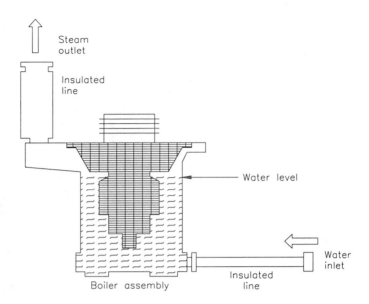

Figure 8.18 Cutaway view of a boiler and tube combination.

The second float switch is operated if the water level drops approximately $\frac{1}{2}$ in below the optimum level. This would be tantamount to a water system failure; the switch would be used to open the control circuit interlocks and remove power from the tube.

For the control box to perform its protective function properly, the water-level mark must be precisely in line with the water-level mark on the boiler. For electrical reasons, the control box generally is mounted some distance from the boiler, and therefore leveling of the two components must be carefully checked during installation. Figure 8.18 shows a cutaway drawing of a classic boiler and tube combination. Figure 8.19 is a cutaway drawing of a control box, showing the position of the float switches and the overflow pipe.

During extended operation, some quantity of water and steam being circulated through the condenser will be lost. The amount is dependent on the size of the system. The water level in the boiler, therefore, will gradually drop. The use of the control box as a reservoir minimizes this effect. In large or multiple-tube installations, an auxiliary reservoir is connected to the control box to increase the ratio of stored water to circulating water and steam. Where it may be necessary to operate multiple tubes at different physical elevations, individual control boxes are required. A multiple-tube system is shown in Figure 8.20.

Equalizer Line

For the control box to duplicate the same pressure conditions that exist in the boiler, the vapor-phase system must be fitted with an equalizer line [1]. This length of tubing connects the steam side of the system with the top of the control box. As partial steam

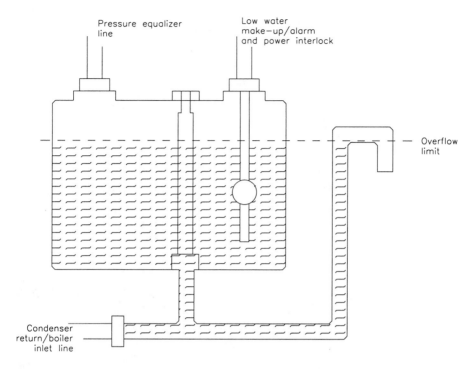

Figure 8.19 Cutaway view of a control box.

pressure begins to build up in the boiler, the equalizer line allows this same pressure to appear in the control box. Although the steam pressure is low, less than 0.5 psi above atmosphere, an error in the operation of the control box would be introduced unless the vessel were equalized.

The fitting used to connect the equalizer line to the steam outlet tube must be constructed to prevent a venturi effect from developing because of the velocity of the vapor. This is best accomplished by directing an elbow within the adapter fitting toward the boiler, as shown in Figure 8.21.

Condenser

Both air-cooled and water-cooled condensers are available for vapor cooling systems [1]. Choose condensers with good reserve capabilities and low pressure drop. Air- and water-cooled condensers may be mounted in any position, provided they allow the condensed water to flow freely by gravity to the boiler return line. Water must not be allowed to stand in the condenser where it might cause back pressure to the entering steam.

The condenser should be mounted above the level of the boiler(s) so that water will drain from it to the boiler return line. Where it is necessary to mount the condenser at a lower physical level than the system water level, an auxiliary pump must be used to re-

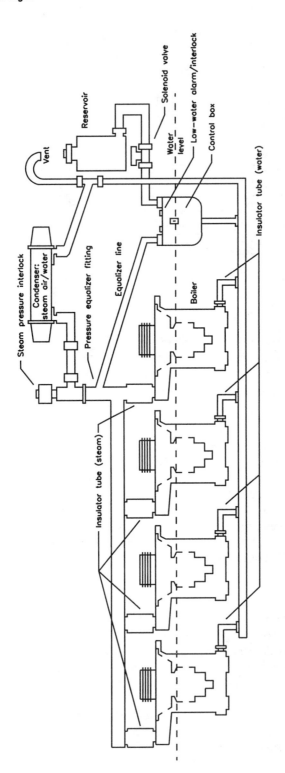

Figure 8.20 Typical four-tube vapor cooling system with a common water supply.

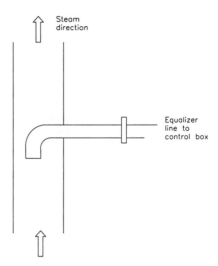

Figure 8.21 Cutaway view of a pressure equalizer fitting.

turn water to the boiler. This arrangement is recommended for the "steam-out-the-bottom" boiler system, as illustrated in Figure 8.22.

Pressure Interlock

Most tube manufacturers recommend the use of a steam pressure interlock switch on the steam or inlet side of the condenser [1]. This switch, set to about 0.5 lbs/in^2, is used as a power interlock that senses abnormal steam pressure resulting from constrictions in the condenser or piping.

Piping

Piping should be of copper or glass throughout the system [1]. The steam piping should be the same diameter as the Pyrex tube from the boiler. The size is dependent on the power level and the volume of generated steam, and it will range from 1-1/4 in at the 8 kW level to 6 in for the 250 kW level of dissipation. The steam path should be as direct as practical and must be sloped to prevent condensate from collecting at a low point where it might cause back pressure. All low spots should be drained back to the inlet water line.

The diameter of water return piping from the condenser to the control box will vary from 3/4 to 1-3/4 in, depending again on the power level. This tubing should be the same diameter as the boiler inlet water fitting. It also should be sloped so that water or vapor pockets are not allowed to form, and it must allow the condensate to return by gravity to the control box and the boiler. A vent to air on the outlet side of the condenser should be incorporated to maintain the water side of the system at atmospheric pressure. Provisions for draining the distilled water should be made at the lowest physical level of the system.

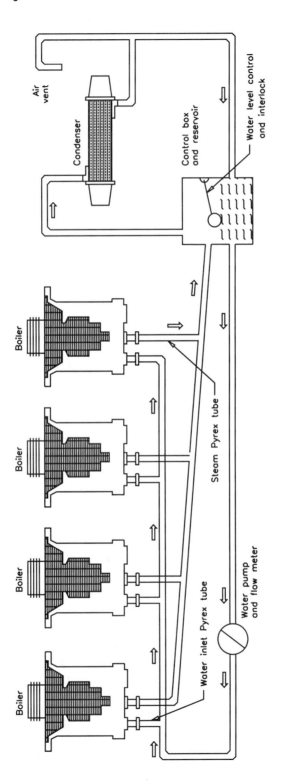

Figure 8.22 Typical four-tube vapor cooling system using "steam-out-the-bottom" boilers.

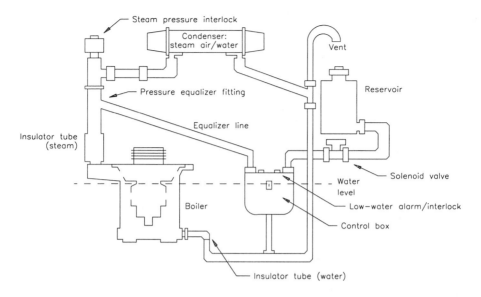

Figure 8.23 Vapor cooling system incorporating a reservoir of distilled water.

The equalizer line also should be sloped from the adapter fitting on the steam line to the top of the control box. This will allow the condensate to return to the control box.

Automatic Refilling System

Figure 8.23 shows a typical vapor cooling system with provisions to allow additional water into the control box [1]. An auxiliary reservoir is connected through a solenoid-operated water valve to the controller. When water loss resulting from evaporation causes the water level in the boiler and the control box to drop about $\frac{1}{4}$ in below normal, the first float switch in the control box closes and actuates the solenoid-controlled valve to permit makeup water to enter the system. When the proper level is restored, the switch opens, the valve closes, and the flow of makeup water is stopped.

Alternative Vapor Cooling Systems

The systems described thus far are "classic" designs in which a separate tube, boiler, condenser, and level control box are used [1]. Variations on this basic scheme are numerous. One such alternative system, offered for use with large power tubes, utilizes a "steam-out-the-bottom" boiler. This configuration makes it possible to keep the steam and water systems, plus the plumbing, below the tubes. This configuration often simplifies the electrical design of the amplifier or oscillator. Figure 8.24 shows a typical boiler used in this cooling technique.

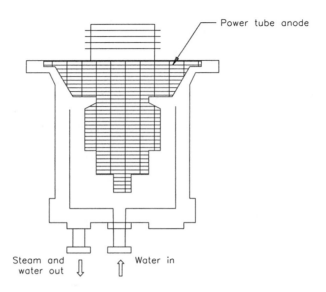

Figure 8.24 Cutaway view of a "steam-out-the-bottom" boiler.

A small water pump circulates a continuous flow of water over a weir—or baffle—in the boiler, maintaining a constant water level. Generated steam is forced under slight pressure out the bottom of the boiler, through an insulator tube in the condenser. Water from the condenser flows into the control box before being pumped back into the boiler. Protective devices include a water-flow interlock and water-level switch in the control box to ensure an adequate supply of coolant.

Maintenance

Maintenance problems associated with circulating water systems are practically eliminated in a vapor cooling system [1]. As mentioned previously, systems can be designed to eliminate all rotating machinery or moving parts. Good engineering practice does, however, dictate periodic maintenance. The glass insulator tubes should be inspected occasionally to be sure they contain no deposits that might cause high-voltage flashover. Water conductivity can be checked by measuring the dc resistance, as in a typical circulating water system. The water should be replaced if the dc resistance drops below 1.0 MΩ-cm.

In practice, a vapor-cooled system will remain cleaner longer than a water-cooled system. In the vapor-cooled boiler, the water is continuously being redistilled, and only pure water is introduced at the bottom of the boiler. Any contaminants will tend to remain in the boiler itself, where they can be removed easily. The periods between equipment shutdown for draining and cleaning will be at least twice as long for a vapor cooling system because of this inherent self-cleaning action.

Each time a tube is removed or replaced, the rubber O-ring between the boiler and the tube should be inspected and, if necessary, replaced. At the same time, the inside of the boiler and the control box can be inspected and cleaned as needed. The electrolytic *target* should be replaced whenever its metallic end is no longer visible in the inlet water line.

8.2.4 Temperature Measurements

Before completing any design employing power tubes, check temperatures in critical areas, such as metal/ceramic seals and (except for water- and vapor-cooled types) the anode just above and below the cooling fins [1]. Another critical area on many medium and large tubes of coaxial design is the central part of the base, which is often recessed and may need special cooling provisions. Electrical contact to the tube terminals is usually by copper-beryllium collets, which should not exceed a temperature of 150°C for any extended period of time. A number of manufacturers make temperature-sensitive paints that are useful for such measurements.[3] "Crayon-type" indicators are also available, in a wide range of temperatures. Conventional temperature measuring-techniques may be affected by the radio frequency energy concentrated at the tube and nearby components. Furthermore, because all power tubes operate at high voltage, there is danger of electric shock with directly connected devices.

When a temperature-sensitive paint or crayon is used for measurement, it must be remembered that many of the areas of concern are vacuum seals involving relatively thin metal. It is important, therefore, that the indicator material be noncorrosive.

Considering the importance of tube element temperatures, every design must be evaluated carefully. All power tubes carry an absolute maximum temperature rating for the seals and envelope, and in the case of an external-anode forced-air-cooled tube, for the anode core itself. Operation above the rated maximum can cause early tube failure. Where long life and consistent performance are important factors, it is normally desirable to maintain tube temperatures comfortably below the rated maximum.

The equipment operator has the ultimate responsibility of making sure that the power tubes used in the system are operating within specified ratings. Most current tube designs use a ceramic envelope and an externally mounted anode or collector. The only visible sign of excessive temperature may be the cosmetic plating beginning to darken or turn black (oxidize). By then, significant damage may have occurred.

"Crayon" Temperature Measurement

When a crystalline solid is heated, a temperature will be reached at which the solid changes sharply to a liquid [1]. This melting point has a definite, reproducible value that is virtually unaffected by ambient conditions that may cause errors in other temperature-sensing methods. For example, fusible temperature indicators are practical in induction heating and in the presence of static electricity or ionized air around

---

3 *Tempilaq* is one such product (Tempil Division, Big Three Industrial Gas and Equipment Co., South Plainfield, NJ).

electric equipment, where electronic means of measuring temperatures often function erratically.

The most popular type of fusible indicator is a temperature-sensitive stick closely resembling a crayon, with a calibrated melting point. These crayon-type indicators are made in 100 or so different specified temperature ratings in the range of 100 to 2500°F. Typically, each crayon has a temperature-indicating accuracy within 1 percent of its rating. The workpiece to be tested is marked with the crayon. When the workpiece attains the predetermined melting point of the crayon, the mark instantly changes from a solid to a liquid phase (liquefies), indicating that the workpiece has reached that temperature.

Under certain circumstances, premarking with crayon is not practical. This is the case if one or more of the following is true:

- A prolonged heating period is experienced.

- The surface is highly polished and does not readily accept a crayon mark.

- The material being marked is one that gradually absorbs the liquid phase of the crayon.

Note that a melted mark, upon cooling, will not solidify at the exact temperature at which it melted. Solidification of a melted crayon mark, therefore, cannot be relied on for exact temperature indication.

Phase-Change Fluid

A phase-changing fluid, or *fusible temperature-indicating lacquer*, offers the greatest flexibility in temperature measurement of a power tube [1]. The lacquer-type fluid contains a solid material of calibrated melting point suspended in an inert, volatile, nonflammable vehicle. As with crayon-type indicators, there are approximately 100 different temperature ratings, covering a range from 100 to 2500°F. Accuracy is typically within 1 percent of the specified value.

The lacquer is supplied in the proper consistency for brushing. If spraying or dipping is desirable as the mode of application, a thinner is available to alter the viscosity without impairing the temperature-indicating accuracy.

Phase-changing lacquers often are used when a smooth or soft surface must be tested, or in situations where the surface is not readily accessible for application of a crayon mark during the heating process. Within seconds after application, the lacquer dries to a dull matte finish, and it responds rapidly when the temperature to be indicated is reached. The response delay of a lacquer mark is only a fraction of a second. This time can be reduced to milliseconds if a mark of minimal thickness is applied.

Upon reaching its rated temperature, the lacquer mark will liquefy. On subsequent cooling, however, the fluid will not revert to its original unmelted appearance but, rather, to a glossy or crystalline coating, which is evidence of its having reached the required temperature. Temperature-indicating lacquers, upon cooling, will not resolidify at the same temperature at which they melted.

If spraying of the phase-change lacquer is to be the sole means of application, the operator may find it more convenient to use an aerosol-packaged form of this material. An aerosol phase-change temperature indicator is identical to the brush-on material in performance and interpretation.

The first commercial form of the fusible indicator was the *pellet*, which continues to be useful in certain applications. Pellets are most frequently employed when extended heating periods are involved, or where a greater bulk of indicator material is necessary. They are also useful when observations must be made from a distance and when air-space temperatures are to be monitored.

Phase-change temperature-indicator pellets are available in flat, $\frac{7}{16}$-in-diameter $\times$ $\frac{1}{8}$-in-thick tablets. For special applications, miniature pellets, $\frac{1}{8}$ in $\times \frac{1}{8}$ in, are also available. The range of 100 to 2500°F covered by the crayon and lacquer-type is, likewise, covered by the pellets. Pellets with coverage extending to 3200°F also may be obtained.

Another variation of the phase-change indicator is the temperature-sensitive label.[4] These self-adhesive-backed monitors consist of one or more heat-sensitive indicators sealed under transparent heat-resistant windows. The centers of the indicator circles turn from white to black at the temperature ratings shown on the label. The change to black is irreversible, representing an absorption of the temperature-sensitive substance into its backing material. After registering the temperature history of the workpiece, the exposed monitor label can be removed and affixed to a service report to remain part of a permanent record.

Selection Process

Fusible temperature indicators and lacquers have at least three major advantages over other methods of determining surface temperature [1]:

- The temperature indications obtained are unquestionably those of the surface being tested. It is not necessary to equilibrate a relatively massive probe with the surface (a probe may conduct heat away from the region being tested), which requires the use of a correction factor to obtain the actual surface temperature.

- There is no delay in obtaining an indication. Because a mark left by a crayon or a lacquer is of extremely small mass, it attains rapid equilibrium with the surface. There is no conduction of heat away from the surface, which would prolong response time and result in erroneously low temperature readings.

- The technique is simple and economical. Determination of surface temperature by most other means requires technical competence and skill and, in some cases, sophisticated instrumentation. Accurate surface temperature readings can be obtained with fusible indicators and lacquers with little effort, training, and expense.

---

4 *Temp-Plate* is one such product (Williams-Wahl Corp., Santa Monica, CA).

There are numerous instances, particularly in determining heat distribution in complex systems, in which the relative simplicity of fusible indicators and lacquers makes surface temperature investigations feasible.

8.2.5 Air-Handling System

All modern air-cooled PA tubes use an air-system socket and matching chimney for cooling [1]. The chimney is designed to be an integral part of the tube cooling system. Operation without the chimney may significantly reduce airflow through the tube and result in overdissipation of the device. It also is possible that operation without the proper chimney could damage other components in the circuit because of excessive radiated heat. Normally, the tube socket is mounted in a pressurized compartment so that cooling air passes through the socket and is guided to the anode cooling fins, as illustrated in Figure 8.25.

Cooling of the socket assembly is important for proper cooling of the tube base and for cooling the contact rings of the tube itself. The contact fingers used in the collet assembly of a socket typically are made of beryllium copper. If subjected to temperatures above 150°C for an extended period of time, the beryllium copper will lose its temper (springy characteristic) and will no longer make good contact with the base rings of the device. In extreme cases, this type of socket problem may lead to arcing, which can burn through the metal portion of the tube base ring. Such an occurrence ultimately may lead to catastrophic failure of the device because of a loss of the vacuum envelope. Other failure modes for a tube socket include arcing between the collet and tube ring, which can weld a part of the socket and tube together. The result is failure of both the tube and the socket.

8.3   Operating Environment

Long-term reliability of a power vacuum tube requires regular attention to the operating environment. Periodic tests and preventive maintenance are important components of this effort. Optimum performance of the cooling system can be achieved only when all elements of the system are functioning properly.

8.3.1 Air-Handling System

The temperature of the intake air supply is a parameter that is usually under the control of the end user. The preferred cooling air temperature is typically no higher than 75°F, and no lower than the room dew point. The air temperature should not vary because of an oversized air-conditioning system or because of the operation of other pieces of equipment at the facility. Monitoring the PA tube exhaust stack temperature is an effective method of evaluating overall RF system performance. This can be easily accomplished. It also provides valuable data on the cooling system and final stage tuning.

Another convenient method for checking the efficiency of the transmitter cooling system over a period of time involves documenting the back pressure that exists within

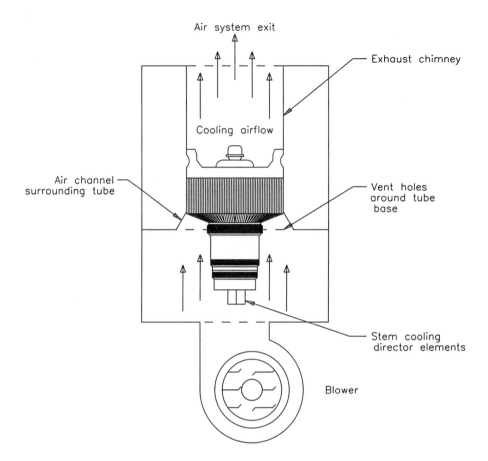

Figure 8.25 Airflow system for an air-cooled power tube.

the PA cavity. This measurement is made with a *manometer*, a simple device that is available from most heating, ventilation, and air-conditioning (HVAC) suppliers. The connection of a simplified manometer to a transmitter PA output compartment is illustrated in Figure 8.26.

By charting the manometer readings, it is possible to accurately measure the performance of the transmitter cooling system over time. Changes resulting from the buildup of small dust particles (*microdust*) may be too gradual to be detected except through back-pressure charting. Deviations from the typical back-pressure value, either higher or lower, could signal a problem with the air-handling system. Decreased PA input or output compartment back pressure could indicate a problem with the blower motor or an accumulation of dust and dirt on the blades of the blower assembly. Increased back pressure, on the other hand, could indicate dirty PA tube anode cooling fins (for the input compartment case) or a buildup of dirt on the PA exhaust ducting (for the output compartment case). Either condition is cause for concern. A system suffering from re-

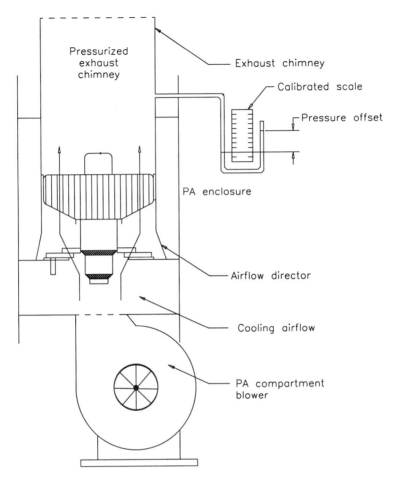

Pressurized
exhaust
chimney

Exhaust chimney

Calibrated scale

Pressure offset

PA enclosure

Airflow director

Cooling airflow

PA compartment
blower

Figure 8.26 A manometer device used for measuring back pressure in the PA compartment of an RF generator.

duced air pressure into the PA compartment must be serviced as soon as possible. Failure to restore the cooling system to proper operation may lead to premature failure of the PA tube or other components in the input or output compartments. Cooling problems do not improve with time; they always get worse.

Failure of the PA compartment air-interlock switch to close reliably may be an early indication of impending trouble in the cooling system. This situation could be caused by normal mechanical wear or vibration of the switch assembly, or it may signal that the PA compartment air pressure has dropped. In such a case, documentation of manometer readings will show whether the trouble is caused by a failure of the air pressure switch or a decrease in the output of the air-handling system.

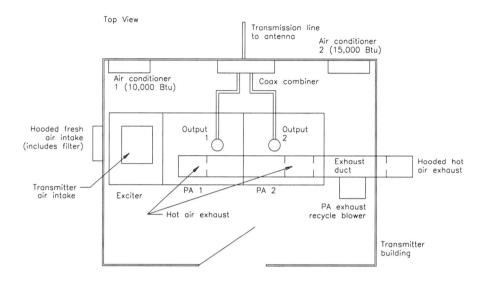

Figure 8.27 A typical heating and cooling arrangement for a high-power transmitter installation. Ducting of PA exhaust air should be arranged so that it offers minimum resistance to airflow.

8.3.2 Air Cooling System Design

Cooling system performance in an RF generator is not necessarily related to airflow volume. The cooling capability of air is a function of its mass, not its volume. The designer must determine an appropriate airflow rate within the equipment and establish the resulting resistance to air movement. A specified static pressure that should be present within the ducting of the transmitter can be a measure of airflow. For any given combination of ducting, filters, heat sinks, RFI honeycomb shielding, tubes, tube sockets, and other elements in the transmitter, a specified system resistance to airflow can be determined. It is important to realize that any changes in the position or number of restricting elements within the system will change the system resistance and, therefore, the effectiveness of the cooling. The altitude of operation is also a consideration in cooling system design. As altitude increases, the density (and cooling capability) of air decreases. A calculated increase in airflow is required to maintain the cooling effectiveness that the system was designed to achieve.

Figure 8.27 shows a typical high-power transmitter plant. The building is oriented so that the cooling activity of the blowers is aided by normal wind currents during the summer months. Air brought in from the outside for cooling is filtered in a hooded air-intake assembly. The building includes a heater and air conditioner.

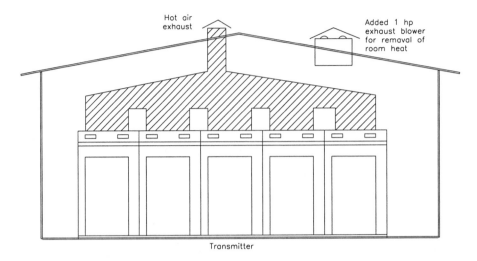

Figure 8.28 Case study in which excessive summertime heating was eliminated through the addition of a 1 hp exhaust blower to the building.

The layout of a transmitter room HVAC system can have a significant impact on the life of the PA tube(s) and the ultimate reliability of the RF generator. Air intake and output ports must be designed with care to avoid airflow restrictions and back-pressure problems. This process, however, is not as easy as it may seem. The science of airflow is complex and generally requires the advice of a qualified HVAC consultant.

To help illustrate the importance of proper cooling system design and the real-world problems that some facilities have experienced, consider the following examples taken from actual case histories:

Case 1

A fully automatic building ventilation system (Figure 8.28) was installed to maintain room temperature at 20°C during the fall, winter, and spring. During the summer, however, ambient room temperature would increase to as much as 60°C. A field survey showed that the only building exhaust route was through the transmitter. Therefore, air entering the room was heated by test equipment, people, solar radiation on the building, and radiation from the transmitter itself before entering the transmitter. The problem was solved through the addition of an exhaust fan (3000 cfm). The 1 hp fan lowered room temperature by 20°C.

Case 2

A simple remote installation was constructed with a heat-recirculating feature for the winter (Figure 8.29). Outside supply air was drawn by the transmitter cooling system blowers through a bank of air filters, and hot air was exhausted through the roof. A

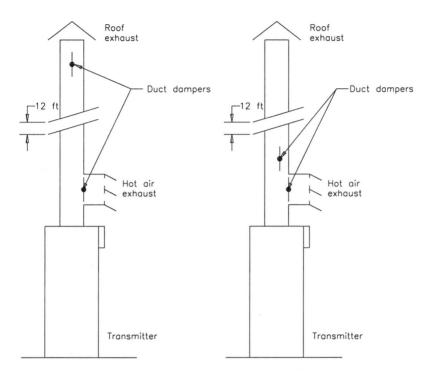

Figure 8.29 Case study in which excessive back pressure to the PA cavity was experienced during winter periods, when the rooftop damper was closed. The problem was eliminated by repositioning the damper as shown.

small blower and damper were installed near the roof exit point. The damper allowed hot exhaust air to blow back into the room through a tee duct during the winter months. For summer operation, the roof damper was switched open and the room damper closed. For winter operation, the arrangement was reversed. The facility, however, experienced short tube life during winter operation, even though the ambient room temperature during winter was not excessive.

The solution involved moving the roof damper 12 ft down to just above the tee. This eliminated the stagnant "air cushion" above the bottom heating duct damper and significantly improved airflow in the region. Cavity back pressure was, therefore, reduced. With this relatively simple modification, the problem of short tube life disappeared.

Case 3

An inconsistency regarding test data was discovered within a transmitter manufacturer's plant. Units tested in the engineering lab typically ran cooler than those at the manufacturing test facility. Figure 8.30 shows the test station difference, a 4-ft exhaust stack that was used in the engineering lab. The addition of the stack increased

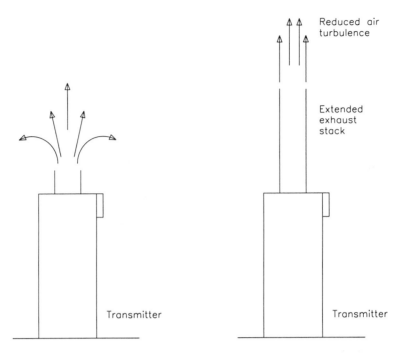

Figure 8.30 Case study in which air turbulence at the exhaust duct resulted in reduced airflow through the PA compartment. The problem was eliminated by adding a 4-ft extension to the output duct.

airflow by up to 20 percent because of reduced air turbulence at the output port, resulting in a 20°C decrease in tube temperature.

These examples point out how easily a cooling problem can be caused during HVAC system design. All power delivered to the transmitter either is converted to RF energy and sent to the antenna or becomes heated air (or water). Proper design of a cooling system, therefore, is a part of transmitter installation that should not be taken lightly.

8.3.3 Site Design Guidelines

There are any number of physical plant designs that will provide for reliable operation of high-power RF systems [3]. One constant, however, is the requirement for tight temperature control. Cooling designs can be divided into three broad classifications:

- Closed site design
- Open site design
- Hybrid design

If the equipment user is to provide adequately for hot air exhaust and fresh air intake, the maximum and minimum environmental conditions in which the equipment will operate must be known. In addition, the minimum cooling requirements of the equipment must be provided by the manufacturer. The following parameters should be considered:

- Site altitude
- Maximum expected outside air temperature
- Minimum expected outside air temperature
- Total airflow through the transmitter
- Air temperature rise through the transmitter
- Air exhaust area

Keep in mind that the recommended cooling capacity for a given RF system applies only to cooling the transmitter; any additional cooling load in the building must be considered separately when selecting the air system components. The transmitter exhaust should not be the only exhaust port in the room because heat from the peripheral equipment would then be forced to go out through the transmitter.

The *sensible-heat load*, then, is the sum of all additional heat loads, including:

- Solar radiation
- Heat gains from equipment and lights
- Heat gains from personnel in the area that is to be cooled

Closed Site Design

Figure 8.31 illustrates a site layout that works well in most climates, as long as the transmitter is small and the building is sealed and well insulated [3]. In fact, this closed configuration will ensure the longest possible transmitter life and lowest maintenance cost. No outside air laden with moisture and contaminants circulates through the transmitter.

At sites using this arrangement, it has been observed that periodic transmitter cleaning is seldom required. In a typical closed system, the air conditioner is set to cool when the room temperature reaches 75 to 80°F. The closed system also uses a louvered emergency intake blower, which is set by its own thermostat to pull in outside air if the room temperature reaches excessive levels (above 90°F). This blower is required in a closed configuration to prevent the possibility of thermal runaway if the air conditioner fails. Without such an emergency ventilation system, the transmitter would recirculate its own heated air, further heating the room. System failure probably would result.

During winter months, the closed system is self-heating (unless the climate is harsh, or the transmitter power output is low), because the transmitter exhaust is not ducted outside but simply empties into the room. Also during these months, the emergency intake blower can be used to draw cold outside air into the room instead of using the air conditioner, although this negates some of the cleanliness advantages inherent to the closed system.

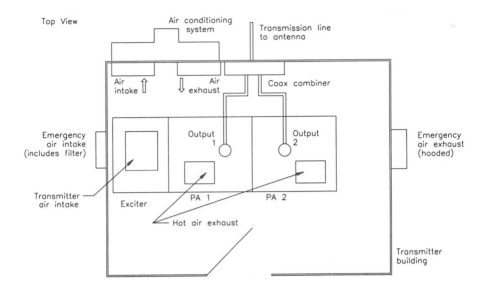

Figure 8.31 Closed site ventilation design, with a backup inlet/outlet system.

An exhaust blower should not be substituted for an intake blower, because positive room pressure is desired for venting the room. This ensures that all air in the building has passed through the intake air filter. Furthermore, the louvered emergency exhaust vent(s) should be mounted high in the room, so that hot air is pushed out of the building first.

The closed system usually makes economic sense only if the transmitter exhaust heat load is relatively small.

Periodic maintenance of a closed system involves the following activities:

- Checking and changing the air conditioner filter periodically

- Cleaning the transmitter air filter as needed

- Checking that the emergency vent system works properly

- Keeping the building sealed from insects and rodents

Open Site Design

Figure 8.32 depicts a site layout that is the most economical to construct and operate [3]. The main attribute of this approach is that the transmitter air supply is not heated or cooled, resulting in cost savings. This is not a closed system; outdoor air is pumped into the transmitter room through air filters. The transmitter then exhausts the hot air. If the duct work is kept simple and the transmitter has a dedicated exhaust port, such a direct exhaust system works adequately. Many transmitters do not lend themselves to

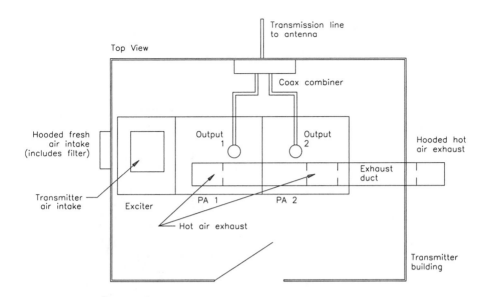

Figure 8.32 Open site ventilation design, using no air conditioner.

a direct exhaust connection, however, and a hood mounted over the transmitter may be required to collect the hot air, as illustrated in Figure 8.33. With a hooded arrangement, it may be necessary to install a booster fan in the system, typically at the wall or roof exit, to avoid excessive back pressure.

When building an open air-circulation system, there are some important additional considerations. The room must be positively pressurized so that only filtered air, which has come through the intake blower and filter, is available to the transmitter. With a negatively pressurized room, air will enter through every hole and crack in the building, and will not necessarily be filtered. Under negative pressure, the transmitter blower also will have to work harder to exhaust air.

The intake blower should be a "squirrel-cage" type rather than a fan type. A fan is meant to move air within a pressurized environment; it cannot compress the air. A squirrel-cage blower will not only move air, but also will pressurize the area into which the air is directed. Furthermore, the intake air blower must be rated for more cubic feet per minute (cfm) airflow than the transmitter will exhaust outside. A typical 20 kW FM broadcast transmitter, for example, will exhaust 500 cfm to 1000 cfm. A blower of 1200 cfm, therefore, would be an appropriate size to replenish the transmitter exhaust and positively pressurize the room.

In moderate and warm climates, the intake blower should be located on a north-facing outside wall. If the air intake is on the roof, it should be elevated so that it does not pick up air heated by the roof surface.

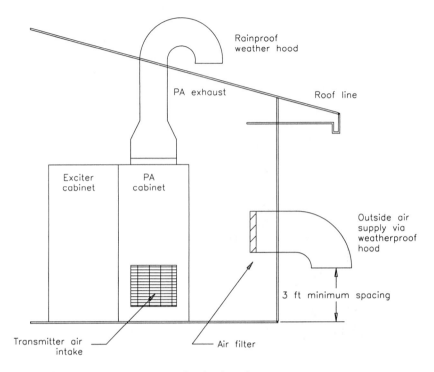

Figure 8.33 Transmitter exhaust-collection hood.

High-quality pleated air filters are recommended. Home-style fiberglass filters are not sufficient. Local conditions may warrant using a double filtration system, with coarse and fine filters in series.

Secure the advice of a knowledgeable HVAC shop when designing filter boxes. For a given cfm requirement, the larger the filtration area, the lower the required air velocity through the filters. This lower velocity results in better filtration than forcing more air through a small filter. In addition, the filters will last longer. Good commercial filtration blowers designed for outdoor installation are available from industrial supply houses and are readily adapted to RF facility use.

The transmitter exhaust may be ducted through a nearby wall or through the roof. Avoid ducting straight up, however. Many facilities have suffered water damage to transmitters in such cases because of the inevitable deterioration of roofing materials. Normally, ducting the exhaust through an outside wall is acceptable. The duct work typically is bent downward a foot or two outside the building to keep direct wind from creating back pressure in the exhaust duct. Minimize all bends in the duct work. If a 90° bend must be made, it should be a large-radius bend with curved *helper* vanes inside the duct to minimize turbulence and back pressure. A 90° L- or T-bend is not recommended, unless oversized and equipped with internal vanes to assist the turning airflow.

For moderate and cool climates, an automatic damper can be employed in the exhaust duct to direct a certain amount of hot exhaust air back into the transmitter build-

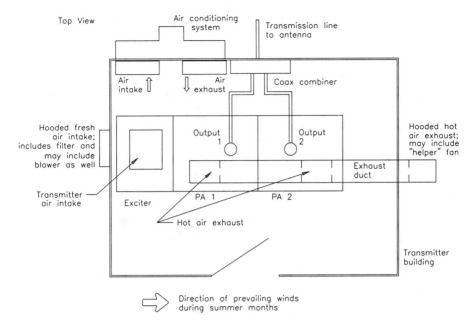

Figure 8.34 Hybrid site ventilation design, using an air conditioner in addition to filtered outside air.

ing as needed for heating. This will reduce outside air requirements, providing clean, dry, heated air to the transmitter during cold weather.

Hybrid Design

Figure 8.34 depicts a hybrid site layout that is often used for high-power transmitters or sites supporting multiple transmitters [3]. As with the layout in Figure 8.32, the room is positively pressurized with clean, filtered air from the outside. A portion of this outside air is then drawn through the air conditioner for cooling before delivery to the transmitter area. Although not all of the air in the room goes through the air conditioner, enough does to make a difference in the room temperature. In humid areas, much of the moisture is removed by the air conditioner. The transmitter exhaust is directed outside.

Such a hybrid system is often the choice for larger transmitter sites where a closed system would prove too costly, but an unconditioned system would run excessively hot during the summer months.

Some closed or hybrid systems use two parallel air conditioners. Most of the time, only one is in use. The thermostats of the units are staggered so that if one cannot keep the air below the ideal operating point, the other will turn on to assist.

8.3.4 Water/Vapor Cooling System Maintenance

The cooling system is vital to any RF generator. In a klystron-based unit, for example, the cooling system may dissipate a large percentage of the input ac power in the form of waste heat in the collector. Pure, distilled water should be used. Deionized water is required unless the isolation of high voltages is unnecessary, as in the case of the collector of a klystron (the collector typically operates at or near ground potential).

Any impurities in the cooling system water eventually will find their way into the water jacket and cause corrosion of the plate or collector. It is essential to use high-purity water with low conductivity, and to replace the water in the cooling system as needed. Efficient energy transfer from the heated surface into the water is necessary for long vacuum tube life. Oil, grease, soldering flux residue, and pipe sealant containing silicone compounds must be excluded from the cooling system. This applies to both vapor- and liquid-conduction cooling systems, although it is usually more critical in the vapor-phase type. The sight glass in a vapor-phase water jacket provides a convenient checkpoint for coolant condition.

In general, greater flows and greater pressures are inherent in liquid-cooled vs. vapor-phase systems, and, when a leak occurs, large quantities of coolant can be lost before the problem is discovered. The condition of gaskets, seals, and fittings, therefore, should be checked regularly. Many liquid-cooled tubes use a distilled water and ethylene glycol mixture. (Do not exceed a 50:50 mix by volume.) The heat transfer of the mixture is lower than that of pure water, requiring the flow to be increased, typically by 25 percent. Greater coolant flow suggests closer inspection of the cooling system after adding the glycol.

The action of heat and air on ethylene glycol causes the formation of acidic products. The acidity of the coolant can be checked with litmus paper. Buffers can be added to the glycol mixture. Buffers are alkaline salts that neutralize acid forms and prevent corrosion. Because they are ionizable chemical salts, the buffers cause the conductivity of the coolant to increase. Consult the power tube manufacturer before adding buffers.

The following general guidelines are recommended for power tube cooling systems:

- The resistivity of the water must be maintained above the minimum level established by the power tube manufacturer. Although recommended values vary, generally speaking, resistivity should be maintained at or above 1 MΩ-cm (at 30°C) for power tubes operating at a high-voltage, referenced to ground. A resistivity value of 30 kΩ-cm or greater (at 30°C) is usually sufficient for tubes whose collector operates at or near ground potential (such as a klystron).

- The pH factor must be within the range of 6.0 to 8.0.

- The particulate matter size must not be greater than 50 microns (325 mesh).

- The inlet water temperature must not exceed 70°C, and this temperature should be regulated to ± 5°C.

When the water in the cooling system fails to satisfy any one of these requirements, prompt action is necessary to correct the deficiency. If the water is contaminated, the system must be flushed and replaced with clean water.

Tests for Purity

The recommended method for measuring resistivity, pH, and particulate matter is by use of laboratory instruments. However, the following simple tests should provide sufficient information to determine the general condition of the coolant.

- The appearance of the water is a good general indicator of coolant condition. If the water looks turbid or tastes or smells brackish, it is good practice to change the water and flush the system.

- The pH factor is easily checked by using pH paper or litmus paper. Sold under a variety of trade names, pH paper is available at laboratory supply stores. Directions for use usually are printed on the package.

- Particulate matter and impurities can be checked by using the "foaming test" (described in the next section).

- Resistivity can be measured accurately with a resistance bridge. Generally speaking, if the water does not pass the foaming test, the resistivity will be low as well.

Foaming Test for Water Purity

One of the many factors affecting the life and operating efficiency of vapor-cooled tubes is the purity of the water in the cooling system. If impurities are present, foaming may occur that will inhibit heat transfer, thereby lowering the cooling efficiency of the system. The impurities that most frequently produce foaming are:

- Cleaning-compound residue

- Detergents

- Joint-sealing compounds

- Oily rust preventives in pumps and other components

- Valve-stem packing

- Impurities in tap water

Tests should be conducted periodically to confirm the quality of the coolant in a vapor-phase system. Conduct such tests after each water change, system cleaning, or modification. The following items are required:

- A ½-in × 4-in glass test tube with rubber stopper

- A 1-pint glass or polypropylene bottle with cap

The procedure is as follows:

- Fill the cooling system with water, and circulate it until thoroughly mixed (about 30 minutes).

- Drain a sample of water into the bottle, and cool to room temperature.

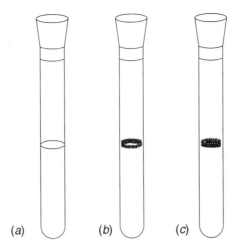

Figure 8.35 Results of the "foaming test" for water purity: (*a*) pure sample, (*b*) sample that is acceptable on a short-term basis, (*c*) unacceptable sample.

- If the water sample stands for more than 1 hour, slowly invert the capped bottle about 10 times. Avoid shaking the bottle because this will create air bubbles in the water. (When the water is static, foaming impurities tend to collect at the surface. This step mixes the sample without generating foam.)

- Using the sample water, rinse the test tube and stopper three times.

- Fill the test tube to the halfway mark with the sample water.

- Shake the test tube vigorously for 15 s.

- Let the sample stand for 15 s.

- Observe the amount of foam remaining on top of the water and compare it with the drawings shown in Figure 8.35.

A completely foam-free water surface and test tube, as illustrated in (*a*), indicates no foam-producing impurities. If the water surface and test tube wall are partially covered with foam, but a circle of clear water appears in the center, the impurity level is temporarily acceptable (*b*). A second test should be made in about one week. If the foam layer completely bridges the inside of the test tube (*c*), the system should be flushed and cleaned.

8.4   References

1. Laboratory Staff, *The Care and Feeding of Power Grid Tubes*, Varian Associates, San Carlos, CA, 1984.
2. Besch, David F., "Thermal Properties," in The Electronics Handbook, Jerry C. Whitaker (ed.), CRC Press, Boca Raton, FL, pp. 127–134, 1996.

3. Harnack, Kirk, "Airflow and Cooling in RF Facilities," *Broadcast Engineering*, Intertec Publishing, Overland Park, KS, pp. 33–38, November 1992.

8.5   Bibliography

"Cleaning and Flushing Klystron Water and Vapor Cooling Systems," Application Engineering Bulletin no. AEB-32, Varian Associates, Palo Alto, CA, February 1982.

Crutchfield, E. B. (ed.), *NAB Engineering Handbook*, 8th Ed., National Association of Broadcasters, Washington, DC, pp. 547–550, 1992.

"Foaming Test for Water Purity," Application Engineering Bulletin no. AEB-26, Varian Associates, Palo Alto, CA, September 1971.

Harnack, Kirk, "Airflow and Cooling in RF Facilities," *Broadcast Engineering*, Intertec Publishing, Overland Park, KS, pp. 33–38, November 1992.

Harper, Charles A., *Electronic Packaging and Interconnection Handbook*, McGraw-Hill, New York, NY, 1991.

"High Power Transmitting Tubes for Broadcasting and Research," Philips Technical Publication, Eindhoven, the Netherlands, 1988.

Integral Cavity Klystrons for UHF-TV Transmitters, Varian Associates, Palo Alto, CA.

Kimmel, E., "Temperature Sensitive Indicators: How They Work, When to Use Them," *Heat Treating*, March 1983.

"Power Grid Tubes for Radio Broadcasting," Thomson-CSF Publication #DTE-115, Thomson-CSF, Dover, NJ, 1986.

"Protecting Integral-Cavity Klystrons Against Water Leakage," Technical Bulletin no. 3834, Varian Associates, Palo Alto, CA, July 1978.

"Temperature Measurements with Eimac Power Tubes," Application Bulletin no. 20, Varian Associates, San Carlos, CA, January 1984.

"Water Purity Requirements in Liquid Cooling Systems," Application Bulletin no. 16, Varian Associates, San Carlos, CA, July 1977.

"Water Purity Requirements in Water and Vapor Cooling Systems," Application Engineering Bulletin no. AEB-31, Varian Associates, Palo Alto, CA, September 1991.

Whitaker, Jerry C., *Maintaining Electronic Systems*, CRC Press, Boca Raton, FL, 1992.

Whitaker, Jerry C., *Radio Frequency Transmission Systems: Design and Operation*, McGraw-Hill, New York, NY, 1990.

Wozniak, Joe, "Using Tetrode Power Amplifiers in High Power UHF-TV Transmitters," *NAB 1992 Broadcast Engineering Conference Proceedings*, National Association of Broadcasters, Washington, DC, pp. 209–211, 1992.

Reliability Considerations

9.1 Introduction

The ultimate goal of any design engineer or maintenance department is zero down-time. This is an elusive goal, but one that can be approximated by examining the vulnerable areas of plant operation and taking steps to prevent a sequence of events that could result in system failure. In cases where failure prevention is not practical, a reliability assessment should encompass the stocking of spare parts, circuit boards, or even entire systems. A large facility may be able to cost-justify the purchase of backup gear that can be used as spares for the entire complex. Backup hardware is expensive, but so is downtime. Because of the finite lifetime of power vacuum tubes, these considerations are of special significance.

Failures can, and do, occur in electronic systems. The goal of product quality assurance at every step in the manufacturing and operating chain is to ensure that failures do not produce a systematic or repeatable pattern. The ideal is to eliminate failures altogether. Short of that, the secondary goal is to end up with a random distribution of failure modes. This indicates that the design of the system is fundamentally optimized and that failures are caused by random events that cannot be predicted. In an imperfect world, this is often the best that end users can hope for. Reliability and maintainability must be built into products or systems at every step in the design, construction, and maintenance process. They cannot be treated as an afterthought.

9.1.1 Terminology

To understand the principles of reliability engineering, the following basic terms must be defined:

- **Availability**. The probability that a system subject to repair will operate satisfactorily on demand.

- **Average life**. The mean value for a normal distribution of product or component lives. This term is generally applied to mechanical failures resulting from "wear-out."

- **Burn-in**. The initially high failure rate encountered when a component is placed on test. Burn-in failures usually are associated with manufacturing defects and the debugging phase of early service.

- **Defect**. Any deviation of a unit or product from specified requirements. A unit or product may contain more than one defect.

- **Degradation failure**. A failure that results from a gradual change, over time, in the performance characteristics of a system or part.

- **Downtime**. Time during which equipment is not capable of doing useful work because of malfunction. This does not include preventive maintenance time. Downtime is measured from the occurrence of a malfunction to its correction.

- **Failure**. A detected cessation of ability to perform a specified function or functions within previously established limits. A failure is beyond adjustment by the operator by means of controls normally accessible during routine operation of the system.

- **Failure mode and effects analysis (FMEA)**. An iterative documented process performed to identify basic faults at the component level and determine their effects at higher levels of assembly.

- **Failure rate**. The rate at which failure occurs during an interval of time as a function of the total interval length.

- **Fault tree analysis (FTA)**. An iterative documented process of a systematic nature performed to identify basic faults, determine their causes and effects, and establish their probabilities of occurrence.

- **Lot size**. A specific quantity of similar material or a collection of similar units from a common source; in inspection work, the quantity offered for inspection and acceptance at any one time. This may be a collection of raw material, parts, subassemblies inspected during production, or a consignment of finished products to be sent out for service.

- **Maintainability**. The probability that a failure will be repaired within a specified time after it occurs.

- **Mean time between failure (MTBF)**. The measured operating time of a single piece of equipment divided by the total number of failures during the measured period of time. This measurement normally is made during that period between early life and wear-out failures.

- **Mean time to repair (MTTR)**. The measured repair time divided by the total number of failures of the equipment.

- **Mode of failure**. The physical description of the manner in which a failure occurs and the operating condition of the equipment or part at the time of the failure.

- **Part failure rate**. The rate at which a part fails to perform its intended function.

- **Quality assurance (QA)**. All those activities, including surveillance, inspection, control, and documentation, aimed at ensuring that a product will meet its performance specifications.

- **Reliability**. The probability that an item will perform satisfactorily for a specified period of time under a stated set of use conditions.

- **Reliability growth**. Actions taken to move a hardware item toward its reliability potential, during development, subsequent manufacturing, or operation.

- **Reliability predictions**. Compiled failure rates for parts, components, subassemblies, assemblies, and systems. These generic failure rates are used as basic data to predict a value for reliability.

- **Sample**. One or more units selected at random from a quantity of product to represent that product for inspection purposes.

- **Sequential sampling**. Sampling inspection in which, after each unit is inspected, the decision is made to accept, reject, or inspect another unit. (Note: Sequential sampling as defined here is sometimes called *unit sequential sampling* or *multiple sampling*.)

- **System**. A combination of parts, assemblies, and sets joined together to perform a specific operational function or functions.

- **Test to failure**. Testing conducted on one or more items until a predetermined number of failures have been observed. Failures are induced by increasing electrical, mechanical, and/or environmental stress levels, usually in contrast to *life tests*, in which failures occur after extended exposure to predetermined stress levels. A life test can be considered a test to failure using age as the stress.

9.2 Quality Assurance

Electronic component and system manufacturers design and implement quality assurance procedures for one fundamental reason: Nothing is perfect. The goal of a QA program is to ensure, for both the manufacturer and the customer, that all but some small, mutually acceptable percentage of devices or systems shipped will be as close to perfection as economics and the state of the art allow. There are tradeoffs in this process. It would be unrealistic, for example, to perform extensive testing to identify potential failures if the cost of that testing exceeded the cost savings that would be realized by not having to replace the devices later in the field.

The focal points of any QA effort are *quality* and *reliability*. These terms are not synonymous. They are related, but they do not provide the same measure of a product:

- Quality is the measure of a product's performance relative to some established criteria.

- Reliability is the measure of a product's life expectancy.

Stated from a different perspective, quality answers the question of whether the product meets applicable specifications *now*; reliability answers the question of *how long* the product will continue to meet its specifications.

9.2.1 Inspection Process

Quality assurance for components normally is performed through sampling rather than through 100 percent inspection. The primary means used by QA departments for controlling product quality at the various processing steps include:

- *Gates.* A mandatory sampling of every lot passing through a critical production stage. Material cannot move on to the next operation until QA has inspected and accepted the lot.

- *Monitor points.* A periodic sampling of some attribute of the component. QA personnel sample devices at a predetermined frequency to verify that machines and operators are producing material that meets preestablished criteria.

- *Quality audit.* An audit carried out by a separate group within the QA department. This group is charged with ensuring that all production steps throughout the manufacturer's facility are in accordance with current specifications.

- *Statistical quality control.* A technique, based on computer modeling, that incorporates data accumulated at each gate and monitor point to construct statistical profiles for each product, operation, and piece of equipment within the plant. Analysis of this data over time allows QA engineers to assess trends in product performance and failure rates.

Quality assurance for a finished subassembly or system may range from a simple go/no-go test to a thorough operational checkout that may take days to complete.

9.2.2 Reliability Evaluation

Reliability prediction is the process of quantitatively assessing the reliability of a component or system during development, before large-scale fabrication and field operation. During product development, predictions serve as a guide by which design alternatives can be judged for reliability. To be effective, the prediction technique must relate engineering variables to reliability variables.

A prediction of reliability is obtained by determining the reliability of each critical item at the lowest system level and proceeding through intermediate levels until an estimate of overall reliability is obtained. This prediction method depends on the availability of accurate evaluation models that reflect the reliability of lower-level components. Various formal prediction procedures are used, based on theoretical and statistical concepts.

Parts-Count Method

Although the parts-count method does not relate directly to vacuum tube design, it is worthwhile to discuss the technique briefly in the context of overall system reliability. It is fair to point out that this technique has considerable validity when comparisons of vacuum tube vs. solid-state amplifier designs are being considered. To achieve high power levels, far more individual components are necessary for a solid-state design than for a vacuum tube-based system

The parts-count approach to reliability prediction provides an estimate of reliability based on a count by part type (ICs, transistors, vacuum tube devices, resistors, capacitors, and other components). This method is useful during the early design stage of a product, when the amount of available detail is limited. The technique involves counting the number of components of each type, multiplying that number by a generic failure rate for each part type, and summing the products to obtain the failure rate of each functional circuit, subassembly, assembly, and/or block depicted in the system block diagram. The parts-count method is useful in the design phase because it provides rapid estimates of reliability, permitting assessment of the feasibility of a given concept.

Stress-Analysis Method

The stress-analysis technique is similar to the parts-count method, but utilizes a detailed parts model plus calculation of circuit stress values for each part before determining the failure rate. Each part is evaluated in its electric circuit and mechanical assembly application based on an electrical and thermal stress analysis. After part failure rates have been established, a combined failure rate for each functional block is determined.

9.2.3 Failure Analysis

Failure mode and effects analysis can be performed with data taken from actual failure modes observed in the field, or from hypothetical failure modes derived from one of the following:

- Design analysis

- Reliability prediction activities

- Experience with how specific parts fail

In the most complete form of FMEA, failure modes are identified at the component level. Failures are induced analytically into each component, and failure effects are evaluated and noted, including the severity and frequency (or probability) of occurrence. Using this approach, the probability of various system failures can be calculated, based on the probability of lower-level failure modes.

Fault tree analysis is a tool commonly used to analyze failure modes found during design, factory test, or field operations. The approach involves several steps, including the development of a detailed logic diagram that depicts basic faults and events that can lead to system failure and/or safety hazards. These data are used to formulate corrective suggestions that, when implemented, will eliminate or minimize faults considered critical. An example FTA chart is shown in Figure 9.1.

9.2.4 Standardization

Standardization and reliability go hand in hand. Standardization of electronic components began with military applications in mind; the first recorded work was performed by the U.S. Navy with vacuum tubes. The navy recognized that some control at the component level was essential to the successful incorporation of electronics into naval systems.

Standardization and reliability are closely related, although there are many aspects of standardization whose reliability implications are subtle. The primary advantages of standardization include:

- *Total product interchangeability*. Standardization ensures that products of the same part number provide the same physical and electrical characteristics. There have been innumerable instances of a replacement device bearing the same part number as a failed device, but not functioning identically to it. In many cases, the differences in performance were so great that the system would not function at all with the new device.

- *Consistency and configuration control*. Component manufacturers constantly redefine their products to improve yields and performance. Consistency and configuration control assure the user that product changes will not affect the interchangeability of the part.

- *Efficiencies of volume production*. Standardization programs usually result in production efficiencies that reduce the costs of parts, relative to components with the same level of reliability screening and control.

- *Effective spares management*. The use of standardized components makes the stocking of spare parts a much easier task. This aspect of standardization is not a minor consideration. For example, the costs of placing, expediting, and receiving material against one Department of Defense purchase order may range from $300 to $1100. Accepting the lowest estimate, the conversion of 10 separate part numbers to one standardized component could effect immediate savings of $3000 just in purchasing and receiving costs.

- *Multiple product sources*. Standardization encourages second-sourcing. Multiple sources help hold down product costs and encourage manufacturers to strive for better product performance.

9.3 Reliability Analysis

The science of reliability and maintainability matured during the 1960s with the development of sophisticated computer systems and complex military and spacecraft electronics. Components and systems never fail without a reason. That reason may be

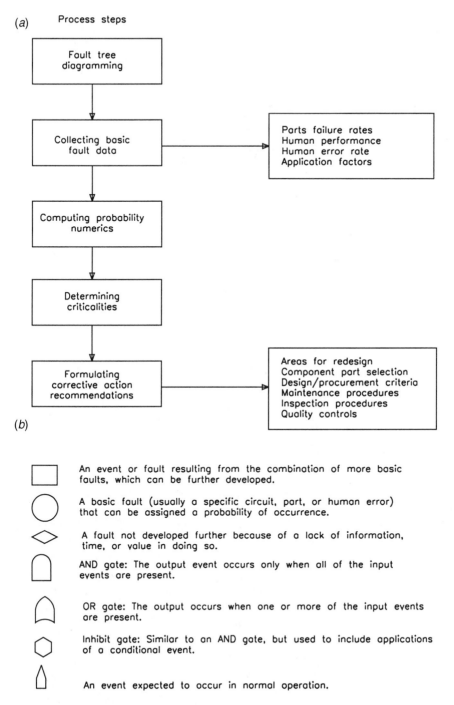

Figure 9.1 Example fault tree analysis diagram: (a) process steps, (b) fault tree symbols, (c, next page) example diagram.

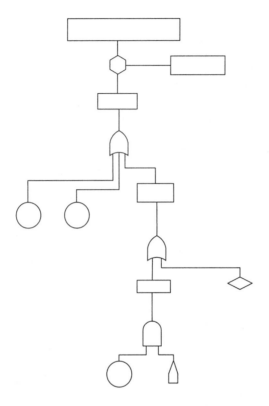

Figure 9.1c

difficult to find, but determination of failure modes and weak areas in system design or installation is fundamental to increasing the reliability of any component or system, whether it is a power vacuum tube, integrated circuit, aircraft autopilot, or broadcast transmitter.

All equipment failures are logical; some are predictable. A system failure usually is related to poor-quality components or to abuse of the system or a part within, either because of underrating or environmental stress. Even the best-designed components can be badly manufactured. A process can go awry, or a step involving operator intervention may result in an occasional device that is substandard or likely to fail under normal stress. Hence, the process of screening and/or *burn-in* to weed out problem parts is a universally accepted quality control tool for achieving high reliability.

9.3.1 Statistical Reliability

Figure 9.2 illustrates what is commonly known as the *bathtub curve*. It divides the expected lifetime of a class of parts into three segments: *infant mortality*, *useful life*, and *wear-out*. A typical burn-in procedure consists of the following steps:

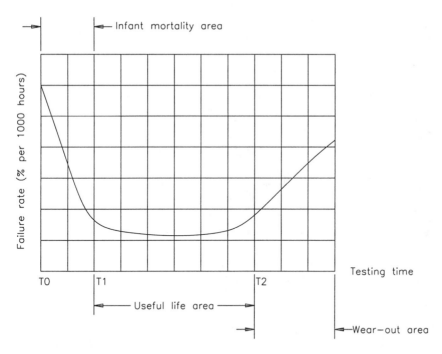

Figure 9.2 The statistical distribution of equipment or component failures vs. time for electronic systems and devices.

- The parts are electrically biased and loaded; that is, they are connected in a circuit representing a typical application.

- The parts are cycled on and off (power applied, then removed) for a predetermined number of times. The number of cycles may range from 10 to several thousand during the burn-in period, depending on the component under test.

- The components under load are exposed to high operating temperatures for a selected time (typically 72 to 168 hours). This constitutes an accelerated life test for the part.

An alternative approach involves temperature shock testing, in which the component product is subjected to temperature extremes, with rapid changes between the *hot-soak* and *cold-soak* conditions. After the stress period, the components are tested for adherence to specifications. Parts meeting the established specifications are accepted for shipment to customers. Parts that fail to meet them are discarded.

Figure 9.3 illustrates the benefits of temperature cycling to product reliability. The charts compare the distribution of component failures identified through steady-state high-temperature burn-in vs. temperature cycling. Note that cycling screened out a significant number of failures. The distribution of failures under temperature cycling usually resembles the distribution of field failures. Temperature cycling simulates

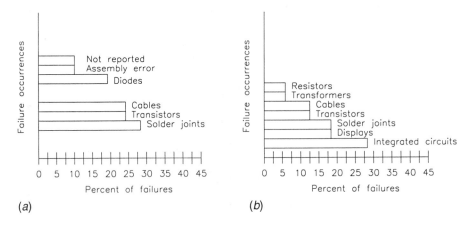

Figure 9.3 Distribution of component failures identified through burn-in testing: (*a*) steady-state high-temperature burn-in, (*b*) temperature cycling.

real-world conditions more closely than steady-state burn-in. The goal of burn-in testing is to ensure that the component lot is advanced beyond the infant mortality stage (*T1* on the bathtub curve). This process is used not only for individual components, but for entire systems as well.

Such a systems approach to reliability is effective, but not foolproof. The burn-in period is a function of statistical analysis; there are no absolute guarantees. The natural enemies of electronic parts are heat, vibration, and excessive voltage. Figure 9.4 documents failures vs. hours in the field for a piece of avionics equipment. The conclusion is made that a burn-in period of 200 hours or more will eliminate 60 percent of the expected failures. However, the burn-in period for another system using different components may well be a different number of hours.

The goal of burn-in testing is to catch system problems and potential faults before the device or unit leaves the manufacturer. The longer the burn-in period, the greater the likelihood of catching additional failures. The problems with extended burn-in, however, are time and money. Longer burn-in translates to longer delivery delays and additional costs for the equipment manufacturer, which are likely to be passed on to the end user. The point at which a product is shipped is based largely on experience with similar components or systems and the financial requirement to move products to customers.

Roller-Coaster Hazard Rate

The bathtub curve has been used for decades to represent the failure rate of an electronic system. More recent data, however, has raised questions regarding the accuracy of the curve shape. A growing number of reliability scientists now believe that the probability of failure, known in the trade as the *hazard rate*, is more accurately represented as a roller-coaster track, as illustrated in Figure 9.5. Hazard rate calculations

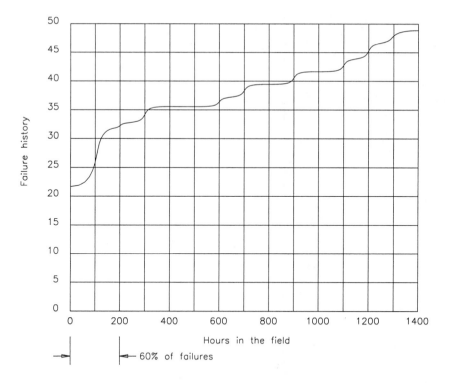

Figure 9.4 The failure history of a piece of avionics equipment vs. time. Note that 60 percent of the failures occurred within the first 200 hours of service. (*After* [1].)

require analysis of the number of failures of the system under test, as well as the number of survivors. Advocates of this approach point out that previous estimating processes and averaging tended to smooth the roller-coaster curve so that the humps were less pronounced, leading to an incorrect conclusion insofar as the hazard rate was concerned. The testing environment also has a significant effect on the shape of the hazard curve, as illustrated in Figure 9.6. Note that at the higher operating temperature (greater environmental stress), the roller-coaster hump has moved to an earlier age.

9.3.2 Environmental Stress Screening

The science of reliability analysis is rooted in the understanding that there is no such thing as a random failure; every failure has a cause. For reasonably designed and constructed electronic equipment, failures not caused by outside forces result from built-in flaws or *latent defects*. Because different flaws are sensitive to different stresses, a variety of environmental forces must be applied to a unit under test to identify any latent defects. This is the underlying concept behind *environmental stress screening* (ESS).

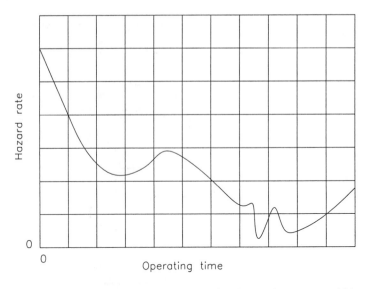

Figure 9.5 The roller-coaster hazard rate curve for electronic systems. (*After* [2].)

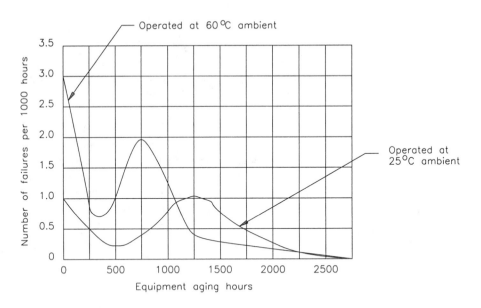

Figure 9.6 The effects of environmental conditions on the roller-coaster hazard rate curve. (*After* [2].)

ESS, which has come into widespread use for aeronautics and military products, takes the burn-in process a step further by combining two of the major environmental factors that cause parts or units to fail: heat and vibration. Qualification testing for

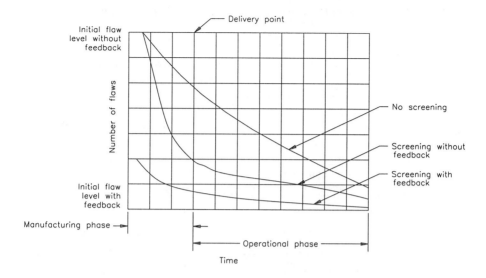

Figure 9.7 The effects of environmental stress screening on the reliability *bathtub* curve. (*After* [3].)

products at a factory practicing ESS involves a carefully planned series of checks for each unit off the assembly line. Units are subjected to random vibration and temperature cycling during production (for subassemblies and discrete components) and upon completion (for systems). The goal is to catch product defects at the earliest possible stage of production. ESS also can lead to product improvements in design and manufacture if feedback from the qualification stage to the design and manufacturing stages is implemented. Figure 9.7 illustrates the improvement in reliability that typically can be achieved through ESS over simple burn-in screening, and through ESS with feedback to earlier production stages. Significant reductions in equipment failures in the field can be gained. Table 9.1 compares the merits of conventional reliability testing and ESS.

Designing an ESS procedure for a given product is no easy task. The environmental stresses imposed on the product must be great enough to cause fallout of marginal components during qualification testing. The stresses must not be so great, however, as to cause failures in good products. Any unit that is stressed beyond its design limits eventually will fail. The proper selection of stress parameters—generally, random vibration on a vibration generator and temperature cycling in an environmental chamber—can, in minutes, uncover product flaws that might take weeks or months to manifest themselves in the field. The result is greater product reliability for the user.

The ESS concept requires that every product undergo qualification testing before integration into a larger system for shipment to an end user. The flaws uncovered by ESS vary from one unit to the next, but types of failures tend to respond to particular en-

Table 9.1 Comparison of Conventional Reliability Testing and Environmental Stress Screening (*After* [2].)

Parameter	Conventional Testing	Environmental Stress Screening
Test condition	Simulates operational equipment profile	Accelerated stress condition
Test sample size	Small	100 percent of production
Total test time	Limited	High
Number of failures	Small	Large
Reliability growth	Potential for gathering useful data small	Potential for gathering useful data good

vironmental stresses. Available data clearly demonstrate that the burn-in screens must match the flaws sought; otherwise, the flaws will probably not be found.

The concept of flaw-stimulus relationships can be presented in Venn diagram form. Figure 9.8 shows a Venn diagram for a hypothetical, but specific, product. The required screen would be different for a different product. For clarity, not all stimuli are shown. Note that there are many latent defects that will not be uncovered by any one stimulus. For example, a solder splash that is just barely clinging to a circuit element probably would not be broken loose by high-temperature burn-in or voltage cycling, but vibration or thermal cycling probably would break the particle loose. Remember also that the defect may be observable only during stimulation and not during a static bench test.

The levels of stress imposed on a product during ESS should be greater than the stress to which the product will be subjected during its operational lifetime, but still be below the maximum design parameters. This rule of thumb is pushed to the limits under an *enhanced screening* process. Enhanced screening places the component or system at well above the expected field environmental levels. This process has been found to be useful and cost-effective for many programs and products. Enhanced screening, however, requires the development of screens that are carefully identified during product design and development so that the product can survive the qualification tests. Enhanced screening techniques often are required for cost-effective products on a cradle-to-grave basis; that is, early design changes for screenability save tremendous costs over the lifetime of the product.

The types of products that can be checked economically through ESS break down into two categories: high-dollar items and mass-produced items. Units that are physically large in size, such as RF generators, are difficult to test in the finished state. Still, qualification tests using more primitive methods, such as cam-driven truck-bed shakers, are practical. Because most large systems generate a large amount of heat, subjecting the equipment to temperature extremes also may be accomplished. Sophisticated ESS for large systems, however, must rely on qualification testing at the subassembly stage.

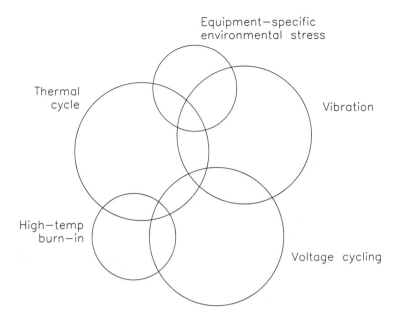

Figure 9.8 Venn diagram representation of the relationship between flaw precipitation and applied environmental stress. (*After* [4].)

The basic hardware complement for an ESS test station includes a thermal chamber shaker and controller/monitor. A typical test sequence includes 10 minutes of exposure to random vibration, followed by 10 cycles between temperature minimum and maximum. To save time, the two tests may be performed simultaneously.

9.3.3 Latent Defects

The cumulative failure rate observed during the early life of an electronic system is dominated by the latent defect content of the product, not its inherent failure rate. Product design is the major determinant of inherent failure rate. A product design will show a higher-than-expected inherent rate if the system contains components that are marginally overstressed, have inadequate functional margin, or contain a subpopulation of components that exhibit a wear-out life shorter than the useful life of the product. Industry has grown to expect the high instantaneous failure rate observed when a new product is placed into service. The burn-in process, whether ESS or conventional, is aimed at shielding customers from the detrimental effects of infant mortality. The key to reducing early-product-life failures lies in reducing the number of latent defects.

A latent defect is some abnormal characteristic of the product or its parts that is likely to result in failure at some point, depending on:

- The degree of abnormality

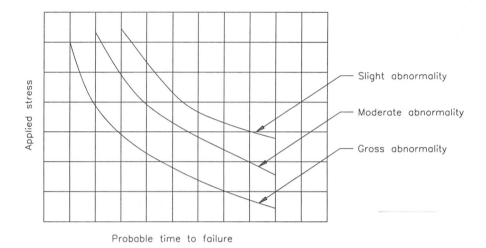

Figure 9.9 Estimation of the probable time to failure from an abnormal solder joint. (*After* [5].)

- The magnitude of applied stress
- The duration of applied stress

For example, consider a solder joint on the connecting pin of a vacuum tube device. One characteristic of the joint is the degree to which the pin hole is filled with solder, characterized as "percent fill." All other characteristics being acceptable, a joint that is 100 percent filled offers the maximum mechanical strength, minimum resistance, and greatest current carrying capacity. Conversely, a joint that is zero percent filled has no mechanical strength, and only if the lead is touching the barrel does it have any significant electrical properties. Between these two extreme cases are degrees of abnormality. For a fixed magnitude of applied stress:

- A grossly abnormal solder joint probably will fail in a short time.
- A moderately abnormal solder joint probably will fail, but after a longer period of time than a grossly abnormal joint.
- A mildly abnormal solder joint probably will fail, but after a much longer period of time than in either of the preceding conditions.

Figure 9.9 illustrates this concept. A similar time-stress relationship holds for a fixed degree of abnormality and variable applied stress.

A latent defect eventually will advance to a *patent defect* when exposed to environmental, or other, stimuli. A patent defect is a flaw that has advanced to the point at which an abnormality actually exists. To carry on the solder example, a cold solder joint represents a flaw (latent defect). After vibration and/or thermal cycling, the joint (it is assumed) will crack. The joint will now have become a detectable (patent) defect. Some latent defects can be stimulated into patent defects by thermal cycling, some by vibration, and some by voltage cycling. Not all flaws respond to all stimuli.

There is strong correlation between the total number of physical and functional defects found per unit of product during the manufacturing process, and the average latent defect content of shipping product. Product- and process-design changes aimed at reducing latent defects not only improve the reliability of shipping product, but also result in substantial manufacturing cost savings.

9.3.4 Operating Environment

The operating environment of an electronic system, either because of external environmental conditions or unintentional component underrating, may be significantly more stressful than the system manufacturer or the component supplier anticipated. Unintentional component underrating represents a design fault, but unexpected environmental conditions are possible for many applications, particularly in remote locations.

Conditions of extreme low or high temperatures, high humidity, and vibration during transportation may have a significant impact on long-term reliability of the system. For example, it is possible—and more likely, probable—that the vibration stress of the truck ride to a remote transmitting site will represent the worst-case vibration exposure of the transmitter and all components within it during the lifetime of the product.

Manufacturers report that most of the significant vibration and shock problems for land-based products arise from the shipping and handling environment. Shipping tends to be an order of magnitude more severe than the operating environment with respect to vibration and shock. Early testing for these problems involved simulation of actual shipping and handling events, such as end-drops, truck trips, side impacts, and rolls over curbs and cobblestones. Although unsophisticated by today's standards, these tests are capable of improving product resistance to shipping-induced damage.

9.3.5 Failure Modes

Latent failures aside, the circuit elements most vulnerable to failure in any piece of electronic hardware are those exposed to the outside world. In most systems, the greatest threat typically involves one or more of the following components or subsystems:

- The ac-to-dc power supply
- Sensitive signal-input circuitry
- High-power output stages and devices
- Circuitry operating into an unpredictable load, or into a load that may be exposed to lightning and other transient effects (such as an antenna)

Derating of individual components is a key factor in improving the overall reliability of a given system. The goal of derating is the reduction of electrical, mechanical, thermal, and other environmental stresses on a component to decrease the degradation rate and prolong expected life. Through derating, the margin of safety between the operating stress level and the permissible stress level for a given part is increased. This adjustment provides added protection from system overstress, unforeseen during design.

9.3.6 Maintenance Considerations

The reliability and operating costs over the lifetime of an RF system can be affected significantly by the effectiveness of the preventive maintenance program designed and implemented by the engineering staff. In the case of a *critical-system* unit that must be operational continuously or during certain periods, maintenance can have a major impact—either positive or negative—on downtime.

The reliability of any electronic system may be compromised by an *enabling event phenomenon*. This is an event that does not cause a failure by itself, but sets up (or enables) a second event that can lead to failure of the system. Such a phenomenon is insidious because the enabling event may not be self-revealing. Examples include the following:

- A warning system that has failed or has been disabled for maintenance

- One or more controls that are set incorrectly, providing false readouts for operations personnel

- Redundant hardware that is out of service for maintenance

- Remote metering that is out of calibration

Common-Mode Failure

A *common-mode failure* is one that can lead to the failure of all paths in a redundant configuration. In the design of redundant systems, therefore, it is important to identify and eliminate sources of common-mode failures, or to increase their reliability to at least an order of magnitude above the reliability of the redundant system. Common-mode failure points in a high-power RF system include the following:

- Switching circuits that activate standby or redundant hardware

- Sensors that detect a hardware failure

- Indicators that alert personnel to a hardware failure

- Software that is common to all paths in a redundant system

The concept of software reliability in control and monitoring has limited meaning in that a good program will always run, and copies of a good program will always run. On the other hand, a program with one or more errors will always fail, and so will the copies, given the same input data. The reliability of software, unlike hardware, cannot be improved through redundancy if the software in the parallel path is identical to that in the primary path.

Spare Parts

The spare parts inventory is a key aspect of any successful equipment maintenance program. Having adequate replacement components on hand is important not only to correct equipment failures, but to identify those failures as well. Many parts—particularly in the high-voltage power supply and RF chain—are difficult to test under static conditions. The only reliable way to test the component may be to substitute one of known quality. If the system returns to normal operation, then the original component is defective. Substitution is also a valuable tool in troubleshooting intermittent failures caused by component breakdown under peak power conditions.

9.4 Vacuum Tube Reliability

The end user should keep an accurate record of performance for each tube at the facility. Shorter-than-normal tube life could point to a problem in the RF amplifier or related hardware. The average life that may be expected from a power tube is a function of many operational parameters, including:

- Filament voltage

- Ambient operating temperature

- RF power output

- Operating frequency

- Operating efficiency

The best estimate of life expectancy for a given system at a particular location comes from on-site experience. As a general rule of thumb, however, at least 4000 hours of service can be expected from most power grid tubes. Klystrons and other microwave devices typically provide in excess of 15,000 hours. Possible causes of short tube life include the following:

- Improper power stage tuning

- Inaccurate panel meters or external wattmeter, resulting in more demand from the tube than is actually required

- Poor filament voltage regulation

- Insufficient cooling system airflow

- Improper stage neutralization

An examination of the data sheet for a given vacuum tube will show that a number of operating conditions are possible, depending upon the class of service required by the application. As long as the maximum ratings of the device are not exceeded, a wide choice of operating parameters is possible. When studying the characteristic curves of each tube, remember that they represent the performance of a *typical* device. All electronic products have some tolerance among devices of a single type. Operation of a given device in a particular system may be different than that specified on the data sheet. This effect is more pronounced at VHF and above.

9.4.1 Thermal Cycling

Most power tube manufacturers recommend a warm-up period between the application of *filament-on* and *plate-on* commands; about 5 minutes is typical. The minimum warm-up time is 2 minutes. Some RF generators include a time-delay relay to prevent the application of a plate-on command until a predetermined warm-up cycle is completed.

Most manufacturers also specify a recommended cool-down period between the application of *plate-off* and *filament-off* commands. This cool-down, generally about 10 minutes, is designed to prevent excessive temperatures on the PA tube surfaces when the cooling air is shut off. Large vacuum tubes contain a significant mass of metal, which stores heat effectively. Unless cooling air is maintained at the base of the tube and through the anode cooling fins, excessive temperature rise can occur. Again, the result can be shortened tube life, or even catastrophic failure because of seal cracks caused by thermal stress.

Most tube manufacturers suggest that cooling air continue to be directed toward the tube base and anode cooling fins after filament voltage has been removed to further cool the device. Unfortunately, however, not all control circuits are configured to permit this mode of operation.

9.4.2 Tube-Changing Procedure

Plug-in power tubes must be seated firmly in their sockets, and the connections to the anodes of the tubes must be tight. Once in place, the device typically requires no maintenance during its normal operating lifetime.

Insertion of a replacement tube must be performed properly. Gently rock and slightly rotate the tube as it is being inserted into the socket. This will help prevent bending and breaking the fingerstock. Be certain to apply sufficient force to seat the tube completely into the socket. Never use a lever or other tool on the tube to set it in place. Manual pressure should be adequate.

An intermediate point is reached when the grid contact fingerstock slides up the tube sides and encounters the contact area. It is important that the tube is fully inserted in the socket beyond this initial point of resistance.

Some sockets have adjustable stops that are set so that the tube grid contacts rest in the middle of the contact area when the tube is fully inserted. This positioning can be checked by inserting, then removing, a new tube. The scratch marks on the grid contacts will indicate the position of the tube relative to the socket elements. Figure 9.10 shows the minute scoring in the base contact rings of a power grid tube. A well-maintained socket will score the tube contacts when the device is inserted. If all fingers do not make contact, more current will flow through fewer contact fingers, causing overheating and burning, as shown in Figure 9.11.

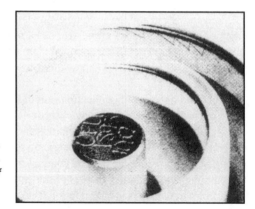

Figure 9.10 Proper fingerstock scoring of filament and control grid contacts on a power tube. (*Courtesy of Varian.*)

Figure 9.11 Damage to the grid contact of a power tube caused by insufficient fingerstock pressure. (*Courtesy of Varian.*)

Poor contact between a tube and its socket will always lead to problems. If the socket connectors fail to provide the proper electrical contact with the cylindrical tube elements, concentration of RF charging currents will result, leading to local overheating that may be destructive. After the connector loses its spring action, the heating is aggravated, and damage to the tube is likely. Some socket designs include a wire-wound spring encircling the outer circumference of the fingerstock to increase the contact pressure of the individual fingers. These springs should be replaced if they break or lose tension.

When a tube is removed from its socket, the socket should be blown or wiped clean and carefully inspected. Any discoloration caused by overheating of the socket fingerstock could signal impending tube/socket failure.

Many industrial and low-frequency tubes are not socketed, but installed by bolted or clamped connections. Stainless steel has a much lower coefficient of thermal expansion than copper. Therefore, clamped stainless steel anode connections should have some method of strain relief to prevent excess pressure from collapsing the tube anode as it heats.

All bolted or screwed connections should be tight. Verify that the clamps are snug and provide a good electrical contact around the entire circumference of the device. Because of the RF fields present, all clamps and bolts should be made from nonmagnetic materials. Copper, brass, and nonmagnetic series 300 stainless steel fasteners are preferred. Stainless steel is not a good conductor of electricity, so, although it may be used for clamping, it should not be part of the current path.

Before a new tube is installed in a system, it should be inspected for cracks or loose connections (in the case of tubes that do not socket-mount). Manufacturers typically recommend that after a new tube—or one that has been on the shelf for some time—is installed in the transmitter, it should be run with *filaments only* for at least 30 minutes, after which plate voltage may be applied. Next, the drive (modulation) is slowly brought up, in the case of an amplitude-modulated system. Residual gas inside the tube may cause an interelectrode arc (usually indicated as a plate overload) unless it is burned off in such a warm-up procedure.

9.4.3 Power Tube Conditioning

Large power tubes are subjected to rigorous processing during exhaust pumping at the time of manufacture [6]. Active elements are processed at temperatures several hundred degrees Celsius higher than those expected during normal operation. This procedure is designed to drive off surface and subsurface gas from the metals in the device to lower the possibility of these gases being released during the service life of the tube. Free gas molecules, however, always will be present to some degree in a fully processed tube. Gas, particularly oxygen-containing compounds, can combine chemically with cathode material to either permanently or temporarily destroy the electron emission capability of the cathode. Free gas molecules, when struck by electrons moving from the cathode to the anode, also may be ionized by having one or more electrons knocked from their systems. If enough of these ions, plus freed electrons, are present, a conduction path will be established that is not subject to control by the grid. This can result in runaway arcing that may involve all elements of the device. Current then will be limited primarily by the source voltage and impedance. (The space charge, to some degree, is neutralized by the presence of both electrons and positive ions.)

Electrons from sources other than the heated cathode provide low-current paths between elements when the voltage gradient is sufficiently high at the negative element for pure field emission. The voltage gradient at the negative element is governed by the following parameters:

- The applied voltage between elements
- The spacing between elements

- The surface contour of the negative element

A large voltage gradient can exist in the following areas:

- In front of a point on the negative element

- In front of a particle adhering to the negative element

- In front of a clump of gas molecules on the surface of the negative element

Field emission occurs readily from a cold surface if the operating conditions provide the necessary voltage gradient.

Ionization of free gas can result from bombardment by field-emitted electrons. Arcing is likely to occur as a result of field emission in operating equipment because the plate voltage is maximum during that part of the signal cycle when ordinary plate current from the cathode is shut off by the control grid. For this reason, an important part of tube processing is high-voltage conditioning to remove sharp points or small particles from tube elements. This part of tube processing may, and sometimes should, be repeated in the field after shipment or storage if the tube is intended for use at a plate voltage in excess of 10 kV.

High-voltage conditioning (sometimes called *spot-knocking* or *debarnacling*) consists of applying successively higher voltages between tube elements, permitting the tube to spark internally at each voltage level until stable (no arcing). The voltage then is raised to the next higher potential until the tube is stable at a voltage approximately 15 percent higher than the peak signal voltage it will encounter in service.

The equipment used for tube conditioning is simple but specialized. It may provide dc or ac voltage, or both. The current required is small. The voltage should be continuously variable from practically zero to the highest value required for proper conditioning of the tube. The energy per spark is controlled by the internal resistance of the supply plus any external series resistor used. In dc conditioning, a milliammeter typically is included to measure the level of field emission current prior to sparking, or simply to determine whether the field emission current is within the specified range for a tube being tested. Also in dc conditioning, a capacitor may be placed across the tube under test to closely control the energy released for each spark. If the conditioning is to proceed effectively, the amount of energy delivered by the capacitor must be great enough to condition the surfaces, but not so large as to cause permanent damage.

Power Supply

The power supply is the heart of any high-voltage tube conditioning station [6]. Current-limiting resistance is desirable; the power supply may have an internal resistance on the order of 100 kΩ. In addition to the power supply, a current-limiting resistor of 100 to 300 kΩ should be provided in the circuit to the tube being conditioned. In dc processing, a storage capacitor of 1000 to 3000 pF can be placed directly across the device under test (DUT), with a high-voltage noninductive resistor of approximately 50 Ω connected between the tube and the capacitor. Naturally, this shunt capacitor must be capable of withstanding the full applied potential of the power supply.

Using both polarities on the tube elements is advantageous in high-voltage conditioning, suggesting that an ac supply might be the best choice for field tube conditioning. The advantage of the dc system, on the other hand, is that exact levels of field emission current and breakdown voltage may be determined for comparison with known "clean" tubes, and with the manufacturer's specification. The polarity of the applied voltage may be reversed by mounting the DUT on an insulating frame so that the positive and negative high-voltage leads may be interchanged during processing. Also, the power supply rectifier and dc meters may be bypassed with suitable switching so that ac is obtained from the supply for initial tube cleanup. The storage capacitor across the tube should be disconnected if ac is used.

Conditioning Procedure

Manufacturers of large power tubes typically specify an upper limit for field emission between the filament, control grid, and screen grid connected together, and the anode [6]. Such devices should be allowed to spark until stable at the voltage where sparking first occurs. The voltage then is increased in steps determined by the voltage that initiates sparking again. This process is continued until the tube is stable at the maximum test voltage. It is advisable to reverse the polarity of the supply if dc testing is employed.

Usually, no field emission specifications exist for applied voltage between grids and between grid and filament, but these elements normally should hold off 10 kV internally. The need for conditioning between these elements is indicated if internal sparking occurs at a significantly lower voltage. Consult the tube manufacturer for specific recommendations on conditioning voltages.

During high-voltage processing, particularly between grids and between grid and filament, some of the redistributed gas molecules may be deposited on the cold filament, causing a temporary loss of emission. If this happens, operate the tube for approximately 1 hour, with normal filament power only, to drive off the volatile material. Normal electron emission from the filament will be reestablished by this procedure. It is recommended that routine practice include operation of the filament for approximately 1 hour as an immediate follow-up to high-voltage conditioning.

After conditioning, the tube should be stored in the same position (orientation) as that used for high-voltage conditioning and installation in the equipment. Movement of the tube after conditioning should be minimized to prevent any redistribution of particles within the device.

Considerations for Very Large Tubes

High-voltage ac conditioning is not recommended for very large tubes, such as the 8971/X2177, 8972/X2176, 8973/X2170, 8974/2159, 4CM400000A, and similar devices [6]. High-potential 50 or 60 Hz voltage can excite natural mechanical resonances in the large filament and grid structures of these tube types. Excitation of these resonances can break the filament and/or grids, thus destroying the device. If ac conditioning must be used, the following precautions should be taken:

- Connect all filament terminals together (external to the tube).

- Connect all terminals of the tube to a terminal of the high-voltage power supply. For example, if high voltage is to be applied between the screen and anode, the filament, grid, and screen should be connected together, then to one terminal of the supply. If high voltage is to be applied between the filament and grid, connect the grid, screen, and anode together and to one terminal of the supply. No terminal of the tube should be allowed to float electrically during high-voltage conditioning.

- As the test voltage is increased, listen carefully for vibration inside the tube. If any is detected, immediately remove the ac high voltage.

These precautions will reduce the risk of damage from ac conditioning of very large tubes. However, because of the special procedures required and the risk of damage resulting from mechanical resonances, it is recommended that only dc conditioning be used for such devices.

Safety

Personnel protection dictates that the tube being processed, as well as the power supply, be enclosed in a metal cabinet with access doors interlocked so that no high voltage can be applied until the doors are closed [6]. The enclosure material, wall thickness, and dimensions must be chosen to limit X-ray radiation from the DUT to safe levels. A steel-wall thickness of $\frac{3}{32}$ in has been found adequate for dc testing up to approximately 70 kV. For conditioning up to 100 kV dc, it is necessary to add 0.05 in of lead sheet as a lining to the steel enclosure.

The field of health standards regarding x-radiation dosage for a given period of time is complex and should be discussed with experts on the subject. Calibrated film badges or other reliable measuring devices normally are used to determine human exposure to X rays over time.

9.4.4 Filament Voltage

A *true reading* rms voltmeter is required for accurate measurement of filament voltage. The measurement is taken directly from the tube socket connections. A true-reading rms meter, instead of the more common *average responding* rms meter, is suggested because it can accurately measure a voltage despite an input waveform that is not a pure sine wave. Some filament voltage regulators use silicon controlled rectifiers (SCRs) to regulate the output voltage. Saturated core transformers, which often deliver nonsine waveforms, also are commonly used to regulate filament input power.

Long tube life requires filament voltage regulation. A tube whose filament voltage is allowed to vary along with the primary line voltage will not achieve the life expectancy possible with a tightly regulated supply. This problem is particularly acute at mountain-top installations, where utility regulation is generally poor.

To extend tube life, some end-users leave the filaments on at all times, not shutting down at sign-off. If the sign-off period is 3 hours or less, this practice can be beneficial. Filament voltage regulation is a must in such situations because the primary line voltages may vary substantially from the carrier-on to carrier-off value.

When a cold filament is turned-on, damage can be caused by two effects [7]:

- Current inrush into the cold filament, which can be up to 10 times the normal operating current if the filament supply is of low impedance.

- Grain reorientation, which occurs at about 600 to 700 °C, known as the *Miller-Larson effect*. This grain reorientation will result in a momentary plastic state that can cause the filament wire to grow in length and, therefore, become thinner.

The Miller-Larson effect is aggravated by variations in the cross-sectional area of the filament wires along their length. This will cause hot spots in the thinner sections because of the greater current densities there. The higher temperatures at the hot spots cause increased growth during warmup through the 600 °C temperature range, when they have a higher current density than the rest of the wire, and a much higher power dissipation per unit length as a result of the higher resistance. Each time the filament is turned-on, the wire becomes thinner until the hot spot temperature enters a runaway condition.

Filament voltage for a klystron should not be left on for a period of more than 2 hours if no beam voltage is applied. The net rate of evaporation of emissive material from the cathode surface of a klystron is greater without beam voltage. Subsequent condensation of the material on gun components may lead to voltage holdoff problems and an increase in body current.

Black Heat

One practice used occasionally to extend the life of the thoriated tungsten power tube filament is to operate the device at a low percentage of rated voltage during sign-off periods [7]. This technique, known as *black heat*, involves operating the filament at 25 to 40 percent of the rated filament voltage. This range produces a temperature high enough to maintain the filament above the 600 °C region where the Miller-Larson effect occurs, and sometimes low enough so that forced cooling is not required during stand-by.

Operating histories of several black heat installations indicate that the trauma of raising the filament voltage from 25 percent up to the operating voltage is still potentially damaging. For this reason, a higher stand-by voltage of about 80 to 90 percent of the full rated value (*orange heat*) has been used with some success.

The voltage range of 40 to 80 percent should be avoided because the filament can act as a getter within this range, and absorb gases that will raise the work function of the thoriated tungsten filament. For orange heat operation, cooling system operation is required. Before attempting to implement a black heat or orange heat filament management program, consult with the tube and transmitter manufacturers.

9.4.5 Filament Voltage Management

As outlined in Chapter 3, the thoriated-tungsten power grid tube lends itself to life extension through management of the applied filament voltage. Accurate adjustment of the filament voltage can extend the useful life of the device considerably, sometimes to twice the normal life expectancy. The following procedure typically is recommended:

- Install the tube, and tune for proper operation.

- Operate the filament at its full rated voltage for the first 200 hours following installation.

- Following the burn-in period, reduce the filament voltage by 0.1 V per step until system power output begins to fall (for frequency-modulated systems) or until modulating waveform distortion begins to increase (for amplitude-modulated systems).

- When the *emissions floor* has been reached, increase the filament voltage approximately 0.2 V peak-to-peak (P-P).

Long-term operation at this voltage can result in a substantial extension of the useful life of the tube, as illustrated in Figure 9.12.

The filament voltage should not, in any event, be operated at or below 90 percent of its rated value. At regular intervals, about every 3 months, the filament voltage is checked and increased if power output begins to fall or waveform distortion begins to rise. Filament voltage never should be increased to more than 105 percent of the rated voltage.

Note that some tube manufacturers place the minimum operating point at 94 percent. Others recommend that the tube be set for 100 percent filament voltage and left there. The choice is left to the user, after consultation with the manufacturer.

The filament current should be checked when the tube is installed, and at annual intervals thereafter, to ensure that the filament draws the expected current. Tubes can fail early in life because of an open filament bar—a problem that could be discovered during the warranty period if a current check is made upon installation.

For 1 week of each year of tube operation, the filament should be run at full rated voltage. This will operate the *getter* and clean the tube of gas.

Klystron Devices

Filament voltage is an equally important factor in achieving long life in a klystron. The voltage recommended by the manufacturer must be accurately set and checked on a regular basis. Attempts have been made to implement a filament voltage management program—as outlined previously for a thoriated-tungsten tube—on a klystron, with varying degrees of success. If filament management is attempted, the following guidelines should be observed:

- Install the tube, and tune for proper operation.

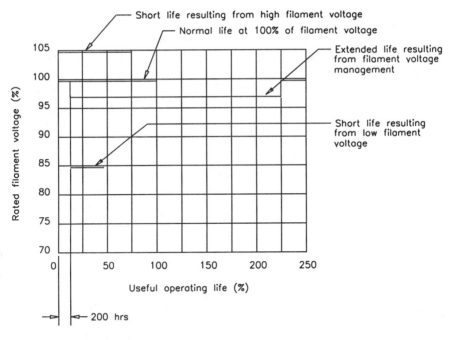

(a)

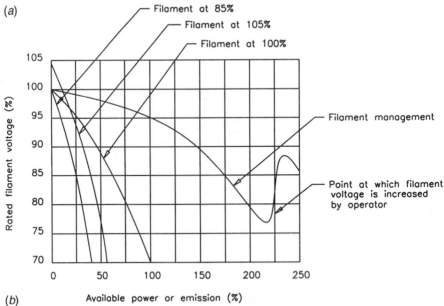

(b)

Figure 9.12 The effects of filament voltage management on a thoriated-tungsten filament power tube: (*a*) useful life as a function of filament voltage, (*b*) available power as a function of filament voltage. Note the dramatic increase in emission hours when filament voltage management is practiced.

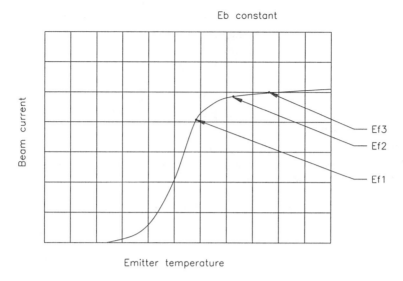

Figure 9.13 The effect of klystron emitter temperature on beam current.

- Operate the filament at its full rated voltage for the first 200 hours following installation.

- Following the burn-in period, reduce the filament voltage by 0.1 V per step until collector current begins to decrease. After each filament voltage change, wait at least 5 minutes to allow for cathode cooling.

- When the emissions floor has been reached, increase the filament voltage approximately 0.25 V P-P.

Figure 9.13 charts the relationship between emitter temperature and beam current. The filament should be adjusted to a point between E_{f1} and E_{f2}, which will vary with each klystron. Note that continued operation below the E_{f1} level, or at a point that causes emission instability, can result in beam defocusing, leading to increased modulating anode current and body current.

For operation at a reduced filament voltage, it may be necessary to allow a longer warm-up period when bringing the tube up from a cold start. If loss of emission is noted at turn-on, it will be necessary to either extend the warm-up time or raise the filament voltage to overcome the loss.

9.4.6 PA Stage Tuning

The PA stage of an RF generator may be tuned in several ways. Experience is the best teacher when it comes to adjusting for peak efficiency and performance. Compromises often must be made among various operating parameters.

Tuning can be affected by any number of changes in the PA stage. Replacing the final tube in a low- to medium-frequency RF generator usually does not significantly alter stage tuning. It is advisable, however, to run through a touch-up tuning procedure just to be sure. Replacing a tube in a high-frequency RF generator, on the other hand, can significantly alter stage tuning. At high frequencies, normal tolerances and variations in tube construction result in changes in element capacitance and inductance. Likewise, replacing a component in the PA stage may cause tuning changes because of normal device tolerances.

Stability is one of the primary objectives of tuning. Adjust for broad peaks or dips, as required. Tune so that the system is stable from a cold startup to normal operating temperature. Readings should not vary measurably after the first minute of operation.

Tuning is adjusted not only for peak efficiency, but also for peak performance. These two elements, unfortunately, do not always coincide. Tradeoffs must sometimes be made to ensure proper operation of the system.

9.4.7 Fault Protection

An arc is a self-sustained discharge of electricity between electrodes with a voltage drop at the cathode on the order of the minimum ionizing potential of the environment [8]. The arc supports large currents by providing its own mechanism of electron emission from the negative terminal, plus space-charge neutralization.

All power vacuum tubes operate at voltages that can cause severe damage in the event of an internal arc unless the tube is properly protected. This damage can be relatively minor in tubes operating from low-energy power supplies, but it will become catastrophic in cases where a large amount of stored energy or follow-on current is involved. Because any high-voltage vacuum device may arc at one time or another, some means of protection is advised in all cases; it is mandatory in those instances where destructive quantities of energy are involved. In addition to protecting the tube in the event of a tube or circuit malfunction, such measures also protect the external circuitry.

Most power grid tubes employing thoriated-tungsten filaments are capable of withstanding relatively high energy arcs. However, while employing substantial materials, such devices usually incorporate fine wire grids that must be protected. In large tube applications, it is often found that the screen grid or bias supply, in addition to the anode supply, is capable of delivering destructive energy to the tube structure in the event of an internal arc. Consequently, protection must be provided in all cases where potentially destructive energy is involved. Large tubes usually are capable of withstanding up to 50 joules (J) total energy without permanent damage.

In the design of a power amplifier, the stage should be tested to ensure that the 50 J limit is not exceeded by any of the power supplies feeding the tube. Each supply, whether anode or grid, should be short-circuit-tested through a 6-in length of 0.255-mm-diameter (no. 30 AWG) soft copper wire. A similar test should be made from plate-to-screen grid to ensure that the tube is protected in the event of a plate-to-screen arc. A protective spark gap between the screen grid and cathode is recommended so that, if an arc occurs, the resulting spark from screen to cathode is external to the tube structure. The test wire will remain intact if the energy dissipated is less than 50 J. The

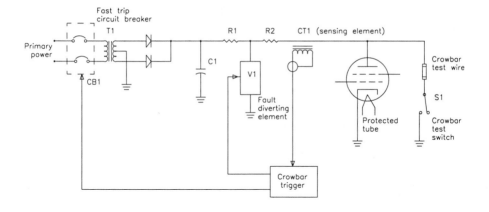

Figure 9.14 Arc protection circuit utilizing an energy-diverting element.

50 J test includes total energy delivered from the energy storage device and/or from the follow-on current from the power supply prior to the opening of the primary circuit.

Each power supply in the system must pass this test, or some form of energy-diverting or energy-limiting circuit must be employed. Figure 9.14 shows an energy-diverting circuit for a high-power tube. Energy diverters at high-power levels are complex. They typically employ large high-speed diverting elements such as thyratrons, ignitrons, triggered spark gaps, and, in some cases, a high-speed vacuum switch. These devices may be required to handle thousands of amperes of peak current. Resistor R1 in Figure 9.14 is used to limit the peak current into the energy-diverting switch V1, as well as to limit peak current demand on the transformer and rectifiers in the power supply. Resistor R2 must be large enough to keep the voltage across V1 from dropping so low under load-arc conditions as to inhibit the proper firing of the fault-diverting elements. R1 and R2 must also damp circuit ringing sufficiently to prevent current reversal in V1, which could cause V1 to open before the fault conditions are cleared. R1 and R2 are sized to minimize average power losses and to handle the required voltage and power transients while properly performing their functions. The sensing element CT1 detects any abnormal peak-current levels in the system. Signals from the current transformer are used to trigger V1 in the event of an overcurrent condition. A crowbar test switch, S1, can be used to perform the wire test outlined previously, to ensure that the total energy dissipated in an arc is less than 50 J. Typically, S1 is a remote-controlled high-voltage switch of either air or vacuum construction.

In a typical test, when S1 is closed, current transformer CT1 senses the fast rising current and fires the energy-diverting element V1; it simultaneously sends a fast trip signal to the main circuit breaker. Most energy diverters fire within a few microseconds of the time the fault is detected; all voltages and currents are removed from the protected device in tens of microseconds.

By using this technique, it is possible to limit the energy delivered during an arc to much less than the recommended maximum. In a properly operating system, essentially no damage will occur within the tube even under severe arcing conditions. The test wire must be connected at the tube itself to take into account all stored energy (L and C) in the system.

9.4.8 Vacuum Tube Life

The vacuum tube suffers wear-out because of a predictable chemical reaction. Life expectancy is one of the most important factors to be considered in the use of vacuum tubes. In general, manufacturers specify maximum operating parameters for power grid tubes so that operation within the ratings will provide for a minimum useful life of 4000 hours. Considerable variation can be found with microwave power tubes, but 15,000 hours is probably a fair average useful life. It should be noted that some microwave devices, such as certain klystrons, have an expected useful life of 30,000 hours or more.

The cathode is the heart of a power tube. The device is said to *wear out* when filament emissions are inadequate for full power output or acceptable waveform distortion. In the case of the common thoriated-tungsten filament tube, three primary factors determine the number of hours a device will operate before reaching wear-out:

- The rate of evaporation of thorium from the cathode

- The quality of the tube vacuum

- The operating temperature of the filament

In the preparation of thoriated tungsten, 1 to 2 percent of thorium oxide (thoria) is added to the tungsten powder before it is sintered and drawn into wire form. After being mounted in the tube, the filament usually is *carburized* by being heated to a temperature of about 2000 K in a low-pressure atmosphere of hydrocarbon gas or vapor until its resistance increases by 10 to 25 percent. This process allows the reduction of the thoria to metallic thorium. The life of the filament as an emitter is increased because the rate of evaporation of thorium from the carburized surface is several times smaller than from a surface of pure tungsten.

Despite the improved performance obtained by carburization of a thoriated-tungsten filament, the element is susceptible to deactivation by the action of positive ions. Although the deactivation process is negligible for anode voltages below a critical value, a trace of residual gas pressure too small to affect the emission from a pure tungsten filament can cause rapid deactivation of a thoriated-tungsten filament. This restriction places stringent requirements on vacuum processing of the tube.

These factors, taken together, determine the wear-out rate of the device. Catastrophic failures due to interelectrode short circuits or failure of the vacuum envelope are considered abnormal and are usually the result of some external influence. Catastrophic failures not the result of the operating environment usually are caused by a defect in the manufacturing process. Such failures generally occur early in the life of the component.

The design of the RF equipment can have a substantial impact on the life expectancy of a vacuum tube. Protection circuitry must remove applied voltages rapidly to prevent damage to the tube in the event of a failure external to the device. The filament turn-on circuit also can affect PA tube life expectancy. The surge current of the filament circuit must be maintained below a certain level to prevent thermal cycling problems. This consideration is particularly important in medium- and high-power PA tubes. When the heater voltage is applied to a cage-type cathode, the tungsten wires expand immediately because of their low thermal inertia. However, the cathode support, which is made of massive parts (relative to the tungsten wires), expands more slowly. The resulting differential expansion can cause permanent damage to the cathode wires. It also can cause a modification of the tube operating characteristics and, occasionally, arcs between the cathode and the control grid.

Catastrophic Failures

Catastrophic failures can be divided into two primary categories:

- The short-circuiting of broken or warped elements within the tube

- A loss of vacuum in the device

Air in the tube causes a loss of dielectric standoff between the internal elements. It is not possible to distinguish between an interelectrode short circuit and a loss of vacuum in the operating equipment.

Catastrophic failures that occur during initial installation are usually the result of broken elements. Failures that occur after an initial burn-in period are more likely to be caused by a loss of vacuum. In either case, continued efforts to operate the tube can result in considerable damage to the device and to other components in the system.

Note that a loss of vacuum can result from any of several anomalies. The two most common are:

- A latent manufacturing defect in one of the ceramic-to-metal seals

- Circulating currents resulting from parasitic oscillations and/or improper tuning

Intermittent overloads are the most difficult to identify. They can be caused by circuit operating conditions or internal tube failures. A broken or warped filament element, for example, can move around inside the device and occasionally short circuit to the grid, causing a loss of grid bias and subsequent plate overload. Intermittent overloads also can be caused by a short circuit or high VSWR across the load.

9.4.9 Examining Tube Performance

Examination of a power tube after it has been removed from a transmitter or other type of RF generator can reveal a great deal about how well the equipment-tube combination is working. Contrast the appearance of a new power tube, shown in Figure 9.15, with a component at the end of its useful life. If a power tube fails prematurely,

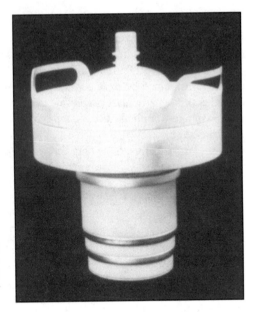

Figure 9.15 A new, unused 4CX15000A tube. Contrast the appearance of this device with the tubes shown in the following figures. (*Courtesy of Varian.*)

the device should be examined to determine whether an abnormal operating condition exists within the transmitter. Consider the following examples:

- Figure 9.16. Two 4CX15000A power tubes with differing anode heat-dissipation patterns. Tube (*a*) experienced excessive heating because of a lack of PA compartment cooling air or excessive dissipation because of poor tuning. Tube (*b*) shows a normal thermal pattern for a silver-plated 4CX15000A. Nickel-plated tubes do not show signs of heating because of the high heat resistance of nickel.

- Figure 9.17. Base-heating patterns on two 4CX15000A tubes. Tube (*a*) shows evidence of excessive heating because of high filament voltage or lack of cooling air directed toward the base of the device. Tube (*b*) shows a typical heating pattern with normal filament voltage.

- Figure 9.18. A 4CX5000A tube with burning on the screen-to-anode ceramic. Exterior arcing of this type generally indicates a socketing problem, or another condition external to the tube.

- Figure 9.19. The stem portion of a 4CX15000A tube that had gone down to air while the filament was on. Note the deposits of tungsten oxide formed when the filament burned up. The grids are burned and melted because of the ionization arcs that subsequently occurred. A failure of this type will trip overload breakers in the RF generator. It is indistinguishable from a short-circuited tube in operation.

- Figure 9.20. A 4CX15000A tube that experienced arcing typical of a bent finger-stock, or exterior arcing caused by components other than the tube.

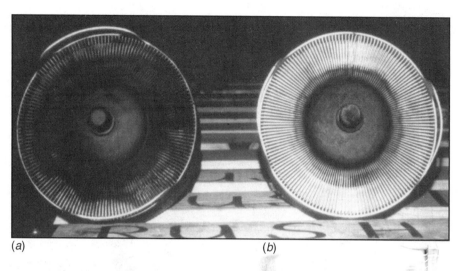

(a)   (b)

Figure 9.16 Anode dissipation patterns on two 4CX15000A tubes: (a) excessive heating, (b) normal wear. (*Courtesy of Econco Broadcast Service.*)

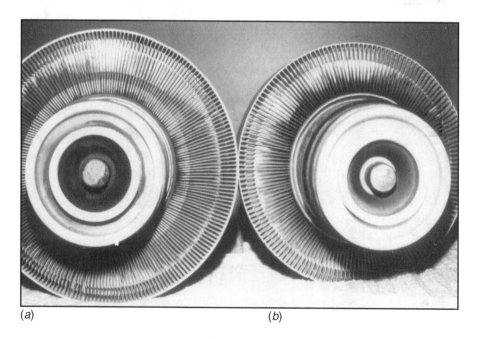

(a)   (b)

Figure 9.17 Base heating patterns on two 4CX15000A tubes: (a) excessive heating, (b) normal wear. (*Courtesy of Econco.*)

Figure 9.18 A 4CX5000A tube that appears to have suffered socketing problems. (*Courtesy of Econco.*)

Figure 9.19 The interior elements of a 4CX15000A tube that had gone to air while the filament was lit. (*Courtesy of Econco.*)

Note that the previous examples apply to silver-plated power grid tubes.

The external metal elements of a tube typically are plated with nickel or silver. Tubes that go into sockets are usually silver-plated. The soft silver provides a better contact interface than the much harder nickel, and it deforms slightly under contact pressure, providing greater contact area. Silver plating has a dull, whitish cast, whereas nickel has a hard, metallic appearance.

Nickel is resistant to discoloration due to heat at normal tube operating temperatures, but silver will tarnish easily. As demonstrated in this section, the heat patterns on

Figure 9.20 A 4CX15000A tube showing signs of external arcing. (*Courtesy of Econco.*)

silver-plated tubes are helpful in problem analysis. If a nickel-plated tube shows signs of heat discoloration, a significant cooling or operating problem exists. Nickel will not discolor until it reaches a temperature much higher than a tube normally achieves under typical operating conditions.

9.4.10 Shipping and Handling Vacuum Tubes

Because of their fragile nature, vacuum tubes are packaged for shipment in foam-filled or spring-supported shipping containers. When it is necessary to transport a tube from one location to another, use the original packing material and box or crate. The tube should be removed whenever the RF equipment is relocated. Never leave a tube installed in its socket during an equipment move.

During installation or service, always keep the tube in its typical operating position (most commonly horizontal). Never allow a tube to roll along a surface. Damage to the filament or to other elements in the device may result.

It is occasionally necessary for an RF generator to be moved from one location to another within a plant. Equipment used in this manner must be equipped with air-filled casters; do not use solid casters.

When it is necessary or desirable to store a tube for an extended period of time, place it in a plastic bag to protect it from moisture, and return the device to its shipping box. Maintenance engineers are discouraged from placing markings on a power tube. Some engineers note service dates using ink or pencil on tubes. This practice is not recommended. Instead, attach a separate service record to the tube using a tie-wrap or similar removable device. Do not use self-adhesive labels for this purpose. When removed, adhesive labels usually leave some residue on the attachment surface.

Although it is not recommended that a power tube be stored indefinitely, most modern devices utilizing metal and ceramic envelopes can be stored for long periods of time without deterioration. It usually is not necessary to rotate a spare tube through an operating socket to degas the device.

Older vacuum tubes utilizing glass as the insulating medium tend to leak gas over time. The glass is not the problem, but, instead, the Kovar alloy used to seal the glass-to-metal parts of the tube. Kovar also is subject to rusting when moisture is present. Glass-insulated vacuum tubes should be sealed in a plastic bag for storage and rotated through the equipment at least once every 12 months. The larger the physical size of the tube, the greater the problem of gassing.

9.5 Klystron Reliability

Klystron amplifiers are well known as reliable, long-life microwave devices that provide the user thousands of hours of trouble-free service. Many of the guidelines discussed in Section 9.4 for power vacuum tubes in general also apply to klystrons and related microwave devices. However, several considerations are specific to the klystron, including:

- Cooling system maintenance

- Ceramic element cleanings

- Gun element reconditioning

- Focusing electromagnet maintenance

- Protection during prime power interruption

- Shutoff precautions

9.5.1 Cleaning and Flushing the Cooling System

One of the major problems encountered in field operation of a klystron cooling system is the formation of corrosion products [9]. Corrosion frequently occurs in the cooling channels of the klystron body on the surface of the water/vapor-cooled collector, and in the cooling channels of the electromagnet. It is the result of a chemical reaction between free oxygen-laden coolant and the hot copper channel wall. Corrosion, as described here, is different from *scale formation*, which can be caused by the use of a coolant other than clean distilled water, or may be the by-product of electrolysis between dissimilar metals in the heat-exchange system. Scale formation can be controlled through the use of on-line purification loops.

New transmitter water lines frequently contain contaminants, including:

- Solder

- Soldering or brazing flux

- Oils

- Metal chips or burrs

- Pieces of Teflon sealing tape

When the water lines are installed, these contaminants must be flushed and cleaned from the system before the klystron and magnet are connected.

The following guidelines are suggested for cleaning the cooling passages of water- and vapor-cooled klystrons and associated RF generation equipment. The procedures should be performed before a klystron is installed, then repeated on a periodic basis while the tube is in service.

Transmitter Flushing

Proper cleaning of the circulating water system in an RF generator is vital to long-term reliability. Although each design is unique, the following general procedures are appropriate for most installations [9]:

- Disconnect the tube and the magnet.

- Add jumper hoses between the input and output of the klystron and the electro-magnet water lines.

- Disconnect or bypass the pump motor.

- Fill the system with hot tap water, if available.

- Open the drain in the transmitter cabinet and flush for 15 minutes or until it is clean.

- Flush the water lines between the tank and pump separately with hot tap water, if available, until they are clean.

- Connect all water lines, and fill the system with hot tap water, if available, and one cup of nonsudsing detergent. Trisodium phosphate is recommended.

- Operate the water system with hot tap water for 30 minutes. An immersion heater may be used to maintain hot water in the system.

- Drain the system, and flush with hot tap water for 30 minutes.

- Remove and clean the filter element.

- Refill the water system with tap water (at ambient temperature).

- Operate the water system. Maintain the water level while draining and flushing the system until no detergent, foam, or foreign objects or particles are visible in the drained or filtered element. To test for detergents in the water, drain a sample of water into a small glass test tube, and allow it to stand for 5 minutes. To generate foam, shake the test tube vigorously for 15 s, then allow it to stand 15 s. A completely foam-free surface indicates no foam-producing impurities.

- Repeat the two previous steps if detergent foam is still present.

- Drain the system and refill with distilled water when the tube and transmitter water lines are clean.

- Remove, clean, and replace the filter before using.

Flushing Klystron Water Lines

Although most klystrons are shipped from the manufacturer with water passages clean and dry, it is good engineering practice to flush all cooling passages before installing the tube. Occasionally, tubes that have been in service for some time develop scale on the collector and also must be flushed. Contaminated water further contributes to dirty water lines, requiring flushing to clear. The following general procedure is recommended for tubes with contaminated water lines, corrosion, scale, or blocked passages [9]:

- Remove the input water fitting (Hansen type), and add a straight pipe extension, approximately 1 to 2 in long, to the tube.

- Attach a hose to this fitting, using a hose clamp to secure it. This connection will serve as the drain line; it should be discharged into a convenient outlet.

- On some tube types, the normal body-cooling output line is fed to the base of the vapor-phase boiler. Remove the hose at the base of the boiler. Do not damage this fitting; it must be reused and must seal tightly.

- Attach a 2 to 3 in-long extension pipe that will fit the small hose at one end and a garden hose at the other end. Secure the extension with hose clamps.

- Connect the garden hose to a tap water faucet. Hot tap water is preferable, if available.

- Back-flush the klystron-body cooling passages for 10 to 15 minutes at full pressure until they are clean.

- Reconnect the input and output water lines to the klystron.

- If scale is present on the vapor-phase collector, use a solution of trisodium phosphate for the first cleaning. Perform this step after cleaning the transmitter water system (as described previously).

- Connect the klystron to the transmitter water lines, and fill the system with tap water (hot, if available) and one cup of nonsudsing detergent.

- Operate the water system for 15 minutes. Make certain that the water level completely covers the collector core.

- Drain the system, and flush for 30 minutes or until no detergent foam is present.

- Remove and clean the filter element.

- Fill the system with distilled water.

- Drain the system, and refill with distilled water.

Cleaning Klystron Water Lines

If heavy scaling has formed on the vapor-phase collector and/or blocked water passages, they may be cleaned using a stronger cleaning agent. A solution of *Ty-Sol*[1] may be used to remove scale and corrosion. The following procedure is recommended [9]:

- Clean the transmitter water system (as described previously).

- Connect the tube to the transmitter water system.

- Fill the system with hot tap water (if available) and add 2 gallons of cleaning solution for every 50 gallons of water.

- Operate the water system for 15 minutes, or until scale has been removed and the collector has a clean copper color.

- Make certain that the water level covers the top of the collector during cleaning. Evidence of cleaning action may be observed in the flow meters.

- Drain the system, and flush with tap water.

- Remove and clean the filter element, then replace.

- Refill the system with distilled water and flush for 30 minutes to 1 hour.

- Check for detergent foam at the end of the flushing period; also check the pH factor. Continue flushing with distilled water if foam is present or if the pH factor is not within the specified range.

- Refill the system with clean distilled water.

Chlorine present in common tap water is harmful to the water passages of the klystron. Thorough flushing with distilled water will remove all chlorine traces. Never use tap water for final refill or for make-up water.

Flushing and Cleaning Magnet Water Lines

The magnet water lines may be flushed and cleaned in the same manner as described for the klystron.

General Cleaning

Two remaining items must be clean to achieve efficient operation: the sight glass and float of the water-flow indicators. The water-flow indicators often become contaminated during use. This contamination collects on the sight glass and on the float, making flow-reading difficult. If too much contamination is present on the glass and float, sticking or erroneous readings may result. The detergent and cleaning solutions may

---

1 Chemical Research Association, Chicago, IL.

not remove all of this contamination. If this is the case, remove and clean the flow meter using a brush to clean the glass surface.

9.5.2 Cleaning Ceramic Elements

Foreign deposits on the ceramic elements of a klystron should be removed periodically to prevent ceramic heating or arc paths that can lead to klystron failure. The recommended procedure is as follows:

- Remove the klystron from the RF generator, and mount it to a stand or other suitable fixture.

- Remove loose debris with a soft-fiber bristle brush.

- Clean the ceramic with a mild abrasive cleanser, preferably one that includes a bleaching agent. (Household cleansers such as Comet and Ajax are acceptable.) While cleaning, do not apply excessive force to the ceramic. Normal hand pressure is sufficient.

- Flush the ceramic with clean water to remove the cleanser.

- Flush a second time with clean isopropyl alcohol.

- Air dry the device. Be certain that all moisture has evaporated from the tube before returning it to service.

9.5.3 Reconditioning Klystron Gun Elements

During the operational life of a klystron, voltage standoff problems may be experienced when the operating mode of the device is changed from one collector voltage to a higher voltage, or from continuous to pulsed service [10]. In both of these cases, the difference of potential between the internal elements is increased over values used in the previous mode of operation. In some instances, this elevated difference of potential can lead to internal arcing as a result of long-term buildup of electrical leakage between elements. This is a part of the normal aging process in klystrons. In many cases, however, this particular process can be reversed through high-voltage conditioning. Maintenance engineers have found it useful to clear any residual leakage between the various elements before pulsing an older klystron, or when increasing the operating potentials applied to an older klystron.

Klystron electron gun leakage can be checked, and, in many cases, the gun elements reconditioned, by applying a high dc potential between these elements. The process is termed *hi-potting*. It requires a current-limited (5 mA maximum), continuously-variable dc supply with a range of 0 to 30 kV. The procedure is as follows:

- Turn off the heater, and allow the cathode sufficient time to cool completely (approximately 1 to 2 hours).

- Remove all electrical connections from the klystron, except for the body and collector grounds.

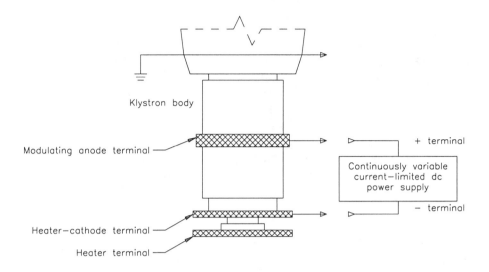

Figure 9.21 Electrical connections for reconditioning the gun elements of a klystron.

- Connect the negative high-voltage output of the hi-potter to the heater-cathode terminal, and connect the positive ground lead to the mod-anode as shown in Figure 9.21. Increase the voltage output gradually, noting the leakage current.

- If high levels of current are observed, reduce the voltage and allow it to stand at the 1 to 2 mA level until the leakage current reduces (burns off).

- Continue to raise the hi-potter output until the voltage applied is equal to the normal operating value for the klystron being checked. The typical leakage level is 200 to 400 µA.

- Check the mod-anode-to-body ground leakage; reduce, if necessary, in the same fashion. For this purpose, the negative high-voltage output is connected to the mod-anode terminal and the positive ground lead to the klystron body.

The leakage level indicated as optimum for this procedure (200 to 400 µA) is a typical value. Tubes still should operate normally above this level, but they may be more susceptible to internal arcing.

9.5.4 Focusing Electromagnet Maintenance

The focusing electromagnet of a klystron, given sufficient preventive maintenance, will provide many years of service [11]. Maintenance personnel can extend the lifetime of the device by reducing the operating temperature of the electromagnet windings and by minimizing thermal cycling. Field experience has shown that the major failure modes for a focusing electromagnet are short-circuited turns in the coil windings and/or short circuits from the windings to the coolant tubes.

Short-circuited turns result in lower magnetic flux density available to focus the klystron beam and are indicated by higher klystron body current. To counteract this effect, the transmitter operator often will increase the electromagnet current to a point above the rating necessary to obtain the magnetic field strength needed for rated klystron body current. This increases the heat dissipation in the device, which may lead to further problems. Such failures are the result of insulation breakdown between turns in the windings, and/or between the windings and the coolant tubes. Insulation breakdown usually is caused by excessive operating temperatures in the individual windings. Excessive temperature is the result of one or more of the following operational conditions:

- Excessive inlet coolant temperature
- Insufficient coolant flow
- Excessive ambient temperature
- Excessive operating current

Insulation breakdown also may result from excessive thermal cycling of the coil windings. Abrasion of internal insulation may occur during expansion and contraction of the winding assembly. Generally speaking, the life of a focusing electromagnet can be doubled for every 10°C drop in operating temperature (up to a point).

To minimize the effects of thermal cycling, the electromagnet temperature should be maintained at or below its operating value during standby periods by running at rated or reduced current, and at rated coolant flow. If neither running nor reduced current is available during standby conditions, heated coolant should be circulated through the electromagnet.

9.5.5 Power Control Considerations

Under certain conditions, high-voltage transients that exceed the specified maximum ratings of a klystron can occur in an RF generator. These transients can cause high-voltage arcing external to the tube across the klystron modulating anode and cathode seals. In some cases, such arcs can puncture sealing rings, letting the tube down to air. High-voltage transient arcing is not common to all installations, but when it does occur, it can result in a costly klystron failure. Overvoltage transients in power circuits can be traced to any of several causes, including lightning.

Collection of moisture on the tube is another potential cause of arc-induced failure. Integral-cavity klystrons, which are installed with the cathode down, may collect moisture in the high-voltage (cathode) area because of water leakage or condensation, leading to external arcing and loss of vacuum.

To provide a measure of protection against these types of failures, arc shields may be used on some tubes to protect the gun components that are the most likely to be damaged by arcs. The shield typically consists of an aluminum cover fitted on a portion of the tube. Although arc shields provide considerable protection, they cannot prevent all failures; the magnitude of a potential arc is an unknown quantity. It must be understood that an arc shield does not eliminate the cause of arcing, which should be addressed separately by the maintenance engineer. It is, instead, intended as a stop-gap measure to prevent destruction of the tube until the root cause of the arcing is eliminated.

Installation of a gas-filled spark-gap transient voltage suppressor also can serve to protect the klystron in the event of an arc or other transient discharge. The rating of the suppressor is determined by the type of tube being protected. The device is connected between ground and the cathode, installed as close as possible to the cathode. The suppressor will prevent the application of transient voltages that exceed the standoff rating of the klystron ceramic.

Power-line suppression components are useful for protecting transmission equipment from incoming line transients and other disturbances, but may not be effective in protecting against a system-induced arc within the transmitter or power supply that could damage or destroy the klystron. Therefore, protection of the klystron itself is recommended.

Primary Power Interruption

Unexpected interruptions of primary power, which occur seasonally during winter storms and summer lightning strikes, can damage or destroy a klystron unless adequate protection measures are taken [12]. There are three main factors involved in such failures:

- Mechanical and/or electrical malfunction of the main circuit breaker of the transmitter. This condition results when the breaker remains closed after the primary power is interrupted. With the loss of primary power, all excitation to the transmitter is removed. If the main circuit breaker remains closed when the power is restored, high voltage can be applied to the klystron before the magnet supply can provide adequate focusing. This allows full beam power dissipation in the body of the klystron, which can melt the drift tube.

- Loss of focusing during shutdown resulting from a power interruption at the transmitter in a system that does not employ magnet power supply filtering. When a loss of primary power occurs, the various power supplies start to discharge. The time constant of the associated filter circuit determines the rate of discharge. If there is no filter capacitance in the magnet supply, it will decay almost instantaneously while the beam filtering allows the high voltage to decay comparatively slowly. This permits defocusing of the electron beam and dissipation of the beam power in the drift tube region. Melting of the drift tube will result.

- Loss of one phase of the primary power supply. In some transmitters, the magnet supply and the protection circuits are fed from a common leg of the 3-phase incoming power line. A loss of this phase will cause a loss of focusing and disable the protection circuits, including body-current protection. If the low-voltage-sensing relay in the main breaker is misadjusted or nonoperational, the high-

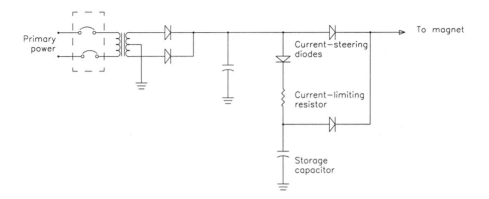

Figure 9.22 Diode-steering filtered power supply for a klystron focusing electromagnet.

voltage supply will remain on (although somewhat reduced) and, without focusing or protection, may melt the tube body.

To prevent these types of failures, the following steps are recommended:

- Conduct regular tests of the main circuit breaker in accordance with the manufacturer's instructions. Note that turning the breaker on and off using the transmitter control circuitry checks only half of the system. Tests also should be conducted by interrupting the primary feed to the breaker.

- Compute or measure the discharge time constant of the filtering circuits of both the beam and magnet supplies. Capacitance should be added to the magnet supply if the time constant is less than that of the beam supply. Be careful, however, not to increase the magnet supply filtering too much because it may also increase the ramp-up time of the magnet voltage. This could result in the beam supply being applied before the magnet supply is at its full potential. A simple diode-steering arrangement, such as the one shown in Figure 9.22, will prevent such problems in most cases.

- Consider installing a multipole vacuum relay at the high-voltage input power lines (between the main breaker and the high-power supply) to prevent instantaneous application of high voltage in the event of a malfunction of the main breaker. The primary or low-voltage circuit configuration of the vacuum relays should be such that it will prevent reenergizing of the relays before the magnetic field has built up when primary power is restored after a line loss.

- Install a solid-state phase-loss sensor if this protection is not already provided.

- Install an electromagnet undercurrent protection circuit that will remove the high voltage if the magnet current falls below a preset level.

An end user who is contemplating any of these changes should consult the original equipment manufacturer before proceeding.

9.6 References

1. Capitano, J., and J. Feinstein, "Environmental Stress Screening Demonstrates Its Value in the Field," *Proceedings of the IEEE Reliability and Maintainability Symposium*, IEEE, New York, 1986.
2. Wong, Kam L., "Demonstrating Reliability and Reliability Growth with Environmental Stress Screening Data," *Proceedings of the IEEE Reliability and Maintainability Symposium*, IEEE, New York, 1990.
3. Tustin, Wayne, "Recipe for Reliability: Shake and Bake," *IEEE Spectrum*, IEEE, New York, December 1986.
4. Hobbs, Gregg K., "Development of Stress Screens," *Proceedings of the IEEE Reliability and Maintainability Symposium*, IEEE, New York, 1987.
5. Smith, William B., "Integrated Product and Process Design to Achieve High Reliability in Both Early and Useful Life of the Product," *Proceedings of the IEEE Reliability and Maintainability Symposium*, IEEE, New York, 1987.
6. "Conditioning of Large Radio-Frequency Power Tubes," Application Bulletin no. 21, Varian/Eimac, San Carlos, CA, 1984.
7. Johnson, Walter, "Techniques to Extend the Service Life of High-Power Vacuum Tubes," *NAB Engineering Conference Proceedings*, National Association of Broadcasters, Washington, DC, pp. 285–293, 1994.
8. "Fault Protection," Application Bulletin no. 17, Varian/Eimac, San Carlos, CA, 1987.
9. "Cleaning and Flushing Klystron Water and Vapor Cooling Systems," Application Engineering Bulletin no. AEB-32, Varian Associates, Palo Alto, CA, 1982.
10. "Reconditioning of UHF-TV Klystron Electron Gun Elements," Technical Bulletin no. 4867, Varian Associates, Palo Alto, CA, July 1985.
11. "UHF-TV Klystron Focusing Electromagnets," Technical Bulletin no. 3122, Varian Associates, Palo Alto, CA, November 1973.
12. "Recommendations for Protection of 4KM Series Klystrons During Prime Power Interruptions," Technical Bulletin no. 3542, Varian Associates, Palo Alto, CA.

9.7 Bibliography

Anderson, Leonard R., *Electric Machines and Transformers*, Reston Publishing Company, Reston, VA.
Brauer, D. C., and G. D. Brauer, "Reliability-Centered Maintenance," *Proceedings of the IEEE Reliability and Maintainability Symposium*, IEEE, New York, 1987.
Cameron, D., and R. Walker, "Run-In Strategy for Electronic Assemblies," *Proceedings of the IEEE Reliability and Maintainability Symposium*, IEEE, New York, 1986.
Clark, Richard J., "Electronic Packaging and Interconnect Technology Working Group Report (IDA/OSD R&M Study)," IDA Record Document D-39, Institute of Defense Analysis, August 1983.

"Cleaning UHF-TV Klystron Ceramics," Technical Bulletin no. 4537, Varian Associates, Palo Alto, CA, May 1982.

DesPlas, Edward, "Reliability in the Manufacturing Cycle," *Proceedings of the IEEE Reliability and Maintainability Symposium*, IEEE, New York, 1986.

Devaney, John, "Piece Parts ESS in Lieu of Destructive Physical Analysis," *Proceedings of the IEEE Reliability and Maintainability Symposium*, IEEE, New York, 1986.

Dick, Brad, "Caring for High-Power Tubes," *Broadcast Engineering*, Intertec Publishing, Overland Park, KS, pp. 48–58, November 1991.

Doyle, Edgar, Jr., "How Parts Fail," *IEEE Spectrum*, IEEE, New York, October 1981.

"Extending Transmitter Tube Life," Application Bulletin no. 18, Varian/Eimac, San Carlos, CA, 1990.

"55 kW Integral Cavity Klystron Arc Shields," Technical Bulletin no. 4441, Varian Associates, Palo Alto, CA, April 1981.

Fink, D., and D. Christiansen (eds.), *Electronics Engineers' Handbook*, 3rd Ed., McGraw-Hill, New York, 1989.

Fortna, H., R. Zavada, and T. Warren, "An Integrated Analytic Approach for Reliability Improvement," *Proceedings of the IEEE Reliability and Maintainability Symposium*, IEEE, New York, 1990.

Frey, R., G. Ratchford, and B. Wendling, "Vibration and Shock Testing for Computers," *Proceedings of the IEEE Reliability and Maintainability Symposium*, IEEE, New York, 1990.

Griffin, P., "Analysis of the F/A-18 Hornet Flight Control Computer Field Mean Time Between Failure," *Proceedings of the IEEE Reliability and Maintainability Symposium*, IEEE, New York, 1985.

Hansen, M. D., and R. L. Watts, "Software System Safety and Reliability," *Proceedings of the IEEE Reliability and Maintainability Symposium*, IEEE, New York, 1988.

Hermason, Sue E., Major, USAF. Letter dated December 2, 1988. From: Yates, W., and D. Shaller, "Reliability Engineering as Applied to Software," *Proceedings of the IEEE Reliability and Maintainability Symposium*, IEEE, New York, 1990.

High Power Transmitting Tubes for Broadcasting and Research, Philips Technical Publication, Eindhoven, the Netherlands, 1988.

Horn, R., and F. Hall, "Maintenance Centered Reliability," *Proceedings of the IEEE Reliability and Maintainability Symposium*, IEEE, New York, 1983.

Irland, Edwin A., "Assuring Quality and Reliability of Complex Electronic Systems: Hardware and Software," *Proceedings of the IEEE*, Vol. 76, no. 1, IEEE, New York, January 1988.

Jordan, Edward C. (ed.), *Reference Data for Engineers: Radio, Electronics, Computers, and Communications*, 7th Ed., Howard W. Sams, Indianapolis, IN, 1985.

Kenett, R., and M. Pollak, "A Semi-Parametric Approach to Testing for Reliability Growth, with Application to Software Systems," *IEEE Transactions on Reliability*, IEEE, New York, August 1986.

Laboratory Staff, *The Care and Feeding of Power Grid Tubes*, Varian EIMAC, San Carlos, CA, 1984.

Maynard, Egbert, "OUSDRE Working Group, VHSIC Technology Working Group Report (IDA/OSD R&M Study)," Document D-42, Institute of Defense Analysis, November 1983.

Meeldijk, Victor, "Why Do Components Fail?" *Electronic Servicing and Technology*, Intertec Publishing, Overland Park, KS, November 1986.

Nenoff, Lucas, "Effect of EMP Hardening on System R&M Parameters," *Proceedings of the 1986 Reliability and Maintainability Symposium*, IEEE, New York, 1986.

Neubauer, R. E., and W. C. Laird, "Impact of New Technology on Repair," *Proceedings of the IEEE Reliability and Maintainability Symposium*, IEEE, New York, 1987.

Pearman, Richard, *Power Electronics*, Reston Publishing Company, Reston, VA, 1980.

Powell, Richard, "Temperature Cycling vs. Steady-State Burn-In," *Circuits Manufacturing*, Benwill Publishing, September 1976.

"Power Grid Tubes for Radio Broadcasting," Publication no. DTE-115, Thomson-CSF, Dover, NJ, 1986.

"Reduced Heater Voltage Operation," Technical Bulletin no. 4863, Varian Associates, Palo Alto, CA, June 1985.

Robinson, D., and S. Sauve, "Analysis of Failed Parts on Naval Avionic Systems," Report no. D180-22840-1, Boeing Company, October 1977.

Spradlin, B. C., "Reliability Growth Measurement Applied to ESS," *Proceedings of the IEEE Reliability and Maintainability Symposium*, IEEE, New York, 1986.

Svet, Frank A., "Factors Affecting On-Air Reliability of Solid State Transmitters," *Proceedings of the SBE Broadcast Engineering Conference*, Society of Broadcast Engineers, Indianapolis, IN, October 1989.

Technical Staff, Military/Aerospace Products Division, *The Reliability Handbook*, National Semiconductor, Santa Clara, CA, 1987.

Tube Topics: A Guide for Vacuum Tube Users in a Transistorized World, Econco Broadcast Service, Woodland, CA, 1990.

Whitaker, Jerry C., *Maintaining Electronic Systems*, CRC Press, Boca Raton, FL, 1992.

Whitaker, Jerry C., *Radio Frequency Transmission Systems: Design and Operation*, McGraw-Hill, New York, 1990.

Wilson, M. F., and M. L. Woodruff, "Economic Benefits of Parts Quality Knowledge," *Proceedings of the IEEE Reliability and Maintainability Symposium*, IEEE, New York, 1985.

Wilson, Sam, "What Do You Know About Electronics," *Electronic Servicing and Technology*, Intertec Publishing, Overland Park, KS, September 1985.

Wong, K. L., I. Quart, L. Kallis, and A. H. Burkhard, "Culprits Causing Avionics Equipment Failures," *Proceedings of the IEEE Reliability and Maintainability Symposium*, IEEE, New York, 1987.

Worm, Charles M., "The Real World: A Maintainer's View," *Proceedings of the IEEE Reliability and Maintainability Symposium*, IEEE, New York, 1987.

10

Device Performance Criteria

10.1 Introduction

Many shortcomings in the received audio, video, and/or data stream of a transmission resulting from noise, reflections, and terrain limitations cannot be controlled by the equipment design engineer or the operating maintenance engineer. However, other problems can be prevented through proper application of vacuum tube devices and proper equipment adjustment. The critical performance parameters for a transmission system include:

- Output power

- Transmission line/antenna VSWR

- Frequency stability/accuracy

- Occupied bandwidth

- Distortion mechanisms, including *total harmonic distortion* (THD), *intercarrier phase distortion, and quadrature distortion*

- Differential phase and gain

- Frequency response errors

- Envelope delay

The adjustment schedule required to maintain a satisfactory level of performance depends upon a number of factors, including:

- Operating environment

- Equipment quality

- Age of the equipment

- Type of service being performed

10.2 Measurement Parameters

Most measurements in the RF field involve characterizing fundamental parameters. These include signal level, phase, and frequency. Most other tests consist of measuring these fundamental parameters and displaying the results in combination by using some convenient format. For example, signal-to-noise ratio (S/N) consists of a pair of level measurements made under different conditions and expressed as a logarithmic ratio. Modern test instruments perform the necessary manipulation of the measured parameters to yield a direct readout of the quantity of interest.

When characterizing a vacuum tube device or stage, it is common to view it as a box with input and output terminals. In normal use, a signal is applied to the input of the box, and the signal—modified in some way—appears at the output. Some measurements are *one-port* tests, such as impedance or noise level, and are not concerned with input/output signals, but with only one or the other.

Measurements are made on equipment to check performance under specified conditions and to assess suitability for use in a particular application. The measurements may be used to verify specified system performance or as a way of comparing several pieces of equipment for use in a system. Measurements also may be used to identify components in need of adjustment or repair.

Many parameters are important in vacuum tube devices and merit attention in the measurement process. Some common measurements include frequency response, gain or loss, harmonic distortion, intermodulation distortion, noise level, phase response, and transient response.

Measurement of level is fundamental to most RF specifications. Level can be measured either in absolute terms or in relative terms. Power output is an example of an absolute level measurement; it does not require any reference. S/N and gain or loss are examples of relative, or ratio, measurements; the result is expressed as a ratio of two measurements. Although it may not appear so at first, frequency response is also a relative measurement. It expresses the gain of the device under test as a function of frequency, with the midband gain (typically) as a reference.

Distortion measurements are a way of quantifying the unwanted components added to a signal by a piece of equipment. The most common technique is total harmonic distortion, but others are often used. Distortion measurements express the unwanted signal components relative to the desired signal, usually as a percentage or decibel value. This is another example of multiple level measurements that are combined to give a new measurement figure.

10.2.1 Power Measurements

The simplest definition of a level measurement is the alternating current amplitude at a particular place in the system under test. However, in contrast to direct current measurements, there are many ways of specifying ac voltage in a circuit. The most common methods include:

- Average response

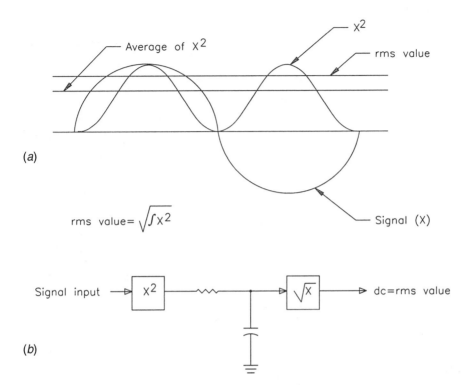

Figure 10.1 Root-mean-square (rms) voltage measurements: (a) the relationship of rms and average values, (b) the rms measurement circuit.

- Root mean square (rms)
- Peak

Strictly speaking, the term *level* refers to a logarithmic, or decibel, measurement. However, common parlance employs the term for an ac amplitude measurement, and that convention will be followed in this chapter.

Root Mean Square

The root-mean-square technique measures the effective power of the ac signal. It specifies the value of the dc equivalent that would dissipate the same power if either were applied to a load resistor. This process is illustrated in Figure 10.1 for voltage measurements. The input signal is squared, and the average value is found. This is equivalent to finding the average power. The square root of this value is taken to transfer the signal from a power value back to a voltage. For the case of a sine wave, the rms value is 0.707 of its maximum value.

Assume that the signal is no longer a sine wave but rather a sine wave and several harmonics. If the rms amplitude of each harmonic is measured individually and added,

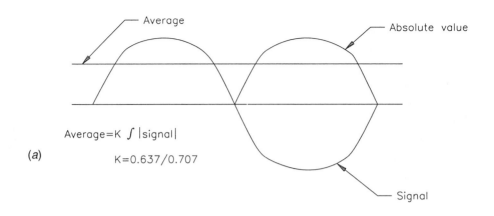

(a)

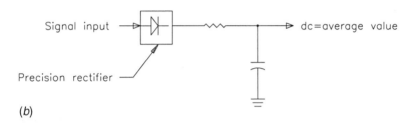

(b)

Figure 10.2 Average voltage measurements: (a) illustration of average detection, (b) average measurement circuit.

the resulting value will be the same as an rms measurement of the signals together. Because rms voltages cannot be added directly, it is necessary to perform an rms addition. Each voltage is squared, and the squared values are added as follows:

$$V_{rms\ total} = \sqrt{V_{rms\ 1}^2 + V_{rms\ 2}^2 + \ldots V_{rms\ n}^2} \tag{10.1}$$

Note that the result is not dependent on the phase relationship of the signal and its harmonics. The rms value is determined completely by the amplitude of the components. This mathematical predictability is useful in practical applications of level measurement, enabling correlation of measurements made at different places in a system. It is also important in correlating measurements with theoretical calculations.

Average-Response Measurement

The average-responding voltmeter measures ac voltage by rectifying it and filtering the resulting waveform to its average value, as shown in Figure 10.2. This results in a dc voltage that can be read on a standard dc voltmeter. As shown in the figure, the average value of a sine wave is 0.637 of its maximum amplitude. Average-responding

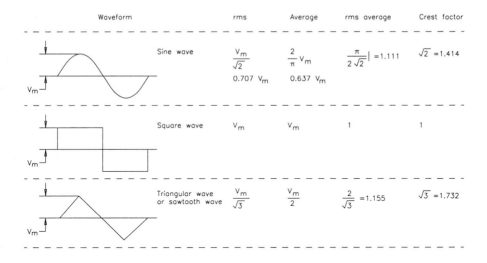

Waveform		rms	Average	rms average	Crest factor
Sine wave		$\dfrac{V_m}{\sqrt{2}}$	$\dfrac{2}{\pi}V_m$	$\dfrac{\pi}{2\sqrt{2}} = 1.111$	$\sqrt{2} = 1.414$
		$0.707\ V_m$	$0.637\ V_m$		
Square wave		V_m	V_m	1	1
Triangular wave or sawtooth wave		$\dfrac{V_m}{\sqrt{3}}$	$\dfrac{V_m}{2}$	$\dfrac{2}{\sqrt{3}} = 1.155$	$\sqrt{3} = 1.732$

Figure 10.3 Comparison of rms and average voltage characteristics.

meters usually are calibrated to read the same as an rms meter for the case of a single sine wave signal. This results in the measurement being scaled by a constant K of 0.707/0.637, or 1.11. Meters of this type are called *average-responding, rms calibrated.* For signals other than sine waves, the response will be different and difficult to predict.

If multiple sine waves are applied, the reading will depend on the phase shift between the components and will no longer match the rms measurement. A comparison of rms and average-response measurements is made in Figure 10.3 for various waveforms. If the average readings are adjusted as described previously to make the average and rms values equal for a sine wave, all the numbers in the *average* column would be increased by 11.1 percent, and the *rms-average* numbers would be reduced by 11.1 percent.

Peak-Response Measurement

Peak-responding meters measure the maximum value that the ac signal reaches as a function of time. This approach is illustrated in Figure 10.4. The signal is full-wave-rectified to find its absolute value, then passed through a diode to a storage capacitor. When the absolute value of the voltage rises above the value stored on the capacitor, the diode will conduct and increase the stored voltage. When the voltage decreases, the capacitor will maintain the old value. Some method of discharging the capacitor is required so that a new peak value can be measured. In a true peak detector, this is accomplished by a solid-state switch. Practical peak detectors usually include a large-

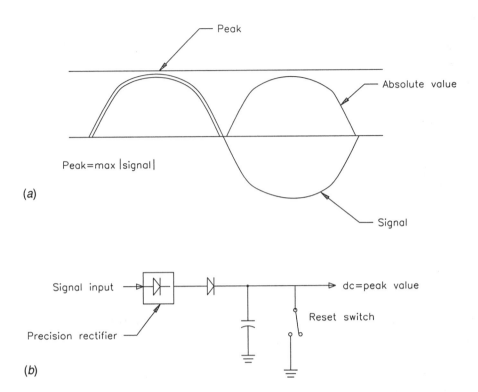

Figure 10.4 Peak voltage measurements: (a) illustration of peak detection, (b) peak measurement circuit.

value resistor to discharge the capacitor gradually after the user has had a chance to read the meter.

The ratio of the true peak to the rms value is called the *crest factor*. For any signal but an ideal square wave, the crest factor will be greater than 1, as demonstrated in Figure 10.5. As the measured signal becomes more peaked, the crest factor increases.

By introducing a controlled charge and discharge time, a *quasi-peak* detector is achieved. The charge and discharge times may be selected, for example, to simulate the transmission pattern for a digital carrier. The gain of a quasi-peak detector is normally calibrated so that it reads the same as an rms detector for sine waves.

The *peak-equivalent sine* is another method of specifying signal amplitude. This value is the rms level of a sine wave having the same peak-to-peak amplitude as the signal under consideration. This is the peak value of the waveform scaled by the correction factor 1.414, corresponding to the peak-to-rms ratio of a sine wave. Peak-equivalent sine is useful when specifying test levels of waveforms in distortion measurements. If the distortion of a device is measured as a function of amplitude, a point will be reached at which the output level cannot increase any further. At this point the peaks of the waveform will be clipped, and the distortion will rise rapidly with further increases in level. If another signal is used for distortion testing on the same device, it is desirable

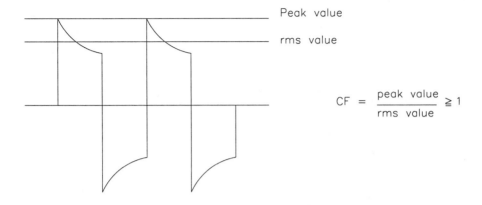

$$CF = \frac{peak \ value}{rms \ value} \geq 1$$

Figure 10.5 Illustration of the crest factor in voltage measurements.

that the levels at which clipping is reached correspond. Signal generators are normally calibrated in this way to allow changing between waveforms without clipping or readjusting levels.

Measurement Bandwidth

The bandwidth of the level-measuring instrument can have a significant effect on the accuracy of the reading. For a meter with a single-pole rolloff (one bandwidth-limiting component in the signal path), significant measurement errors can occur. Such a meter with a specified bandwidth of 1 MHz, for example, will register a 10 percent error in the measurement of signals at 500 kHz. To obtain 1 percent accurate measurements (disregarding other error sources in the meter), the signal frequency must be less than 100 kHz.

Figure 10.6 illustrates another problem associated with limited-bandwidth measuring devices. In the figure, a distorted sine wave is measured by two meters with different bandwidths. The meter with the narrower bandwidth does not respond to all the harmonics and gives a lower reading. The severity of this effect varies with the frequency being measured and the bandwidth of the meter; it can be especially severe in the measurement of wideband noise. Peak measurements are particularly sensitive to bandwidth effects. Systems with restricted low-frequency bandwidth will produce *tilt* in a square wave, and bumps in the high-frequency response will produce an *overshoot*. The effect of either will be an increase in the peak reading.

Meter Accuracy

Accuracy is a measure of how well an instrument quantifies a signal at a midband frequency. This sets a basic limit on the performance of the meter in establishing the absolute amplitude of a signal. It is also important to look at the *flatness* specification to

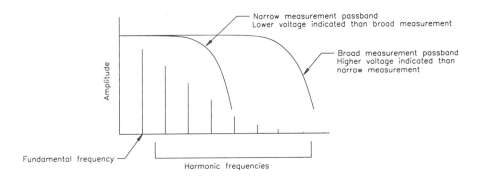

Figure 10.6 The effects of instrument bandwidth on voltage measurements.

see how well this performance is maintained with changes in frequency. Flatness describes how well the measurements at any other frequency track those at the reference. If a meter has an accuracy of 2 percent at 1 MHz and a flatness of 1 dB (10 percent) from 20 kHz to 20 MHz, the inaccuracy can be as great as 12 percent at 20 MHz.

Meters often have a specification of accuracy that changes with voltage range, being most accurate only in the range in which the instrument was calibrated. A meter with 1 percent accuracy on the 2 V range and 1 percent accuracy per step would be 3 percent accurate on the 200 V scale. Using the flatness specification given previously, the overall accuracy for a 100 V, 20 MHz sine wave is 14 percent. In many instruments, an additional accuracy derating is given for readings as a percentage of full scale, making readings at less than full scale less accurate.

However, the accuracy specification is not normally as important as the flatness. When performing frequency response or gain measurements, the results are relative and are not affected by the absolute voltage used. When measuring gain, however, the attenuator accuracy of the instrument is a direct error source. Similar comments apply to the accuracy and flatness specifications for signal generators. Most are specified in the same manner as voltmeters, with the inaccuracies adding in much the same manner.

RF Power Measurement

Measurement of RF power output typically is performed with a directional wattmeter, calibrated for use over a specified range of frequencies and power levels. Various grades of accuracy are available, depending upon the requirements of the application. The most accurate measurements usually are made by determining the temperature rise of the cooling through a calibrated dummy load. The power absorbed by the coolant, usually water, can be calculated from the following equation:

$$P = K \times Q \times \Delta T \tag{10.2}$$

Where:

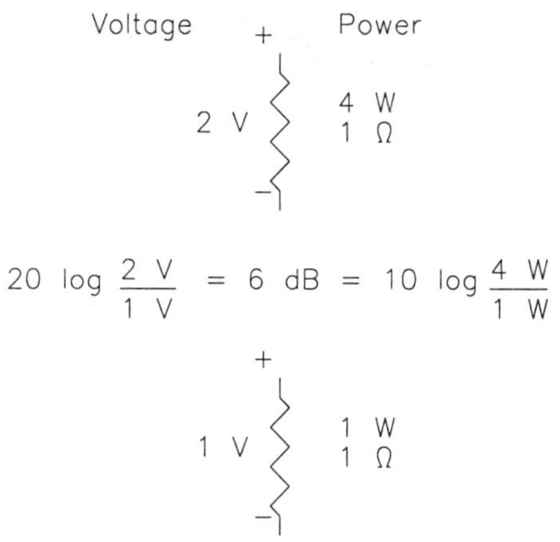

Voltage   +   Power

2 V   4 W
      1 Ω

$$20 \ \log \ \frac{2 \ V}{1 \ V} \ = \ 6 \ dB \ = \ 10 \ \log \ \frac{4 \ W}{1 \ W}$$

+

1 V   1 W
      1 Ω

Figure 10.7 Example of the equivalence of voltage and power decibels.

P = power dissipated in kilowatts
K = a constant, determined by the coolant (for pure water at 30°C, $K = 0.264$)
Q = coolant flow in gallons per minute
ΔT = difference between inlet and outlet water temperature in degrees Celsius

This procedure often is used to verify the accuracy of in-line directional wattmeters, which are more convenient to use but typically offer less accuracy.

10.2.2 Decibel Measurement

Measurements in RF work often are expressed in decibels. Radio frequency signals span a wide range of levels, too wide to be accommodated on a linear scale. The decibel is a logarithmic unit that compresses this wide range down to one that is easier to handle. Order-of-magnitude (factor of 10) changes result in equal increments on a decibel scale. A decibel may be defined as the logarithmic ratio of two power measurements or as the logarithmic ratio of two voltages:

$$dB = 20 \ \log \left\{ \frac{E_1}{E_2} \right\} \tag{10.3}$$

$$dB = 10 \ \log \left\{ \frac{P_1}{P_2} \right\} \tag{10.4}$$

Decibel values from power measurements and decibel values from voltage measurements are equal if the impedances are equal. In both equations, the denominator variable is usually a stated reference, as illustrated by the example in Figure 10.7. Whether the decibel value is computed from the power-based equation or from the voltage-based equation, the same result is obtained.

Table 10.1 Common Decibel Values and Conversion Ratios

dB Value	Voltage Ratio	Power Ratio
0	1	1
+1	1.122	1.259
+2	1.259	1.586
+3	1.412	1.995
+6	1.995	3.981
+10	3.162	10
+20	10	100
+40	100	10,000
−1	0.891	0.794
−2	0.794	0.631
−3	0.707	0.501
−6	0.501	0.251
−10	0.3163	0.1
−20	0.1	0.01
−40	0.01	0.0001

A doubling of voltage will yield a value of 6.02 dB, and a doubling of power will yield 3.01 dB. This is true because doubling of voltage results in an increase in power by a factor of 4. Table 10.1 lists the decibel values for some common voltage and power ratios.

RF engineers often express the decibel value of a signal relative to some standard reference instead of another signal. The reference for decibel measurements may be predefined as a power level, as in dBk (decibels above 1 kW), or it may be a voltage reference. Often, it is desirable to specify levels in terms of a reference transmission level somewhere in the system under test. These measurements are designated dBr, where the reference point or level must be separately conveyed.

10.2.3 Noise Measurement

Noise measurements are specialized level measurements. Noise may be expressed as an absolute level by simply measuring the voltage at the desired point in the system. This approach, however, is often not very meaningful. Specifying the noise performance as the signal-to-noise ratio is a better approach. S/N is a decibel measurement of the noise level using the signal level measured at the same point as a reference. This makes measurements at different points in a system or in different systems directly comparable. A signal with a given S/N can be amplified with a perfect amplifier or attenuated with no change in the S/N. Any degradation in S/N at a later point in the system is the result of limitations of the equipment that follows.

Noise performance is an important parameter in the operation of any power vacuum tube amplifier or oscillator. All electric conductors contain free electrons that are in continuous random motion. It can be expected that, by pure chance, more electrons will be moving in one direction than in another at any instant. The result is that a voltage will be developed across the terminals of the conductor if it is an open circuit, or a current will be delivered to any connected circuit. Because this voltage (or current) varies in a random manner, it represents noise energy distributed throughout the frequency spectrum, from the lowest frequencies well into the microwave range. This effect is commonly referred to as *thermal-agitation noise*, because the motion of electrons results from thermal action. It is also referred to as *resistance noise*. The magnitude of the noise depends upon the following:

- The resistance across which the noise is developed

- Absolute temperature of the resistance

- Bandwidth of the system involved

Random noise, similar in character to that produced in a resistance, is generated in vacuum tubes as a result of irregularities in electron flow. Tube noise can be divided into the following general classes:

- *Shot effect*, representing random variations in the rate of electron emission from the cathode

- *Partition noise*, arising from chance variations in the division of current between two or more positive electrodes

- *Induced grid noise*, produced as a result of variations in the electron stream passing adjacent to a grid

- *Gas noise*, generated by random variations in the rate of ion production by collision

- *Secondary emission noise*, arising from random variations in the rate of production of secondary electrons

- *Flicker effect*, a low-frequency variation in emission that occurs with oxide-coated cathodes

Shot effect in the presence of space charge, partition noise, and induced grid noise is the principal source of tube noise that must be considered in RF work.

10.2.4 Phase Measurement

When a signal is applied to the input of a device, the output will appear at some later point in time. For sine wave excitation, this delay between input and output may be expressed as a proportion of the sine wave cycle, usually in degrees. One cycle is 360°, one half-cycle is 180°, and so on. This measurement is illustrated in Figure 10.8. The

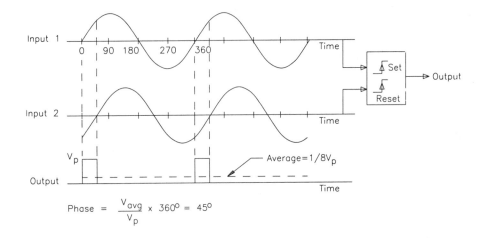

Figure 10.8 Illustration of the measurement of a phase shift between two signals.

phasemeter input signal number 2 is delayed from, or is said to be *lagging*, input number 1 by 45°.

Most RF test equipment checks phase directly by measuring the proportion of one signal cycle between zero crossings of the signals. Phase typically is measured and recorded as a function of frequency over a specified range. For most vacuum tube devices, phase and amplitude responses are closely coupled. Any change in amplitude that varies with frequency will produce a corresponding phase shift.

Relation to Frequency

When dealing with complex signals, the meaning of phase can become unclear. Viewing the signal as the sum of its components according to Fourier theory, a different value of phase shift is found at each frequency. With a different phase value on each component, the one to be used as the reference is unclear. If the signal is periodic and the waveshape is unchanged passing through the device under test, a phase value still may be defined. This may be done by using the shift of the zero crossings as a fraction of the waveform period. Indeed, most commercial phase-measuring instruments will display this value. However, in the case of differential phase shift with frequency, the waveshape will be changed. It is then impossible to define any phase-shift value, and phase must be expressed as a function of frequency.

Group delay is another useful expression of the phase characteristics of an RF device. Group delay is the slope of the phase response. It expresses the relative delay of the spectral components of a complex waveform. If the group delay is flat, all components will arrive at a given point together. A peak or rise in the group delay indicates that those components will arrive later by the amount of the peak or rise. Group delay is computed by taking the derivative of the phase response vs. frequency. Mathematically:

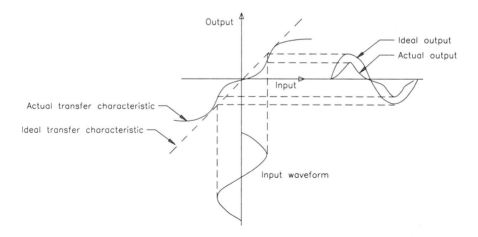

Figure 10.9 Illustration of total harmonic distortion (THD) measurement of an amplifier transfer characteristic.

$$Group\ delay = -\frac{(phase\ at\ f_2 - phase\ at\ f_1)}{f_2 - f_1} \qquad (10.5)$$

This definition requires that phase be measured over a range of frequencies to give a curve that can be differentiated. It also requires that the phase measurements be performed at frequencies sufficiently close to provide a smooth and accurate derivative.

10.2.5 Nonlinear Distortion

Distortion is a measure of signal impurity, a deviation from ideal performance of a device, stage, or system. Distortion usually is expressed as a percentage or decibel ratio of the und3sired components to the desired components of a signal. Distortion of a vacuum tube device or stage is measured by inputting one or more sine waves of various amplitudes and frequencies. In simplistic terms, any frequencies at the output that were not present at the input are distortion. However, strictly speaking, components caused by power line interference or another spurious signal are not distortion but, rather, noise. Many methods of measuring distortion are in common use, including harmonic distortion and several types of intermodulation distortion.

Harmonic Distortion

The transfer characteristic of a typical vacuum tube amplifier is shown in Figure 10.9. The transfer characteristic represents the output voltage at any point in the signal waveform for a given input voltage; ideally this is a straight line. The output waveform is the projection of the input sine wave on the device transfer characteristic. A

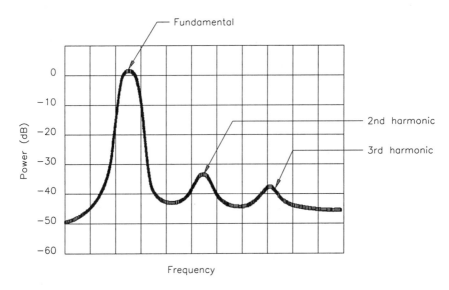

Figure 10.10 Example of reading THD from a spectrum analyzer.

change in the input produces a proportional change in the output. Because the actual transfer characteristic is nonlinear, a distorted version of the input waveshape appears at the output.

Harmonic distortion measurements excite the device under test with a sine wave and measure the spectrum of the output. Because of the nonlinearity of the transfer characteristic, the output is not sinusoidal. By using Fourier series, it can be shown that the output waveform consists of the original input sine wave plus sine waves at integer multiples (harmonics) of the input frequency. The spectrum of a distorted signal is shown in Figure 10.10. The harmonic amplitudes are proportional to the amount of distortion in the device under test. The percentage of harmonic distortion is the rms sum of the harmonic amplitudes divided by the rms amplitude of the fundamental.

Harmonic distortion may be measured with a spectrum analyzer or a distortion test set. Figure 10.11 shows the setup for a spectrum analyzer. As shown in Figure 10.10, the fundamental amplitude is adjusted to the 0 dB mark on the display. The amplitudes of the harmonics are then read and converted to linear scale. The rms sum of these values is taken, which represents the THD.

A simpler approach to the measurement of harmonic distortion can be found in the notch-filter distortion analyzer, illustrated in Figure 10.12. This device, commonly referred to as simply a distortion analyzer, removes the fundamental of the signal to be investigated and measures the remainder. Figure 10.13 shows the notch-filter approach applied to a spectrum analyzer for distortion measurements.

The correct method of representing percentage distortion is to express the level of the harmonics as a fraction of the fundamental level. However, many commercial distortion analyzers use the total signal level as the reference voltage. For small amounts of distortion, these two quantities are essentially equivalent. At large values of distortion,

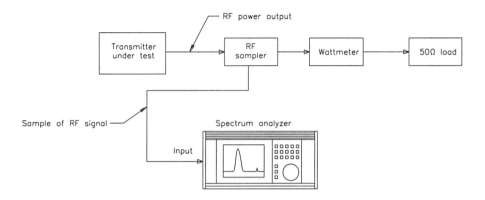

Figure 10.11 Common test setup to measure harmonic distortion with a spectrum analyzer.

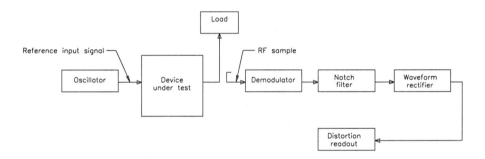

Figure 10.12 Simplified block diagram of a harmonic distortion analyzer.

however, the total signal level will be greater than the fundamental level. This makes distortion measurements on such units lower than the actual value. The relationship between the measured distortion and true distortion is given in Figure 10.14. The errors are not significant until about 20 percent measured distortion.

Because of the notch-filter response, any signal other than the fundamental will influence the results, not just harmonics. Some of these interfering signals are illustrated in Figure 10.15. Any practical signal contains some noise, and the distortion analyzer will include this noise in the reading. Because of these added components, the correct term for this measurement is *total harmonic distortion and noise* (THD+N). Additional filters are included on most distortion analyzers to reduce unwanted noise and permit a more accurate reading.

The use of a sine wave test signal and a notch-type distortion analyzer provides the distinct advantage of simplicity in both design and use. This simplicity has an addi-

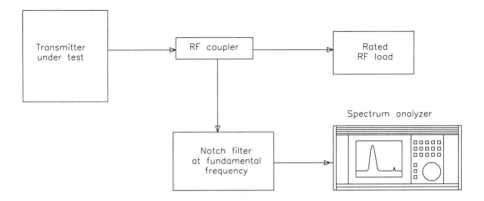

Figure 10.13 Typical test setup for measuring the harmonic and spurious output of a transmitter. The notch filter is used to remove the fundamental frequency to prevent overdriving the spectrum analyzer input and to aid in evaluation of distortion components.

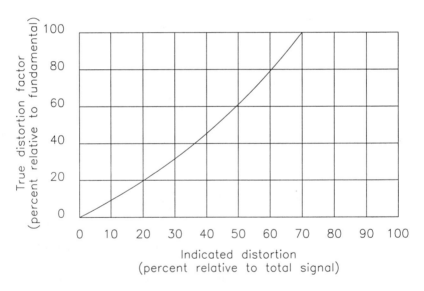

Figure 10.14 Conversion graph for indicated distortion and true distortion.

tional benefit in ease of interpretation. The shape of the output waveform from a notch-type analyzer indicates the slope of the nonlinearity. Displaying the residual components on the vertical axis of an oscilloscope and the input signal on the horizontal axis provides a plot of the deviation of the transfer characteristic from a best-fit straight line. This technique is diagrammed in Figure 10.16. The trace will be a horizontal line for a perfectly linear device. If the transfer characteristic curves upward on positive input voltages, the trace will bend upward at the right-hand side.

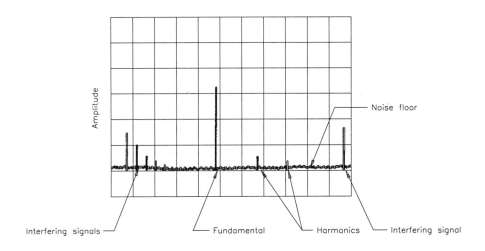

Figure 10.15 Example of interference sources in distortion and noise measurements.

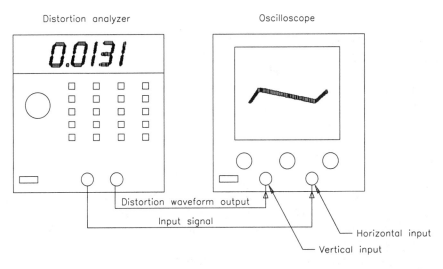

Figure 10.16 Transfer-function monitoring configuration using an oscilloscope and distortion analyzer.

Examination of the distortion components in real time on an oscilloscope allow observation of oscillation on the peaks of the signal and clipping. This is a valuable tool in the design and development of RF circuits, and one that no other distortion measurement method can fully match. Viewing the residual components in the frequency do-

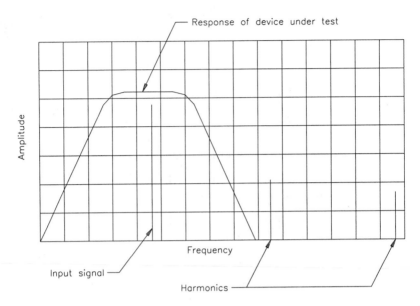

Figure 10.17 Illustration of problems that occur when measuring harmonic distortion in band-limited systems.

main by using a spectrum analyzer also reveals much information about the distortion mechanism inside the device or stage under test.

When measuring distortion at high frequencies, bandwidth limitations are an important consideration, as illustrated in Figure 10.17. Because the components being measured are harmonics of the input frequency, they may fall outside the passband of the device under test.

Intermodulation Distortion

Intermodulation distortion (IMD) describes the presence of unwanted signals that are caused by interactions between two or more desired signals. IMD, as it applies to an RF amplifier, refers to the generation of spurious products in a nonlinear amplifying stage. IMD can be measured using a distortion monitor or a spectrum analyzer.

Nonlinearities in a circuit can cause it to act like a mixer, generating the sum and difference frequencies of two signals that are present. These same imperfections generate harmonics of the signals, which then can be mixed with other fundamental or harmonic frequencies. Figure 10.18 shows the relationship of these signals in the frequency domain. The *order* of a particular product of IMD is defined as the number of "steps" between it and a single fundamental signal. Harmonics add steps. For example, the second harmonic $= 2f_1 = f_1 + f_1$, which is two steps, making it a *second-order* product. The product resulting from $2f_1 - f_2$ is, then, a *third-order* product. Some of the IMD products resulting from two fundamental frequencies include:

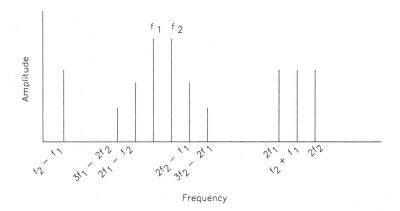

Figure 10.18 Frequency-domain relationships of the desired signals and low-order IMD products.

second-order $= 2f_1$ (10.6)

$= 2f_2$ (10.7)

$= f_1 + f_2$ (10.8)

$= f_1 - f_2$ (10.9)

third-order $= 2f_1 + f_2$ (10.10)

$= 2f_1 - f_2$ (10.11)

$= 2f_2 + f_1$ (10.12)

$= 2f_2 - f_1$ (10.13)

and so on.

The order is important because, in general, the amplitudes of IMD products fall off as the order increases. Therefore, low-order IMD products have the greatest potential to cause problems if they fall within the band of interest, or on frequencies used by other services or equipment. Odd-order IMD products are particularly troublesome because they fall closest to the signals that cause them, usually within the operating frequency band.

Measurement Techniques

A number of techniques have been devised to measure the intermodulation (IM) of two or more signals passing through a device simultaneously. The most common of

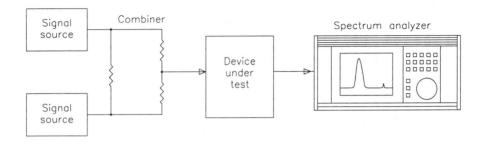

Figure 10.19 Test setup for measuring the intermodulation distortion of an RF amplifier using a spectrum analyzer.

these involves the application of a test waveform consisting of a low-frequency signal and a high-frequency signal mixed in a specified amplitude ratio. The signal is applied to the device under test, and the output waveform is examined for modulation of the upper frequency by the low-frequency signal. The amount by which the low-frequency signal modulates the high-frequency signal indicates the degree of nonlinearity. As with harmonic distortion measurement, this test may be done with a spectrum analyzer or a dedicated distortion analyzer. The test setup for a spectrum analyzer is shown in Figure 10.19. Two independent signal sources are connected using a power combiner to drive the device under test. The sources are set at the same output level, but at different frequencies. A typical spectrum analyzer display of the two-tone distortion test is shown in Figure 10.20. As shown, the third-order products that fall close to the original two tones are measured. This measurement is a common, and important, one because such products are difficult to remove by filtering.

The mechanisms of intermodulation as applied to interfering radiated signals are discussed later in this chapter.

Addition and Cancellation of Distortion Components

The addition and cancellation of distortion components in test equipment or the device under test is an often-overlooked problem in measurement. Consider the examples in Figure 10.21 and 10.22. Assume that one device under test has a transfer characteristic similar to that diagrammed at the top left of Figure 10.21a, and another has the characteristic diagrammed at the center. If the devices are cascaded, the resulting transfer-characteristic nonlinearity will be magnified as shown. The effect on sine waves in the time domain is illustrated in Figure 10.21b. The distortion components generated by each nonlinearity are in phase and will sum to a component of twice the original magnitude. However, if the second device under test has a complementary transfer characteristic, as shown in Figure 10.22, quite a different result is obtained. When the devices are cascaded, the effects of the two curves cancel, yielding a straight line for the transfer characteristic (Figure 10.22a). The corresponding distor-

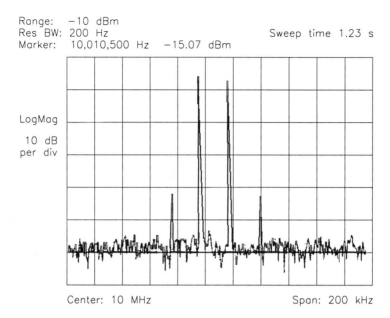

Figure 10.20 Typical two-tone IMD measurement, which evaluates the third-order products within the bandpass of the original tones.

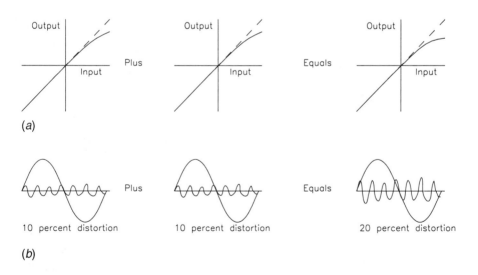

Figure 10.21 Illustration of the addition of distortion components: (a) addition of transfer-function nonlinearities, (b) addition of distortion components.

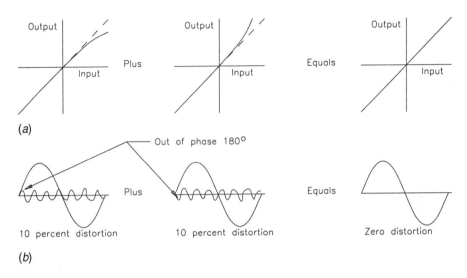

Figure 10.22 Illustration of the cancellation of distortion components: (a) cancellation of transfer-characteristic nonlinearities, (b) cancellation of the distortion waveform.

tion products are out of phase, resulting in no measured distortion components in the final output (Figure 10.22b).

This problem is common at low levels of distortion, especially between the test equipment and the device under test. For example, if the test equipment has a residual of 0.002 percent when connected to itself, and readings of 0.001 percent are obtained from the circuit under test, cancellations are occurring. It is also possible for cancellations to occur in the test equipment itself, with the combined analyzer and signal generator giving readings lower than the sum of their individual residuals. If the distortion is the result of even-order (asymmetrical) nonlinearity, reversing the phase of the signal between the offending devices will change a cancellation to an addition. If the distortion is the result of an odd-order (symmetrical) nonlinearity, phase inversions will not affect the cancellation.

Intermodulation Precorrection Techniques

In a practical system, amplifiers cannot be absolutely linear; all linear amplifiers exhibit intermodulation distortion to some extent, relative to the amplitudes of the parent signals [1]. Adequate linearity is judged by the application, but in general, the in-band odd-order products should be as low as the state of the art permits because they are in-band and thus cannot be filtered out.

All intermodulation products are the result of undesirable time domain multiplication of the matrix of input signals to a linear amplifier. All harmonic components are the result of the same process, only here, an individual signal is multiplied by itself. For

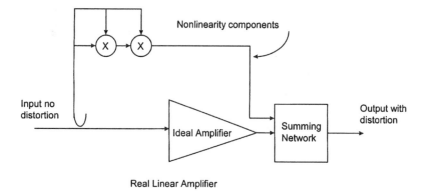

Figure 10.23 Basic model of an intermodulation correction circuit. (*After* [1].)

second order intermodulation, the ideal linear amplifier is modified by modeling a single mixer feeding a time domain multiplied signal around the amplifier. For third order intermodulation, the ideal linear amplifier is modified by modeling two mixers feeding a time domain multiplied signal around the amplifier, as shown in Figure 10.23.

Because conventional amplitude modulation contains at least three parent signals (the carrier plus two sidebands), AM amplification has always been fraught with in-band intermodulation distortion. A common technique to correct for these distortion products is to precorrect the power amplifier using the following process:

- Intentionally generate an anti-intermodulation signal by mixing the three input signals together

- Adjust the amplitude of the anti-IM product to match that of the real IM product

- Invert the phase

- Delay the correction signal or the parent signals in time so as to permit proper alignment

- Add the correction signal into the main parent signal path to cause cancellation in the output circuit of the amplifier

This is a well-known technique used in countless applications.

Each intermodulation product may have its own anti-intermodulation product intentionally generated to cause targeted cancellation. Even if one or more parent signals contains modulation of some type—AM, FM, or PM—the legitimate modulation signal is often included in the intended mixing circuit so that even the sidebands produced around an intermodulation distortion product are canceled.

The precorrection circuit typically consists of mixer devices with a local oscillator (LO) port, an RF port, and an IF port. The LO port is fed with the parent signal, such as the carrier, which contains no modulation. Its purpose is to hard-switch the mixer di-

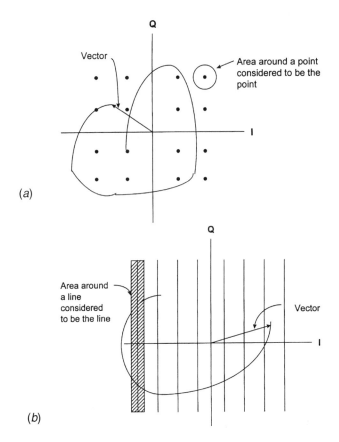

Figure 10.24 Digital decoding process: (*a*) 16-QAM, (*b*) 8-VSB. (*After* [1].)

odes, causing frequency translation of the signal fed to the RF port to the sum and difference frequencies of the LO and RF input signals exiting at the IF port. The intermodulation product to be canceled is either a result of the sum or the difference signal so that the product of no concern is filtered out. The resulting intended mixer product is fed to the IF port of a second mixer where LO injection is from the third parent signal. The third order intended mixing product is then available at the second mixer RF port, and will be at the frequency of the offending IM and coherent with it.

Digital signals suitable for radio transmission actually remain analog, but are termed *digital* because of the nature of the information that they carry. A modern digital radio transmission using quadrature amplitude modulation (QAM) or *n*-level vestigial sideband modulation (e.g., 8-VSB) consists of a vector signal constantly moving in phase angle and magnitude relative to some reference. The vector signal is smooth and without discontinuity so that it is well contained within a finite bandwidth. Its digital nature stems from decoding the proper point or spatial area around a point (QAM) of a

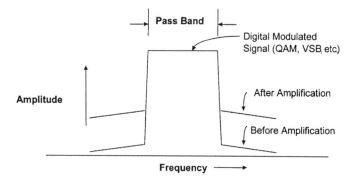

Figure 10.25 Digitally modulated RF signal spectrum before and after amplification. (*After* [1].)

finite number of legitimate points, or the proper line (VSB) or spatial area around a line of a finite number of legitimate lines, that the vector is passing though. Figure 10.24 illustrates this process. In either case, the area is assigned a discrete digital value, which represents a portion of the information being transmitted. Spatial location is generally expressed in terms of rectangular coordinates with I (in phase) and Q (quadrature phase) components.

Because the vector appears to move at random from point to point, especially resulting from the use of an intentional randomizer circuit, the frequency distribution within the passband of the signal is constant, flat, and favors no particular frequency. Figure 10.25 illustrates the frequency distribution with the sidebands shown before and after amplification. As a result, the transmitted signal is said to be "noise-like." As such, it takes on the nature of noise and need not be periodic.

For this case, there is no identifiable set of parent components upon which to obtain signals for the purpose of mixing to generate an anti-intermodulation signal for cancellation purposes. There is no carrier signal to feed into the LO port of a mixer for hard switching to generate sum and difference frequencies. Because the signal is noise-like, there is no correlation of any piece of passband spectrum with any other. Generating a precorrection signal, thus, is not possible using the process described for the discrete signal analog case.

10.3 Vacuum Tube Operating Parameters

Direct current input power to the final RF output stage is a function of the applied plate or collector voltage and the plate or collector current of the final tube(s) [2]. Test points or built-in meters usually are provided for checking the dc voltage and current levels of the final amplifier.

Transmitter deviation or modulation must be set properly. A deviation level set too high or too low adversely affects radio range. If the deviation is set too low, the signal-to-noise ratio of the overall system is diminished. If the deviation is set too high, the transmitter signal may exceed the receiver's modulation acceptance bandwidth, causing severe distortion on modulation peaks and out-of-band spurious radiation.

The difference between the positive and negative excursions of the deviated waveform is defined as *modulation dissymmetry*. Theoretically, positive and negative excursions should be perfectly symmetrical. In practice, however, there is usually some dissymmetry. In a multichannel phase-modulated transmitter, modulation dissymmetry may be worse on some channels than others as a result of modulator tuning. If the modulator tuning is peaked on one channel, for example, response on another channel may be compromised.

10.3.1 Stage Tuning

A power output stage can be tuned in any number of ways [2]. Determining factors include the type of device, circuit design, and operational objectives. A PA can be tuned to optimize bandwidth, linearity, efficiency, and other parameters. It is often necessary to compromise one parameter to enhance another. This tradeoff is dictated by the application.

Power Grid Tubes

When optimizing a tetrode or pentode RF amplifier for proper excitation and loading, the general tuning procedure varies depending upon whether the screen voltage is taken from a fixed supply or a dropping resistor supply with poor regulation [2]. If both the screen supply and grid bias are from fixed sources with good regulation, the plate current is almost entirely controlled by RF excitation. The loading is then varied until the maximum power output is obtained. Following these adjustments, the excitation is trimmed, along with the loading, until the desired control and screen grid currents are obtained.

In the case of an RF amplifier where both the screen and grid bias are taken from sources with poor regulation, the stage will tune in a manner similar to a triode RF power amplifier. The plate current will be controlled principally by varying the loading, and the excitation trimmed to yield the desired control grid current. In this case, the screen current will be almost entirely set by the choice of the dropping resistor. Excitation and loading will vary the screen voltage considerably, and these should be trimmed to yield a desired screen voltage.

The grounded-grid amplifier has been used for many years, and, with the advent of high-power zero-bias triodes, this configuration has become more common. Adjustment of the excitation and loading of a grounded-grid RF amplifier requires a slightly different procedure. The plate voltage (plate and screen voltage in the case of a tetrode or pentode) must be applied before the excitation. If this precaution is not followed, the control grid may sustain damage. The loading is increased as the excitation is increased. When the desired plate current is reached, note the power output. The loading can be reduced slightly and the excitation increased until the plate current is the same as before. If the power output is less than before, a check can be made with increased loading and less excitation. By proper trimming, the proper grid current, plate current, and optimum power output can be obtained. In a grounded-grid circuit, the cathode or input circuit is in series with the plate circuit. Because of this arrangement, any change made in the plate circuit will affect the input circuit. Therefore, the driver amplifier does not see its designed load until the driven stage is up to full plate current.

Typical operating values provide a starting point for tuning efforts. As long as the maximum limits are not exceeded, a wide range of tuning options is available.

Screen Voltage

Typical operating values given on the data sheet for tetrode and pentode devices include recommended screen voltages [2]. These values are not critical for most applications; they are chosen as a convenient value consistent with low driving power and reasonable screen dissipation. If lower values of screen voltage are used, more driving voltage will be required on the grid to obtain the same plate current. If higher values of screen voltage are used, less driving voltage will be required. Thus, high power gain can be achieved, provided the circuit exhibits adequate stability. Be careful not to exceed the screen dissipation limit. The value of screen voltage can be chosen to suit available power supplies or amplifier conditions.

The published characteristic curves of tetrodes and pentodes are shown for commonly used screen voltages. Occasionally, it is desirable to operate the tetrode or pentode at some screen voltage other than that shown on the characteristic curves. It is a relatively simple matter to convert the published curves to corresponding curves at a different screen voltage. This conversion is based on the fact that if all interelectrode voltages are either raised or lowered by the same relative amount, the shape of the voltage field pattern is not altered, and neither is the current distribution. The current lines simply take on new proportionate values in accordance with the *three-halves power law*. This method fails only where insufficient cathode emission or high secondary emission affects the current values.

For example, if the characteristic curves are shown at a screen voltage of 250 V and it is desirable to determine conditions at 500 V on the screen, all voltage scales should be multiplied by the same factor that is applied to the screen voltage (2 in this case). The 1 kV plate voltage line thus becomes 2 kV, the 50 V grid voltage line becomes 100 V, and so on. All current lines then assume new values in accordance with the $\frac{3}{2}$ power law. Because the voltage was increased by a factor of 2, the current lines all will be increased in value by a factor of $2^{\frac{3}{2}}$ or 2.8. All current values are then multiplied by the factor of 2.8. The 100 mA line thus becomes a 280 mA line, and so on.

Likewise, if the screen voltage given on the characteristic curve is higher than the conditions desired, the voltage should be reduced by the same factor that is used to obtain the desired screen voltage. Correspondingly, the current values all will be reduced by an amount equal to the $\frac{3}{2}$ power of this factor. Commonly used factors are given in Table 10.2.

Table 10.2 Three-Halves Current Values of Commonly Used Factors

Voltage Factor	Current Factor
0.25	0.125
0.50	0.35
0.75	0.65
1.00	1.00
1.25	1.40
1.50	1.84
1.75	2.30
2.00	2.80
2.25	3.40
2.50	4.00
2.75	4.60
3.00	5.20

Back-Heating by Electrons

Back-heating of the cathode is an undesirable action involving the motion of electrons within the tube at VHF and UHF [2]. Because the time of flight of electrons from the cathode through the grid structure to the plate becomes an appreciable part of the cycle at high frequencies, electrons can be stopped in flight and turned back by the rapidly changing grid voltage. Under these conditions, the electrons are deflected from their normal paths and given excess energy with which they bombard the cathode and other portions of the tube structure. This effect can be aggravated by the choice of operating conditions to the extent that destructive effects occur. The tube can be destroyed within a few minutes under severe conditions.

Fortunately, the conditions that tend to minimize this back-bombardment by electrons also characterize minimum driving voltage. The tendency of electrons to turn back in flight is reduced by using the lowest possible RF grid voltage on the tube. This is achieved by using the lowest possible dc grid bias. In tetrodes, this effect is inherently lower because of the action of the screen in accelerating the electrons toward the anode. A tetrode also permits the use of smaller grid voltages. Consequently, under favorable conditions, the number of electrons turned back to heat the cathode and tube structure can be kept to a practical, low level. In addition to low dc grid bias, high screen voltage is desirable.

The plate circuit should always operate with heavy loading (low external plate impedance) so that the minimum instantaneous value of plate voltage will stay sufficiently positive to continue accelerating electrons to the anode. For this reason, the longest tube life is realized when the tetrode amplifier is heavily loaded, as indicated by having small values of dc screen and dc control grid current.

Never operate a tube with light plate loading. If the plate load is removed so that the minimum instantaneous plate voltage tends to fall to values approximating cathode potential (as it must do when the loading is removed completely and excitation is present), the number of electrons turned back can be destructive to the tube. It has been found that under conditions of no loading, the electron bombardment and increased electric field heating of the insulating portion of the tube is often sufficient to cause suck-in of the glass or cracking of the ceramic envelope. Automatic protection should be provided to remove all voltages from the tube if the plate circuit loading becomes too light for the amount of excitation applied.

Note that parasitic oscillations seldom are loaded heavily, as indicated by the high grid currents often observed during such self-oscillation. Thus, excessive RF plate voltages are developed that, at VHF, can be as damaging as unloaded operation on a VHF fundamental frequency. If unloaded VHF parasitic oscillations are present simultaneously with apparently satisfactory operation on the fundamental, unexplained reduction of tube life may result.

Occasionally, an output line circuit may resonate simultaneously to a harmonic frequency as well as to the fundamental frequency. The higher resonant modes of practical line circuits are not normally harmonically related. Sometimes, however, the tuning curve of a mode will cross the fundamental tuning curve and, at that point, the circuit will build up resonant voltages at both the harmonic and the fundamental frequency. The harmonic resonance usually is lightly loaded and the damaging action is similar to that of lightly loaded parasitic or fundamental. Again, while operation of the tube and circuit on the fundamental may appear normal, but with lower than expected efficiency, damage can occur to the tube.

In addition to operating the tube with minimum bias, high screen voltage, and heavy loading on the plate circuit, some degree of compensation for the remaining back-heating of the cathode may be required. This can be accomplished by lowering the filament voltage or heater voltage until the cathode operates at a normal temperature. In the case of tetrodes and pentodes, by taking precautions necessary to minimize back-bombardment by electrons, the required compensation for back-heating of the cathode is not significant and may often be neglected.

10.3.2 Amplifier Balance

In a push-pull RF amplifier, lack of balance in the plate circuit or unequal plate dissipation usually is caused by a lack of symmetry in the RF circuit [2]. Normally, the tubes are similar enough that such imbalance is not associated with the tubes and their characteristics. This can readily be checked by interchanging the tubes in the sockets (provided both tubes have common dc voltages to plate, screen, and grid) and observing whether the unbalanced condition remains at the socket location or moves with the tube. If it remains at the socket location, the circuit requires adjustment. If appreciable imbalance is associated with the tube, it is likely that one tube is abnormal and should be replaced.

The basic indicators of balance are the plate current and plate dissipation of each tube. It is assumed that the circuit applies the same dc plate voltage, dc screen voltage (in the case of a tetrode or pentode), and dc grid bias to each tube from common supplies. Also, it is initially assumed that the plate circuit is mechanically and electrically symmetrical (or approximately so).

Imbalance in a push-pull RF amplifier is usually the result of unequal RF voltages applied to the grids of the tubes, or the RF plate circuit applying unequal RF voltages to the plates of the tubes. Correction of this situation involves first balancing grid excitation until equal dc plate currents flow in each tube. Next, the RF plate circuit is adjusted until equal plate dissipation appears on each tube, or equal RF plate voltage is observed. The balance of plate current is a more important criterion than equality of screen current (in a tetrode or pentode) or grid current. This results from the fact that tubes tend to be more uniform in plate current characteristics. However, the screen current also is sensitive to a lack of voltage balance in the RF plate circuit and may be used as an indicator. If the tubes differ somewhat in screen current characteristics, and the circuit has common dc supply voltages, final trimming of the plate circuit balance may be made by interchanging tubes and adjusting the circuit to give the same screen current for each tube, regardless of its location.

Note that the dc grid current has not been used as an indicator of balance of the RF power amplifier. It is probable that after following the foregoing procedure, the grid currents will be fairly well balanced, but this condition, in itself, is not a safe indicator of grid excitation balance.

10.3.3 Parallel Tube Amplifiers

The previous discussion has been oriented toward the RF push-pull amplifier. The same comments can be directed to parallel tube RF amplifiers [2]. Balance—the condition in which each tube carries its fair share of the load—still must be considered. In low-frequency amplifiers operating in class AB_1 or class AB_2, the idle dc plate current per tube should be balanced by separate bias adjustments for each tube. In many cases, some lack of balance of the plate currents will have negligible effect on the overall performance of the amplifier.

When tubes are operating in the idle position close to cutoff, an exact balance is difficult to achieve because plate current cannot be held to a close percentage tolerance. At this point, the action of the plate and screen voltages is in delicate balance with the opposing negative grid voltage. The state of balance is indicated by the plate current. Minor variations of individual grid wires or variations in the diameter of grid wires can upset the balance. It is practically impossible to control such minor variations in manufacture.

10.3.4 Harmonic Energy

A pulse of plate current delivered by the tube to the output circuit contains components of the fundamental and most harmonic frequencies [2]. To generate output power that is an harmonic of the exciting voltage applied to the control grid, it is merely necessary to resonate the plate circuit to the desired harmonic frequency. To optimize the performance of the amplifier, it is necessary to adjust the angle of plate current flow to maximize the desired harmonic. The shorter the length of the current pulse, in the case of a particular harmonic, the higher the plate efficiency. However, the bias, exciting voltage, and driving power must be increased. Also, if the pulse is too long or too short, the output power will decrease appreciably.

The plate circuit efficiency of tetrode and pentode harmonic amplifiers is quite high. In triode amplifiers, if feedback of the output harmonic occurs, the phase of the voltage feedback usually reduces the harmonic content of the plate pulse, and thereby lowers the plate circuit efficiency. Because tetrodes and pentodes have negligible feedback, the efficiency of a harmonic amplifier usually is comparable to that of other amplifier types. Also, the high amplification factor of a tetrode or pentode causes the plate voltage to have little effect on the flow of plate current. A well-designed tetrode or pentode also permits large RF voltages to be developed in the plate circuit while still passing high peaks of plate current in the RF pulse. These two factors help to further increase plate efficiency.

Controlling Harmonics

The previous discussion of harmonics has been for the case where harmonic power in the load is the objective. The generation and radiation of harmonic energy must be kept to a minimum in a *fundamental frequency* RF amplifier [2]. The ability of the tetrode and pentode to isolate the output circuit from the input circuit over a wide range of frequencies is important in avoiding feedthrough of harmonic voltages from the grid. An important part of this shielding is the result of the basic physical design of the devices. The following steps permit reduction of unwanted harmonic energy in the output circuit:

- Minimize the circuit impedance between plate and cathode. This objective may be achieved by having some or all of the tuning capacitance of the resonant circuit close to the tube.

- Completely shield the input and output compartments of the stage.

- Use inductive output coupling from the resonant plate circuit, and possibly a capacitive or Faraday shield between the coupling coil and the tank coil. A high-frequency attenuating circuit, such as a *Pi* or *Pi-L* network, also may be used.

- Incorporate low-pass filters on all supply leads and wires leading to the output and input compartments.

- Use resonant traps for particular frequencies as required.

- Install a low-pass filter in series with the output transmission line.

10.3.5 Klystron Tuning Considerations

The output power stability of a klystron is sensitive to changes in the principal operating parameters, including:

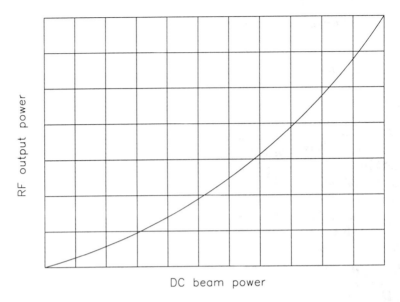

Figure 10.26 RF output power variation as a function of dc beam input power for a klystron.

- Beam input power
- RF drive power
- RF drive frequency
- Device tuning
- Magnetic fields in the vicinity of the tube
- VSWR of the load

Varying any of these parameters will affect the RF output power response.

Figure 10.26 illustrates how changes in the dc beam input power affect the RF output power of a klystron under constant RF drive conditions. Small changes in dc input power produce marked changes in RF output power. Figure 10.27 shows RF output power as a function of RF drive power applied to the tube. From this curve, it can be seen that when the RF drive power is low, the RF output power is low. As the level of RF drive power increases, RF output power increases until an optimum operating point is reached. Beyond this point, further increases in RF drive power result in reduced RF output power.

Klystrons are said to be *saturated* at the point where a further increase in RF drive only decreases the RF output power. To obtain maximum RF output from a klystron, sufficient RF drive must be applied to the tube to reach the point of saturation. Operating at RF drive levels beyond saturation will only overdrive the klystron, decrease RF

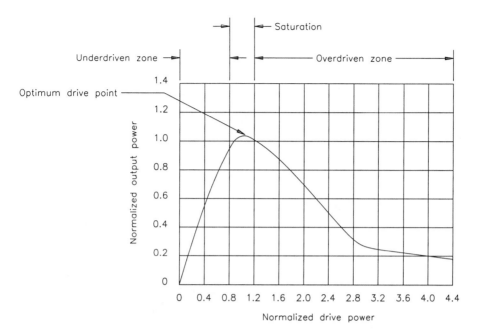

Figure 10.27 RF output power of a klystron as a function of RF drive power.

output power, and increase the amount of beam interception (body current) at the drift tube.

Tuning of any cavity of a multicavity klystron will affect the overall shape of the amplitude-response curve. There are two common methods of tuning klystrons:

- High-efficiency narrowband tuning

- Broadband tuning, yielding lower gain and efficiency

Figure 10.28 shows how RF output power changes with various levels of RF drive applied to a klystron under different tuning conditions. The RF power at point A represents the drive saturation point for a synchronously tuned tube. Point B shows a new point of saturation that is reached by tuning the *penultimate* (next to the last) cavity to a somewhat higher frequency. By tuning the penultimate cavity still further, point C is reached. At point D, increasing the penultimate cavity frequency no longer increases RF output power; instead, it reduces the power delivered by the tube (curve E).

10.3.6 Intermodulation Distortion

Intermodulation distortion products in linear power amplifier circuits can be caused by either amplitude gain nonlinearity or phase shift with change in input signal level. Intermodulation distortion products appear when the RF signal has a varying envelope amplitude. A single continuous-frequency wave will be amplified by a fixed

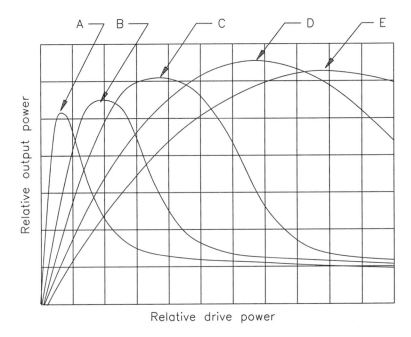

Figure 10.28 Output power variation of a klystron as a function of drive power under different tuning conditions.

amount and shifted in phase by a fixed amount. The nonlinearity of the amplifier will produce only harmonics of the input wave. As the input RF wave changes, the nonlinearity of the amplifier will cause undesirable intermodulation distortion.

When an RF signal with varying amplitude is passed through a nonlinear device, many new products are produced. The frequency and amplitude of each component can be determined mathematically. The nonlinear device can be represented by a power series expanded about the zero-signal operating point. Intermodulation distortion energy is wasted and serves no purpose other than to cause interference to adjacent channels. Intermodulation distortion in a power amplifier tube is principally the result of its transfer characteristics. An ideal transfer characteristic curve is shown in Figure 10.29.

Even-order products do not contribute to the intermodulation distortion problem because they usually fall outside the amplifier passband. Therefore, if the transfer characteristic produces an even-order curvature at the small-signal end of the curve (from point A to point B) and the remaining portion of the curve (point B to point C) is linear, the tube is considered to have an ideal transfer characteristic. If the operating point of the amplifier is set at point 0 (located midway horizontally between point A and point B), there will be no distortion in a class AB amplifier. However, no tube has this idealized transfer characteristic. It is possible, by manipulation of the electron ballistics within a given tube structure, to alter the transfer characteristic and minimize the distortion products.

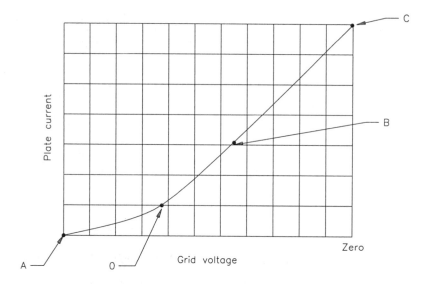

Figure 10.29 Ideal grid-plate transfer curve for class AB operation.

Generation of RF Intermodulation Products

When two or more transmitters are coupled to each other, new spectral components are produced by mixing of the fundamental and harmonic terms of each of the desired output frequencies.[1] The most common form of coupling results from transmitting antennas that are physically closely spaced. If sufficient filtering is not provided at the output stages and/or on one or both of the transmitter feedlines, energy from one transmitter may travel back to the final power amplifier of the other. In the PA stage, the desired and undesired signals mix, producing spurious intermodulation products.

Consider the following example involving two transmitters. The third-order intermodulation products (IM3) are generated in the following manner:

- The output of transmitter 1 (f_1) is coupled into the nonlinear output stage of transmitter 2 (f_2). This condition is the result of insufficient isolation between the two output signals.

- Signal f_1 mixes with the second harmonic of f_2, producing an in-band third-order term with a frequency of $2f_2 - f_1$ at the output of transmitter 2.

- In a similar manner, another third-order term will be produced at a frequency of $2f_1 - f_2$ at the output of transmitter 1.

---

1   Portions of this section were contributed by Geoffrey N. Mendenhall, P. E., Cincinnati, OH.

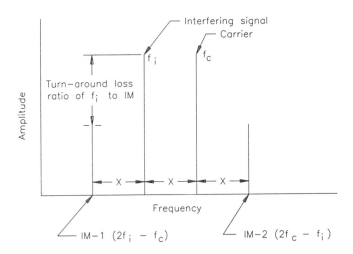

Figure 10.30 Frequency spectrum of third-order IM with the interfering level equal to the carrier level.

This implies that the second-harmonic content within the output stage of each transmitter, along with the specific nonlinear characteristics of the output stage, will have an effect on the value of the mixing loss.

It is also possible to generate these same third-order terms in another way. If the difference frequency between the two transmitters $(f_2 - f_1)$, which is an out-of-band frequency, remixes with either f_1 or f_2, the same third-order intermodulation frequencies will be produced.

Empirical measurements [3] indicate that the $2f_2 - f_1$ type of mechanism is the dominant mode generating third-order IM products in RF generators using a tuned cavity for the output network. Figure 10.30 illustrates the mechanism of third-order IM for the two-signal condition.

IM As a Function of Turnaround Loss[2]

Turnaround or *mixing loss* describes the phenomenon whereby the interfering signal mixes with the fundamental and its harmonics within the nonlinear output device. This mixing occurs with a net conversion loss; hence, the term *turnaround loss* has become widely used to quantify the ratio of the interfering level to the resulting IM3 level. A turn around loss of 10 dB means that the IM3 product fed back to the antenna system will be 10 dB below the interfering signal fed into the transmitter output stage.

Turnaround loss will increase if the interfering signal falls outside the passband of the transmitter output circuit, varying with frequency separation of the desired signal

---

2 Mendenhall, Geoffrey N., "A Study of RF Intermodulation Between FM Broadcast Transmitters Sharing Filterplexed or Colocated Antenna Systems, Broadcast Electronics, Quincy, Il, 1983.

and the interfering signal. This is because the interfering signal is first attenuated by the output circuit selectivity going into the nonlinear device, then the IM3 product is further attenuated as it comes back through the frequency-selective circuit. Turnaround loss, thus, can be divided into three principal components:

- The basic in-band conversion loss of the nonlinear device

- Attenuation of the out-of-band interfering signal resulting from the selectivity of the output stage

- Attenuation of the resulting out-of-band IM3 products resulting from the selectivity of the output circuit

As the turnaround loss increases, the level of undesirable intermodulation products is reduced, and the amount of isolation required between the transmitters also is reduced.

Transmitter output circuit loading directly affects the power amplifier source impedance and, therefore, also affects the efficiency of coupling the interfering signal into the output circuit where it mixes with the other frequencies present to yield IM3 products. Light loading reduces the amount of interference that enters the output circuit with a resulting increase in turnaround loss. In addition, output loading will change the output circuit bandwidth (loaded Q) and, therefore, also affect the amount of attenuation that out-of-band signals will encounter in passing into and out of the output circuit.

Second-harmonic traps or low-pass filters in the transmission line of either transmitter have little effect on the generation of IM products. This is because the harmonic content of the interfering signal entering the output circuit of the transmitter has much less effect on IM3 generation than the harmonic content within the nonlinear device itself. The interfering signal and the resulting IM3 products thus fall within the passband of the low-pass filters and outside the reject band of the second-harmonic traps; these devices typically offer little, if any, attenuation to RF intermodulation products.

Site-related intermodulation problems and typical solutions are discussed later in this chapter.

10.3.7 VSWR

Among all the possible load measurements, VSWR is the most common. An in-line directional wattmeter commonly is used to measure both the forward and reflected power in a feedline. It is the comparison of reflected power (P_r) to forward power (P_f) that provides a qualitative measure of the match between a transmitter and its feedline/load.

The standing wave ratio (SWR) on a transmission line is related to the ratio of the forward to reflected power by the following:

$$SWR = \frac{1 + \sqrt{P_r/P_f}}{1 - \sqrt{P_r/P_f}} \tag{10.14}$$

The standing wave is the result of the presence of two components of power: one traveling toward the load and the other reflected by a load mismatch, traveling back toward the generator. These components are defined by the following:

$$P_f = \frac{E_f{}^2}{Z_o} = I_f{}^2 (Z_o) \tag{10.15}$$

$$P_r = \frac{E_r{}^2}{Z_o} = I_r{}^2 (Z_o) \tag{10.16}$$

$$P_n = (P_f - P_r) \tag{10.17}$$

Where:
P_f = forward power
P_r = reflected power
E_f = forward voltage
E_r = reflected voltage
I_f = forward current
I_r = reflected current
Z_o = characteristic impedance of the line
P_n = net power absorbed by the load (transmission line loss and antenna radiation)

Because the forward and reflected voltages and currents are traveling in opposite directions, they will add in phase at some point along the line to produce a voltage maximum. At $\frac{1}{4}$ wavelength along the line in either direction from this maximum, the forward and reflected components will be out of phase, and will produce a voltage minimum. The forward and reflected components of current also add vectorally to produce a current standing wave. The magnitude of this wave is defined by:

$$SWR = \frac{E_{max}}{E_{min}} = \frac{I_{max}}{I_{min}} \tag{10.18}$$

The ratio of the highest voltage point on the line to the lowest voltage point on the line commonly is used to evaluate system performance. This property is known as the *voltage standing wave ratio* (VSWR). A VSWR of 1.0:1.0 means that a perfect match has been achieved. A VSWR of 2.0:1.0 means that a mismatch is causing power to be reflected back from the load, approximately 11 percent in this particular case.

At the point of reflection (the load mismatch), the phase of the reflected current is reversed 180° from the forward current. The reflected voltage does not exhibit this phase reversal. This displaces the voltage and current standing waves by 90° along the line so that the E_{max} and I_{min} occur at the same points, while E_{min} and I_{max} occur 90° in either direction from E_{max} and I_{min}.

The fact that the reflected current is reversed in phase makes it possible to measure forward and reflected power separately using a directional coupler device. A small volt-

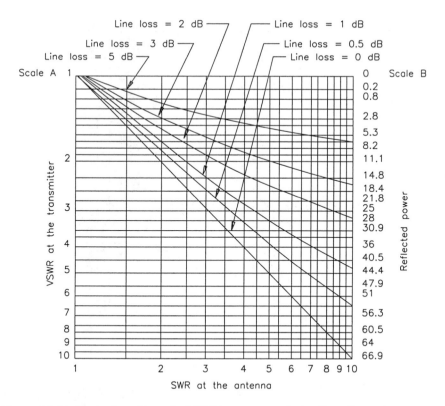

Figure 10.31 Illustration of how the VSWR on a transmission line affects the attenuation performance of the line. Scale *A* represents the VSWR at the transmitter; scale *B* represents the corresponding reflected power in watts for an example system.

age is obtained by inductive coupling that represents the current in the transmission line. To this signal is added a sample of the voltage across the line, simultaneously obtained by capacitive coupling. These two samples are adjusted to be exactly equal when the line is terminated with its characteristic impedance. The two RF samples, when added, yield a resultant RF voltage proportional to the forward components of voltage and current. Reflected power is measured by installing a second coupling section physically turned in the opposite direction. The phase of the current sample is reversed, and the reflected components add while the forward components balance out.

On a theoretical, lossless feedline, VSWR is the same at all points along the line. Actually, all feedlines have some attenuation distributed uniformly along their entire lengths. Any attenuation between the reflection point and the source reduces VSWR at the source. This concept is illustrated in Figure 10.31. Standing waves on the feedline increase feedline attenuation to a level greater than normal attenuation under matched-line conditions. The amount of feedline attenuation increase depends on the VSWR and the extent of losses under matched-line conditions. The higher the matched-line attenuation, the greater the additional attenuation caused by a given

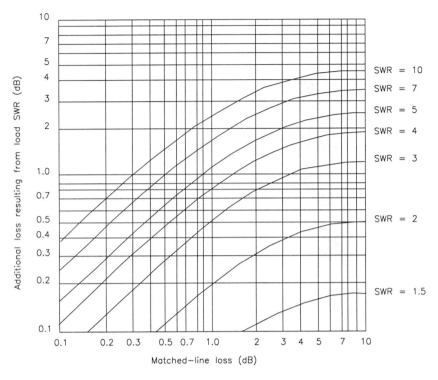

Figure 10.32 Illustration of how VSWR causes additional loss in a length of feedline.

VSWR. Figure 10.32 correlates additional attenuation with matched-line attenuation for various VSWR levels.

When a feedline is terminated by an impedance equal to the feedline *characteristic impedance*, the feedline impedance at any point is equal to the feedline characteristic impedance. When the feedline is terminated by an impedance not equal to the feedline characteristic impedance, the feedline input impedance will vary with changes in feedline length. The greater the impedance mismatch at the load, the wider the feedline input impedance variation with changes in line length. The largest feedline impedance change occurs with the worst mismatches: open circuit or short circuit.

A short circuit at the end of a feedline appears as an open circuit at a point ¼ wavelength down the feedline. It is the greatest impedance transformation that can occur on a feedline. Less severe mismatches cause smaller impedance transformations. At a point ½ wavelength down the line, the impedance equals the load impedance. Feedlines with high VSWR sometimes are called *resonant* or *tuned feedlines*. Matched lines—those with little VSWR—are referred to as *nonresonant* or *untuned feedlines*. This phenomenon means that high feedline VSWR can cause the feedline input to present a load impedance to the output tube that differs greatly from the feedline characteristic impedance. Some RF generators incorporate an output matching network that provides a *conjugate match* to a broad range of impedances.

10.4 RF System Performance

Assessing the performance of an RF generator has been greatly simplified by recent advancements in system design. The overall performance of most transmitters can be assessed by measuring the audio, video, or data performance of the system as a whole. If these "intelligence signal" tests indicate good performance, chances are good that all of the RF parameters also meet specifications. Both AM and FM transmission systems have certain limitations that prevent them from ever being a completely transparent medium. Through proper system design and regular equipment maintenance, however, overall performance can be maintained at acceptable levels in most operational environments.

10.4.1 Key System Measurements

The procedures for measuring transmitter performance vary widely, depending on the type of equipment being used. Certain basic measurements, however, can be applied to characterize the overall performance of most systems:

- Frequency response: the actual deviation from a constant amplitude across a given span of frequencies.

- Total harmonic distortion: the creation, by a nonlinear device, of spurious signals harmonically related to the applied waveform. THD is sensitive to the noise floor of the system under test. If the system has a signal-to-noise ratio of 60 dB, the distortion analyzer's best possible reading will be higher than 0.1 percent (60 dB = 0.001 = 0.1 percent).

- Intermodulation distortion: the creation, by a nonlinear device, of spurious signals not harmonically related to the audio waveform. These distortion components are *sum-and-difference* (beat note) mixing products. The IM measurement is relatively impervious to the noise floor of the system under test.

- Signal-to-noise ratio (S/N): the amplitude difference, expressed in decibels, between a reference-level audio signal and the residual noise and hum of the overall system.

- Crosstalk: the amplitude difference, expressed in decibels, between two or more channels of a multichannel transmission system.

10.4.2 Synchronous AM in FM Systems[3]

The output spectrum of an FM transmitter contains many sideband frequency components, theoretically an infinite number. The practical considerations of transmitter de-

---

3 Mendenhall, Geoffrey N., "Techniques for Measuring Synchronous AM Noise in FM Transmitters," Broadcast Electronics, Quincy, IL, 1988.

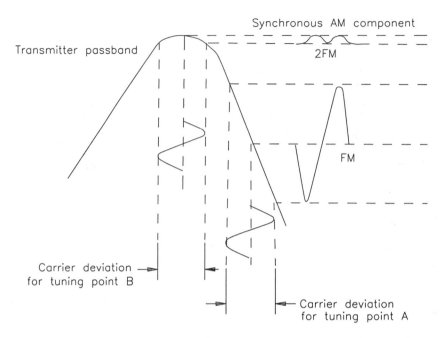

Figure 10.33 The generation of synchronous AM in a bandwidth-limited FM system. Note that minimum synchronous AM occurs when the system operates in the center of its passband.

sign and frequency allocation make it necessary to restrict the bandwidth of FM signals. Bandwidth restriction brings with it a number of undesirable side effects, including:

- Phase shifts through the transmission chain

- Generation of *synchronous AM* components

- Distortion in the demodulated output of certain receivers

In most medium- and high-power FM transmitters, the primary out-of-band filtering is performed in the output cavity of the final stage. Previous stages in the transmitter (the exciter and IPA) are designed to be broadband, or at least more broadband than the PA. As the bandwidth of an FM transmission system is reduced, synchronous amplitude modulation increases for a given carrier deviation (modulation). Synchronous AM is produced as tuned circuits with finite bandwidth are swept by the frequency of modulation. The amount of synchronous AM generated is dependent on tuning, which determines (to a large extent) the bandwidth of the system. Figure 10.33 illustrates how synchronous AM is generated through modulation of the FM carrier.

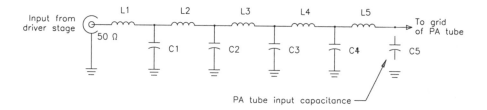

Figure 10.34 IPA plate-to-PA-grid matching network designed for wide bandwidth operation.

Bandwidth Limiting

The design goal of most medium- and high-power FM transmission equipment is to limit the bandwidth of the transmitted RF signal at one stage only. In this way, control over system bandwidth can be tightly maintained, and the tradeoffs required in any practical FM system can be optimized. If, on the other hand, more than one narrowband stage exists within the transmitter, adjustment for peak efficiency and performance can be a difficult proposition. The following factors can affect the bandwidth of an FM system:

- Total number of tuned circuits in the system

- Amplitude and phase response of the total combination of all tuned circuits in the RF path

- Amount of drive to each RF stage (saturation effects)

- Nonlinear transfer function within each RF stage

The following techniques can be used to improve the bandwidth of a given system:

- Maintain wide bandwidth until the final RF stage. This can be accomplished through the use of a broadband exciter and broadband (usually solid-state) IPA.

- Minimize the number of interactive-tuned networks within the system. Through proper matching network design, greater bandwidth and simplified tuning can be accomplished.

Bandwidth affects the gain and efficiency of any FM RF amplifier. The bandwidth is determined by the load resistance across the tuned circuit and the output or input capacitance of the RF stage. Because grid-driven PA tubes typically exhibit high input capacitance, the PA input circuit generally has the greatest effect on bandwidth limiting in an FM transmitter. The grid circuit can be broadened through resistive swamping in the input circuit, or through the use of a broadband input impedance matching network. One approach involves a combination of series inductor and shunt capacitor elements implemented on a printed wiring board, as illustrated in Figure 10.34. In this design, the

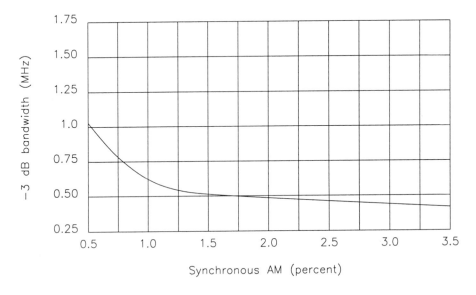

Figure 10.35 Synchronous AM content of an example FM system as a function of system bandwidth.

inductors and capacitors are etched into the copper-clad laminate of the PWB. This approach provides the necessary impedance transformation from a 50 Ω solid-state driver to a high-impedance PA grid in a series of small steps, rather than more conventional *L*, *Pi*, or *Tee* matching networks.

Effects of Synchronous AM

The effect of bandwidth on synchronous AM performance for an example FM RF amplifier is plotted in Figure 10.35. Notice that as the –3 dB points of the RF system passband are narrowed, a dramatic increase in synchronous AM occurs. The effect shown in the figure is applicable to any bandwidth-limited FM system.

Bandwidth also affects the distortion floor of the demodulated signal of an FM system, as plotted in Figure 10.36. The data shown apply to a test setup involving a single-tuned circuit fed by an FM generator with a deviation of ±75 kHz at a modulating frequency of 15 kHz (no de-emphasis is applied to the output signal). Remember that FM is a nonlinear process and that interpretation of distortion numbers for multiple-carrier signals is not accurate. The example shown, however, illustrates how the bandwidth of the RF channel can set a minimum performance limit on system total harmonic distortion. By understanding the mechanics of synchronous AM, adjustment for optimum performance of an FM system can be accomplished.

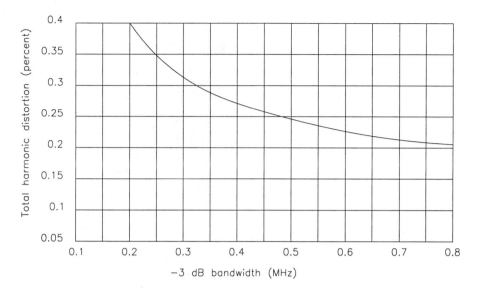

Figure 10.36 Total harmonic distortion (THD) content of an example demodulated FM signal as a function of transmission-channel bandwidth.

10.4.3 Incidental Phase Modulation

Incidental phase modulation (IPM) is phase modulation produced by an AM transmitter as a result of amplitude modulation. In other words, as an AM transmitter develops the amplitude-modulated signal, it also produces a phase-modulated, or PM, version of the input intelligence. In theory, IPM is of little consequence for classic AM applications because FM and PM do not affect the carrier amplitude. An envelope detector will, in theory at least, ignore IPM. However, a number of applications utilize manipulation of the carrier phase to convey information in addition to amplitude modulation of the carrier. Receivers designed to decode such AM/PM waveforms will not perform properly if excessive IPM is produced by the transmitter. Although the effects of IPM vary from one system to another, optimum performance is realized when IPM is minimized.

As a general rule, because IPM is a direct result of the modulation process, it can be generated in any stage that is influenced by modulation. The most common cause of IPM in plate-modulated and pulse-modulated transmitters is improper neutralization of the final RF amplifier. Adjusting the transmitter for minimum IPM is an accurate way of achieving proper neutralization. The reverse is not always true, however, because some neutralization methods will not necessarily result in the lowest IPM.

Improper tuning of the IPA stage is the second most common cause of IPM. As modulation changes the driver loading to the PA grid, the driver output also may vary. The circuits that feed the driver stage usually are sufficiently isolated from the PA that they do not produce IPM. An exception is if the power supply for the IPA is influenced

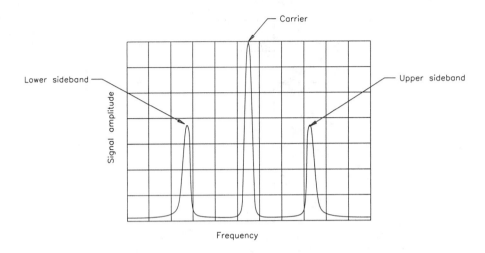

Figure 10.37 The spectrum of an ideal AM transmitter. THD measures 0 percent; there are no harmonic sidebands.

by modulation. Such a problem could be caused by a lack of adequate capacitance in the filters of the high-voltage power supply.

Bandwidth Considerations

When a perfect AM transmitter is modulated with a tone, two sidebands are created, one above the carrier and one below it. Phase modulation of the same carrier, however, produces an infinite number of sidebands at intervals above and below the carrier. The number of significant sidebands depends on the magnitude of the modulation. With moderate amounts of IPM, the transmitted spectrum can be increased to the point at which the legal channel limits are exceeded.

Corrective Procedures

The spectrum analyzer is generally the best instrument for identifying IPM components. The transmitter is first modulated with a pure sine wave, and a THD measurement is taken with a distortion analyzer. These data will provide an idea of what the spectrum analyzer display should look like when it is connected to the transmitter. If the transmitter has an excessive amount of harmonic distortion, it will be impossible to tell the difference between the normal distortion sidebands and the IPM sidebands.

Figure 10.37 shows the trace of an ideal transmitted spectrum on a spectrum analyzer when modulated with a 1 kHz tone at 100 percent modulation. Notice that there are just three components: one carrier and two sidebands. This transmitter would measure 0 percent distortion, because no harmonic sidebands are present. Also, no IPM is present, because it would show up as extra sidebands in the spectrum.

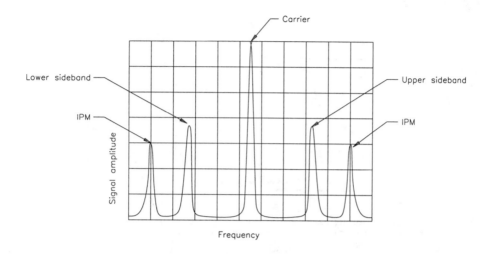

Figure 10.38 The spectrum of an ideal transmitter with the addition of IPM.

If the transmitter had an IPM problem, the spectrum analyzer display might look like Figure 10.38. The distortion analyzer still would read 0 percent, but the sidebands on the spectrum analyzer would not agree. The key is to start with a low harmonic distortion reading on the transmitter so that any sidebands will be recognizable immediately as the result of IPM.

10.4.4 Carrier Amplitude Regulation

Carrier amplitude regulation (*carrier level shift*) is the effective shift in apparent carrier level as a result of the amplitude modulation process [4]. Carrier level shift can be caused by the following:

- Poor power supply regulation

- Modulation-related even order harmonic distortion that generates a dc offset component in the modulated RF envelope

- A combination of the previous two mechanisms

Carrier level shift can be either positive or negative, although is usually negative because power supply regulation is most often the major source of carrier level shift, and power supply regulation is generally negative in sign (i.e., lower voltage output at higher current loads). Strictly speaking, carrier level shift is defined as the shift in effective carrier voltage or current resulting from modulation. (See. Figure 10.39.)

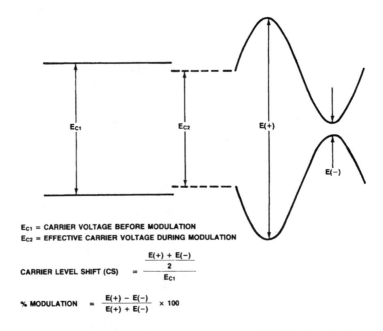

$$E_{C1} = \text{CARRIER VOLTAGE BEFORE MODULATION}$$
$$E_{C2} = \text{EFFECTIVE CARRIER VOLTAGE DURING MODULATION}$$

$$\text{CARRIER LEVEL SHIFT (CS)} = \frac{\dfrac{E(+) + E(-)}{2}}{E_{C1}}$$

$$\% \text{ MODULATION} = \frac{E(+) - E(-)}{E(+) + E(-)} \times 100$$

Figure 10.39 Carrier amplitude regulation (carrier level shift) defined. (*From* [4]. *Used with permission.*)

10.4.5 Site-Related Intermodulation Products

As discussed previously in this chapter, an intermodulation product (intermod) is a signal created when two or more signals are unintentionally mixed to produce a new signal or group of signals. Mixing can occur in transmitter circuitry, in receiver circuitry, or elsewhere. The apparent effect is the same: one or more low-level products strong enough to interfere with desired signals and prevent proper reception. Sometimes the intermod falls on unused frequencies. Sometimes it falls on a channel in use or near enough to the channel to cause interference.

It is possible to calculate the intermod that may result from frequencies in use at a given transmission site. Normally produced by a computerized analysis, this *intermod study* is helpful in several ways. If the study is conducted as one or more new frequencies are under consideration for future use at an existing site, it may reveal which frequencies are more likely to produce a harmful intermod and which are not. If the study is conducted as part of an investigation into an interference problem, it may help in identifying transmitters that play a role in generating the intermod.

Intermod can be generated by any nonlinear device or circuit. The most common place for intermod to be generated is in the power amplifier of an FM transmitter. Such an amplifier does not need to preserve waveform integrity; the waveform is restored by the tuned output circuit. Thus, the amplifier is operated in an efficient but nonlinear

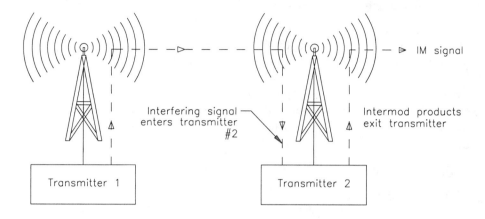

Figure 10.40 The mechanics of intermod generation at a two-transmitter installation.

mode, class C. In contrast, many amplifiers for AM service operate in a linear mode. Such amplifiers normally do not act as mixers.

Generally, solid-state amplifiers are better mixers than their electron tube counterparts, a fact that may explain why intermod may appear suddenly when older equipment is replaced. A mix occurs when RF energy from one transmitter (Tx 1) finds its way down the transmission line and into the final amplifier of another transmitter (Tx 2), as illustrated in Figure 10.40. The energy mixes in the final amplifier with the prefiltered second harmonic of Tx 2 to produce another in-band signal, a third-order intermod product. The reverse mix also may occur in the final amplifier of Tx 1, producing another intermod signal.

After the cause is discovered, a mix generated in a transmitter final amplifier is relatively easy to eliminate: Increase the isolation between the offending transmitters. This may be accomplished by increasing the space between the antennas, or by placing an isolator at the output of each transmitter.

Receiver Intermod

Intermod also may be generated in the front end of a receiver. High-level RF energy from transmitters at a site can generate a bias voltage in the receiver sufficient to place the receiver RF amplifier into a nonlinear state. In such a mode, the stage acts as a mixer and creates unwanted frequencies. Filters may be used to solve this problem. The filters should notch out the offending transmitted signals. If another transmitter is added to the site, another notch filter usually must be added to the receiver transmission line.

Hardware Intermod

At a large transmission site, the entire area—including towers, guy wires, antennas, clamps, and obstruction lighting hardware—is saturated by high-level RF. Any loose or moving hardware can cause broadband noise. Corroded connections, tower joints, or clamps can form semiconductor junctions that may generate harmful intermod. Because the antenna system is exposed to a changing external environment, the degree of semiconductivity can vary with moisture or temperature. Changing semiconductivity causes intermittent problems, at changing levels of severity.

10.5 References

1. Hulick, Timothy P., "Very Simple Out-of-Band IMD Correctors for Adjacent Channel NTSC/DTV Transmitters," *Digital Television '98*, Intertec Publishing, Overland Park, KS, 1998.
2. Laboratory Staff, *Care and Feeding of Power Grid Tubes*, Varian/Eimac, San Carlos, CA, 1984.
3. Mendenhall, Geoffrey N., M. Shrestha, and E. Anthony, "FM Broadcast Transmitters," *NAB Engineering Handbook*, 8th Ed., E. B. Crutchfield (ed.), National Association of Broadcasters, Washington, DC, pp. 463 - 464, 1992.
4. George Woodard, "AM Transmitters," in *NAB Engineering Handbook*, 9th Ed., Jerry C. Whitaker (ed.), National Association of Broadcasters, Washington, DC, 1998.

10.6 Bibliography

Benson, K. B., and Jerry C. Whitaker, *Television and Audio Handbook for Engineers and Technicians*, McGraw-Hill, New York, 1989.
Bordonaro, Dominic, "Reducing IPM in AM Transmitters," *Broadcast Engineering*, Intertec Publishing, Overland Park, KS, May 1988.
Crutchfield, E. B. (ed.), *NAB Engineering Handbook*, 8th Ed., National Association of Broadcasters, Washington, DC, 1992.
Dolgosh, Al, and Ron Jakubowski, "Controlling Interference at Communications Sites," *Mobile Radio Technology*, Intertec Publishing, Overland Park, KS, February 1990.
Hershberger, David, and Robert Weirather, "Amplitude Bandwidth, Phase Bandwidth, Incidental AM, and Saturation Characteristics of Power Tube Cavity Amplifiers for FM," application note, Harris Corporation, Quincy, IL, 1982.
Kinley, Harold, "Base Station Feedline Performance Testing," *Mobile Radio Technology*, Intertec Publishing, Overland Park, KS, April 1989.
Mendenhall, Geoffrey N., "A Study of RF Intermodulation Between FM Broadcast Transmitters Sharing Filterplexed or Co-located Antenna Systems," Broadcast Electronics, Quincy, IL, 1983.
Mendenhall, Geoffrey N., "A Systems Approach to Improving Subcarrier Performance," *Proceedings of the 1986 NAB Engineering Conference*, National Association of Broadcasters, Washington, DC, 1986.

Mendenhall, Geoffrey N., "Fine Tuning FM Final Stages," *Broadcast Engineering*, Intertec Publishing, Overland Park, KS, May 1987.

Mendenhall, Geoffrey N., "Improving FM Modulation Performance by Tuning for Symmetrical Group Delay," Broadcast Electronics, Quincy IL, 1991.

Mendenhall, Geoffrey N., "Techniques for Measuring Synchronous AM Noise in FM Transmitters," Broadcast Electronics, Quincy IL, 1988.

Mendenhall, Geoffrey N., "Testing Television Transmission Systems for Multichannel Sound Compatibility," Broadcast Electronics, Quincy, IL, 1985.

Terman, F. E., *Radio Engineering*, 3rd Ed., McGraw-Hill, New York, 1947.

Whitaker, Jerry C., *Maintaining Electronic Systems*, CRC Press, Boca Raton, FL, 1992.

Whitaker, Jerry C., *Radio Frequency Transmission Systems: Design and Operation*, McGraw-Hill, New York, 1990.

Witte, Robert A., "Distortion Measurements Using a Spectrum Analyzer," *RF Design*, Cardiff Publishing, Denver, CO, pp. 75 - 84, September 1992.

Safe Handling of Vacuum Tube Devices

11.1 Introduction

Electrical safety is important when working with any type of electronic hardware. Because vacuum tubes operate at high voltages and currents, safety is doubly important. The primary areas of concern, from a safety standpoint, include:

- Electric shock

- Nonionizing radiation

- Beryllium oxide (BeO) ceramic dust

- Hot surfaces of vacuum tube devices

- Polychlorinated biphenyls (PCBs)

11.2 Electric Shock

Surprisingly little current is required to injure a person. Studies at Underwriters Laboratories (UL) show that the electrical resistance of the human body varies with the amount of moisture on the skin, the muscular structure of the body, and the applied voltage. The typical hand-to-hand resistance ranges between 500 Ω and 600 kΩ, depending on the conditions. Higher voltages have the capability to break down the outer layers of the skin, which can reduce the overall resistance value. UL uses the lower value, 500 Ω, as the standard resistance between major extremities, such as from the hand to the foot. This value is generally considered the minimum that would be encountered and, in fact, may not be unusual because wet conditions or a cut or other break in the skin significantly reduces human body resistance.

Table 11.1 The Effects of Current on the Human Body

Current	Effect
1 mA or less	No sensation, not felt
More than 3 mA	Painful shock
More than 10 mA	Local muscle contractions, sufficient to cause "freezing" to the circuit for 2.5 percent of the population
More than 15 mA	Local muscle contractions, sufficient to cause "freezing" to the circuit for 50 percent of the population
More than 30 mA	Breathing is difficult, can cause unconsciousness
50 mA to 100 mA	Possible ventricular fibrillation
100 mA to 200 mA	Certain ventricular fibrillation
More than 200 mA	Severe burns and muscular contractions; heart more apt to stop than to go into fibrillation
More than a few amperes	Irreparable damage to body tissue

11.2.1 Effects on the Human Body

Table 11.1 lists some effects that typically result when a person is connected across a current source with a hand-to-hand resistance of 2.4 kΩ. The table shows that a current of approximately 50 mA will flow between the hands, if one hand is in contact with a 120 V ac source and the other hand is grounded. The table indicates that even the relatively small current of 50 mA can produce *ventricular fibrillation* of the heart, and perhaps death. Medical literature describes ventricular fibrillation as rapid, uncoordinated contractions of the ventricles of the heart, resulting in loss of synchronization between heartbeat and pulse beat. The electrocardiograms shown in Figure 11.1 compare a healthy heart rhythm with one in ventricular fibrillation. Unfortunately, once ventricular fibrillation occurs, it will continue. Barring resuscitation techniques, death will ensue within a few minutes.

The route taken by the current through the body has a significant effect on the degree of injury. Even a small current, passing from one extremity through the heart to another extremity, is dangerous and capable of causing severe injury or electrocution. There are cases where a person has contacted extremely high current levels and lived to tell about it. However, usually when this happens, the current passes only through a single limb and not through the body. In these instances, the limb is often lost, but the person survives.

Current is not the only factor in electrocution. Figure 11.2 summarizes the relationship between current and time on the human body. The graph shows that 100 mA flowing through a human adult body for 2 s will cause death by electrocution. An important factor in electrocution, the *let-go range*, also is shown on the graph. This range is described as the amount of current that causes "freezing", or the inability to let go of the conductor. At 10 mA, 2.5 percent of the population will be unable to let go of a "live" conductor. At 15 mA, 50 percent of the population will be unable to let go of an energized conductor. It is apparent from the graph that even a small amount of current can

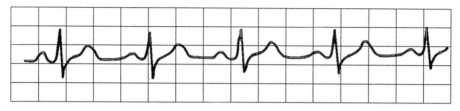

(a)

(b)

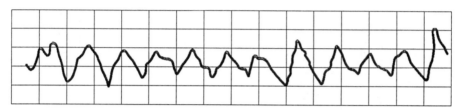

Figure 11.1 Electrocardiogram of a human heartbeat: (a) healthy rhythm, (b) ventricular fibrillation.

"freeze" someone to a conductor. The objective for those who must work around electric equipment is how to protect themselves from electric shock. Table 11.2 lists required precautions for personnel working around high voltages.

11.2.2 Circuit Protection Hardware

The typical primary panel or equipment circuit breaker or fuse will not protect a person from electrocution. In the time it takes a fuse or circuit breaker to blow, someone could die. However, there are protection devices that, properly used, may help prevent electrocution. The *ground-fault current interrupter* (GFCI), shown in Figure 11.3, works by monitoring the current being applied to the load. The GFI uses a differential transformer and looks for an imbalance in load current. If a current (5 mA, ±1 mA) begins to flow between the neutral and ground or between the hot and ground leads, the differential transformer detects the leakage and opens up the primary circuit within 2.5 ms.

GFIs will not protect a person from every type of electrocution. If the victim becomes connected to both the neutral and the hot wire, the GFI will not detect an imbalance.

11.2.3 Working with High Voltage

Rubber gloves are commonly used by engineers working on high-voltage equipment. These gloves are designed to provide protection from hazardous voltages or RF when

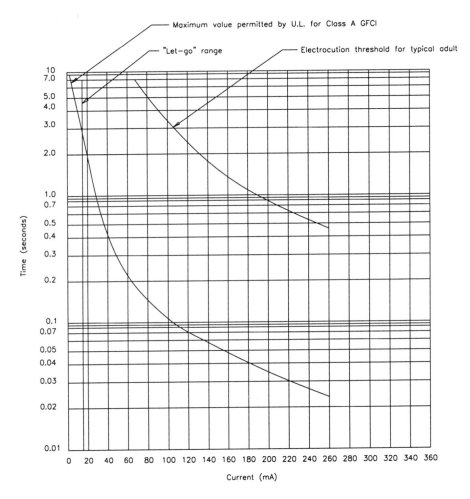

Figure 11.2 Effects of electric current and time on the human body. Note the "let-go" range.

the wearer is working on "hot" ac or RF circuits. Although the gloves may provide some protection from these hazards, placing too much reliance on them can have disastrous consequences. There are several reasons why gloves should be used with a great deal of caution and respect. A common mistake made by engineers is to assume that the gloves always provide complete protection. The gloves found in many facilities may be old or untested. Some may show signs of user repair, perhaps with electrical tape. Few tools could be more hazardous than such a pair of gloves.

Another mistake is not knowing the voltage rating of the gloves. Gloves are rated differently for both ac and dc voltages. For example, a *class 0* glove has a minimum dc breakdown voltage of 35 kV; the minimum ac breakdown voltage, however, is only 6 kV. Furthermore, high-voltage rubber gloves are not usually tested at RF frequencies,

Table 11.2 Required Safety Practices for Engineers Working Around High-Voltage Equipment

High-Voltage Precautions
✓ Remove all ac power from the equipment. Do not rely on internal contactors or SCRs to remove dangerous ac.
✓ Trip the appropriate power distribution circuit breakers at the main breaker panel.
✓ Place signs as needed to indicate that the circuit is being serviced.
✓ Switch the equipment being serviced to the *local control* mode as provided.
✓ Discharge all capacitors using the discharge stick provided by the manufacturer.
✓ Do not remove, short circuit, or tamper with interlock switches on access covers, doors, enclosures, gates, panels, or shields.
✓ Keep away from live circuits.

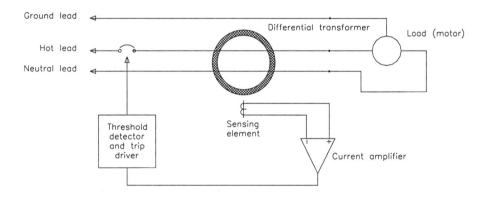

Figure 11.3 Basic design of a ground-fault interrupter (GFI).

and RF can burn a hole in the best of them. It is possible to develop dangerous working habits by assuming that gloves will offer the required protection.

Gloves alone may not be enough to protect an individual in certain situations. Recall the axiom of keeping one hand in a pocket while working around a device with current flowing? That advice is actually based on simple electricity. It is not the "hot" connection that causes the problem, but the ground connection that lets the current begin to flow. Studies have shown that more than 90 percent of electric equipment fatalities occurred when the grounded person contacted a live conductor. Line-to-line electrocution accounted for less than 10 percent of the deaths.

When working around high voltages, always look for grounded surfaces. Keep hands, feet, and other parts of the body away from any grounded surface. Even concrete

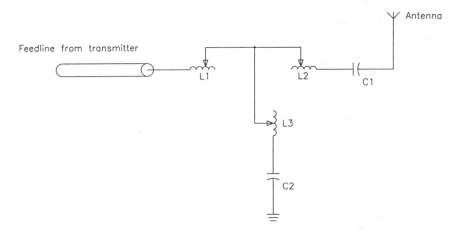

Figure 11.4 Example of how high voltages can be generated in an RF load matching unit.

can act as a ground if the voltage is sufficiently high. If work must be performed in "live" cabinets, then consider using, in addition to rubber gloves, a rubber floor mat, rubber vest, and rubber sleeves. Although this may seem to be a lot of trouble, consider the consequences of making a mistake. Of course, the best troubleshooting methodology is never to work on any circuit without being certain that no hazardous voltages are present. In addition, any circuits or contactors that normally contain hazardous voltages should be firmly grounded before work begins.

RF Considerations

Engineers often rely on electrical gloves when making adjustments to live RF circuits. This practice, however, can be extremely dangerous. Consider the typical load matching unit shown in Figure 11.4. In this configuration, disconnecting the coil from either L2 or L3 places the full RF output literally at the engineer's fingertips. Depending on the impedances involved, the voltages can become quite high, even in a circuit that normally is relatively tame.

In the Figure 11.4 example, assume that the load impedance is approximately 106 $+j202$ Ω. With 1 kW feeding into the load, the rms voltage at the matching output will be approximately 700 V. The peak voltage (which determines insulating requirements) will be close to 1 kV, and perhaps more than twice that if the carrier is being amplitude-modulated. At the instant the output coil clip is disconnected, the current in the shunt leg will increase rapidly, and the voltage easily could more than double.

11.2.4 First Aid Procedures

All engineers working around high-voltage equipment should be familiar with first aid treatment for electric shock and burns. Always keep a first aid kit on hand at the facility. Figure 11.5 illustrates the basic treatment for victims of electric shock. Copy

If the victim is not responsive, follow the A–B–Cs of basic life support.

A   AIRWAY: If the victim is unconscious, open airway.

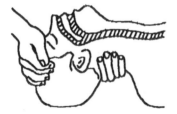

1. Lift up neck
2. Push forehead back
3. Clear out mouth if necessary
4. Observe for breathing

B   BREATHING: If the victim is not breathing, begin artificial breathing.

1. Tilt head	Check carotid pulse. If pulse
2. Pinch nostrils	is absent, begin artificial circulation.
3. Make airtight seal	Remember that mouth–to–mouth
4. Provide four quick full breaths	resuscitation must be commenced as soon as possible.

C   CIRCULATION: Depress the sternum 1.2 to 2 inches.

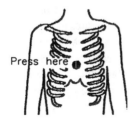 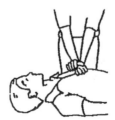

Press here ●

For situations in which there is one rescuer, provide 15 compressions and then 2 quick breaths. The approximate rate of compressions should be 80 per minute.

For situations in which there are two rescuers, provide 5 compressions and then 1 breath. The approximate rate of compressions should be 60 per minute.

Do not interrupt the rhythm of compressions when a second person is giving breaths.

If the victim is responsive, keep warm and quiet, loosen clothing, and place in a reclining position. Call for medical assistance as soon as possible.

Figure 11.5 Basic first aid treatment for electric shock.

the information, and post it in a prominent location. Better yet, obtain more detailed information from the local heart association or Red Cross chapter. Personalized instruction on first aid usually is available locally.

11.3 Operating Hazards

A number of potential hazards exist in the operation and maintenance of high-power vacuum tube RF equipment. Maintenance personnel must exercise extreme care around such hardware. Consider the following guidelines:

- Use caution around the high-voltage stages of the equipment. Many power tubes operate at voltages high enough to kill through electrocution. Always break the primary ac circuit of the power supply, and discharge all high-voltage capacitors.

- Minimize exposure to RF radiation. Do not permit personnel to be in the vicinity of open, energized RF generating circuits, RF transmission systems (waveguides, cables, or connectors), or energized antennas. High levels of radiation can result in severe bodily injury, including blindness. Cardiac pacemakers may also be affected.

- Avoid contact with beryllium oxide (BeO) ceramic dust and fumes. BeO ceramic material may be used as a thermal link to carry heat from a tube to the heat sink. Do not perform any operation on any BeO ceramic that might produce dust or fumes, such as grinding, grit blasting, or acid cleaning. Beryllium oxide dust and fumes are highly toxic, and breathing them can result in serious injury or death. BeO ceramics must be disposed of as prescribed by the device manufacturer.

- Avoid contact with hot surfaces within the equipment. The anode portion of most power tubes is air-cooled. The external surface normally operates at a high temperature (up to 250°C). Other portions of the tube also may reach high temperatures, especially the cathode insulator and the cathode/heater surfaces. All hot surfaces may remain hot for an extended time after the tube is shut off. To prevent serious burns, avoid bodily contact with these surfaces during tube operation and for a reasonable cool-down period afterward. Table 11.3 lists basic first aid procedures for burns.

11.3.1 OSHA Safety Considerations

The U.S. government has taken a number of steps to help improve safety within the workplace under the auspices of the Occupational Safety and Health Administration (OSHA). The agency helps industries monitor and correct safety practices. OSHA has developed a number of guidelines designed to help prevent accidents. OSHA records show that electrical standards are among the most frequently violated of all safety standards. Table 11.4 lists 16 of the most common electrical violations, including exposure of live conductors, improperly labeled equipment, and faulty grounding.

Table 11.3 Basic First Aid Procedures for Burns (More detailed information can be obtained from any Red Cross office.)

	Extensively Burned and Broken Skin
✓	Cover affected area with a clean sheet or cloth.
✓	Do not break blisters, remove tissue, remove adhered particles of clothing, or apply any salve or ointment.
✓	Treat victim for shock as required.
✓	Arrange for transportation to a hospital as quickly as possible.
✓	If arms or legs are affected, keep them elevated.
✓	If medical help will not be available within an hour and the victim is conscious and not vomiting, prepare a weak solution of salt and soda: 1 level teaspoon of salt and $\frac{1}{2}$ level teaspoon of baking soda to each quart of tepid water. Allow the victim to sip slowly about 4 ounces (half a glass) over a period of 15 minutes. Discontinue fluid intake if vomiting occurs. (Do not offer alcohol.)
	Less Severe Burns (First and Second Degree)
✓	Apply cool (not ice-cold) compresses using the cleanest available cloth article.
✓	Do not break blisters, remove tissue, remove adhered particles of clothing, or apply salve or ointment.
✓	Apply clean, dry dressing if necessary.
✓	Treat victim for shock as required.
✓	Arrange for transportation to a hospital as quickly as possible.
✓	If arms or legs are affected, keep them elevated.

Protective Covers

Exposure of live conductors is a common safety violation. All potentially dangerous electric conductors should be covered with protective panels. The danger is that someone may come into contact with the exposed current-carrying conductors. It is also possible for metallic objects such as ladders, cable, or tools to contact a hazardous voltage, creating a life-threatening condition. Open panels also present a fire hazard.

Identification and Marking

Circuit breakers and switch panels should be properly identified and labeled. Labels on breakers and equipment switches may be many years old and may no longer reflect the equipment actually in use. This is a safety hazard. Casualties or unnecessary damage can be the result of an improperly labeled circuit panel if no one who understands the system is available in an emergency. If a number of devices are connected to a single disconnect switch or breaker, a diagram should be provided for clarification. Label with brief phrases, and use clear, permanent, and legible markings.

Equipment marking is a closely related area of concern. This is not the same thing as equipment identification. Marking equipment means labeling the equipment breaker

Table 11.4 Sixteen Common OSHA Violations (*After*[1].)

Fact Sheet	Subject	NEC[1] Reference
1	Guarding of live parts	110-17
2	Identification	110-22
3	Uses allowed for flexible cord	400-7
4	Prohibited uses of flexible cord	400-8
5	Pull at joints and terminals must be prevented	400-10
6.1	Effective grounding, Part 1	250-51
6.2	Effective grounding, Part 2	250-51
7	Grounding of fixed equipment, general	250-42
8	Grounding of fixed equipment, specific	250-43
9	Grounding of equipment connected by cord and plug	250-45
10	Methods of grounding, cord and plug-connected equipment	250-59
11	AC circuits and systems to be grounded	250-5
12	Location of overcurrent devices	240-24
13	Splices in flexible cords	400-9
14	Electrical connections	110-14
15	Marking equipment	110-21
16	Working clearances about electric equipment	110-16
[1] National Electrical Code		

panels and ac disconnect switches according to device rating. Breaker boxes should contain a nameplate showing the manufacturer, rating, and other pertinent electrical factors. The intent is to prevent devices from being subjected to excessive loads or voltages.

Grounding

OSHA regulations describe two types of grounding: *system grounding* and *equipment grounding*. System grounding actually connects one of the current-carrying conductors (such as the terminals of a supply transformer) to ground. (See Figure 11.6.) Equipment grounding connects all of the noncurrent-carrying metal surfaces together and to ground. From a grounding standpoint, the only difference between a grounded electrical system and an ungrounded electrical system is that the *main bonding jumper* from the service equipment ground to a current-carrying conductor is omitted in the ungrounded system. The system ground performs two tasks:

- It provides the final connection from equipment-grounding conductors to the grounded circuit conductor, thus completing the ground-fault loop.

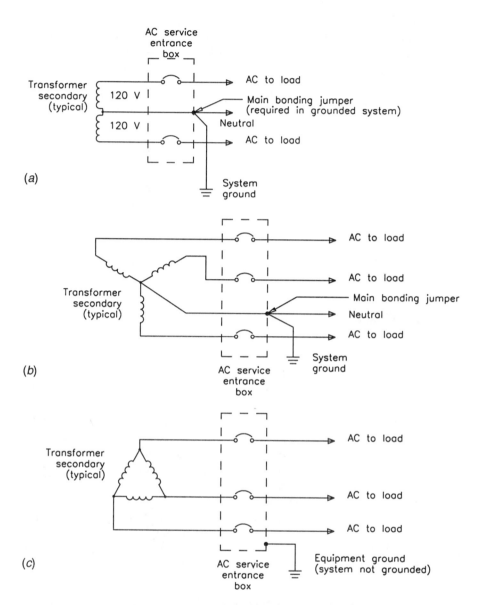

Figure 11.6 AC service entrance bonding requirements: (*a*) 120 V phase-to-neutral (240 V phase-to-phase), (*b*) 3-phase 208 V wye (120 V phase-to-neutral), (*c*) 3-phase 240 V (or 480 V) delta. Note that the main bonding jumper is required in only two of the designs.

- It solidly ties the electrical system and its enclosures to their surroundings (usually earth, structural steel, and plumbing). This prevents voltages at any source from rising to harmfully high voltage-to-ground levels.

Note that equipment grounding—bonding all electric equipment to ground—is required whether or not the system is grounded. Equipment grounding serves two important tasks:

- It bonds all surfaces together so that there can be no voltage difference among them.

- It provides a ground-fault current path from a fault location back to the electrical source, so that if a fault current develops, it will rise to a level high enough to operate the breaker or fuse.

The National Electrical Code (NEC) is complex and contains numerous requirements concerning electrical safety. The fact sheets listed in Table 11.4 are available from OSHA.

11.3.2 Beryllium Oxide Ceramics

Some tubes, both power grid and microwave, contain beryllium oxide (BeO) ceramics, typically at the output waveguide window or around the cathode. Never perform any operations on BeO ceramics that produce dust or fumes, such as grinding, grit blasting, or acid cleaning. Beryllium oxide dust and fumes are highly toxic, and breathing them can result in serious personal injury or death.

If a broken window is suspected on a microwave tube, carefully remove the device from its waveguide, and seal the output flange of the tube with tape. Because BeO warning labels may be obliterated or missing, maintenance personnel should contact the tube manufacturer before performing any work on the device. Some tubes have BeO internal to the vacuum envelope.

Take precautions to protect personnel working in the disposal or salvage of tubes containing BeO. All such personnel should be made aware of the deadly hazards involved and the necessity for great care and attention to safety precautions. Some tube manufacturers will dispose of tubes without charge, provided they are returned to the manufacturer prepaid, with a written request for disposal.

11.3.3 Corrosive and Poisonous Compounds

The external output waveguides and cathode high-voltage bushings of microwave tubes are sometimes operated in systems that use a dielectric gas to impede microwave or high-voltage breakdown. If breakdown does occur, the gas may decompose and combine with impurities, such as air or water vapor, to form highly toxic and corrosive compounds. Examples include Freon gas, which may form lethal *phosgene*, and sulfur hexafluoride (SF_6) gas, which may form highly toxic and corrosive sulfur or fluorine compounds such as *beryllium fluoride*. When breakdown does occur in the presence of these gases, proceed as follows:

- Ventilate the area to outside air
- Avoid breathing any fumes or touching any liquids that develop
- Take precautions appropriate for beryllium compounds and for other highly toxic and corrosive substances

If a coolant other than pure water is used, follow the precautions supplied by the coolant manufacturer.

11.3.4 FC-75 Toxic Vapor

The decomposition products of FC-75 are highly toxic. Decomposition may occur as a result of any of the following:

- Exposure to temperatures above 200°C
- Exposure to liquid fluorine or alkali metals (lithium, potassium, or sodium)
- Exposure to ionizing radiation

Known thermal decomposition products include *perfluoroisobutylene* (PFIB; $[CF_3]_2$ $C = CF_2$), which is highly toxic in small concentrations.

If FC-75 has been exposed to temperatures above 200°C through fire, electric heating, or prolonged electric arcs, or has been exposed to alkali metals or strong ionizing radiation, take the following steps:

- Strictly avoid breathing any fumes or vapors.
- Thoroughly ventilate the area.
- Strictly avoid any contact with the FC-75.

Under such conditions, promptly replace the FC-75 and handle and dispose of the contaminated FC-75 as a toxic waste.

11.3.5 Nonionizing Radiation

Nonionizing radio frequency radiation (RFR) resulting from high-intensity RF fields is a growing concern to engineers who must work around high-power transmission equipment. The principal medical concern regarding nonionizing radiation involves heating of various body tissues, which can have serious effects, particularly if there is no mechanism for heat removal. Recent research has also noted, in some cases, subtle psychological and physiological changes at radiation levels below the threshold for heat-induced biological effects. However, the consensus is that most effects are thermal in nature.

High levels of RFR can affect one or more body systems or organs. Areas identified as potentially sensitive include the ocular (eye) system, reproductive system, and the immune system. Nonionizing radiation also is thought to be responsible for metabolic effects on the central nervous system and cardiac system.

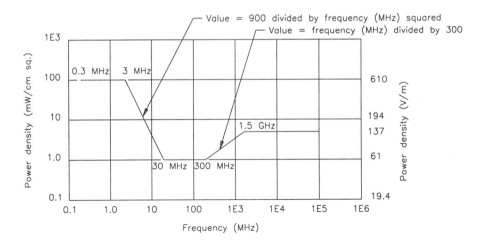

Figure 11.7 The power density limits for nonionizing radiation exposure for humans.

In spite of these studies, many of which are ongoing, there is still no clear evidence in Western literature that exposure to medium-level nonionizing radiation results in detrimental effects. Russian findings, on the other hand, suggest that occupational exposure to RFR at power densities above 1.0 mW/cm^2 does result in symptoms, particularly in the central nervous system.

Clearly, the jury is still out as to the ultimate biological effects of RFR. Until the situation is better defined, however, the assumption must be made that potentially serious effects can result from excessive exposure. Compliance with existing standards should be the minimum goal, to protect members of the public as well as facility employees.

NEPA Mandate

The National Environmental Policy Act of 1969 required the Federal Communications Commission to place controls on nonionizing radiation. The purpose was to prevent possible harm to the public at large and to those who must work near sources of the radiation. Action was delayed because no hard and fast evidence existed that low- and medium-level RF energy is harmful to human life. Also, there was no evidence showing that radio waves from radio and TV stations did not constitute a health hazard.

During the delay, many studies were carried out in an attempt to identify those levels of radiation that might be harmful. From the research, suggested limits were developed by the American National Standards Institute (ANSI) and stated in the document known as ANSI C95.1-1982. The protection criteria outlined in the standard are shown in Figure 11.7.

The energy-level criteria were developed by representatives from a number of industries and educational institutions after performing research on the possible effects of nonionizing radiation. The projects focused on absorption of RF energy by the human body, based upon simulated human body models. In preparing the document, ANSI attempted to determine those levels of incident radiation that would cause the body to absorb less than 0.4 W/kg of mass (averaged over the whole body) or peak absorption values of 8 W/kg over any 1 gram of body tissue.

From the data, the researchers found that energy would be absorbed more readily at some frequencies than at others. The absorption rates were found to be functions of the size of a specific individual and the frequency of the signal being evaluated. It was the result of these absorption rates that culminated in the shape of the *safe curve* shown in the figure. ANSI concluded that no harm would come to individuals exposed to radio energy fields, as long as specific values were not exceeded when averaged over a period of 0.1 hour. It was also concluded that higher values for a brief period would not pose difficulties if the levels shown in the standard document were not exceeded when averaged over the 0.1-hour time period.

The FCC adopted ANSI C95.1-1982 as a standard that would ensure adequate protection to the public and to industry personnel who are involved in working around RF equipment and antenna structures.

Revised Guidelines

The ANSI C95.1-1982 standard was intended to be reviewed at 5-year intervals. Accordingly, the 1982 standard was due for reaffirmation or revision in 1987. The process was indeed begun by ANSI, but was handed off to the Institute of Electrical and Electronics Engineers (IEEE) for completion. In 1991, the revised document was completed and submitted to ANSI for acceptance as ANSI/IEEE C95.1-1992.

The IEEE standard incorporated changes from the 1982 ANSI document in four major areas:

- An additional safety factor was provided in certain situations. The most significant change was the introduction of new *uncontrolled* (public) exposure guidelines, generally established at one-fifth of the *controlled* (occupational) exposure guidelines. Figure 11.8 illustrates the concept for the microwave frequency band.

- For the first time, guidelines were included for body currents; examination of the electric and magnetic fields were determined to be insufficient to determine compliance.

- Minor adjustments were made to occupational guidelines, including relaxation of the guidelines at certain frequencies and the introduction of *breakpoints* at new frequencies.

- Measurement procedures were changed in several aspects, most notably with respect to spatial averaging and to minimum separation from reradiating objects and structures at the site.

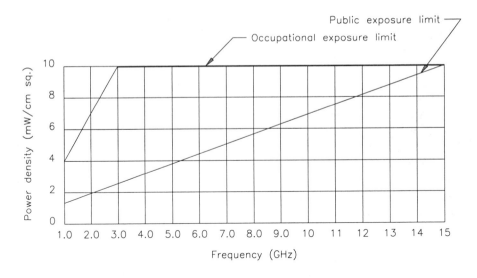

Figure 11.8 ANSI/IEEE exposure guidelines for microwave frequencies.

The revised guidelines are complex and beyond the scope of this handbook. Refer to the ANSI/IEEE document for details.

Multiple-User Sites

At a multiple-user site, the responsibility for assessing the RFR situation—although officially triggered by either a new user or the license renewal of all site tenants—is, in reality, the joint responsibility of all the site tenants. In a multiple-user environment involving various frequencies, and various protection criteria, compliance is indicated when the fraction of the RFR limit within each pertinent frequency band is established and added to the sum of all the other fractional contributions. The sum must not be greater than 1.0. Evaluating the multiple-user environment is not a simple matter, and corrective actions, if indicated, may be quite complex.

Operator Safety Considerations

RF energy must be contained properly by shielding and transmission lines. All input and output RF connections, cables, flanges, and gaskets must be RF leakproof. The following guidelines should be followed at all times:

- Never operate a power tube without a properly matched RF energy absorbing load attached.

- Never look into or expose any part of the body to an antenna or open RF generating tube, circuit, or RF transmission system that is energized.

- Monitor the RF system for radiation leakage at regular intervals and after servicing.

11.3.6 X-Ray Radiation Hazard

The voltages typically used in microwave tubes are capable of producing dangerous X rays. As voltages increase beyond 15 kV, metal-body tubes are capable of producing progressively more dangerous radiation. Adequate X-ray shielding must be provided on all sides of such tubes, particularly at the cathode and collector ends, as well as at the modulator and pulse transformer tanks (as appropriate). High-voltage tubes never should be operated without adequate X-ray shielding in place. The X-ray radiation of the device should be checked at regular intervals and after servicing.

11.3.7 Implosion Hazard

Because of the high internal vacuum in power grid and microwave tubes, the glass or ceramic output window or envelope can shatter inward (implode) if struck with sufficient force or exposed to sufficient mechanical shock. Flying debris could result in bodily injury, including cuts and puncture wounds. If the device is made of beryllium oxide ceramic, implosion may produce highly toxic dust or fumes.

In the event of such an implosion, assume that toxic BeO ceramic is involved unless confirmed otherwise.

11.3.8 Hot Coolant and Surfaces

Extreme heat occurs in the electron collector of a microwave tube and the anode of a power grid tube during operation. Coolant channels used for water or vapor cooling also can reach high temperatures (boiling—100°C—and above), and the coolant is typically under pressure (as high as 100 psi). Some devices are cooled by boiling the coolant to form steam.

Contact with hot portions of the tube or its cooling system can scald or burn. Carefully check that all fittings and connections are secure, and monitor back pressure for changes in cooling system performance. If back pressure is increased above normal operating values, shut the system down and clear the restriction.

For a device whose anode or collector is air-cooled, the external surface normally operates at a temperature of 200 to 300°C. Other parts of the tube also may reach high temperatures, particularly the cathode insulator and the cathode/heater surfaces. All hot surfaces remain hot for an extended time after the tube is shut off. To prevent serious burns, take care to avoid bodily contact with these surfaces during operation and for a reasonable cool-down period afterward.

11.3.9 Polychlorinated Biphenyls

PCBs belong to a family of organic compounds known as *chlorinated hydrocarbons*. Virtually all PCBs in existence today have been synthetically manufactured. PCBs

Table 11.5 Commonly Used Names for PCB Insulating Material

PCB Trade Names					
Apirolio	Abestol	Askarel[1]	Aroclor B	Chlorexto	Chlophen
Chlorinol	Clorphon	Diaclor	DK	Dykanol	EEC-18
Elemex	Eucarel	Fenclor	Hyvol	Inclor	Inerteen
Kanechlor	No-Flamol	Phenodlor	Pydraul	Pyralene	Pyranol
Pyroclor	Sal-T-Kuhl	Santothern FR	Santovac	Solvol	Thermin

[1] Generic name used for nonflammable dielectric fluids containing PCBs.

have a heavy oil-like consistency, high boiling point, a high degree of chemical stability, low flammability, and low electrical conductivity. These characteristics resulted in the widespread use of PCBs in high-voltage capacitors and transformers. Commercial products containing PCBs were widely distributed between 1957 and 1977 under several trade names including:

- Aroclor

- Pyroclor

- Sanotherm

- Pyranol

- Askarel

Askarel is also a generic name used for nonflammable dielectric fluids containing PCBs. Table 11.5 lists some common trade names used for Askarel. These trade names typically will be listed on the nameplate of a PCB transformer or capacitor.

PCBs are harmful because once they are released into the environment, they tend not to break apart into other substances. Instead, PCBs persist, taking several decades to slowly decompose. By remaining in the environment, they can be taken up and stored in the fatty tissues of all organisms, from which they are slowly released into the bloodstream. Therefore, because of the storage in fat, the concentration of PCBs in body tissues can increase with time, even though PCB exposure levels may be quite low. This process is called *bioaccumulation*. Furthermore, as PCBs accumulate in the tissues of simple organisms, and as they are consumed by progressively higher organisms, the concentration increases. This process is called *biomagnification*. These two factors are especially significant because PCBs are harmful even at low levels. Specifically, PCBs have been shown to cause chronic (long-term) toxic effects in some species of animals and aquatic life. Well-documented tests on laboratory animals show that various levels of PCBs can cause reproductive effects, gastric disorders, skin lesions, and cancerous tumors.

PCBs may enter the body through the lungs, the gastrointestinal tract, and the skin. After absorption, PCBs are circulated in the blood throughout the body and stored in fatty tissues and a variety of organs, including the liver, kidneys, lungs, adrenal glands, brain, heart, and skin.

The health risk from PCBs lies not only in the PCB itself, but also in the chemicals that develop when PCBs are heated. Laboratory studies have confirmed that PCB by-products, including *polychlorinated dibenzofurans* (PCDFs) and *polychlorinated dibenzo-p-dioxins* (PCDDs), are formed when PCBs or *chlorobenzenes* are heated to temperatures ranging from approximately 900 to 1300°F. Unfortunately, these products are more toxic than PCBs themselves.

Governmental Action

The U.S. Congress took action to control PCBs in October 1975 by passing the Toxic Substances Control Act (TSCA). A section of this law specifically directed the EPA to regulate PCBs. Three years later the Environmental Protection Agency (EPA) issued regulations to implement the congressional ban on the manufacture, processing, distribution, and disposal of PCBs. Since that time, several revisions and updates have been issued by the EPA. One of these revisions, issued in 1982, specifically addressed the type of equipment used in industrial plants and transmitting stations. Failure to properly follow the rules regarding the use and disposal of PCBs has resulted in high fines and even jail sentences.

Although PCBs are no longer being produced for electrical products in the United States, there are throusands of PCB transformers and millions of small PCB capacitors still in use or in storage. The threat of widespread contamination from PCB fire-related incidents is one reason behind the EPA's efforts to reduce the number of PCB products in the environment. The users of high-power equipment are affected by the regulations primarily because of the widespread use of PCB transformers and capacitors. These components usually are located in older (pre-1979) systems, so this is the first place to look for them. However, some facilities also maintain their own primary power transformers. Unless these transformers are of recent vintage, it is quite likely that they too contain a PCB dielectric. Table 11.6 lists the primary classifications of PCB devices.

PCB Components

The two most common PCB components are transformers and capacitors. A PCB transformer is one containing at least 500 ppm (parts per million) PCBs in the dielectric fluid. An Askarel transformer generally has 600,000 ppm or more. A PCB transformer may be converted to a *PCB-contaminated device* (50 to 500 ppm) or a *non-PCB device* (less than 50 ppm) by having it drained, refilled, and tested. The testing must not take place until the transformer has been in service for a minimum of 90 days. Note that this is *not* something a maintenance technician can do. It is the exclusive domain of specialized remanufacturing companies.

PCB transformers must be inspected quarterly for leaks. If an impervious dike is built around the transformer sufficient to contain all of the liquid material, the inspec-

Table 11.6 Definition of PCB Terms as Identified by the EPA

Term	Definition	Examples
PCB	Any chemical substance that is limited to the biphenyl molecule that has been chlorinated to varying degrees, or any combination of substances that contain such substances.	PCB dielectric fluids, PCB heat-transfer fluids, PCB hydraulic fluids, 2,2',4-trichlorobiphenyl
PCB article	Any manufactured article, other than a PCB container, that contains PCBs and whose surface has been in direct contact with PCBs.	Capacitors, transformers, electric motors, pumps, pipes
PCB container	A device used to contain PCBs or PCB articles, and whose surface has been in direct contact with PCBs.	Packages, cans, bottles, bags, barrels, drums, tanks
PCB article container	A device used to contain PCB articles or equipment, and whose surface has not been in direct contact with PCBs.	Packages, cans, bottles, bags, barrels, drums, tanks
PCB equipment	Any manufactured item, other than a PCB container or PCB article container, that contains a PCB article or other PCB equipment.	Microwave systems, fluorescent light ballasts, electronic equipment
PCB item	Any PCB article, PCB article container, PCB container, or PCB equipment that deliberately or unintentionally contains, or has as a part of it, any PCBs.	
PCB transformer	Any transformer that contains PCBs in concentrations of 500 ppm or greater.	
PCB contaminated	Any electric equipment that contains more than 50, but less than 500 ppm of PCBs. (Oil-filled electric equipment other than circuit breakers, reclosers, and cable whose PCB concentration is unknown must be assumed to be PCB-contaminated electric equipment.)	Transformers, capacitors, contaminated circuit breakers, reclosers, voltage regulators, switches, cable, electromagnets

tions can be conducted yearly. Similarly, if the transformer is tested and found to contain less than 60,000 ppm, a yearly inspection is sufficient. Failed PCB transformers cannot be repaired; they must be properly disposed of.

If a leak develops, it must be contained and daily inspections begun. A cleanup must be initiated as soon as possible, but no later than 48 hours after the leak is discovered. Adequate records must be kept of all inspections, leaks, and actions taken for 3 years after disposal of the component. Combustible materials must be kept a minimum of 5 m from a PCB transformer and its enclosure.

As of October 1, 1990, the use of PCB transformers (500 ppm or greater) was prohibited in or near commercial buildings when the secondary voltages are 480 V ac or higher.

The EPA regulations also require that the operator notify others of the possible dangers. All PCB transformers (including PCB transformers in storage for reuse) must be registered with the local fire department. The following information must be supplied:

- The location of the PCB transformer(s).

- Address(es) of the building(s) and, for outdoor PCB transformers, the location.

- Principal constituent of the dielectric fluid in the transformer(s).

- Name and telephone number of the contact person in the event of a fire involving the equipment.

Any PCB transformers used in a commercial building must be registered with the building owner. All building owners within 30 m of such PCB transformers also must be notified. In the event of a fire-related incident involving the release of PCBs, the Coast Guard National Spill Response Center (800-424-8802) must be notified immediately. Appropriate measures also must be taken to contain and control any possible PCB release into water.

Capacitors are divided into two size classes, *large* and *small*. A PCB small capacitor contains less than 1.36 kg (3 lbs) of dielectric fluid. A capacitor having less than 100 in^3 also is considered to contain less than 3 lb of dielectric fluid. A PCB large capacitor has a volume of more than 200 in^3 and is considered to contain more than 3 lb of dielectric fluid. Any capacitor having a volume between 100 and 200 in^3 is considered to contain 3 lb of dielectric, provided the total weight is less than 9 lb. A PCB *large high-voltage capacitor* contains 3 lb or more of dielectric fluid and operates at voltages of 2 kV or greater. A *large low-voltage capacitor* also contains 3 lb or more of dielectric fluid but operates below 2 kV.

The use and servicing of PCB small capacitors is not restricted by the EPA unless there is a leak. In that event, the leak must be repaired or the capacitor disposed of. Disposal may be handled by an approved incineration facility, or the component may be placed in a specified container and buried in an approved chemical waste landfill. Items such as capacitors that are leaking oil greater than 500 ppm PCBs should be taken to an EPA-approved PCB disposal facility.

PCB Liability Management

Properly managing the PCB risk is not particularly difficult; the keys are understanding the regulations and following them carefully. Any program should include the following steps:

- Locate and identify all PCB devices. Check all stored or spare devices.

- Properly label PCB transformers and capacitors according to EPA requirements.

- Perform the required inspections and maintain an accurate log of PCB items, their location, inspection results, and actions taken. These records must be maintained for 3 years after disposal of the PCB component.

- Complete the annual report of PCBs and PCB items by July 1 of each year. This report must be retained for 5 years.

- Arrange for necessary disposal through a company licensed to handle PCBs. If there are any doubts about the company's license, contact the EPA.

- Report the location of all PCB transformers to the local fire department and to the owners of any nearby buildings.

The importance of following the EPA regulations cannot be overstated.

11.4 References

1. National Electrical Code, NFPA #70.

11.5 Bibliography

"Current Intelligence Bulletin #45," National Institute for Occupational Safety and Health, Division of Standards Development and Technology Transfer, February 24, 1986.

Code of Federal Regulations, 40, Part 761.

"Electrical Standards Reference Manual," U.S. Department of Labor, Washington, DC.

Hammett, William F., "Meeting IEEE C95.1-1991 Requirements," *NAB 1993 Broadcast Engineering Conference Proceedings*, National Association of Broadcasters, Washington, D.C., pp. 471 - 476, April 1993.

Hammar, Willie, *Occupational Safety Management and Engineering*, Prentice Hall, New York.

Markley, Donald, "Complying with RF Emission Standards," *Broadcast Engineering*, Intertec Publishing, Overland Park, KS, May 1986.

"Occupational Injuries and Illnesses in the United States by Industry," OSHA Bulletin 2278, U.S. Department of Labor, Washington, DC, 1985.

OSHA, "Electrical Hazard Fact Sheets," U.S. Department of Labor, Washington, DC, January 1987.

OSHA, "Handbook for Small Business," U.S. Department of Labor, Washington, DC.

Pfrimmer, Jack, "Identifying and Managing PCBs in Broadcast Facilities," *1987 NAB Engineering Conference Proceedings*, National Association of Broadcasters, Washington, DC, 1987.

"Safety Precautions," Publication no. 3386A, Varian Associates, Palo Alto, CA, March 1985.

Smith, Milford K., Jr., "RF Radiation Compliance," *Proceedings of the Broadcast Engineering Conference*, Society of Broadcast Engineers, Indianapolis, IN, 1989.

"Toxics Information Series," Office of Toxic Substances, July 1983.

Whitaker, Jerry C., *AC Power Systems*, 2nd Ed., CRC Press, Boca Raton, FL, 1998.

Whitaker, Jerry C., G. DeSantis, and C. Paulson, *Interconnecting Electronic Systems*, CRC Press, Boca Raton, FL, 1993.

Whitaker, Jerry C., *Maintaining Electronic Systems*, CRC Press, Boca Raton, FL, 1991.

Whitaker, Jerry C., *Radio Frequency Transmission Systems: Design and Operation*, McGraw-Hill, New York, 1990.

12

Reference Data

Standard Units

Name	Symbol	Quantity
ampere	A	electric current
ampere per meter	A/m	magnetic field strength
ampere per square meter	A/m^2	current density
Becquerel	Bq	activity (of a radionuclide)
candela	cd	luminous intensity
coulomb	C	electric charge
coulomb per kilogram	C/kg	exposure (x- and gamma-rays)
coulomb per sq. meter	C/m^2	electric flux density
cubic meter	m^3	volume
cubic meter per kilogram	m^3/kg	specific volume
degree Celsius	°C	Celsius temperature
farad	F	capacitance
farad per meter	F/m	permittivity
henry	H	inductance
henry per meter	H/m	permeability
hertz	Hz	frequency
joule	J	energy, work, quantity of heat
joule per cubic meter	J/m^3	energy density
joule per kelvin	J/K	heat capacity
joule per kilogram K	J/(kg-K)	specific heat capacity
joule per mole	J/mol	molar energy
kelvin	K	thermodynamic temperature
kilogram	kg	mass
kilogram per cubic meter	kg/m^3	density, mass density
lumen	lm	luminous flux
lux	lx	luminance
meter	m	length
meter per second	m/s	speed, velocity
meter per second sq.	m/s^2	acceleration

(Continued on next page)

Standard Units Continued

Name	Symbol	Quantity
mole	mol	amount of substance
pascal	Pa	pressure, stress
pascal second	Pa•s	dynamic viscosity
radian	rad	plane angle
radian per second	rad/s	angular velocity
radian per second squared	rad/s^2	angular acceleration
second	s	time
siemens	S	electric conductance
square meter	m^2	area
steradian	sr	solid angle
tesla	T	magnetic flux density
volt	V	electrical potential
volt per meter	V/m	electric field strength
watt	W	power, radiant flux
watt per meter kelvin	W/(m•K)	thermal conductivity
watt per square meter	W/m^2	heat (power) flux density
weber	Wb	magnetic flux

Standard Prefixes

Multiple	Prefix	Symbol
10^{18}	exa	E
10^{15}	peta	P
10^{12}	tera	T
10^9	giga	G
10^6	mega	M
10^3	kilo	k
10^2	hecto	h
10	deka	da
10^{-1}	deci	d
10^{-2}	centi	c
10^{-3}	milli	m
10^{-6}	micro	μ
10^{-9}	nano	n
10^{-12}	pico	p
10^{-15}	femto	f
10^{-18}	atto	a

Common Standard Units for Electronics Work

Unit	Symbol
centimeter	cm
cubic centimeter	cm^3
cubic meter per second	m^3/s
gigahertz	GHz
gram	g
kilohertz	kHz
kilohm	kΩ
kilojoule	kJ
kilometer	km
kilovolt	kV
kilovoltampere	kVA
kilowatt	kW
megahertz	MHz
megavolt	MV
megawatt	MW
megohm	MΩ
microampere	μA
microfarad	μF
microgram	μg
microhenry	μH
microsecond	μs
microwatt	μW
milliampere	mA
milligram	mg
millihenry	mH
millimeter	mm
millisecond	ms
millivolt	mV
milliwatt	mW
nanoampere	nA
nanofarad	nF
nanometer	nm
nanosecond	ns
nanowatt	nW
picoampere	pA
picofarad	pF
picosecond	ps
picowatt	pW

Power Conversion Factors (Decibels to Watts)

dBm	dBw	Watts	Multiple	Prefix
+150	+120	1,000,000,000,000	10^{12}	1 terawatt
+140	+110	100,000,000,000	10^{11}	100 gigawatts
+130	+100	10,000,000,000	10^{10}	10 gigawatts
+120	+90	1,000,000,000	10^{9}	1 gigawatt
+110	+80	100,000,000	10^{8}	100 megawatts
+100	+70	10,000,000	10^{7}	10 megawatts
+90	+60	1,000,000	10^{6}	1 megawatt
+80	+50	100,000	10^{5}	100 kilowatts
+70	+40	10,000	10^{4}	10 kilowatts
+60	+30	1000	10^{3}	1 kilowatt
+50	+20	100	10^{2}	1 hectowatt
+40	+10	10	10	1 decawatt
+30	0	1	1	1 watt
+20	−10	0.1	10^{-1}	1 deciwatt
+10	−20	0.01	10^{-2}	1 centiwatt
0	−30	0.001	10^{-3}	1 milliwatt
−10	−40	0.0001	10^{-4}	100 microwatts
−20	−50	0.00001	10^{-5}	10 microwatts
−30	−60	0.000,001	10^{-6}	1 microwatt
−40	−70	0.0,000,001	10^{-7}	100 nanowatts
−50	−80	0.00,000,001	10^{-8}	10 nanowatts
−60	−90	0.000,000,001	10^{-9}	1 nanowatt
−70	−100	0.0,000,000,001	10^{-10}	100 picowatts
−80	−110	0.00,000,000,001	10^{-11}	10 picowatts
−90	−120	0.000,000,000,001	10^{-12}	1 picowatt

Voltage Standing Wave Ratio and Key Parameters

VSWR	Reflection Coefficient	Return Loss	Power Ratio	Percent Reflected
1.01:1	0.0050	46.1 dB	0.00002	0.002
1.02:1	0.0099	40.1 dB	0.00010	0.010
1.04:1	0.0196	34.2 dB	0.00038	0.038
1.06:1	0.0291	30.7 dB	0.00085	0.085
1.08:1	0.0385	28.3 dB	0.00148	0.148
1.10:1	0.0476	26.4 dB	0.00227	0.227
1.20:1	0.0909	20.8 dB	0.00826	0.826
1.30:1	0.1304	17.7 dB	0.01701	1.7
1.40:1	0.1667	15.6 dB	0.02778	2.8
1.50:1	0.2000	14.0 dB	0.04000	4.0
1.60:1	0.2308	12.7 dB	0.05325	5.3
1.70:1	0.2593	11.7 dB	0.06722	6.7
1.80:1	0.2857	10.9 dB	0.08163	8.2
1.90:1	0.3103	10.2 dB	0.09631	9.6
2.00:1	0.3333	9.5 dB	0.11111	11.1
2.20:1	0.3750	8.5 dB	0.14063	14.1
2.40:1	0.4118	7.7 dB	0.16955	17.0
2.60:1	0.4444	7.0 dB	0.19753	19.8
2.80:1	0.4737	6.5 dB	0.22438	22.4
3.00:1	0.5000	6.0 dB	0.25000	25.0
3.50:1	0.5556	5.1 dB	0.30864	30.9
4.00:1	0.6000	4.4 dB	0.36000	36.0
4.50:1	0.6364	3.9 dB	0.40496	40.5
5.00:1	0.6667	3.5 dB	0.44444	44.4
6.00:1	0.7143	2.9 dB	0.51020	51.0
7.00:1	0.7500	2.5 dB	0.56250	56.3
8.00:1	0.7778	2.2 dB	0.60494	60.5
9.00:1	0.8000	1.9 dB	0.64000	64.0
10.00:1	0.8182	1.7 dB	0.66942	66.9
15.00:1	0.8750	1.2 dB	0.76563	76.6
20.00:1	0.9048	0.9 dB	0.81859	81.9
30.00:1	0.9355	0.6 dB	0.87513	97.5
40.00:1	0.9512	0.4 dB	0.90482	90.5
50.00:1	0.9608	0.3 dB	0.92311	92.3

Specifications of Standard Copper Wire

Wire Size AWG	Diam. in mils	Circular mil Area	Turns Per Linear Inch[1] enam.	SCE	DCC	Ohms Per 100 ft[2]	Current-Carrying Capacity[3]	Diam. in mm
1	289.3	83810	-	-	-	0.1239	119.6	7.348
2	257.6	05370	-	-	-	0.1563	94.8	6.544
3	229.4	62640	-	-	-	0.1970	75.2	5.827
4	204.3	41740	-	-	-	0.2485	59.6	5.189
5	181.9	33100	-	-	-	0.3133	47.3	4.621
6	162.0	26250	-	-	-	0.3951	37.5	4.115
7	144.3	20820	-	-	-	0.4982	29.7	3.665
8	128.5	16510	7.6	-	7.1	0.6282	23.6	3.264
9	114.4	13090	8.6	-	7.8	0.7921	18.7	2.906
10	101.9	10380	9.6	9.1	8.9	0.9989	14.8	2.588
11	90.7	8234	10.7	-	9.8	1.26	11.8	2.305
12	80.8	6530	12.0	11.3	10.9	1.588	9.33	2.063
13	72.0	5178	13.5	-	12.8	2.003	7.40	1.828
14	64.1	4107	15.0	14.0	13.8	2.525	5.87	1.628
15	57.1	3257	16.8	-	14.7	3.184	4.65	1.450
16	50.8	2583	18.9	17.3	16.4	4.016	3.69	1.291
17	45.3	2048	21.2	-	18.1	5.064	2.93	1.150
18	40.3	1624	23.6	21.2	19.8	6.386	2.32	1.024
19	35.9	1288	26.4	-	21.8	8.051	1.84	0.912
20	32.0	1022	29.4	25.8	23.8	10.15	1.46	0.812
21	28.5	810	33.1	-	26.0	12.8	1.16	0.723
22	25.3	642	37.0	31.3	30.0	16.14	0.918	0.644
23	22.6	510	41.3	-	37.6	20.36	0.728	0.573
24	20.1	404	46.3	37.6	35.6	25.67	0.577	0.511
25	17.9	320	51.7	-	38.6	32.37	0.458	0.455
26	15.9	254	58.0	46.1	41.8	40.81	0.363	0.406
27	14.2	202	64.9	-	45.0	51.47	0.288	0.361
28	12.6	160	72.7	54.6	48.5	64.9	0.228	0.321
29	11.3	127	81.6	-	51.8	81.83	0.181	0.286
30	10.0	101	90.5	64.1	55.5	103.2	0.144	0.255
31	8.9	50	101	-	59.2	130.1	0.114	0.227
32	8.0	63	113	74.1	61.6	164.1	0.090	0.202
33	7.1	50	127	-	66.3	206.9	0.072	0.180
34	6.3	40	143	86.2	70.0	260.9	0.057	0.160
35	5.6	32	158	-	73.5	329.0	0.045	0.143
36	5.0	25	175	103.1	77.0	414.8	0.036	0.127
37	4.5	20	198	-	80.3	523.1	0.028	0.113
38	4.0	16	224	116.3	83.6	659.6	0.022	0.101
39	3.5	12	248	-	86.6	831.8	0.018	0.090

1. Based on 25.4 mm.
2. Ohms per 1000 ft measured at 20°C.
3. Current-carrying capacity at 700 circular mils per amp.

Celsius-to-Fahrenheit Conversion

°Celsius	°Fahrenheit	°Celsius	°Fahrenheit
−50	−58	125	257
−45	−49	130	266
−40	−40	135	275
−35	−31	140	284
−30	−22	145	293
−25	−13	150	302
−20	4	155	311
−15	5	160	320
−10	14	165	329
−5	23	170	338
0	32	175	347
5	41	180	356
10	50	185	365
15	59	190	374
20	68	195	383
25	77	200	392
30	86	205	401
35	95	210	410
40	104	215	419
45	113	220	428
50	122	225	437
55	131	230	446
60	140	235	455
65	149	240	464
70	158	245	473
75	167	250	482
80	176	255	491
85	185	260	500
90	194	265	509
95	203	270	518
100	212	275	527
105	221	280	536
110	230	285	545
115	239	290	554
120	248	295	563

Inch-to-Millimeter Conversion

Inch	0	1/8	1/4	3/8	1/2	5/8	3/4	7/8	Inch
0	0.0	3.18	6.35	9.52	12.70	15.88	19.05	22.22	0
1	25.40	28.58	31.75	34.92	38.10	41.28	44.45	47.62	1
2	50.80	53.98	57.15	60.32	63.50	66.68	69.85	73.02	2
3	76.20	79.38	82.55	85.72	88.90	92.08	95.25	98.42	3
4	101.6	104.8	108.0	111.1	114.3	117.5	120.6	123.8	4
5	127.0	130.2	133.4	136.5	139.7	142.9	146.0	149.2	5
6	152.4	155.6	158.8	161.9	165.1	168.3	171.4	174.6	6
7	177.8	181.0	184.2	187.3	190.5	193.7	196.8	200.0	7
8	203.2	206.4	209.6	212.7	215.9	219.1	222.2	225.4	8
9	228.6	231.8	235.0	238.1	241.3	244.5	247.6	250.8	9
10	254.0	257.2	260.4	263.5	266.7	269.9	273.0	276.2	10
11	279	283	286	289	292	295	298	302	11
12	305	308	311	314	317	321	324	327	12
13	330	333	337	340	343	346	349	352	13
14	356	359	362	365	368	371	375	378	14
15	381	384	387	391	394	397	400	403	15
16	406	410	413	416	419	422	425	429	16
17	432	435	438	441	445	448	451	454	17
18	457	460	464	467	470	473	476	479	18
19	483	486	489	492	495	498	502	505	19
20	508	511	514	518	521	524	527	530	20

Millimeters-to-Decimal Inches Conversion

mm	Inches	mm	Inches	mm	Inches
1	0.039370	46	1.811020	91	3.582670
2	0.078740	47	1.850390	92	3.622040
3	0.118110	48	1.889760	93	3.661410
4	0.157480	49	1.929130	94	3.700780
5	0.196850	50	1.968500	95	3.740150
6	0.236220	51	2.007870	96	3.779520
7	0.275590	52	2.047240	97	3.818890
8	0.314960	53	2.086610	98	3.858260
9	0.354330	54	2.125980	99	3.897630
10	0.393700	55	2.165350	100	3.937000
11	0.433070	56	2.204720	105	4.133848
12	0.472440	57	2.244090	110	4.330700
13	0.511810	58	2.283460	115	4.527550
14	0.551180	59	2.322830	120	4.724400
15	0.590550	60	2.362200	125	4.921250
16	0.629920	61	2.401570	210	8.267700
17	0.669290	62	2.440940	220	8.661400
18	0.708660	63	2.480310	230	9.055100
19	0.748030	64	2.519680	240	9.448800
20	0.787400	65	2.559050	250	9.842500
21	0.826770	66	2.598420	260	10.236200
22	0.866140	67	2.637790	270	10.629900
23	0.905510	68	2.677160	280	11.032600
24	0.944880	69	2.716530	290	11.417300
25	0.984250	70	2.755900	300	11.811000
26	1.023620	71	2.795270	310	12.204700
27	1.062990	72	2.834640	320	12.598400
28	1.102360	73	2.874010	330	12.992100
29	1.141730	74	2.913380	340	13.385800
30	1.181100	75	2.952750	350	13.779500
31	1.220470	76	2.992120	360	14.173200
32	1.259840	77	3.031490	370	14.566900
33	1.299210	78	3.070860	380	14.960600
34	1.338580	79	3.110230	390	15.354300
35	1.377949	80	3.149600	400	15.748000
36	1.417319	81	3.188970	500	19.685000
37	1.456689	82	3.228340	600	23.622000
38	1.496050	83	3.267710	700	27.559000
39	1.535430	84	3.307080	800	31.496000
40	1.574800	85	3.346450	900	35.433000
41	1.614170	86	3.385820	1000	39.370000
42	1.653540	87	3.425190	2000	78.740000
43	1.692910	88	3.464560	3000	118.110000
44	1.732280	89	3.503903	4000	157.480000
45	1.771650	90	3.543300	5000	196.850000

Conversion of Common Fractions to Decimal and Millimeter Units

Common Fractions[1]	Decimal Fractions[1]	mm (approx.)	Common Fractions	Decimal Fractions	mm (appox.)
1/128	0.008	0.20	1/2	0.500	12.70
1/64	0.016	0.40	33/64	0.516	13.10
1/32	0.031	0.79	17/32	0.531	13.49
3/64	0.047	1.19	35/64	0.547	13.89
1/16	0.063	1.59	9/16	0.563	14.29
5/64	0.078	1.98	37/64	0.578	14.68
3/32	0.094	2.38	19/32	0.594	15.08
7/64	0.109	2.78	39/64	0.609	15.48
1/8	0.125	3.18	5/8	0.625	15.88
9/64	0.141	3.57	41/64	0.641	16.27
5/32	0.156	3.97	21/32	0.656	16.67
11/64	0.172	4.37	43/64	0.672	17.07
3/16	0.188	4.76	11/16	0.688	17.46
13/64	0.203	5.16	45/64	0.703	17.86
7/32	0.219	5.56	23/32	0.719	18.26
15/64	0.234	5.95	47/64	0.734	18.65
1/4	0.250	6.35	3/4	0.750	19.05
17/64	0.266	6.75	49/64	0.766	19.45
9/32	0.281	7.14	25/32	0.781	19.84
19/64	0.297	7.54	51/64	0.797	20.24
5/16	0.313	7.94	13/16	0.813	20.64
21/64	0.328	8.33	53/64	0.828	21.03
11/32	0.344	8.73	27/32	0.844	21.43
23/64	0.359	9.13	55/64	0.859	21.83
3/8	0.375	9.53	7/8	0.875	22.23
25/64	0.391	9.92	57/64	0.891	22.62
13/32	0.406	10.32	29/32	0.906	23.02
27/64	0.422	10.72	59/64	0.922	23.42
7/16	0.438	11.11	15/16	0.938	23.81
29/64	0.453	11.51	61/64	0.953	24.21
15/32	0.469	11.91	31/32	0.969	24.61
31/64	0.484	12.30	63/64	0.984	25.00

[1] Units in inches.

Conversion Ratios for Length

Known Quantity	Multiply By	Quantity to Find
inches (in)	2.54	centimeters (cm)
feet (ft)	30	centimeters (cm)
yards (yd)	0.9	meters (m)
miles (mi)	1.6	kilometers (km)
millimeters (mm)	0.04	inches (in)
centimeters (cm)	0.4	inches (in)
meters (m)	3.3	feet (ft)
meters (m)	1.1	yards (yd)
kilometers (km)	0.6	miles (mi)
centimeters (cm)	10	millimeters (mm)
decimeters (dm)	10	centimeters (cm)
decimeters (dm)	100	millimeters (mm)
meters (m)	10	decimeters (dm)
meters (m)	1000	millimeters (mm)
dekameters (dam)	10	meters (m)
hectometers (hm)	10	dekameters (dam)
hectometers (hm)	100	meters (m)
kilometers (km)	10	hectometers (hm)
kilometers (km)	1000	meters (m)

Conversion Ratios for Area

Known Quantity	Multiply By	Quantity to Find
square inches (in^2)	6.5	square centimeters (cm^2)
square feet (ft^2)	0.09	square meters (m^2)
square yards (yd^2)	0.8	square meters (m^2)
square miles (mi^2)	2.6	square kilometers (km^2)
acres	0.4	hectares (ha)
square centimeters (cm^2)	0.16	square inches (in^2)
square meters (m^2)	1.2	square yards (yd^2)
square kilometers (km^2)	0.4	square miles (mi^2)
hectares (ha)	2.5	acres
square centimeters (cm^2)	100	square millimeters (mm^2)
square meters (m^2)	10,000	square centimeters (cm^2)
square meters (m^2)	1,000,000	square millimeters (mm^2)
ares (a)	100	square meters (m^2)
hectares (ha)	100	ares (a)
hectares (ha)	10,000	square meters (m^2)
square kilometers (km^2)	100	hectares (ha)
square kilometers (km^2)	1,000	square meters (m^2)

Conversion Ratios for Mass

Known Quantity	Multiply By	Quantity to Find
ounces (oz)	28	grams (g)
pounds (lb)	0.45	kilograms (kg)
tons	0.9	tonnes (t)
grams (g)	0.035	ounces (oz)
kilograms (kg)	2.2	pounds (lb)
tonnes (t)	100	kilograms (kg)
tonnes (t)	1.1	tons
centigrams (cg)	10	milligrams (mg)
decigrams (dg)	10	centigrams (cg)
decigrams (dg)	100	milligrams (mg)
grams (g)	10	decigrams (dg)
grams (g)	1000	milligrams (mg)
dekagram (dag)	10	grams (g)
hectogram (hg)	10	dekagrams (dag)
hectogram (hg)	100	grams (g)
kilograms (kg)	10	hectograms (hg)
kilograms (kg)	1000	grams (g)
metric tons (t)	1000	kilograms (kg)

Conversion Ratios for Volume

Known Quantity	Multiply By	Quantity to Find
milliliters (mL)	0.03	fluid ounces (fl oz)
liters (L)	2.1	pints (pt)
liters (L)	1.06	quarts (qt)
liters (L)	0.26	gallons (gal)
gallons (gal)	3.8	liters (L)
quarts (qt)	0.95	liters (L)
pints (pt)	0.47	liters (L)
cups (c)	0.24	liters (L)
fluid ounces (fl oz)	30	milliliters (mL)
teaspoons (tsp)	5	milliliters (mL)
tablespoons (tbsp)	15	milliliters (mL)
liters (L)	100	milliliters (mL)

Conversion Ratios for Cubic Measure

Known Quantity	Multiply By	Quantity to Find
cubic meters (m^3)	35	cubic feet (ft^3)
cubic meters (m^3)	1.3	cubic yards (yd^3)
cubic yards (yd^3)	0.76	cubic meters (m^3)
cubic feet (ft^3)	0.028	cubic meters (m^3)
cubic centimeters (cm^3)	1000	cubic millimeters (mm^3)
cubic decimeters (dm^3)	1000	cubic centimeters (cm^3)
cubic decimeters (dm^3)	1,000,000	cubic millimeters (mm^3)
cubic meters (m^3)	1000	cubic decimeters (dm^3)
cubic meters (m^3)	1	steres
cubic feet (ft^3)	1728	cubic inches (in^3)
cubic feet (ft^3)	28.32	liters (L)
cubic inches (in^3)	16.39	cubic centimeters (cm^3)
cubic meters (m^3)	264	gallons (gal)
cubic yards (yd^3)	27	cubic feet (ft^3)
cubic yards (yd^3)	202	gallons (gal)
gallons (gal)	231	cubic inches (in^3)

Conversion Ratios for Electrical Quantities

Known Quantity	Multiply By	Quantity to Find
Btu per minute	0.024	horsepower (hp)
Btu per minute	17.57	watts (W)
horsepower (hp)	33,000	foot-pounds per min (ft-lb/min)
horsepower (hp)	746	watts (W)
kilowatts (kW)	57	Btu per minute
kilowatts (kW)	1.34	horsepower (hp)

Basic Electrical Units

Unit	Symbol	Definition
Ampere	A	One ampere is the constant current flowing in two parallel conductors one meter apart that would produce a force of 2×10^{-7} newtons per meter of length.
Coulomb	C	One coulomb is the charge that is moved in one second by a current of one ampere.
Farad	F	One farad is the capacitance of a capacitor that produces a potential of one volt between the plates when charged to one coulomb.
Henry	H	One henry is the inductance of a coil that has one volt induced in it when the current varies uniformly at one ampere per second.
Joule	J	One joule is the work done by a force of one newton acting over a distance of one meter.
Ohm	Ω	One ohm is the resistance between two points of a conductor that produces a current of one ampere when one volt is applied between these points.
Siemens	S	One siemens is the conductance between two points of a conductor that produces a current of one ampere when one volt is applied between these points. Reciprocal of the ohm. (Formerly mho).
Volt	V	One volt is the potential difference between two points in a wire carrying one ampere when the power dissipated between these points is one watt.
Watt	W	One watt is the power that produces energy at one joule per second.
Weber	Wb	One weber is the magnetic flux that produces an electromotive force of one volt in a circuit of one turn as the flux changes uniformly from maximum to zero in one second.

A

absolute delay The amount of time a signal is delayed. The delay may be expressed in time or number of pulse events.

absolute zero The lowest temperature theoretically possible, −273.16°C. *Absolute zero* is equal to zero degrees Kelvin.

absorption The transference of some or all of the energy contained in an electromagnetic wave to the substance or medium in which it is propagating or upon which it is incident.

absorption auroral The loss of energy in a radio wave passing through an area affected by solar auroral activity.

ac coupling A method of coupling one circuit to another through a capacitor or transformer so as to transmit the varying (ac) characteristics of the signal while blocking the static (dc) characteristics.

ac/dc coupling Coupling between circuits that accommodates the passing of both ac and dc signals (may also be referred to as simply dc coupling).

accelerated life test A special form of reliability testing performed by an equipment manufacturer. The unit under test is subjected to stresses that exceed those typically experienced in normal operation. The goal of an *accelerated life test* is to improve the reliability of products shipped by forcing latent failures in components to become evident before the unit leaves the factory.

accelerating electrode The electrode that causes electrons emitted from an electron gun to accelerate in their journey to the screen of a cathode ray tube.

accelerating voltage The voltage applied to an electrode that accelerates a beam of electrons or other charged particles.

acceptable reliability level The maximum number of failures allowed per thousand operating hours of a given component or system.

acceptance test The process of testing newly purchased equipment to ensure that it is fully compliant with contractual specifications.

access The point at which entry is gained to a circuit or facility.

acquisition time In a communication system, the amount of time required to attain synchronism.

active Any device or circuit that introduces gain or uses a source of energy other than that inherent in the signal to perform its function.

adapter A fitting or electrical connector that links equipment that cannot be connected directly.

adaptive A device able to adjust or react to a condition or application, as an *adaptive circuit*. This term usually refers to filter circuits.

adaptive system A general name for a system that is capable of reconfiguring itself to meet new requirements.

adder A device whose output represents the sum of its inputs.

adjacent channel interference Interference to communications caused by a transmitter operating on an adjacent radio channel. The sidebands of the transmitter mix with the carrier being received on the desired channel, resulting in noise.

admittance A measure of how well alternating current flows in a conductor. It is the reciprocal of *impedance* and is expressed in *siemens*. The real part of admittance is *conductance*; the imaginary part is *susceptance*.

AFC (automatic frequency control) A circuit that automatically keeps an oscillator on frequency by comparing the output of the oscillator with a standard frequency source or signal.

air core An inductor with no magnetic material in its core.

algorithm A prescribed finite set of well-defined rules or processes for the solution of a problem in a finite number of steps.

alignment The adjustment of circuit components so that an entire system meets minimum performance values. For example, the stages in a radio are aligned to ensure proper reception.

allocation The planned use of certain facilities and equipment to meet current, pending, and/or forecasted circuit- and carrier-system requirements.

alternating current (ac) A continuously variable current, rising to a maximum in one direction, falling to zero, then reversing direction and rising to a maximum in the other direction, then falling to zero and repeating the cycle. Alternating current usually follows a sinusoidal growth and decay curve. Note that the correct usage of the term *ac* is lower case.

alternator A generator that produces alternating current electric power.

ambient electromagnetic environment The radiated or conducted electromagnetic signals and noise at a specific location and time.

ambient level The magnitude of radiated or conducted electromagnetic signals and noise at a specific test location when equipment-under-test is not powered.

ambient temperature The temperature of the surrounding medium, typically air, that comes into contact with an apparatus. Ambient temperature may also refer simply to room temperature.

American National Standards Institute (ANSI) A nonprofit organization that coordinates voluntary standards activities in the U.S.

American Wire Gauge (AWG) The standard American method of classifying wire diameter.

ammeter An instrument that measures and records the amount of current in amperes flowing in a circuit.

amp (A) An abbreviation of the term *ampere*.

ampacity A measure of the current carrying capacity of a power cable. *Ampacity* is determined by the maximum continuous-performance temperature of the insulation, by the heat generated in the cable (as a result of conductor and insulation losses), and by the heat-dissipating properties of the cable and its environment.

ampere (amp) The standard unit of electric current.

ampere per meter The standard unit of magnetic field strength.

ampere-hour The energy that is consumed when a current of one ampere flows for a period of one hour.

ampere-turns The product of the number of turns of a coil and the current in amperes flowing through the coil.

amplification The process that results when the output of a circuit is an enlarged reproduction of the input signal. Amplifiers may be designed to provide amplification of voltage, current, or power, or a combination of these quantities.

amplification factor In a vacuum tube, the ratio of the change in plate voltage to the change in grid voltage that causes a corresponding change in plate current. Amplification factor is expressed by the Greek letter μ (*mu*).

amplifier (1—general) A device that receives an input signal and provides as an output a magnified replica of the input waveform. **(2—audio)** An amplifier designed to cover the normal audio frequency range (20 Hz to 20 kHz). **(3—balanced)** A circuit with two identical connected signal branches that operate in phase opposition, with input and output connections each balanced to ground. **(4—bridging)** An amplifying circuit featuring high input impedance to prevent loading of the source. **(5—broadband)** An amplifier capable of operating over a specified broad band of frequencies with acceptably small amplitude variations as a function of frequency. **(6—buffer)** An amplifier stage used to isolate a frequency-sensitive circuit from variations in the load presented by following stages. **(7—linear)** An amplifier in which the instantaneous output signal is a linear function of the corresponding input signal. **(8—magnetic)** An amplifier incorporating a control device dependent on magnetic saturation. A small dc signal applied to a control circuit triggers a large change in operating impedance and, hence, in the output of the circuit. **(9—microphone)** A circuit that amplifies the low level output from a microphone to make it sufficient to be used as an input signal to a power amplifier or another stage in a modulation circuit. Such a circuit is commonly known as a *preamplifier*. **(10—push-pull)** A balanced amplifier with two similar amplifying units connected in phase opposition in order to cancel undesired harmonics and minimize distortion. **(11—tuned radio frequency)** An amplifier tuned to a particular radio frequency or band so that only selected frequencies are amplified.

amplifier operating class (1—general) The operating point of an amplifying stage. The operating point, termed the operating *class*, determines the period during which current flows in the output. **(2—class A)** An amplifier in which output current flows during the whole of the input current cycle. **(3—class AB)** An amplifier in which the output current flows for more than half but less than the whole of the input cycle. **(4—class B)** An amplifier in which output current is cut off at zero input signal; a half-wave rectified output is produced. **(5—class C)** An amplifier in which output current flows for less than half the input cycle. **(6—class D)** An amplifier operating in a pulse-only mode.

amplitude The magnitude of a signal in voltage or current, frequently expressed in terms of *peak*, *peak-to-peak*, or *root-mean-square* (RMS). The actual amplitude of a quantity at a particular instant often varies in a sinusoidal manner.

amplitude distortion A distortion mechanism occurring in an amplifier or other device when the output amplitude is not a linear function of the input amplitude under specified conditions.

amplitude equalizer A corrective network that is designed to modify the amplitude characteristics of a circuit or system over a desired frequency range.

amplitude-versus-frequency distortion The distortion in a transmission system caused by the nonuniform attenuation or gain of the system with respect to frequency under specified conditions.

analog carrier system A carrier system whose signal amplitude, frequency, or phase is varied continuously as a function of a modulating input.

anode (1 — general) A positive pole or element. **(2—vacuum tube)** The outermost positive element in a vacuum tube, also called the *plate*. **(3—battery)** The positive element of a battery or cell.

anodize The formation of a thin film of oxide on a metallic surface, usually to produce an insulating layer.

antenna (1—general) A device used to transmit or receive a radio signal. An antenna is usually designed for a specified frequency range and serves to couple electromagnetic energy from a transmission line to and/or from the free space through which it travels. Directional antennas concentrate the energy in a particular horizontal or vertical direction. **(2—aperiodic)** An antenna that is not periodic or resonant at particular frequencies, and so can be used over a wide band of frequencies. **(3—artificial)** A device that behaves, so far as the transmitter is concerned, like a proper antenna, but does not radiate any power at radio frequencies. **(4—broadband)** An antenna that operates within specified performance limits over a wide band of frequencies, without requiring retuning for each individual frequency. **(5—Cassegrain)** A double reflecting antenna, often used for ground stations in satellite systems. **(6—coaxial)** A dipole antenna made by folding back on itself a quarter wavelength of the outer conductor of a coaxial line, leaving a quarter wavelength of the inner conductor exposed. **(7—corner)** An antenna within the angle formed by two plane-reflecting surfaces. **(8—dipole)** A center-fed antenna, one half-wavelength long. **(9—directional)** An antenna designed to receive or emit radiation more efficiently in a particular direction. **(10—dummy)** An artificial antenna, designed to accept power from the transmitter but not to radiate it. **(11—ferrite)** A common AM broadcast receive antenna that uses a small coil mounted on a short rod of ferrite material. **(12—flat top)** An antenna in which all the horizontal components are in the same horizontal plane. **(13—folded dipole)** A radiating device consisting of two ordinary half-wave dipoles joined at their outer ends and fed at the center of one of the dipoles. **(14—horn reflector)** A radiator in which the feed horn extends into a parabolic reflector, and the power is radiated through a window in the horn. **(15—isotropic)** A theoretical antenna in free space that transmits or receives with the same efficiency in all directions. **(16—log-periodic)** A broadband directional antenna incorporating an array of dipoles of different lengths, the length and spacing between dipoles increasing logarithmically away from the feeder element. **(17—long wire)** An antenna made up of one or more conductors in a straight line pointing in the required direction with a total length of several wavelengths at the operating frequency. **(18—loop)** An antenna consisting of one or more turns of wire in the same or parallel planes. **(19—nested rhombic)** An assembly of two rhombic antennas, one smaller than the other, so that the complete diamond-shaped antenna fits inside the area occupied by the larger unit. **(20—omnidirectional)** An antenna whose radiating or receiving properties are the same in all horizontal plane directions. **(21—periodic)** A resonant antenna designed for use at a particular frequency. **(22—quarter-wave)** A dipole antenna whose length is equal to one quarter of a wavelength at the operating frequency. **(23—rhombic)** A large diamond-shaped antenna, with sides of the diamond several wavelengths long. The rhombic antenna is fed at one of the corners, with directional efficiency in the direction of the diagonal. (**24—series fed**) A vertical antenna that is fed at its lower end. (**25—shunt fed**) A vertical antenna whose base is grounded, and is fed at a specified point above ground. The point at which the antenna is fed above ground determines the operating impedance. (**26—steerable**) An antenna so constructed that its major lobe may readily be changed in direction. (**27—top-loaded**) A vertical antenna capacitively loaded at its upper end, often by simple enlargement or the attachment of a disc or plate. (**28—turnstile**) An antenna with one or more tiers of horizontal dipoles, crossed at right angles to each other and with excitation of the dipoles in phase quadrature. (**29—whip**) An antenna constructed of a thin semiflexible metal rod or tube, fed at its base. (**30—Yagi**) A directional antenna constructed of a series of dipoles cut to specific lengths. *Director* elements are placed in front of the active dipole and *reflector* elements are placed behind the active element.

antenna array A group of several antennas coupled together to yield a required degree of directivity.

antenna beamwidth The angle between the *half-power* points (3 dB points) of the main lobe of the antenna pattern when referenced to the peak power point of the antenna pattern. *Antenna beamwidth* is measured in degrees and normally refers to the horizontal radiation pattern.

antenna directivity factor The ratio of the power flux density in the desired direction to the average value of power flux density at crests in the antenna directivity pattern in the interference section.

antenna factor A factor that, when applied to the voltage appearing at the terminals of measurement equipment, yields the electrical field strength at an antenna. The unit of antenna factor is volts per meter per measured volt.

antenna gain The ratio of the power required at the input of a theoretically perfect omnidirectional reference antenna to the power supplied to the input of the given antenna to produce the same field at the same distance. When not specified otherwise, the figure expressing the gain of an antenna refers to the gain in the direction of the radiation main lobe. In services using *scattering* modes of propagation, the full gain of an antenna may not be realizable in practice and the apparent gain may vary with time.

antenna gain-to-noise temperature For a satellite earth terminal receiving system, a figure of merit that equals G/T, where G is the gain in dB of the earth terminal antenna at the receive frequency, and T is the equivalent noise temperature of the receiving system in Kelvins.

antenna matching The process of adjusting an antenna matching circuit (or the antenna itself) so that the input impedance of the antenna is equal to the characteristic impedance of the transmission line.

antenna monitor A device used to measure the ratio and phase between the currents flowing in the towers of a directional AM broadcast station.

antenna noise temperature The temperature of a resistor having an available noise power per unit bandwidth equal to that at the antenna output at a specified frequency.

antenna pattern A diagram showing the efficiency of radiation in all directions from the antenna.

antenna power rating The maximum continuous-wave power that can be applied to an antenna without degrading its performance.

antenna preamplifier A small amplifier, usually mast-mounted, for amplifying weak signals to a level sufficient to compensate for down-lead losses.

apparent power The product of the root-mean-square values of the voltage and current in an alternating-current circuit without a correction for the phase difference between the voltage and current.

arc A sustained luminous discharge between two or more electrodes.

arithmetic mean The sum of the values of several quantities divided by the number of quantities, also referred to as the *average*.

armature winding The winding of an electrical machine, either a motor or generator, in which current is induced.

array (1—antenna) An assembly of several directional antennas so placed and interconnected that directivity may be enhanced. **(2—broadside)** An antenna array whose elements are all in the same plane, producing a major lobe perpendicular to the plane. **(3—colinear)** An antenna array whose elements are in the same line, either horizontal or vertical. **(4—end-fire)** An antenna array whose elements are in parallel rows, one behind the other, producing a major lobe perpendicular to the plane in which individual elements are placed. **(5—linear)** An antenna array whose elements are arranged end-to-end. **(6—stacked)** An antenna array whose elements are stacked, one above the other.

artificial line An assembly of resistors, inductors, and capacitors that simulates the electrical characteristics of a transmission line.

assembly A manufactured part made by combining several other parts or subassemblies.

assumed values A range of values, parameters, levels, and other elements assumed for a mathematical model, hypothetical circuit, or network, from which analysis, additional estimates, or calculations will be made. The range of values, while not measured, represents the best engineering judgment and is generally derived from values found or measured in real circuits or networks of the same generic type, and includes projected improvements.

atmosphere The gaseous envelope surrounding the earth, composed largely of oxygen, carbon dioxide, and water vapor. The atmosphere is divided into four primary layers: *troposphere, stratosphere, ionosphere,* and *exosphere.*

atmospheric noise Radio noise caused by natural atmospheric processes, such as lightning.

attack time The time interval in seconds required for a device to respond to a control stimulus.

attenuation The decrease in amplitude of an electrical signal traveling through a transmission medium caused by dielectric and conductor losses.

attenuation coefficient The rate of decrease in the amplitude of an electrical signal caused by attenuation. The *attenuation coefficient* can be expressed in decibels or nepers per unit length. It may also be referred to as the *attenuation constant.*

attenuation distortion The distortion caused by attenuation that varies over the frequency range of a signal.

attenuation-limited operation The condition prevailing when the received signal amplitude (rather than distortion) limits overall system performance.

attenuator A fixed or adjustable component that reduces the amplitude of an electrical signal without causing distortion.

atto A prefix meaning one *quintillionth.*

attraction The attractive force between two unlike magnetic poles (N/S) or electrically charged bodies (+/-).

attributes The characteristics of equipment that aid planning and circuit design.

automatic frequency control (AFC) A system designed to maintain the correct operating frequency of a receiver. Any drift in tuning results in the production of a control voltage, which is used to adjust the frequency of a local oscillator so as to minimize the tuning error.

automatic gain control (AGC) An electronic circuit that compares the level of an incoming signal with a previously defined standard and automatically amplifies or attenuates the signal so it arrives at its destination at the correct level.

autotransformer A transformer in which both the primary and secondary currents flow through one common part of the coil.

auxiliary power An alternate source of electric power, serving as a back-up for the primary utility company ac power.

availability A measure of the degree to which a system, subsystem, or equipment is operable and not in a stage of congestion or failure at any given point in time.

avalanche effect The effect obtained when the electric field across a barrier region is sufficiently strong for electrons to collide with *valence electrons*, thereby releasing more electrons and giving a cumulative multiplication effect in a semiconductor.

average life The mean value for a normal distribution of product or component lives, generally applied to mechanical failures resulting from "wear-out."

B

back emf A voltage induced in the reverse direction when current flows through an inductance. *Back emf* is also known as *counter-emf*.

back scattering A form of wave scattering in which at least one component of the scattered wave is deflected opposite to the direction of propagation of the incident wave.

background noise The total system noise in the absence of information transmission, independent of the presence or absence of a signal.

backscatter The deflection or reflection of radiant energy through angles greater than 90° with respect to the original angle of travel.

backscatter range The maximum distance from which backscattered radiant energy can be measured.

backup A circuit element or facility used to replace an element that has failed.

backup supply A redundant power supply that takes over if the primary power supply fails.

balance The process of equalizing the voltage, current, or other parameter between two or more circuits or systems.

balanced A circuit having two sides (conductors) carrying voltages that are symmetrical about a common reference point, typically ground.

balanced circuit A circuit whose two sides are electrically equal in all transmission respects.

balanced line A transmission line consisting of two conductors in the presence of ground capable of being operated in such a way that when the voltages of the two conductors at all transverse planes are equal in magnitude and opposite in polarity with respect to ground, the currents in the two conductors are equal in magnitude and opposite in direction.

balanced modulator A modulator that combines the information signal and the carrier so that the output contains the two sidebands without the carrier.

balanced three-wire system A power distribution system using three conductors, one of which is balanced to have a potential midway between the potentials of the other two.

balanced-to-ground The condition when the impedance to ground on one wire of a two-wire circuit is equal to the impedance to ground on the other wire.

balun (balanced/unbalanced) A device used to connect balanced circuits with unbalanced circuits.

band A range of frequencies between a specified upper and lower limit.

band elimination filter A filter having a single continuous attenuation band, with neither the upper nor lower cut-off frequencies being zero or infinite. A *band elimination filter* may also be referred to as a *band-stop*, *notch*, or *band reject* filter.

bandpass filter A filter having a single continuous transmission band with neither the upper nor the lower cut-off frequencies being zero or infinite. A bandpass filter permits only a specific band of frequencies to pass; frequencies above or below are attenuated.

bandwidth The range of signal frequencies that can be transmitted by a communications channel with a defined maximum loss or distortion. Bandwidth indicates the information-carrying capacity of a channel.

bandwidth expansion ratio The ratio of the necessary bandwidth to the baseband bandwidth.

bandwidth-limited operation The condition prevailing when the frequency spectrum or bandwidth, rather than the amplitude (or power) of the signal, is the limiting factor in communication capability. This condition is reached when the system distorts the shape of the waveform beyond tolerable limits.

bank A group of similar items connected together in a specified manner and used in conjunction with one another.

bare A wire conductor that is not enameled or enclosed in an insulating sheath.

baseband The band of frequencies occupied by a signal before it modulates a carrier wave to form a transmitted radio or line signal.

baseband channel A channel that carries a signal without modulation, in contrast to a *passband* channel.

baseband signal The original form of a signal, unchanged by modulation.

bath tub The shape of a typical graph of component failure rates: high during an initial period of operation, falling to an acceptable low level during the normal usage period, and then rising again as the components become time-expired.

battery A group of several cells connected together to furnish current by conversion of chemical, thermal, solar, or nuclear energy into electrical energy. A single cell is itself sometimes also called a battery.

bay A row or suite of racks on which transmission, switching, and/or processing equipment is mounted.

Bel A unit of power measurement, named in honor of Alexander Graham Bell. The commonly used unit is one tenth of a Bel, or a decibel (dB). One Bel is defined as a tenfold increase in power. If an amplifier increases the power of a signal by 10 times, the power gain of the amplifier is equal to 1 Bel or 10 *decibels* (dB). If power is increased by 100 times, the power gain is 2 Bels or 20 decibels.

bend A transition component between two elements of a transmission waveguide.

bending radius The smallest bend that may be put into a cable under a stated pulling force. The bending radius is typically expressed in inches.

bias A dc voltage difference applied between two elements of an active electronic device, such as a vacuum tube, transistor, or integrated circuit. Bias currents may or may not be drawn, depending on the device and circuit type.

bidirectional An operational qualification which implies that the transmission of information occurs in both directions.

bifilar winding A type of winding in which two insulated wires are placed side by side. In some components, bifilar winding is used to produce balanced circuits.

bipolar A signal that contains both positive-going and negative-going amplitude components. A bipolar signal may also contain a zero amplitude state.

bleeder A high resistance connected in parallel with one or more filter capacitors in a high voltage dc system. If the power supply load is disconnected, the capacitors discharge through the bleeder.

block diagram An overview diagram that uses geometric figures to represent the principal divisions or sections of a circuit, and lines and arrows to show the path of a signal, or to show program functionalities. It is not a *schematic*, which provides greater detail.

blocking capacitor A capacitor included in a circuit to stop the passage of direct current.

BNC An abbreviation for *bayonet Neill-Concelman*, a type of cable connector used extensively in RF applications (named for its inventor).

Boltzmann's constant 1.38×10^{-23} joules.

bridge A type of network circuit used to match different circuits to each other, ensuring minimum transmission impairment.

bridging The shunting or paralleling of one circuit with another.

broadband The quality of a communications link having essentially uniform response over a given range of frequencies. A communications link is said to be *broadband* if it offers no perceptible degradation to the signal being transported.

buffer A circuit or component that isolates one electrical circuit from another.

burn-in The operation of a device, sometimes under extreme conditions, to stabilize its characteristics and identify latent component failures before bringing the device into normal service.

bus A central conductor for the primary signal path. The term bus may also refer to a signal path to which a number of inputs may be connected for feed to one or more outputs.

busbar A main dc power bus.

bypass capacitor A capacitor that provides a signal path that effectively shunts or bypasses other components.

bypass relay A switch used to bypass the normal electrical route of a signal or current in the event of power, signal, or equipment failure.

C

cable An electrically and/or optically conductive interconnecting device.

cable loss Signal loss caused by passing a signal through a coaxial cable. Losses are the result of resistance, capacitance, and inductance in the cable.

cable splice The connection of two pieces of cable by joining them mechanically and closing the joint with a weather-tight case or sleeve.

cabling The wiring used to interconnect electronic equipment.

calibrate The process of checking, and adjusting if necessary, a test instrument against one known to be set correctly.

calibration The process of identifying and measuring errors in instruments and/or procedures.

capacitance The property of a device or component that enables it to store energy in an electrostatic field and to release it later. A capacitor consists of two conductors separated by an insulating material. When the conductors have a voltage difference between them, a charge will be stored in the electrostatic field between the conductors.

capacitor A device that stores electrical energy. A capacitor allows the apparent flow of alternating current, while blocking the flow of direct current. The degree to which the device permits ac current flow depends on the frequency of the signal and the size of the capacitor. Capacitors are used in filters, delay-line components, couplers, frequency selectors, timing elements, voltage transient suppression, and other applications.

carrier A single frequency wave that, prior to transmission, is modulated by another wave containing information. A carrier may be modulated by manipulating its amplitude and/or frequency in direct relation to one or more applied signals.

carrier frequency The frequency of an unmodulated oscillator or transmitter. Also, the average frequency of a transmitter when a signal is frequency modulated by a symmetrical signal.

cascade connection A tandem arrangement of two or more similar component devices or circuits, with the output of one connected to the input of the next.

cascaded An arrangement of two or more circuits in which the output of one circuit is connected to the input of the next circuit.

cathode ray tube (CRT) A vacuum tube device, usually glass, that is narrow at one end and widens at the other to create a surface onto which images can be projected. The narrow end contains the necessary circuits to generate and focus an electron beam on the luminescent screen at the other end. CRTs are used to display pictures in TV receivers, video monitors, oscilloscopes, computers, and other systems.

cell An elementary unit of communication, of power supply, or of equipment.

Celsius A temperature measurement scale, expressed in degrees C, in which water freezes at 0°C and boils at 100°C. To convert to degrees Fahrenheit, multiply by 0.555 and add 32. To convert to Kelvins add 273 (approximately).

center frequency In frequency modulation, the resting frequency or initial frequency of the carrier before modulation.

center tap A connection made at the electrical center of a coil.

channel The smallest subdivision of a circuit that provides a single type of communication service.

channel decoder A device that converts an incoming modulated signal on a given channel back into the source-encoded signal.

channel encoder A device that takes a given signal and converts it into a form suitable for transmission over the communications channel.

channel noise level The ratio of the channel noise at any point in a transmission system to some arbitrary amount of circuit noise chosen as a reference. This ratio is usually expressed in *decibels above reference noise*, abbreviated *dBrn*.

channel reliability The percent of time a channel is available for use in a specific direction during a specified period.

channelization The allocation of communication circuits to channels and the forming of these channels into groups for higher order multiplexing.

characteristic The property of a circuit or component.

characteristic impedance The impedance of a transmission line, as measured at the driving point, if the line were of infinite length. In such a line, there would be no standing waves. The *characteristic impedance* may also be referred to as the *surge impedance*.

charge The process of replenishing or replacing the electrical charge in a secondary cell or storage battery.

charger A device used to recharge a battery. Types of charging include: (1) *constant voltage charge*, (2) *equalizing charge*, and (3) *trickle charge*.

chassis ground A connection to the metal frame of an electronic system that holds the components in a place. The chassis ground connection serves as the ground return or electrical common for the system.

circuit Any closed path through which an electrical current can flow. In a *parallel circuit*, components are connected between common inputs and outputs such that all paths are parallel to each other. The same voltage appears across all paths. In a *series circuit*, the same current flows through all components.

circuit noise level The ratio of the circuit noise at some given point in a transmission system to an established reference, usually expressed in decibels above the reference.

circuit reliability The percentage of time a circuit is available to the user during a specified period of scheduled availability.

circular mil The measurement unit of the cross-sectional area of a circular conductor. A *circular mil* is the area of a circle whose diameter is one mil, or 0.001 inch.

clear channel A transmission path wherein the full bandwidth is available to the user, with no portions of the channel used for control, framing, or signaling. Can also refer to a classification of AM broadcast station.

clipper A limiting circuit which ensures that a specified output level is not exceeded by restricting the output waveform to a maximum peak amplitude.

clipping The distortion of a signal caused by removing a portion of the waveform through restriction of the amplitude of the signal by a circuit or device.

coax A short-hand expression for *coaxial cable*, which is used to transport high-frequency signals.

coaxial cable A transmission line consisting of an inner conductor surrounded first by an insulating material and then by an outer conductor, either solid or braided. The mechanical dimensions of the cable determine its *characteristic impedance*.

coherence The correlation between the phases of two or more waves.

coherent The condition characterized by a fixed phase relationship among points on an electromagnetic wave.

coherent pulse The condition in which a fixed phase relationship is maintained between consecutive pulses during pulse transmission.

cold joint A soldered connection that was inadequately heated, with the result that the wire is held in place by rosin flux, not solder. A cold joint is sometimes referred to as a *dry joint*.

comb filter An electrical filter circuit that passes a series of frequencies and rejects the frequencies in between, producing a frequency response similar to the teeth of a comb.

common A point that acts as a reference for circuits, often equal in potential to the local ground.

common mode Signals identical with respect to amplitude, frequency, and phase that are applied to both terminals of a cable and/or both the input and reference of an amplifier.

common return A return path that is common to two or more circuits, and returns currents to their source or to ground.

common return offset The dc common return potential difference of a line.

communications system A collection of individual communications networks, transmission systems, relay stations, tributary stations, and terminal equipment capable of interconnection and interoperation to form an integral whole. The individual components must serve a common purpose, be technically compatible, employ common procedures, respond to some form of control, and, in general, operate in unison.

commutation A successive switching process carried out by a commutator.

commutator A circular assembly of contacts, insulated one from another, each leading to a different portion of the circuit or machine.

compatibility The ability of diverse systems to exchange necessary information at appropriate levels of command directly and in usable form. Communications equipment items are compatible if signals can be exchanged between them without the addition of buffering or translation for the specific purpose of achieving workable interface connections, and if the equipment or systems being interconnected possess comparable performance characteristics, including the suppression of undesired radiation.

complex wave A waveform consisting of two or more sinewave components. At any instant of time, a complex wave is the algebraic sum of all its sinewave components.

compliance For mechanical systems, a property which is the reciprocal of stiffness.

component An assembly, or part thereof, that is essential to the operation of some larger circuit or system. A *component* is an immediate subdivision of the assembly to which it belongs.

COMSAT The *Communications Satellite Corporation*, an organization established by an act of Congress in 1962. COMSAT launches and operates the international satellites for the INTELSAT consortium of countries.

concentricity A measure of the deviation of the center conductor position relative to its ideal location in the exact center of the dielectric cross-section of a coaxial cable.

conditioning The adjustment of a channel in order to provide the appropriate transmission characteristics needed for data or other special services.

conditioning equipment The equipment used to match transmission levels and impedances, and to provide equalization between facilities.

conductance A measure of the capability of a material to conduct electricity. It is the reciprocal of *resistance* (ohm) and is expressed in *siemens*. (Formerly expressed as *mho*.)

conducted emission An electromagnetic energy propagated along a conductor.

conduction The transfer of energy through a medium, such as the conduction of electricity by a wire, or of heat by a metallic frame.

conduction band A partially filled or empty atomic energy band in which electrons are free to move easily, allowing the material to carry an electric current.

conductivity The conductance per unit length.

conductor Any material that is capable of carrying an electric current.

configuration A relative arrangement of parts.

connection A point at which a junction of two or more conductors is made.

connector A device mounted on the end of a wire or fiber optic cable that mates to a similar device on a specific piece of equipment or another cable.

constant-current source A source with infinitely high output impedance so that output current is independent of voltage, for a specified range of output voltages.

constant-voltage charge A method of charging a secondary cell or storage battery during which the terminal voltage is kept at a constant value.

constant-voltage source A source with low, ideally zero, internal impedance, so that voltage will remain constant, independent of current supplied.

contact The points that are brought together or separated to complete or break an electrical circuit.

contact bounce The rebound of a contact, which temporarily opens the circuit after its initial *make*.

contact form The configuration of a contact assembly on a relay. Many different configurations are possible from simple *single-make* contacts to complex arrangements involving *breaks* and *makes*.

contact noise A noise resulting from current flow through an electrical contact that has a rapidly varying resistance, as when the contacts are corroded or dirty.

contact resistance The resistance at the surface when two conductors make contact.

continuity A continuous path for the flow of current in an electrical circuit.

continuous wave An electromagnetic signal in which successive oscillations of the waves are identical.

control The supervision that an operator or device exercises over a circuit or system.

control grid The grid in an electron tube that controls the flow of current from the cathode to the anode.

convention A generally acceptable symbol, sign, or practice in a given industry.

Coordinated Universal Time (UTC) The time scale, maintained by the BIH (Bureau International de l'Heure) that forms the basis of a coordinated dissemination of standard frequencies and time signals.

copper loss The loss resulting from the heating effect of current.

corona A bluish luminous discharge resulting from ionization of the air near a conductor carrying a voltage gradient above a certain *critical level*.

corrective maintenance The necessary tests, measurements, and adjustments required to remove or correct a fault.

cosmic noise The random noise originating outside the earth's atmosphere.

coulomb The standard unit of electric quantity or charge. One *coulomb* is equal to the quantity of electricity transported in 1 second by a current of 1 ampere.

Coulomb's Law The attraction and repulsion of electric charges act on a line between them. The charges are inversely proportional to the square of the distance between them, and proportional to the product of their magnitudes. (Named for the French physicist Charles-Augustine de Coulomb, 1736-1806.)

counter-electromotive force The effective electromotive force within a system that opposes the passage of current in a specified direction.

couple The process of linking two circuits by inductance, so that energy is transferred from one circuit to another.

coupled mode The selection of either ac or dc coupling.

coupling The relationship between two components that enables the transfer of energy between them. Included are *direct coupling* through a direct electrical connection, such as a wire; *capacitive coupling* through the capacitance formed by two adjacent conductors; and *inductive coupling* in which energy is transferred through a magnetic field. Capacitive coupling is also called *electrostatic coupling*. Inductive coupling is often referred to as *electromagnetic coupling*.

coupling coefficient A measure of the electrical coupling that exists between two circuits. The *coupling coefficient* is equal to the ratio of the mutual impedance to the square root of the product of the self impedances of the coupled circuits.

cross coupling The coupling of a signal from one channel, circuit, or conductor to another, where it becomes an undesired signal.

crossover distortion A distortion that results in an amplifier when an irregularity is introduced into the signal as it crosses through a zero reference point. If an amplifier is properly designed and biased, the upper half cycle and lower half cycle of the signal coincide at the zero crossover reference.

crossover frequency The frequency at which output signals pass from one channel to the other in a *crossover network*. At the *crossover frequency* itself, the outputs to each side are equal.

crossover network A type of filter that divides an incoming signal into two or more outputs, with higher frequencies directed to one output, and lower frequencies to another.

crosstalk Undesired transmission of signals from one circuit into another circuit in the same system. Crosstalk is usually caused by unintentional capacitive (ac) coupling.

crosstalk coupling The ratio of the power in a disturbing circuit to the induced power in the disturbed circuit, observed at a particular point under specified conditions. Crosstalk coupling is typically expressed in dB.

crowbar A short-circuit or low resistance path placed across the input to a circuit, usually for protective purposes..

CRT (cathode ray tube) A vacuum tube device that produces light when energized by the electron beam generated inside the tube. A CRT includes an electron gun, deflection mechanism, and phosphor-covered faceplate.

crystal A solidified form of a substance that has atoms and molecules arranged in a symmetrical pattern.

crystal filter A filter that uses piezoelectric crystals to create resonant or antiresonant circuits.

crystal oscillator An oscillator using a piezoelectric crystal as the tuned circuit that controls the resonant frequency.

crystal-controlled oscillator An oscillator in which a piezoelectric-effect crystal is coupled to a tuned oscillator circuit in such a way that the crystal pulls the oscillator frequency to its own natural frequency and does not allow frequency drift.

current (1—general) A general term for the transfer of electricity, or the movement of electrons or *holes*. **(2—alternating)** An electric current that is constantly varying in amplitude and periodically reversing direction. **(3—average)** The arithmetic mean of the instantaneous values of current, averaged over one complete half cycle. **(4—charging)** The current that flows in to charge a capacitor when it is first connected to a source of electric potential. **(5—direct)** Electric current that flows in one direction only. **(6—eddy)** A wasteful current that flows in the core of a transformer and produces heat. *Eddy currents* are largely eliminated through the use of laminated cores. **(7—effective)** The ac current that will produce the same effective heat in a resistor as is produced by dc. If the ac is sinusoidal, the *effective current* value is 0.707 times the peak ac value. **(8—fault)** The current that flows between conductors or to ground during a fault condition. **(9—ground fault)** A fault current that flows to ground. **(10—ground return)** A current that returns through the earth. **(11—lagging)** A phenomenon observed in an inductive circuit where alternating current lags behind the voltage that produces it. **(12—leading)** A phenomenon observed in a capacitive circuit where alternating current leads the voltage that produces it. **(13—magnetizing)** The current in a transformer primary winding that is just sufficient to magnetize the core and offset iron losses. **(14—neutral)** The current that flows in the neutral conductor of an unbalanced polyphase power circuit. If correctly balanced, the neutral would carry no net current. **(15—peak)** The maximum value reached by a varying current during one cycle. **(16—pick-up)** The minimum current at which a relay just begins to operate. **(17—plate)** The anode current of an electron tube. **(18—residual)** The vector sum of the currents in the phase wires of an unbalanced polyphase power circuit. **(19—space)** The total current flowing through an electron tube.

current amplifier A low output impedance amplifier capable of providing high current output.

current probe A sensor, clamped around an electrical conductor, in which an induced current is developed from the magnetic field surrounding the conductor. For measurements, the current probe is connected to a suitable test instrument.

current transformer A transformer-type of instrument in which the primary carries the current to be measured and the secondary is in series with a low current ammeter. A current transformer is used to measure high values of alternating current.

current-carrying capacity A measure of the maximum current that can be carried continuously without damage to components or devices in a circuit.

cut-off frequency The frequency above or below which the output current in a circuit is reduced to a specified level.

cycle The interval of time or space required for a periodic signal to complete one period.

cycles per second The standard unit of frequency, expressed in Hertz (one cycle per second).

D

damped oscillation An oscillation exhibiting a progressive diminution of amplitude with time.

damping The dissipation and resultant reduction of any type of energy, such as electromagnetic waves.

dB (decibel) A measure of voltage, current, or power gain equal to 0.1 Bel. Decibels are given by the equations

$$20 \log \frac{V_{out}}{V_{in}}, \ 20 \log \frac{I_{out}}{I_{in}}, \ \text{or } 10 \log \frac{P_{out}}{P_{in}}.$$

dBk A measure of power relative to 1 kilowatt. 0 dBk equals 1 kW.

dBm (decibels above 1 milliwatt) A logarithmic measure of power with respect to a reference power of one milliwatt.

dBmv A measure of voltage gain relative to 1 millivolt at 75 ohms.

dBr The power difference expressed in dB between any point and a reference point selected as the *zero relative transmission level* point. A power expressed in *dBr* does not specify the absolute power; it is a relative measurement only.

dBu A term that reflects comparison between a measured value of voltage and a reference value of 0.775 V, expressed under conditions in which the impedance at the point of measurement (and of the reference source) are not considered.

dbV A measure of voltage gain relative to 1 V.

dBW A measure of power relative to 1 watt. 0 dBW equals 1 W.

dc An abbreviation for *direct current*. Note that the preferred usage of the term *dc* is lower case.

dc amplifier A circuit capable of amplifying dc and slowly varying alternating current signals.

dc component The portion of a signal that consists of direct current. This term may also refer to the average value of a signal.

dc coupled A connection configured so that both the signal (ac component) and the constant voltage on which it is riding (dc component) are passed from one stage to the next.

dc coupling A method of coupling one circuit to another so as to transmit the static (dc) characteristics of the signal as well as the varying (ac) characteristics. Any dc offset present on the input signal is maintained and will be present in the output.

dc offset The amount that the dc component of a given signal has shifted from its correct level.

dc signal bounce Overshoot of the proper dc voltage level resulting from multiple ac couplings in a signal path.

de-energized A system from which sources of power have been disconnected.

deca A prefix meaning *ten*.

decay The reduction in amplitude of a signal on an exponential basis.

decay time The time required for a signal to fall to a certain fraction of its original value.

decibel (dB) One tenth of a Bel. The decibel is a logarithmic measure of the ratio between two powers.

decode The process of recovering information from a signal into which the information has been encoded.

decoder A device capable of deciphering encoded signals. A decoder interprets input instructions and initiates the appropriate control operations as a result.

decoupling The reduction or removal of undesired coupling between two circuits or stages.

deemphasis The reduction of the high-frequency components of a received signal to reverse the preemphasis that was placed on them to overcome attenuation and noise in the transmission process.

defect An error made during initial planning that is normally detected and corrected during the development phase. Note that a *fault* is an error that occurs in an in-service system.

deflection The control placed on electron direction and motion in CRTs and other vacuum tube devices by varying the strengths of electrostatic (electrical) or electromagnetic fields.

degradation In susceptibility testing, any undesirable change in the operational performance of a test specimen. This term does not necessarily mean malfunction or catastrophic failure.

degradation failure A failure that results from a gradual change in performance characteristics of a system or part with time.

delay The amount of time by which a signal is delayed or an event is retarded.

delay circuit A circuit designed to delay a signal passing through it by a specified amount.

delay distortion The distortion resulting from the difference in phase delays at two frequencies of interest.

delay equalizer A network that adjusts the velocity of propagation of the frequency components of a complex signal to counteract the delay distortion characteristics of a transmission channel.

delay line A transmission network that increases the propagation time of a signal traveling through it.

delta connection A common method of joining together a three-phase power supply, with each phase across a different pair of the three wires used.

delta-connected system A 3-phase power distribution system where a single-phase output can be derived from each of the adjacent pairs of an equilateral triangle formed by the service drop transformer secondary windings.

demodulator Any device that recovers the original signal after it has modulated a high-frequency carrier. The output from the unit may be in baseband composite form.

demultiplexer (demux) A device used to separate two or more signals that were previously combined by a compatible multiplexer and are transmitted over a single channel.

derating factor An operating safety margin provided for a component or system to ensure reliable performance. A *derating allowance* also is typically provided for operation under extreme environmental conditions, or under stringent reliability requirements.

desiccant A drying agent used for drying out cable splices or sensitive equipment.

design A layout of all the necessary equipment and facilities required to make a special circuit, piece of equipment, or system work.

design objective The desired electrical or mechanical performance characteristic for electronic circuits and equipment.

detection The rectification process that results in the modulating signal being separated from a modulated wave.

detectivity The reciprocal of *noise equivalent power*.

detector A device that converts one type of energy into another.

device A functional circuit, component, or network unit, such as a vacuum tube or transistor.

dewpoint The temperature at which moisture will condense out.

diagnosis The process of locating errors in software, or equipment faults in hardware.

diagnostic routine A software program designed to trace errors in software, locate hardware faults, or identify the cause of a breakdown.

dielectric An insulating material that separates the elements of various components, including capacitors and transmission lines. Dielectric materials include air, plastic, mica, ceramic, and Teflon. A dielectric material must be an insulator. (*Teflon* is a registered trademark of Du Pont.)

dielectric constant The ratio of the capacitance of a capacitor with a certain dielectric material to the capacitance with a vacuum as the dielectric. The *dielectric constant* is considered a measure of the capability of a dielectric material to store an electro-static charge.

dielectric strength The potential gradient at which electrical breakdown occurs.

differential amplifier An input circuit that rejects voltages that are the same at both input terminals but amplifies any voltage difference between the inputs. Use of a differential amplifier causes any signal present on both terminals, such as common mode hum, to cancel itself.

differential dc The maximum dc voltage that can be applied between the differential inputs of an amplifier while maintaining linear operation.

differential gain The difference in output amplitude (expressed in percent or dB) of a small high frequency sinewave signal at two stated levels of a low frequency signal on which it is superimposed.

differential phase The difference in output phase of a small high frequency sinewave signal at two stated levels of a low frequency signal on which it is superimposed.

differential-mode interference An interference source that causes a change in potential of one side of a signal transmission path relative to the other side.

diffuse reflection The scattering effect that occurs when light, radio, or sound waves strike a rough surface.

diffusion The spreading or scattering of a wave, such as a radio wave.

diode A semiconductor or vacuum tube with two electrodes that passes electric current in one direction only. Diodes are used in rectifiers, gates, modulators, and detectors.

direct coupling A coupling method between stages that permits dc current to flow between the stages.

direct current An electrical signal in which the direction of current flow remains constant.

discharge The conversion of stored energy, as in a battery or capacitor, into an electric current.

discontinuity An abrupt nonuniform point of change in a transmission circuit that causes a disruption of normal operation.

discrete An individual circuit component.

discrete component A separately contained circuit element with its own external connections.

discriminator A device or circuit whose output amplitude and polarity vary according to how much the input signal varies from a standard or from another signal. A discriminator can be used to recover the modulating waveform in a frequency modulated signal.

dish An antenna system consisting of a parabolic shaped reflector with a signal feed element at the focal point. Dish antennas commonly are used for transmission and reception from microwave stations and communications satellites.

dispersion The wavelength dependence of a parameter.

display The representation of text and images on a cathode-ray tube, an array of light-emitting diodes, a liquid-crystal readout, or another similar device.

display device An output unit that provides a visual representation of data.

distortion The difference between the wave shape of an original signal and the signal after it has traversed a transmission circuit.

distortion-limited operation The condition prevailing when the shape of the signal, rather than the amplitude (or power), is the limiting factor in communication capability. This condition is reached when the system distorts the shape of the waveform beyond tolerable limits. For linear systems, *distortion-limited* operation is equivalent to *bandwidth-limited* operation.

disturbance The interference with normal conditions and communications by some external energy source.

disturbance current The unwanted current of any irregular phenomenon associated with transmission that tends to limit or interfere with the interchange of information.

disturbance power The unwanted power of any irregular phenomenon associated with transmission that tends to limit or interfere with the interchange of information.

disturbance voltage The unwanted voltage of any irregular phenomenon associated with transmission that tends to limit or interfere with the interchange of information.

diversity receiver A receiver using two antennas connected through circuitry that senses which antenna is receiving the stronger signal. Electronic gating permits the stronger source to be routed to the receiving system.

documentation A written description of a program. *Documentation* can be considered as any record that has permanence and can be read by humans or machines.

down-lead A lead-in wire from an antenna to a receiver.

downlink The portion of a communication link used for transmission of signals from a satellite or airborne platform to a surface terminal.

downstream A specified signal modification occurring after other given devices in a signal path.

downtime The time during which equipment is not capable of doing useful work because of malfunction. This does not include preventive maintenance time. In other words, *downtime* is measured from the occurrence of a malfunction to the correction of that malfunction.

drift A slow change in a nominally constant signal characteristic, such as frequency.

drift-space The area in a klystron tube in which electrons drift at their entering velocities and form electron *bunches*.

drive The input signal to a circuit, particularly to an amplifier.

driver An electronic circuit that supplies an isolated output to drive the input of another circuit.

drop-out value The value of current or voltage at which a relay will cease to be operated.

dropout The momentary loss of a signal.

dropping resistor A resistor designed to carry current that will make a required voltage available.

duplex separation The frequency spacing required in a communications system between the *forward* and *return* channels to maintain interference at an acceptably low level.

duplex signaling A configuration permitting signaling in both transmission directions simultaneously.

duty cycle The ratio of operating time to total elapsed time of a device that operates intermittently, expressed in percent.

dynamic A situation in which the operating parameters and/or requirements of a given system are continually changing.

dynamic range The maximum range or extremes in amplitude, from the lowest to the highest (noise floor to system clipping), that a system is capable of reproducing. The dynamic range is expressed in dB against a reference level.

dynamo A rotating machine, normally a dc generator.

dynamotor A rotating machine used to convert dc into ac.

E

earth A large conducting body with no electrical potential, also called *ground*.

earth capacitance The capacitance between a given circuit or component and a point at ground potential.

earth current A current that flows to earth/ground, especially one that follows from a fault in the system. *Earth current* may also refer to a current that flows in the earth, resulting from ionospheric disturbances, lightning, or faults on power lines.

earth fault A fault that occurs when a conductor is accidentally grounded/earthed, or when the resistance to earth of an insulator falls below a specified value.

earth ground A large conducting body that represents *zero level* in the scale of electrical potential. An *earth ground* is a connection made either accidentally or by design between a conductor and earth.

earth potential The potential taken to be the arbitrary zero in a scale of electric potential.

effective ground A connection to ground through a medium of sufficiently low impedance and adequate current-carrying capacity to prevent the buildup of voltages that might be hazardous to equipment or personnel.

effective resistance The increased resistance of a conductor to an alternating current resulting from the *skin effect*, relative to the direct-current resistance of the conductor. Higher frequencies tend to travel only on the outer skin of the conductor, whereas dc flows uniformly through the entire area.

efficiency The useful power output of an electrical device or circuit divided by the total power input, expressed in percent.

electric Any device or circuit that produces, operates on, transmits, or uses electricity.

electric charge An excess of either electrons or protons within a given space or material.

electric field strength The magnitude, measured in volts per meter, of the electric field in an electromagnetic wave.

electric flux The amount of electric charge, measured in coulombs, across a dielectric of specified area. *Electric flux* may also refer simply to electric lines of force.

electricity An energy force derived from the movement of negative and positive electric charges.

electrode An electrical terminal that emits, collects, or controls an electric current.

electrolysis A chemical change induced in a substance resulting from the passage of electric current through an electrolyte.

electrolyte A nonmetallic conductor of electricity in which current is carried by the physical movement of ions.

electromagnet An iron or steel core surrounded by a wire coil. The core becomes magnetized when current flows through the coil but loses its magnetism when the current flow is stopped.

electromagnetic compatibility The capability of electronic equipment or systems to operate in a specific electromagnetic environment, at designated levels of efficiency and within a defined margin of safety, without interfering with itself or other systems.

electromagnetic field The electric and magnetic fields associated with radio and light waves.

electromagnetic induction An electromotive force created with a conductor by the relative motion between the conductor and a nearby magnetic field.

electromagnetism The study of phenomena associated with varying magnetic fields, electromagnetic radiation, and moving electric charges.

electromotive force (EMF) An electrical potential, measured in volts, that can produce the movement of electrical charges.

electron A stable elementary particle with a negative charge that is mainly responsible for electrical conduction. Electrons move when under the influence of an electric field. This movement constitutes an *electric current.*

electron beam A stream of emitted electrons, usually in a vacuum.

electron gun A hot cathode that produces a finely focused stream of fast electrons, which are necessary for the operation of a vacuum tube, such as a cathode ray tube. The gun is made up of a hot cathode electron source, a control grid, accelerating anodes, and (usually) focusing electrodes.

electron lens A device used for focusing an electron beam in a cathode ray tube. Such focusing can be accomplished by either magnetic forces, in which external coils are used to create the proper magnetic field within the tube, or electrostatic forces, where metallic plates within the tube are charged electrically in such a way as to control the movement of electrons in the beam.

electron volt The energy acquired by an electron in passing through a potential difference of one volt in a vacuum.

electronic A description of devices (or systems) that are dependent on the flow of electrons in electron tubes, semiconductors, and other devices, and not solely on electron flow in ordinary wires, inductors, capacitors, and similar passive components.

Electronic Industries Association (EIA) A trade organization, based in Washington, DC, representing the manufacturers of electronic systems and parts, including communications systems. The association develops standards for electronic components and systems.

electronic switch A transistor, semiconductor diode, or a vacuum tube used as an on/off switch in an electrical circuit. Electronic switches can be controlled manually, by other circuits, or by computers.

electronics The field of science and engineering that deals with electron devices and their utilization.

electroplate The process of coating a given material with a deposit of metal by electrolytic action.

electrostatic The condition pertaining to electric charges that are at rest.

electrostatic field The space in which there is electric stress produced by static electric charges.

electrostatic induction The process of inducing static electric charges on a body by bringing it near other bodies that carry high electrostatic charges.

element A substance that consists of atoms of the same atomic number. Elements are the basic units in all chemical changes other than those in which *atomic changes,* such as fusion and fission, are involved.

EMI (electromagnetic interference) Undesirable electromagnetic waves that are radiated unintentionally from an electronic circuit or device into other circuits or devices, disrupting their operation.

emission (1—radiation) The radiation produced, or the production of radiation by a radio transmitting system. The emission is considered to be a *single emission* if the modulating signal and other characteristics are the same for every transmitter of the radio transmitting system and the spacing between antennas is not more than a few wavelengths. **(2—cathode)** The release of electrons from the cathode of a vacuum tube. **(3—parasitic)** A spurious radio frequency emission unintentionally generated at frequencies that are independent of the carrier frequency being amplified or modulated. **(4—secondary)** In an electron tube, emission of electrons by a plate or grid because of bombardment by *primary emission* electrons from the cathode of the tube. **(5—spurious)** An emission outside the radio frequency band authorized for a transmitter. **(6—thermonic)** An emission from a cathode resulting from high temperature.

emphasis The intentional alteration of the frequency-amplitude characteristics of a signal to reduce the adverse effects of noise in a communication system.

empirical A conclusion not based on pure theory, but on practical and experimental work.

emulation The use of one system to imitate the capabilities of another system.

enable To prepare a circuit for operation or to allow an item to function.

enabling signal A signal that permits the occurrence of a specified event.

encode The conversion of information from one form into another to obtain characteristics required by a transmission or storage system.

encoder A device that processes one or more input signals into a specified form for transmission and/or storage.

energized The condition when a circuit is switched on, or powered up.

energy spectral density A frequency-domain description of the energy in each of the frequency components of a pulse.

envelope The boundary of the family of curves obtained by varying a parameter of a wave.

envelope delay The difference in absolute delay between the fastest and slowest propagating frequencies within a specified bandwidth.

envelope delay distortion The maximum difference or deviation of the envelope-delay characteristic between any two specified frequencies.

envelope detection A demodulation process that senses the shape of the modulated RF envelope. A diode detector is one type of envelop detection device.

environmental An equipment specification category relating to temperature and humidity.

EQ (equalization) network A network connected to a circuit to correct or control its transmission frequency characteristics.

equalization (EQ) The reduction of frequency distortion and/or phase distortion of a circuit through the introduction of one or more networks to compensate for the difference in attenuation, time delay, or both, at the various frequencies in the transmission band.

equalize The process of inserting in a line a network with complementary transmission characteristics to those of the line, so that when the loss or delay in the line and that in the equalizer are combined, the overall loss or delay is approximately equal at all frequencies.

equalizer A network that corrects the transmission-frequency characteristics of a circuit to allow it to transmit selected frequencies in a uniform manner.

equatorial orbit The plane of a satellite orbit which coincides with that of the equator of the primary body.

equipment A general term for electrical apparatus and hardware, switching systems, and transmission components.

equipment failure The condition when a hardware fault stops the successful completion of a task.

equipment ground A protective ground consisting of a conducting path to ground of noncurrent carrying metal parts.

equivalent circuit A simplified network that emulates the characteristics of the real circuit it replaces. An equivalent circuit is typically used for mathematical analysis.

equivalent noise resistance A quantitative representation in resistance units of the spectral density of a noise voltage generator at a specified frequency.

error A collective term that includes all types of inconsistencies, transmission deviations, and control failures.

excitation The current that energizes field coils in a generator.

expandor A device with a nonlinear gain characteristic that acts to increase the gain more on larger input signals than it does on smaller input signals.

extremely high frequency (EHF) The band of microwave frequencies between the limits of 30 GHz and 300 GHz (wavelengths between 1 cm and 1 mm).

extremely low frequency The radio signals with operating frequencies below 300 Hz (wavelengths longer than 1000 km).

F

fail-safe operation A type of control architecture for a system that prevents improper functioning in the event of circuit or operator failure.

failure A detected cessation of ability to perform a specified function or functions within previously established limits. A *failure* is beyond adjustment by the operator by means of controls normally accessible during routine operation of the system. (This requires that measurable limits be established to define "satisfactory performance".)

failure effect The result of the malfunction or failure of a device or component.

failure in time (FIT) A unit value that indicates the reliability of a component or device. One failure in time corresponds to a failure rate of 10^{-9} per hour.

failure mode and effects analysis (FMEA) An iterative documented process performed to identify basic faults at the component level and determine their effects at higher levels of assembly.

failure rate The ratio of the number of actual failures to the number of times each item has been subjected to a set of specified stress conditions.

fall time The length of time during which a pulse decreases from 90 percent to 10 percent of its maximum amplitude.

farad The standard unit of capacitance equal to the value of a capacitor with a potential of one volt between its plates when the charge on one plate is one coulomb and there is an equal and opposite charge on the other plate. The farad is a large value and is more commonly expressed in *microfarads* or *picofarads*. The *farad* is named for the English chemist and physicist Michael Faraday (1791-1867).

fast frequency shift keying (FFSK) A system of digital modulation where the digits are represented by different frequencies that are related to the baud rate, and where transitions occur at the zero crossings.

fatigue The reduction in strength of a metal caused by the formation of crystals resulting from repeated flexing of the part in question.

fault A condition that causes a device, a component, or an element to fail to perform in a required manner. Examples include a short-circuit, broken wire, or intermittent connection.

fault to ground A fault caused by the failure of insulation and the consequent establishment of a direct path to ground from a part of the circuit that should not normally be grounded.

fault tree analysis (FTA) An iterative documented process of a systematic nature performed to identify basic faults, determine their causes and effects, and establish their probabilities of occurrence.

feature A distinctive characteristic or part of a system or piece of equipment, usually visible to end users and designed for their convenience.

Federal Communications Commission (FCC) The federal agency empowered by law to regulate all interstate radio and wireline communications services originating in the United States, including radio, television, facsimile, telegraph, data transmission, and telephone systems. The agency was established by the Communications Act of 1934.

feedback The return of a portion of the output of a device to the input. *Positive feedback* adds to the input, *negative feedback* subtracts from the input.

feedback amplifier An amplifier with the components required to feed a portion of the output back into the input to alter the characteristics of the output signal.

feedline A transmission line, typically coaxial cable, that connects a high frequency energy source to its load.

femto A prefix meaning *one quadrillionth* (10^{-15}).

ferrite A ceramic material made of powdered and compressed ferric oxide, plus other oxides (mainly cobalt, nickel, zinc, yttrium-iron, and manganese). These materials have low eddy current losses at high frequencies.

ferromagnetic material A material with low relative permeability and high coercive force so that it is difficult to magnetize and demagnetize. Hard ferromagnetic materials retain magnetism well, and are commonly used in permanent magnets.

fidelity The degree to which a system, or a portion of a system, accurately reproduces at its output the essential characteristics of the signal impressed upon its input.

field strength The strength of an electric, magnetic, or electromagnetic field.

filament A wire that becomes hot when current is passed through it, used either to emit light (for a light bulb) or to heat a cathode to enable it to emit electrons (for an electron tube).

film resistor A type of resistor made by depositing a thin layer of resistive material on an insulating core.

filter A network that passes desired frequencies but greatly attenuates other frequencies.

filtered noise White noise that has been passed through a filter. The power spectral density of filtered white noise has the same shape as the transfer function of the filter.

fitting A coupling or other mechanical device that joins one component with another.

fixed A system or device that is not changeable or movable.

flashover An arc or spark between two conductors.

flashover voltage The voltage between conductors at which flashover just occurs.

flat face tube The design of CRT tube with almost a flat face, giving improved legibility of text and reduced reflection of ambient light.

flat level A signal that has an equal amplitude response for all frequencies within a stated range.

flat loss A circuit, device, or channel that attenuates all frequencies of interest by the same amount, also called *flat slope*.

flat noise A noise whose power per unit of frequency is essentially independent of frequency over a specified frequency range.

flat response The performance parameter of a system in which the output signal amplitude of the system is a faithful reproduction of the input amplitude over some range of specified input frequencies.

floating A circuit or device that is not connected to any source of potential or to ground.

fluorescence The characteristic of a material to produce light when excited by an external energy source. Minimal or no heat results from the process.

flux The electric or magnetic lines of force resulting from an applied energy source.

flywheel effect The characteristic of an oscillator that enables it to sustain oscillations after removal of the control stimulus. This characteristic may be desirable, as in the case of a phase-locked loop employed in a synchronous system, or undesirable, as in the case of a voltage-controlled oscillator.

focusing A method of making beams of radiation converge on a target, such as the face of a CRT.

Fourier analysis A mathematical process for transforming values between the frequency domain and the time domain. This term also refers to the decomposition of a time-domain signal into its frequency components.

Fourier transform An integral that performs an actual transformation between the frequency domain and the time domain in Fourier analysis.

frame A segment of an analog or digital signal that has a repetitive characteristic, in that corresponding elements of successive *frames* represent the same things.

free electron An electron that is not attached to an atom and is, thus, mobile when an electromotive force is applied.

free running An oscillator that is not controlled by an external synchronizing signal.

free-running oscillator An oscillator that is not synchronized with an external timing source.

frequency The number of complete cycles of a periodic waveform that occur within a given length of time. Frequency is usually specified in cycles per second (*Hertz*). Frequency is the reciprocal of wavelength. The higher the frequency, the shorter the wavelength. In general, the higher the frequency of a signal, the more capacity it has to carry information, the smaller an antenna is required, and the more susceptible the signal is to absorption by the atmosphere and by physical structures. At microwave frequencies, radio signals take on a *line-of-sight* characteristic and require highly directional and focused antennas to be used successfully.

frequency accuracy The degree of conformity of a given signal to the specified value of a frequency.

frequency allocation The designation of radio-frequency bands for use by specific radio services.

frequency content The band of frequencies or specific frequency components contained in a signal.

frequency converter A circuit or device used to change a signal of one frequency into another of a different frequency.

frequency coordination The process of analyzing frequencies in use in various bands of the spectrum to achieve reliable performance for current and new services.

frequency counter An instrument or test set used to measure the frequency of a radio signal or any other alternating waveform.

frequency departure An unintentional deviation from the nominal frequency value.

frequency difference The algebraic difference between two frequencies. The two frequencies can be of identical or different nominal values.

frequency displacement The end-to-end shift in frequency that may result from independent frequency translation errors in a circuit.

frequency distortion The distortion of a multifrequency signal caused by unequal attenuation or amplification at the different frequencies of the signal. This term may also be referred to as *amplitude distortion*.

frequency domain A representation of signals as a function of frequency, rather than of time.

frequency modulation (FM) The modulation of a carrier signal so that its instantaneous frequency is proportional to the instantaneous value of the modulating wave.

frequency multiplier A circuit that provides as an output an exact multiple of the input frequency.

frequency offset A frequency shift that occurs when a signal is sent over an analog transmission facility in which the modulating and demodulating frequencies are not identical. A channel with frequency offset does not preserve the waveform of a transmitted signal.

frequency response The measure of system linearity in reproducing signals across a specified bandwidth. Frequency response is expressed as a frequency range with a specified amplitude tolerance in dB.

frequency response characteristic The variation in the transmission performance (gain or loss) of a system with respect to variations in frequency.

frequency reuse A technique used to expand the capacity of a given set of frequencies or channels by separating the signals either geographically or through the use of different polarization techniques. Frequency reuse is a common element of the *frequency coordination* process.

frequency selectivity The ability of equipment to separate or differentiate between signals at different frequencies.

frequency shift The difference between the frequency of a signal applied at the input of a circuit and the frequency of that signal at the output.

frequency shift keying (FSK) A commonly-used method of digital modulation in which a one and a zero (the two possible states) are each transmitted as separate frequencies.

frequency stability A measure of the variations of the frequency of an oscillator from its mean frequency over a specified period of time.

frequency standard An oscillator with an output frequency sufficiently stable and accurate that it is used as a reference.

frequency-division multiple access (FDMA) The provision of multiple access to a transmission facility, such as an earth satellite, by assigning each transmitter its own frequency band.

frequency-division multiplexing (FDM) The process of transmitting multiple analog signals by an orderly assignment of frequency slots, that is, by dividing transmission bandwidth into several narrow bands, each of which carries a single communication and is sent simultaneously with others over a common transmission path.

full duplex A communications system capable of transmission simultaneously in two directions.

full-wave rectifier A circuit configuration in which both positive and negative half-cycles of the incoming ac signal are rectified to produce a unidirectional (dc) current through the load.

functional block diagram A diagram illustrating the definition of a device, system, or problem on a logical and functional basis.

functional unit An entity of hardware and/or software capable of accomplishing a given purpose.

fundamental frequency The lowest frequency component of a complex signal.

fuse A protective device used to limit current flow in a circuit to a specified level. The fuse consists of a metallic link that melts and opens the circuit at a specified current level.

fuse wire A fine-gauge wire made of an alloy that overheats and melts at the relatively low temperatures produced when the wire carries overload currents. When used in a fuse, the wire is called a fuse (or fusible) link.

G

gain An increase or decrease in the level of an electrical signal. Gain is measured in terms of decibels or number-of-times of magnification. Strictly speaking, *gain* refers to an increase in level. Negative numbers, however, are commonly used to denote a decrease in level.

gain-bandwidth The gain times the frequency of measurement when a device is biased for maximum obtainable gain.

gain/frequency characteristic The gain-versus-frequency characteristic of a channel over the bandwidth provided, also referred to as *frequency response.*

gain/frequency distortion A circuit defect in which a change in frequency causes a change in signal amplitude.

galvanic A device that produces direct current by chemical action.

gang The mechanical connection of two or more circuit devices so that they can all be adjusted simultaneously.

gang capacitor A variable capacitor with more than one set of moving plates linked together.

gang tuning The simultaneous tuning of several different circuits by turning a single shaft on which ganged capacitors are mounted.

ganged One or more devices that are mechanically coupled, normally through the use of a shared shaft.

gas breakdown The ionization of a gas between two electrodes caused by the application of a voltage that exceeds a threshold value. The ionized path has a low impedance. Certain types of circuit and line protectors rely on gas breakdown to divert hazardous currents away from protected equipment.

gas tube A protection device in which a sufficient voltage across two electrodes causes a gas to ionize, creating a low impedance path for the discharge of dangerous voltages.

gas-discharge tube A gas-filled tube designed to carry current during gas breakdown. The gas-discharge tube is commonly used as a protective device, preventing high voltages from damaging sensitive equipment.

gauge A measure of wire diameter. In measuring wire gauge, the lower the number, the thicker the wire.

Gaussian distribution A statistical distribution, also called the *normal* distribution. The graph of a Gaussian distribution is a bell-shaped curve.

Gaussian noise Noise in which the distribution of amplitude follows a Gaussian model, that is, the noise is random but distributed about a reference voltage of zero.

Gaussian pulse A pulse that has the same form as its own Fourier transform.

generator A machine that converts mechanical energy into electrical energy, or one form of electrical energy into another form.

geosynchronous The attribute of a satellite in which the relative position of the satellite as viewed from the surface of a given planet is stationary. For earth, the geosynchronous position is 22,300 miles above the planet.

getter A metal used in vaporized form to remove residual gas from inside an electron tube during manufacture.

giga A prefix meaning one billion.

gigahertz (GHz) A measure of frequency equal to one billion cycles per second. Signals operating above 1 gigahertz are commonly known as *microwaves*, and begin to take on the characteristics of visible light.

glitch A general term used to describe a wide variety of momentary signal discontinuities.

graceful degradation An equipment failure mode in which the system suffers reduced capability, but does not fail altogether.

graticule A fixed pattern of reference markings used with oscilloscope CRTs to simplify measurements. The graticule may be etched on a transparent plate covering the front of the CRT or, for greater accuracy in readings, may be electrically generated within the CRT itself.

grid (1—general) A mesh electrode within an electron tube that controls the flow of electrons between the cathode and plate of the tube. **(2—bias)** The potential applied to a grid in an electron tube to control its center operating point. **(3—control)** The grid in an electron tube to which the input signal is usually applied. **(4—screen)** The grid in an electron tube, typically held at a steady potential, that screens the control grid from changes in anode potential. **(5—suppressor)** The grid in an electron tube near the anode (plate) that suppresses the emission of secondary electrons from the plate.

ground An electrical connection to earth or to a common conductor usually connected to earth.

ground clamp A clamp used to connect a ground wire to a ground rod or system.

ground loop An undesirable circulating ground current in a circuit grounded via multiple connections or at multiple points.

ground plane A conducting material at ground potential, physically close to other equipment, so that connections may be made readily to ground the equipment at the required points.

ground potential The point at zero electric potential.

ground return A conductor used as a path for one or more circuits back to the ground plane or central facility ground point.

ground rod A metal rod driven into the earth and connected into a mesh of interconnected rods so as to provide a low resistance link to ground.

ground window A single-point interface between the integrated ground plane of a building and an isolated ground plane.

ground wire A copper conductor used to extend a good low-resistance earth ground to protective devices in a facility.

grounded The connection of a piece of equipment to earth via a low resistance path.

grounding The act of connecting a device or circuit to ground or to a conductor that is grounded.

group delay A condition where the different frequency elements of a given signal suffer differing propagation delays through a circuit or a system. The delay at a lower frequency is different from the delay at a higher frequency, resulting in a time-related distortion of the signal at the receiving point.

group delay time The rate of change of the total phase shift of a waveform with angular frequency through a device or transmission facility.

group velocity The speed of a pulse on a transmission line.

guard band A narrow bandwidth between adjacent channels intended to reduce interference or crosstalk.

H

half-wave rectifier A circuit or device that changes only positive or negative half-cycle inputs of alternating current into direct current.

Hall effect The phenomenon by which a voltage develops between the edges of a current-carrying metal strip whose faces are perpendicular to an external magnetic field.

hard-wired Electrical devices connected through physical wiring.

harden The process of constructing military telecommunications facilities so as to protect them from damage by enemy action, especially *electromagnetic pulse* (EMP) radiation.

hardware Physical equipment, such as mechanical, magnetic, electrical, or electronic devices or components.

harmonic A periodic wave having a frequency that is an integral multiple of the fundamental frequency. For example, a wave with twice the frequency of the fundamental is called the *second harmonic*.

harmonic analyzer A test set capable of identifying the frequencies of the individual signals that make up a complex wave.

harmonic distortion The production of harmonics at the output of a circuit when a periodic wave is applied to its input. The level of the distortion is usually expressed as a percentage of the level of the input.

hazard A condition that could lead to danger for operating personnel.

headroom The difference, in decibels, between the typical operating signal level and a peak overload level.

heat loss The loss of useful electrical energy resulting from conversion into unwanted heat.

heat sink A device that conducts heat away from a heat-producing component so that it stays within a safe working temperature range.

heater In an electron tube, the filament that heats the cathode to enable it to emit electrons.

hecto A prefix meaning 100.

henry The standard unit of electrical inductance, equal to the self-inductance of a circuit or the mutual inductance of two circuits when there is an induced electromotive force of one volt and a current change of one ampere per second. The symbol for inductance is H, named for the American physicist Joseph Henry (1797-1878).

hertz (Hz) The unit of frequency that is equal to one cycle per second. Hertz is the reciprocal of the *period*, the interval after which the same portion of a periodic waveform recurs. Hertz was named for the German physicist Heinrich R. Hertz (1857-1894).

heterodyne The mixing of two signals in a nonlinear device in order to produce two additional signals at frequencies that are the sum and difference of the original frequencies.

heterodyne frequency The sum of, or the difference between, two frequencies, produced by combining the two signals together in a modulator or similar device.

heterodyne wavemeter A test set that uses the heterodyne principle to measure the frequencies of incoming signals.

high-frequency loss Loss of signal amplitude at higher frequencies through a given circuit or medium. For example, high frequency loss could be caused by passing a signal through a coaxial cable.

high Q An inductance or capacitance whose ratio of reactance to resistance is high.

high tension A high voltage circuit.

high-pass filter A network that passes signals of higher than a specified frequency but attenuates signals of all lower frequencies.

homochronous Signals whose corresponding significant instants have a constant but uncontrolled phase relationship with each other.

horn gap A lightning arrester utilizing a gap between two horns. When lightning causes a discharge between the horns, the heat produced lengthens the arc and breaks it.

horsepower The basic unit of mechanical power. One horsepower (hp) equals 550 foot-pounds per second or 746 watts.

hot A charged electrical circuit or device.

hot dip galvanized The process of galvanizing steel by dipping it into a bath of molten zinc.

hot standby System equipment that is fully powered but not in service. A *hot standby* can rapidly replace a primary system in the event of a failure.

hum Undesirable coupling of the 60 Hz power sine wave into other electrical signals and/or circuits.

HVAC An abbreviation for *heating, ventilation, and air conditioning* system.

hybrid system A communication system that accommodates both digital and analog signals.

hydrometer A testing device used to measure specific gravity, particularly the specific gravity of the dilute sulphuric acid in a lead-acid storage battery, to learn the state of charge of the battery.

hygrometer An instrument that measures the relative humidity of the atmosphere.

hygroscopic The ability of a substance to absorb moisture from the air.

hysteresis The property of an element evidenced by the dependence of the value of the output, for a given excursion of the input, upon the history of prior excursions and direction of the input. Originally, *hysteresis* was the name for magnetic phenomena only—the lagging of flux density behind the change in value of the magnetizing flux—but now, the term is also used to describe other inelastic behavior.

hysteresis loop The plot of magnetizing current against magnetic flux density (or of other similarly related pairs of parameters), which appears as a loop. The area within the loop is proportional to the power loss resulting from hysteresis.

hysteresis loss The loss in a magnetic core resulting from hysteresis.

I

I^2R **loss** The power lost as a result of the heating effect of current passing through resistance.

idling current The current drawn by a circuit, such as an amplifier, when no signal is present at its input.

image frequency A frequency on which a carrier signal, when heterodyned with the local oscillator in a superheterodyne receiver, will cause a sum or difference frequency that is the same as the intermediate frequency of the receiver. Thus, a signal on an *image frequency* will be demodulated along with the desired signal and will interfere with it.

impact ionization The ionization of an atom or molecule as a result of a high energy collision.

impedance The total passive opposition offered to the flow of an alternating current. *Impedance* consists of a combination of resistance, inductive reactance, and capacitive reactance. It is the vector sum of resistance and reactance $(R + jX)$ or the vector of magnitude Z at an angle θ.

impedance characteristic A graph of the impedance of a circuit showing how it varies with frequency.

impedance irregularity A discontinuity in an impedance characteristic caused, for example, by the use of different coaxial cable types.

impedance matching The adjustment of the impedances of adjoining circuit components to a common value so as to minimize reflected energy from the junction and to maximize energy transfer across it. Incorrect adjustment results in an *impedance mismatch.*

impedance matching transformer A transformer used between two circuits of different impedances with a turns ratio that provides for maximum power transfer and minimum loss by reflection.

impulse A short high energy surge of electrical current in a circuit or on a line.

impulse current A current that rises rapidly to a peak then decays to zero without oscillating.

impulse excitation The production of an oscillatory current in a circuit by impressing a voltage for a relatively short period compared with the duration of the current produced.

impulse noise A noise signal consisting of random occurrences of energy spikes, having random amplitude and bandwidth.

impulse response The amplitude-versus-time output of a transmission facility or device in response to an impulse.

impulse voltage A unidirectional voltage that rises rapidly to a peak and then falls to zero, without any appreciable oscillation.

in-phase The property of alternating current signals of the same frequency that achieve their peak positive, peak negative, and zero amplitude values simultaneously.

incidence angle The angle between the perpendicular to a surface and the direction of arrival of a signal.

increment A small change in the value of a quantity.

induce To produce an electrical or magnetic effect in one conductor by changing the condition or position of another conductor.

induced current The current that flows in a conductor because a voltage has been induced across two points in, or connected to, the conductor.

induced voltage A voltage developed in a conductor when the conductor passes through magnetic lines of force.

inductance The property of an inductor that opposes any change in a current that flows through it. The standard unit of inductance is the *Henry*.

induction The electrical and magnetic interaction process by which a changing current in one circuit produces a voltage change not only in its own circuit (*self inductance*) but also in other circuits to which it is linked magnetically.

inductive A circuit element exhibiting inductive reactance.

inductive kick A voltage surge produced when a current flowing through an inductance is interrupted.

inductive load A load that possesses a net inductive reactance.

inductive reactance The reactance of a circuit resulting from the presence of inductance and the phenomenon of induction.

inductor A coil of wire, usually wound on a core of high permeability, that provides high inductance without necessarily exhibiting high resistance.

inert An inactive unit, or a unit that has no power requirements.

infinite line A transmission line that appears to be of infinite length. There are no reflections back from the far end because it is terminated in its characteristic impedance.

infra low frequency (ILF) The frequency band from 300 Hz to 3000 Hz.

inhibit A control signal that prevents a device or circuit from operating.

injection The application of a signal to an electronic device.

input The waveform fed into a circuit, or the terminals that receive the input waveform.

insertion gain The gain resulting from the insertion of a transducer in a transmission system, expressed as the ratio of the power delivered to that part of the system following the transducer to the power delivered to that same part before insertion. If more than one component is involved in the input or output, the particular component used must be specified. This ratio is usually expressed in decibels. If the resulting number is negative, an *insertion loss* is indicated.

insertion loss The signal loss within a circuit, usually expressed in decibels as the ratio of input power to output power.

insertion loss-vs.-frequency characteristic The amplitude transfer characteristic of a system or component as a function of frequency. The amplitude response may be stated as actual gain, loss, amplification, or attenuation, or as a ratio of any one of these quantities at a particular frequency, with respect to that at a specified reference frequency.

inspection lot A collection of units of product from which a sample is drawn and inspected to determine conformance with acceptability criteria.

instantaneous value The value of a varying waveform at a given instant of time. The value can be in volts, amperes, or phase angle.

Institute of Electrical and Electronics Engineers (IEEE) The organization of electrical and electronics scientists and engineers formed in 1963 by the merger of the Institute of Radio Engineers (IRE) and the American Institute of Electrical Engineers (AIEE).

instrument multiplier A measuring device that enables a high voltage to be measured using a meter with only a low voltage range.

instrument rating The range within which an instrument has been designed to operate without damage.

insulate The process of separating one conducting body from another conductor.

insulation The material that surrounds and insulates an electrical wire from other wires or circuits. *Insulation* may also refer to any material that does not ionize easily and thus presents a large impedance to the flow of electrical current.

insulator A material or device used to separate one conducting body from another.

intelligence signal A signal containing information.

intensity The strength of a given signal under specified conditions.

interconnect cable A short distance cable intended for use between equipment (generally less than 3 m in length)

interface A device or circuit used to interconnect two pieces of electronic equipment.

interface device A unit that joins two interconnecting systems.

interference emission An emission that results in an electrical signal being propagated into and interfering with the proper operation of electrical or electronic equipment.

interlock A protection device or system designed to remove all dangerous voltages from a machine or piece of equipment when access doors or panels are opened or removed.

intermediate frequency A frequency that results from combining a signal of interest with a signal generated within a radio receiver. In superheterodyne receivers, all incoming signals are converted to a single intermediate frequency for which the amplifiers and filters of the receiver have been optimized.

intermittent A noncontinuous recurring event, often used to denote a problem that is difficult to find because of its unpredictable nature.

intermodulation The production, in a nonlinear transducer element, of frequencies corresponding to the sums and differences of the fundamentals and harmonics of two or more frequencies that are transmitted through the transducer.

intermodulation distortion (IMD) The distortion that results from the mixing of two input signals in a nonlinear system. The resulting output contains new frequencies that represent the sum and difference of the input signals and the sums and differences of their harmonics. IMD is also called *intermodulation noise.*

intermodulation noise In a transmission path or device, the noise signal that is contingent upon modulation and demodulation, resulting from nonlinear characteristics in the path or device.

internal resistance The actual resistance of a source of electric power. The total electromotive force produced by a power source is not available for external use; some of the energy is used in driving current through the source itself.

International Standards Organization (ISO) An international body concerned with worldwide standardization for a broad range of industrial products, including telecommunications equipment. Members are represented by national standards organizations, such as ANSI (American National Standards Institute) in the United States. ISO was established in 1947 as a specialized agency of the United Nations.

International Telecommunications Union (ITU) A specialized agency of the United Nations established to maintain and extend international cooperation for the maintenance, development, and efficient use of telecommunications. The union does this through standards and recommended regulations, and through technical and telecommunications studies.

International Telecommunications Satellite Consortium (Intelsat) A nonprofit cooperative of member nations that owns and operates a satellite system for international and, in many instances, domestic communications.

interoperability The condition achieved among communications and electronics systems or equipment when information or services can be exchanged directly between them or their users, or both.

interpolate The process of estimating unknown values based on a knowledge of comparable data that falls on both sides of the point in question.

interrupting capacity The rating of a circuit breaker or fuse that specifies the maximum current the device is designed to interrupt at its rated voltage.

interval The points or numbers lying between two specified endpoints.

inverse voltage The effective value of voltage across a rectifying device, which conducts a current in one direction during one half cycle of the alternating input, during the half cycle when current is not flowing.

inversion The change in the polarity of a pulse, such as from positive to negative.

inverter A circuit or device that converts a direct current into an alternating current.

ionizing radiation The form of electromagnetic radiation that can turn an atom into an ion by knocking one or more of its electrons loose. Examples of ionizing radiation include X rays, gamma rays, and cosmic rays

IR drop A drop in voltage because of the flow of current (I) through a resistance (R), also called *resistance drop*.

IR loss The conversion of electrical power to heat caused by the flow of electrical current through a resistance.

isochronous A signal in which the time interval separating any two significant instants is theoretically equal to a specified unit interval or to an integral multiple of the unit interval.

isolated ground A ground circuit that is isolated from all equipment framework and any other grounds, except for a single-point external connection.

isolated ground plane A set of connected frames that are grounded through a single connection to a ground reference point. That point and all parts of the frames are insulated from any other ground system in a building.

isolated pulse A pulse uninfluenced by other pulses in the same signal.

isophasing amplifier A timing device that corrects for small timing errors.

isotropic A quantity exhibiting the same properties in all planes and directions.

J

jack A receptacle or connector that makes electrical contact with the mating contacts of a plug. In combination, the plug and jack provide a ready means for making connections in electrical circuits.

jacket An insulating layer of material surrounding a wire in a cable.

jitter Small, rapid variations in a waveform resulting from fluctuations in a supply voltage or other causes.

joule The standard unit of work that is equal to the work done by one newton of force when the point at which the force is applied is displaced a distance of one meter in the direction of the force. The *joule* is named for the English physicist James Prescott Joule (1818-1889).

Julian date A chronological date in which days of the year are numbered in sequence. For example, the first day is 001, the second is 002, and the last is 365 (or 366 in a leap year).

K

Kelvin (K) The standard unit of thermodynamic temperature. Zero degrees Kelvin represents *absolute zero*. Water freezes at 273 K and water boils at 373 K under standard pressure conditions.

kilo A prefix meaning one thousand.

kilohertz (kHz) A unit of measure of frequency equal to 1,000 Hz.

kilovar A unit equal to one thousand volt-amperes.

kilovolt (kV) A unit of measure of electrical voltage equal to 1,000 V.

kilowatt A unit equal to one thousand watts.

Kirchoff's Law At any point in a circuit, there is as much current flowing into the point as there is flowing away from it.

klystron (1—general) A family of electron tubes that function as microwave amplifiers and oscillators. Simplest in form are two-cavity klystrons in which an electron beam passes through a cavity that is excited by a microwave input, producing a velocity-modulated beam which passes through a second cavity a precise distance away that is coupled to a tuned circuit, thereby producing an amplified output of the original input signal frequency. If part of the output is fed back to the input, an oscillator can be the result. **(2—multi-cavity)** An amplifier device for UHF and microwave signals based on velocity modulation of an electron beam. The beam is directed through an input cavity, where the input RF signal polarity initializes a *bunching effect* on electrons in the beam. The bunching effect excites subsequent cavities, which increase the bunching through an energy flywheel concept. Finally, the beam passes to an output cavity that couples the amplified signal to the load (antenna system). The beam falls onto a collector element that forms the return path for the current and dissipates the heat resulting from electron beam bombardment. **(3—reflex)** A klystron with only one cavity. The action is the same as in a two-cavity klystron but the beam is reflected back into the cavity in which it was first excited, after being sent out to a reflector. The one cavity, therefore, acts both as the original exciter (or buncher) and as the collector from which the output is taken.

knee In a response curve, the region of maximum curvature.

ku band Radio frequencies in the range of 15.35 GHz to 17.25 GHz, typically used for satellite telecommunications.

L

ladder network A type of filter with components alternately across the line and in the line.

lag The difference in phase between a current and the voltage that produced it, expressed in electrical degrees.

lagging current A current that lags behind the alternating electromotive force that produced it. A circuit that produces a *lagging current* is one containing inductance alone, or whose effective impedance is inductive.

lagging load A load whose combined inductive reactance exceeds its capacitive reactance. When an alternating voltage is applied, the current lags behind the voltage.

laminate A material consisting of layers of the same or different materials bonded together and built up to the required thickness.

latitude An angular measurement of a point on the earth above or below the equator. The equator represents 0°, the north pole +90°, and the south pole –90°.

layout A proposed or actual arrangement or allocation of equipment.

LC circuit An electrical circuit with both inductance (L) and capacitance (C) that is resonant at a particular frequency.

LC ratio The ratio of inductance to capacitance in a given circuit.

lead An electrical wire, usually insulated.

leading edge The initial portion of a pulse or wave in which voltage or current rise rapidly from zero to a final value.

leading load A reactive load in which the reactance of capacitance is greater than that of inductance. Current through such a load *leads* the applied voltage causing the current.

leakage The loss of energy resulting from the flow of electricity past an insulating material, the escape of electromagnetic radiation beyond its shielding, or the extension of magnetic lines of force beyond their intended working area.

leakage resistance The resistance of a path through which leakage current flows.

level The strength or intensity of a given signal.

level alignment The adjustment of transmission levels of single links and links in tandem to prevent overloading of transmission subsystems.

life cycle The predicted useful life of a class of equipment, operating under normal (specified) working conditions.

life safety system A system designed to protect life and property, such as emergency lighting, fire alarms, smoke exhaust and ventilating fans, and site security.

life test A test in which random samples of a product are checked to see how long they can continue to perform their functions satisfactorily. A form of *stress testing* is used, including temperature, current, voltage, and/or vibration effects, cycled at many times the rate that would apply in normal usage.

limiter An electronic device in which some characteristic of the output is automatically prevented from exceeding a predetermined value.

limiter circuit A circuit of nonlinear elements that restricts the electrical excursion of a variable in accordance with some specified criteria.

limiting A process by which some characteristic at the output of a device is prevented from exceeding a predetermined value.

line loss The total end-to-end loss in decibels in a transmission line.

line-up The process of adjusting transmission parameters to bring a circuit to its specified values.

linear A circuit, device, or channel whose output is directly proportional to its input.

linear distortion A distortion mechanism that is independent of signal amplitude.

linearity A constant relationship, over a designated range, between the input and output characteristics of a circuit or device.

lines of force A group of imaginary lines indicating the direction of the electric or magnetic field at all points along it.

lissajous pattern The looping patterns generated by a CRT spot when the horizontal (X) and vertical (Y) deflection signals are sinusoids. The lissajous pattern is useful for evaluating the delay or phase of two sinusoids of the same frequency.

live A device or system connected to a source of electric potential.

load The work required of an electrical or mechanical system.

load factor The ratio of the average load over a designated period of time to the peak load occurring during the same period.

load line A straight line drawn across a grouping of plate current/plate voltage characteristic curves showing the relationship between grid voltage and plate current for a particular plate load resistance of an electron tube.

logarithm The power to which a base must be raised to produce a given number. Common logarithms are to base 10.

logarithmic scale A meter scale with displacement proportional to the logarithm of the quantity represented.

long persistence The quality of a cathode ray tube that has phosphorescent compounds on its screen (in addition to fluorescent compounds) so that the image continues to glow after the original electron beam has ceased to create it by producing the usual fluorescence effect. Long persistence is often used in radar screens or where photographic evidence is needed of a display. Most such applications, however, have been superseded through the use of digital storage techniques.

longitude The angular measurement of a point on the surface of the earth in relation to the meridian of Greenwich (London). The earth is divided into 360° of longitude, beginning at the Greenwich mean. As one travels west around the globe, the longitude increases.

longitudinal current A current that travels in the same direction on both wires of a pair. The return current either flows in another pair or via a ground return path.

loss The power dissipated in a circuit, usually expressed in decibels, that performs no useful work.

loss deviation The change of actual loss in a circuit or system from a designed value.

loss variation The change in actual measured loss over time.

lossy The condition when the line loss per unit length is significantly greater than some defined normal parameter.

lossy cable A coaxial cable constructed to have high transmission loss so it can be used as an artificial load or as an attenuator.

lot size A specific quantity of similar material or a collection of similar units from a common source; in inspection work, the quantity offered for inspection and acceptance at any one time. The **lot size** may be a collection of raw material, parts, subassemblies inspected during production, or a consignment of finished products to be sent out for service.

low tension A low voltage circuit.

low-pass filter A filter network that passes all frequencies below a specified frequency with little or no loss, but that significantly attenuates higher frequencies.

lug A tag or projecting terminal onto which a wire may be connected by wrapping, soldering, or crimping.

lumped constant A resistance, inductance, or capacitance connected at a point, and not distributed uniformly throughout the length of a route or circuit.

M

mA An abbreviation for *milliamperes* (0.001 A).

magnet A device that produces a magnetic field and can attract iron, and attract or repel other magnets.

magnetic field An energy field that exists around magnetic materials and current-carrying conductors. Magnetic fields combine with electric fields in light and radio waves.

magnetic flux The field produced in the area surrounding a magnet or electric current. The standard unit of flux is the *Weber*.

magnetic flux density A vector quantity measured by a standard unit called the *Tesla*. The *magnetic flux density* is the number of magnetic lines of force per unit area, at right angles to the lines.

magnetic leakage The magnetic flux that does not follow a useful path.

magnetic pole A point that appears from the outside to be the center of magnetic attraction or repulsion at or near one end of a magnet.

magnetic storm A violent local variation in the earth's magnetic field, usually the result of sunspot activity.

magnetism A property of iron and some other materials by which external magnetic fields are maintained, other magnets being thereby attracted or repelled.

magnetization The exposure of a magnetic material to a magnetizing current, field, or force.

magnetizing force The force producing magnetization.

magnetomotive force The force that tends to produce lines of force in a magnetic circuit. The *magnetomotive force* bears the same relationship to a magnetic circuit that voltage does to an electrical circuit.

magnetron A high-power, ultra high frequency electron tube oscillator that employs the interaction of a strong electric field between an anode and cathode with the field of a strong permanent magnet to cause oscillatory electron flow through multiple internal cavity resonators. The magnetron may operate in a continuous or pulsed mode.

maintainability The probability that a failure will be repaired within a specified time after the failure occurs.

maintenance Any activity intended to keep a functional unit in satisfactory working condition. The term includes the tests, measurements, replacements, adjustments, and repairs necessary to keep a device or system operating properly.

malfunction An equipment failure or a fault.

manometer A test device for measuring gas pressure.

margin The difference between the value of an operating parameter and the value that would result in unsatisfactory operation. Typical *margin* parameters include signal level, signal-to-noise ratio, distortion, crosstalk coupling, and/or undesired emission level.

Markov model A statistical model of the behavior of a complex system over time in which the probabilities of the occurrence of various future states depend only on the present state of the system, and not on the path by which the present state was achieved. This term was named for the Russian mathematician Andrei Andreevich Markov (1856-1922).

master clock An accurate timing device that generates a synchronous signal to control other clocks or equipment.

master oscillator A stable oscillator that provides a standard frequency signal for other hardware and/or systems.

matched termination A termination that absorbs all the incident power and so produces no reflected waves or mismatch loss.

matching The connection of channels, circuits, or devices in a manner that results in minimal reflected energy.

matrix A logical network configured in a rectangular array of intersections of input/output signals.

Maxwell's equations Four differential equations that relate electric and magnetic fields to electromagnetic waves. The equations are a basis of electrical and electronic engineering.

mean An arithmetic average in which values are added and divided by the number of such values.

mean time between failures (MTBF) For a particular interval, the total functioning life of a population of an item divided by the total number of failures within the population during the measurement interval.

mean time to failure (MTTF) The measured operating time of a single piece of equipment divided by the total number of failures during the measured period of time. This measurement is normally made during that period between early life and wear-out failures.

mean time to repair (MTTR) The total corrective maintenance time on a component or system divided by the total number of corrective maintenance actions during a given period of time.

measurement A procedure for determining the amount of a quantity.

median A value in a series that has as many readings or values above it as below.

medium An electronic pathway or mechanism for passing information from one point to another.

mega A prefix meaning one million.

megahertz (MHz) A quantity equal to one million Hertz (cycles per second).

megohm A quantity equal to one million ohms.

metric system A decimal system of measurement based on the meter, the kilogram, and the second.

micro A prefix meaning one millionth.

micron A unit of length equal to one millionth of a meter (1/25,000 of an inch).

microphonic(s) Unintended noise introduced into an electronic system by mechanical vibration of electrical components.

microsecond One millionth of a second (0.000001 s).

microvolt A quantity equal to one-millionth of a volt.

milli A prefix meaning one thousandth.

milliammeter A test instrument for measuring electrical current, often part of a *multimeter*.

millihenry A quantity equal to one-thousandth of a henry.

milliwatt A quantity equal to one thousandth of a watt.

minimum discernible signal The smallest input that will produce a discernible change in the output of a circuit or device.

mixer A circuit used to combine two or more signals to produce a third signal that is a function of the input waveforms.

mixing ratio The ratio of the mass of water vapor to the mass of dry air in a given volume of air. The *mixing ratio* affects radio propagation.

mode An electromagnetic field distribution that satisfies theoretical requirements for propagation in a waveguide or oscillation in a cavity.

modified refractive index The sum of the refractive index of the air at a given height above sea level, and the ratio of this height to the radius of the earth.

modular An equipment design in which major elements are readily separable, and which the user may replace, reducing the mean-time-to-repair.

modulation The process whereby the amplitude, frequency, or phase of a single-frequency wave (the *carrier*) is varied in step with the instantaneous value of, or samples of, a complex wave (the *modulating wave*).

modulator A device that enables the intelligence in an information-carrying modulating wave to be conveyed by a signal at a higher frequency. A *modulator* modifies a carrier wave by amplitude, phase, and/or frequency as a function of a control signal that carries intelligence. Signals are *modulated* in this way to permit more efficient and/or reliable transmission over any of several media.

module An assembly replaceable as an entity, often as an interchangeable plug-in item. A *module* is not normally capable of being disassembled.

monostable A device that is stable in one state only. An input pulse causes the device to change state, but it reverts immediately to its stable state.

motor A machine that converts electrical energy into mechanical energy.

motor effect The repulsion force exerted between adjacent conductors carrying currents in opposite directions.

moving coil Any device that utilizes a coil of wire in a magnetic field in such a way that the coil is made to move by varying the applied current, or itself produces a varying voltage because of its movement.

ms An abbreviation for *millisecond* (0.001 s).

multimeter A test instrument fitted with several ranges for measuring voltage, resistance, and current, and equipped with an analog meter or digital display readout. The *multimeter* is also known as a *volt-ohm-milliammeter*, or *VOM*.

multiplex (MUX) The use of a common channel to convey two or more channels. This is done either by splitting of the common channel frequency band into narrower bands, each of which is used to constitute a distinct channel (*frequency division multiplex*), or by allotting this common channel to multiple users in turn to constitute different intermittent channels (*time division multiplex*).

multiplexer A device or circuit that combines several signals onto a single signal.

multiplexing A technique that uses a single transmission path to carry multiple channels. In *time division multiplexing* (TDM), path time is shared. For *frequency division multiplexing* (FDM) or *wavelength division multiplexing* (WDM), signals are divided into individual channels sent along the same path but at different frequencies.

multiplication Signal mixing that occurs within a multiplier circuit.

multiplier A circuit in which one or more input signals are mixed under the direction of one or more control signals. The resulting output is a composite of the input signals, the characteristics of which are determined by the scaling specified for the circuit.

mutual induction The property of the magnetic flux around a conductor that induces a voltage in a nearby conductor. The voltage generated in the secondary conductor in turn induces a voltage in the primary conductor. The inductance of two conductors so coupled is referred to as *mutual inductance*.

mV An abbreviation for *millivolt* (0.001 V).

mW An abbreviation for *milliwatt* (0.001 W).

N

nano A prefix meaning one billionth.

nanometer 1×10^{-9} meter.

nanosecond (ns) One billionth of a second (1×10^{-9} s).

narrowband A communications channel of restricted bandwidth, often resulting in degradation of the transmitted signal.

narrowband emission An emission having a spectrum exhibiting one or more sharp peaks that are narrow in width compared to the nominal bandwidth of the measuring instrument, and are far enough apart in frequency to be resolvable by the instrument.

National Electrical Code (NEC) A document providing rules for the installation of electric wiring and equipment in public and private buildings, published by the National Fire Protection Association. The NEC has been adopted as law by many states and municipalities in the U.S.

National Institute of Standards and Technology (NIST) A nonregulatory agency of the Department of Commerce that serves as a national reference and measurement laboratory for the physical and engineering sciences. Formerly called the *National Bureau of Standards*, the agency was renamed in 1988 and given the additional responsibility of aiding U.S. companies in adopting new technologies to increase their international competitiveness.

negative In a conductor or semiconductor material, an excess of electrons or a deficiency of positive charge.

negative feedback The return of a portion of the output signal from a circuit to the input but 180° out of phase. This type of feedback decreases signal amplitude but stabilizes the amplifier and reduces distortion and noise.

negative impedance An impedance characterized by a decrease in voltage drop across a device as the current through the device is increased, or a decrease in current through the device as the voltage across it is increased.

neutral A device or object having no electrical charge.

neutral conductor A conductor in a power distribution system connected to a point in the system that is designed to be at neutral potential. In a balanced system, the neutral conductor carries no current.

neutral ground An intentional ground applied to the neutral conductor or neutral point of a circuit, transformer, machine, apparatus, or system.

newton The standard unit of force. One *newton* is the force that, when applied to a body having a mass of 1 kg, gives it an acceleration of 1 m/s^2.

nitrogen A gas widely used to pressurize radio frequency transmission lines. If a small puncture occurs in the cable sheath, the nitrogen keeps moisture out so that service is not adversely affected.

node The points at which the current is at minimum in a transmission system in which standing waves are present.

noise Any random disturbance or unwanted signal in a communication system that tends to obscure the clarity or usefulness of a signal in relation to its intended use.

noise factor (NF) The ratio of the noise power measured at the output of a receiver to the noise power that would be present at the output if the thermal noise resulting from the resistive component of the source impedance were the only source of noise in the system.

noise figure A measure of the noise in dB generated at the input of an amplifier, compared with the noise generated by an impedance-method resistor at a specified temperature.

noise filter A network that attenuates noise frequencies.

noise generator A generator of wideband random noise.

noise power ratio (NPR) The ratio, expressed in decibels, of signal power to intermodulation product power plus residual noise power, measured at the baseband level.

noise suppressor A filter or digital signal processing circuit in a receiver or transmitter that automatically reduces or eliminates noise.

noise temperature The temperature, expressed in Kelvin, at which a resistor will develop a particular noise voltage. The noise temperature of a radio receiver is the value by which the temperature of the resistive component of the source impedance should be increased—if it were the only source of noise in the system—to cause the noise power at the output of the receiver to be the same as in the real system.

nominal The most common value for a component or parameter that falls between the maximum and minimum limits of a tolerance range.

nominal value A specified or intended value independent of any uncertainty in its realization.

nomogram A chart showing three or more scales across which a straight edge may be held in order to read off a graphical solution to a three-variable equation.

nonionizing radiation Electromagnetic radiation that does not turn an atom into an ion. Examples of nonionizing radiation include visible light and radio waves.

nonconductor A material that does not conduct energy, such as electricity, heat, or sound.

noncritical technical load That part of the technical power load for a facility not required for minimum acceptable operation.

noninductive A device or circuit without significant inductance.

nonlinearity A distortion in which the output of a circuit or system does not rise or fall in direct proportion to the input.

nontechnical load The part of the total operational load of a facility used for such purposes as general lighting, air conditioning, and ventilating equipment during normal operation.

normal A line perpendicular to another line or to a surface.

normal-mode noise Unwanted signals in the form of voltages appearing in line-to-line and line-to-neutral signals.

normalized frequency The ratio between the actual frequency and its nominal value.

normalized frequency departure The frequency departure divided by the nominal frequency value.

normalized frequency difference The algebraic difference between two normalized frequencies.

normalized frequency drift The frequency drift divided by the nominal frequency value.

normally closed Switch contacts that are closed in their nonoperated state, or relay contacts that are closed when the relay is de-energized.

normally open Switch contacts that are open in their nonoperated state, or relay contacts that are open when the relay is de-energized.

north pole The pole of a magnet that seeks the north magnetic pole of the earth.

notch filter A circuit designed to attenuate a specific frequency band; also known as a *band stop filter*.

notched noise A noise signal in which a narrow band of frequencies has been removed.

ns An abbreviation for *nanosecond*.

null A zero or minimum amount or position.

O

octave Any frequency band in which the highest frequency is twice the lowest frequency.

off-line A condition wherein devices or subsystems are not connected into, do not form a part of, and are not subject to the same controls as an operational system.

offset An intentional difference between the realized value and the nominal value.

ohm The unit of electric resistance through which one ampere of current will flow when there is a difference of one volt. The quantity is named for the German physicist Georg Simon Ohm (1787-1854).

Ohm's law A law that sets forth the relationship between voltage (E), current (I), and resistance (R). The law states that $E = I \times R$, $I = \dfrac{E}{R}$, and $R = \dfrac{E}{I}$. *Ohm's Law* is named for the German physicist Georg Simon Ohm (1787-1854).

ohmic loss The power dissipation in a line or circuit caused by electrical resistance.

ohmmeter A test instrument used for measuring resistance, often part of a *multimeter*.

ohms-per-volt A measure of the sensitivity of a voltmeter.

on-line A device or system that is energized and operational, and ready to perform useful work.

open An interruption in the flow of electrical current, as caused by a broken wire or connection.

open-circuit A defined loop or path that closes on itself and contains an infinite impedance.

open-circuit impedance The input impedance of a circuit when its output terminals are open, that is, not terminated.

open-circuit voltage The voltage measured at the terminals of a circuit when there is no load and, hence, no current flowing.

operating lifetime The period of time during which the principal parameters of a component or system remain within a prescribed range.

optimize The process of adjusting for the best output or maximum response from a circuit or system.

orbit The path, relative to a specified frame of reference, described by the center of mass of a satellite or other object in space, subjected solely to natural forces (mainly gravitational attraction).

order of diversity The number of independently fading propagation paths or frequencies, or both, used in a diversity reception system.

original equipment manufacturer (OEM) A manufacturer of equipment that is used in systems assembled and sold by others.

oscillation A variation with time of the magnitude of a quantity with respect to a specified reference when the magnitude is alternately greater than and smaller than the reference.

oscillator A nonrotating device for producing alternating current, the output frequency of which is determined by the characteristics of the circuit.

oscilloscope A test instrument that uses a display, usually a cathode-ray tube, to show the instantaneous values and waveforms of a signal that varies with time or some other parameter.

out-of-band energy Energy emitted by a transmission system that falls outside the frequency spectrum of the intended transmission.

outage duration The average elapsed time between the start and the end of an outage period.

outage probability The probability that an outage state will occur within a specified time period. In the absence of specific known causes of outages, the *outage probability* is the sum of all outage durations divided by the time period of measurement.

outage threshold A defined value for a supported performance parameter that establishes the minimum operational service performance level for that parameter.

output impedance The impedance presented at the output terminals of a circuit, device, or channel.

output stage The final driving circuit in a piece of electronic equipment.

ovenized crystal oscillator (OXO) A crystal oscillator enclosed within a temperature regulated heater (oven) to maintain a stable frequency despite external temperature variations.

overcoupling A degree of coupling greater than the *critical coupling* between two resonant circuits. *Overcoupling* results in a wide bandwidth circuit with two peaks in the response curve.

overload In a transmission system, a power greater than the amount the system was designed to carry. In a power system, an overload could cause excessive heating. In a communications system, distortion of a signal could result.

overshoot The first maximum excursion of a pulse beyond the 100% level. Overshoot is the portion of the pulse that exceeds its defined level temporarily before settling to the correct level. Overshoot amplitude is expressed as a percentage of the defined level.

P

pentode An electron tube with five electrodes, the cathode, control grid, screen grid, suppressor grid, and plate.

photocathode An electrode in an electron tube that will emit electrons when bombarded by photons of light.

picture tube A cathode-ray tube used to produce an image by variation of the intensity of a scanning beam on a phosphor screen.

pin A terminal on the base of a component, such as an electron tube.

plasma (1—arc) An ionized gas in an arc-discharge tube that provides a conducting path for the discharge. **(2—solar)** The ionized gas at extremely high temperature found in the sun.

plate (1—electron tube) The anode of an electron tube. **(2—battery)** An electrode in a storage battery. **(3—capacitor)** One of the surfaces in a capacitor. **(4—chassis)** A mounting surface to which equipment may be fastened.

propagation time delay The time required for a signal to travel from one point to another.

protector A device or circuit that prevents damage to lines or equipment by conducting dangerously high voltages or currents to ground. Protector types include spark gaps, semiconductors, varistors, and gas tubes.

proximity effect A nonuniform current distribution in a conductor, caused by current flow in a nearby conductor.

pseudonoise In a spread-spectrum system, a seemingly random series of pulses whose frequency spectrum resembles that of continuous noise.

pseudorandom A sequence of signals that appears to be completely random but have, in fact, been carefully drawn up and repeat after a significant time interval.

pseudorandom noise A noise signal that satisfies one or more of the standard tests for statistical randomness. Although it seems to lack any definite pattern, there is a sequence of pulses that repeats after a long time interval.

pseudorandom number sequence A sequence of numbers that satisfies one or more of the standard tests for statistical randomness. Although it seems to lack any definite pattern, there is a sequence that repeats after a long time interval.

pulsating direct current A current changing in value at regular or irregular intervals but which has the same direction at all times.

pulse One of the elements of a repetitive signal characterized by the rise and decay in time of its magnitude. A *pulse* is usually short in relation to the time span of interest.

pulse decay time The time required for the trailing edge of a pulse to decrease from 90 percent to 10 percent of its peak amplitude.

pulse duration The time interval between the points on the leading and trailing edges of a pulse at which the instantaneous value bears a specified relation to the peak pulse amplitude.

pulse duration modulation (PDM) The modulation of a pulse carrier by varying the width of the pulses according to the instantaneous values of the voltage samples of the modulating signal (also called *pulse width modulation*).

pulse edge The leading or trailing edge of a pulse, defined as the 50 percent point of the pulse rise or fall time.

pulse fall time The interval of time required for the edge of a pulse to fall from 90 percent to 10 percent of its peak amplitude.

pulse interval The time between the start of one pulse and the start of the next.

pulse length The duration of a pulse (also called *pulse width*).

pulse level The voltage amplitude of a pulse.

pulse period The time between the start of one pulse and the start of the next.

pulse ratio The ratio of the length of any pulse to the total pulse period.

pulse repetition period The time interval from the beginning of one pulse to the beginning of the next pulse.

pulse repetition rate The number of times each second that pulses are transmitted.

pulse rise time The time required for the leading edge of a pulse to rise from 10 percent to 90 percent of its peak amplitude.

pulse train A series of pulses having similar characteristics.

pulse width The measured interval between the 50 percent amplitude points of the leading and trailing edges of a pulse.

puncture A breakdown of insulation or of a dielectric, such as in a cable sheath or in the insulant around a conductor.

pW An abbreviation for picowatt, a unit of power equal to 10^{-12} W (-90 dBm).

Q

Q (quality factor) A figure of merit that defines how close a coil comes to functioning as a pure inductor. *High Q* describes an inductor with little energy loss resulting from resistance. Q is found by dividing the inductive reactance of a device by its resistance.

quadrature A state of alternating current signals separated by one quarter of a cycle (90°).

quadrature amplitude modulation (QAM) A process that allows two different signals to modulate a single carrier frequency. The two signals of interest amplitude modulate two samples of the carrier that are of the same frequency, but differ in phase by 90°. The two resultant signals can be added and transmitted. Both signals may be recovered at a decoder when they are demodulated 90° apart.

quadrature component The component of a voltage or current at an angle of 90° to a reference signal, resulting from inductive or capacitive reactance.

quadrature phase shift keying (QPSK) A type of phase shift keying using four phase states.

quality The absence of objectionable distortion.

quality assurance (QA) All those activities, including surveillance, inspection, control, and documentation, aimed at ensuring that a given product will meet its performance specifications.

quality control (QC) A function whereby management exercises control over the quality of raw material or intermediate products in order to prevent the production of defective devices or systems.

quantum noise Any noise attributable to the discrete nature of electromagnetic radiation. Examples include shot noise, photon noise, and recombination noise.

quantum-limited operation An operation wherein the minimum detectable signal is limited by quantum noise.

quartz A crystalline mineral that when electrically excited vibrates with a stable period. Quartz is typically used as the frequency-determining element in oscillators and filters.

quasi-peak detector A detector that delivers an output voltage that is some fraction of the peak value of the regularly repeated pulses applied to it. The fraction increases toward unity as the pulse repetition rate increases.

quick-break fuse A fuse in which the fusible link is under tension, providing for rapid operation.

quiescent An inactive device, signal, or system.

quiescent current The current that flows in a device in the absence of an applied signal.

R

rack An equipment rack, usually measuring 19 in (48.26 cm) wide at the front mounting rails.

rack unit (RU) A unit of measure of vertical space in an equipment enclosure. One rack unit is equal to 1.75 in (4.45 cm).

radiate The process of emitting electromagnetic energy.

radiation The emission and propagation of electromagnetic energy in the form of waves. *Radiation* is also called *radiant energy*.

radiation scattering The diversion of thermal, electromagnetic, or nuclear radiation from its original path as a result of interactions or collisions with atoms, molecules, or large particles in the atmosphere or other media between the source of radiation and a point some distance away. As a result of scattering, radiation (especially gamma rays and neutrons) will be received at such a point from many directions, rather than only from the direction of the source.

radio The transmission of signals over a distance by means of electromagnetic waves in the approximate frequency range of 150 kHz to 300 GHz. The term may also be used to describe the equipment used to transmit or receive electromagnetic waves.

radio detection The detection of the presence of an object by radio location without precise determination of its position.

radio frequency interference (RFI) The intrusion of unwanted signals or electromagnetic noise into various types of equipment resulting from radio frequency transmission equipment or other devices using radio frequencies.

radio frequency spectrum Those frequency bands in the electromagnetic spectrum that range from several hundred thousand cycles per second (*very low frequency*) to several billion cycles per second (*microwave frequencies*).

radio recognition In military communications, the determination by radio means of the "friendly" or "unfriendly" character of an aircraft or ship.

random noise Electromagnetic signals that originate in transient electrical disturbances and have random time and amplitude patterns. Random noise is generally undesirable; however, it may also be generated for testing purposes.

rated output power The power available from an amplifier or other device under specified conditions of operation.

RC constant The time constant of a resistor-capacitor circuit. The *RC constant* is the time in seconds required for current in an RC circuit to rise to 63 percent of its final steady value or fall to 37 percent of its original steady value, obtained by multiplying the resistance value in ohms by the capacitance value in farads.

RC network A circuit that contains resistors and capacitors, normally connected in series.

reactance The part of the impedance of a network resulting from inductance or capacitance. The *reactance* of a component varies with the frequency of the applied signal.

reactive power The power circulating in an ac circuit. It is delivered to the circuit during part of the cycle and is returned during the other half of the cycle. The *reactive power* is obtained by multiplying the voltage, current, and the sine of the phase angle between them.

reactor A component with inductive reactance.

received signal level (RSL) The value of a specified bandwidth of signals at the receiver input terminals relative to an established reference.

receiver Any device for receiving electrical signals and converting them to audible sound, visible light, data, or some combination of these elements.

receptacle An electrical socket designed to receive a mating plug.

reception The act of receiving, listening to, or watching information-carrying signals.

rectification The conversion of alternating current into direct current.

rectifier A device for converting alternating current into direct current. A *rectifier* normally includes filters so that the output is, within specified limits, smooth and free of ac components.

rectify The process of converting alternating current into direct current.

redundancy A system design that provides a back-up for key circuits or components in the event of a failure. Redundancy improves the overall reliability of a system.

redundant A configuration when two complete systems are available at one time. If the online system fails, the backup will take over with no loss of service.

reference voltage A voltage used for control or comparison purposes.

reflectance The ratio of reflected power to incident power.

reflection An abrupt change, resulting from an impedance mismatch, in the direction of propagation of an electromagnetic wave. For light, at the interface of two dissimilar materials, the incident wave is returned to its medium of origin.

reflection coefficient The ratio between the amplitude of a reflected wave and the amplitude of the incident wave. For large smooth surfaces, the reflection coefficient may be near unity.

reflection gain The increase in signal strength that results when a reflected wave combines, in phase, with an incident wave.

reflection loss The apparent loss of signal strength caused by an impedance mismatch in a transmission line or circuit. The loss results from the reflection of part of the signal back toward the source from the point of the impedance discontinuity. The greater the mismatch, the greater the loss.

reflectometer A device that measures energy traveling in each direction in a waveguide, used in determining the standing wave ratio.

refraction The bending of a sound, radio, or light wave as it passes obliquely from a medium of one density to a medium of another density that varies its speed.

regulation The process of adjusting the level of some quantity, such as circuit gain, by means of an electronic system that monitors an output and feeds back a controlling signal to constantly maintain a desired level.

regulator A device that maintains its output voltage at a constant level.

relative envelope delay The difference in envelope delay at various frequencies when compared with a reference frequency that is chosen as having zero delay.

relative humidity The ratio of the quantity of water vapor in the atmosphere to the quantity that would cause saturation at the ambient temperature.

relative transmission level The ratio of the signal power in a transmission system to the signal power at some point chosen as a reference. The ratio is usually determined by applying a standard test signal at the input to the system and measuring the gain or loss at the location of interest.

relay A device by which current flowing in one circuit causes contacts to operate that control the flow of current in another circuit.

relay armature The movable part of an electromechanical relay, usually coupled to spring sets on which contacts are mounted.

relay bypass A device that, in the event of a loss of power or other failure, routes a critical signal around the equipment that has failed.

release time The time required for a pulse to drop from steady-state level to zero, also referred to as the *decay time*.

reliability The ability of a system or subsystem to perform within the prescribed parameters of quality of service. *Reliability* is often expressed as the probability that a system or subsystem will perform its intended function for a specified interval under stated conditions.

reliability growth The action taken to move a hardware item toward its reliability potential, during development or subsequent manufacturing or operation.

reliability predictions The compiled failure rates for parts, components, subassemblies, assemblies, and systems. These generic failure rates are used as basic data to predict the reliability of a given device or system.

remote control A system used to control a device from a distance.

remote station A station or terminal that is physically remote from a main station or center but can gain access through a communication channel.

repeater The equipment between two circuits that receives a signal degraded by normal factors during transmission and amplifies the signal to its original level for retransmission.

repetition rate The rate at which regularly recurring pulses are repeated.

reply A transmitted message that is a direct response to an original message.

repulsion The mechanical force that tends to separate like magnetic poles, like electric charges, or conductors carrying currents in opposite directions.

reset The act of restoring a device to its default or original state.

residual flux The magnetic flux that remains after a magnetomotive force has been removed.

residual magnetism The magnetism or flux that remains in a core after current ceases to flow in the coil producing the magnetomotive force.

residual voltage The vector sum of the voltages in all the phase wires of an unbalanced polyphase power system.

resistance The opposition of a material to the flow of electrical current. Resistance is equal to the voltage drop through a given material divided by the current flow through it. The standard unit of resistance is the *ohm*, named for the German physicist Georg Simon Ohm (1787-1854).

resistance drop The fall in potential (volts) between two points, the product of the current and resistance.

resistance-grounded A circuit or system grounded for safety through a resistance, which limits the value of the current flowing through the circuit in the event of a fault.

resistive load A load in which the voltage is in phase with the current.

resistivity The resistance per unit volume or per unit area.

resistor A device the primary function of which is to introduce a specified resistance into an electrical circuit.

resonance A tuned condition conducive to oscillation, when the reactance resulting from capacitance in a circuit is equal in value to the reactance resulting from inductance.

resonant frequency The frequency at which the inductive reactance and capacitive reactance of a circuit are equal.

resonator A resonant cavity.

return A return path for current, sometimes through ground.

reversal A change in magnetic polarity, in the direction of current flow.

reverse current A small current that flows through a diode when the voltage across it is such that normal forward current does not flow.

reverse voltage A voltage in the reverse direction from that normally applied.

rheostat A two-terminal variable resistor, usually constructed with a sliding or rotating shaft that can be used to vary the resistance value of the device.

ripple An ac voltage superimposed on the output of a dc power supply, usually resulting from imperfect filtering.

rise time The time required for a pulse to rise from 10 percent to 90 percent of its peak value.

roll-off A gradual attenuation of gain-frequency response at either or both ends of a transmission pass band.

root-mean-square (RMS) The square root of the average value of the squares of all the instantaneous values of current or voltage during one half-cycle of an alternating current. For an alternating current, the RMS voltage or current is equal to the amount of direct current or voltage that would produce the same heating effect in a purely resistive circuit. For a sinewave, the root-mean-square value is equal to 0.707 times the peak value. RMS is also called the *effective value*.

rotor The rotating part of an electric generator or motor.

RU An abbreviation for *rack unit*.

S

scan One sweep of the target area in a camera tube, or of the screen in a picture tube.

screen grid A grid in an electron tube that improves performance of the device by shielding the control grid from the plate.

self-bias The provision of bias in an electron tube through a voltage drop in the cathode circuit.

shot noise The noise developed in a vacuum tube or photoconductor resulting from the random number and velocity of emitted charge carriers.

slope The rate of change, with respect to frequency, of transmission line attenuation over a given frequency spectrum.

slope equalizer A device or circuit used to achieve a specified slope in a transmission line.

smoothing circuit A filter designed to reduce the amount of ripple in a circuit, usually a dc power supply.

snubber An electronic circuit used to suppress high frequency noise.

solar wind Charged particles from the sun that continuously bombard the surface of the earth.

solid A single wire conductor, as contrasted with a stranded, braided, or rope-type wire.

solid-state The use of semiconductors rather than electron tubes in a circuit or system.

source The part of a system from which signals or messages are considered to originate.

source terminated A circuit whose output is terminated for correct impedance matching with standard cable.

spare A system that is available but not presently in use.

spark gap A gap between two electrodes designed to produce a spark under given conditions.

specific gravity The ratio of the weight of a volume, liquid, or solid to the weight of the same volume of water at a specified temperature.

spectrum A continuous band of frequencies within which waves have some common characteristics.

spectrum analyzer A test instrument that presents a graphic display of signals over a selected frequency bandwidth. A cathode-ray tube is often used for the display.

spectrum designation of frequency A method of referring to a range of communication frequencies. In American practice, the designation is a two or three letter acronym for the name. The ranges are: below 300 Hz, ELF (extremely low frequency); 300 Hz–3000 Hz, ILF (infra low frequency); 3 kHz–30 kHz, VLF (very low frequency); 30 kHz–300 kHz, LF (low frequency); 300 kHz–3000 kHz, MF (medium frequency); 3 MHz–30 MHz, HF (high frequency); 30 MHz–300 MHz, VHF (very high frequency); 300 MHz–3000 MHz, UHF (ultra high frequency); 3 GHz–30 GHz, SHF (super high frequency); 30 GHz–300 GHz, EHF (extremely high frequency); 300 GHz–3000 GHz, THF (tremendously high frequency).

spherical antenna A type of satellite receiving antenna that permits more than one satellite to be accessed at any given time. A spherical antenna has a broader angle of acceptance than a parabolic antenna.

spike A high amplitude, short duration pulse superimposed on an otherwise regular waveform.

split-phase A device that derives a second phase from a single phase power supply by passing it through a capacitive or inductive reactor.

splitter A circuit or device that accepts one input signal and distributes it to several outputs.

splitting ratio The ratio of the power emerging from the output ports of a coupler.

sporadic An event occurring at random and infrequent intervals.

spread spectrum A communications technique in which the frequency components of a narrowband signal are spread over a wide band. The resulting signal resembles white noise. The technique is used to achieve signal security and privacy, and to enable the use of a common band by many users.

spurious signal Any portion of a given signal that is not part of the fundamental waveform. Spurious signals include transients, noise, and hum.

square wave A square or rectangular-shaped periodic wave that alternately assumes two fixed values for equal lengths of time, the transition being negligible in comparison with the duration of each fixed value.

square wave testing The use of a square wave containing many odd harmonics of the fundamental frequency as an input signal to a device. Visual examination of the output signal on an oscilloscope indicates the amount of distortion introduced.

stability The ability of a device or circuit to remain stable in frequency, power level, and/or other specified parameters.

standard The specific signal configuration, reference pulses, voltage levels, and other parameters that describe the input/output requirements for a particular type of equipment.

standard time and frequency signal A time-controlled radio signal broadcast at scheduled intervals on a number of different frequencies by government-operated radio stations to provide a method for calibrating instruments.

standing wave ratio (SWR) The ratio of the maximum to the minimum value of a component of a wave in a transmission line or waveguide, such as the maximum voltage to the minimum voltage.

static charge An electric charge on the surface of an object, particularly a dielectric.

station One of the input or output points in a communications system.

stator The stationary part of a rotating electric machine.

status The present condition of a device.

statute mile A unit of distance equal to 1,609 km or 5,280 ft.

steady-state A condition in which circuit values remain essentially constant, occurring after all initial transients or fluctuating conditions have passed.

steady-state condition A condition occurring after all initial transient or fluctuating conditions have damped out in which currents, voltages, or fields remain essentially constant or oscillate uniformly without changes in characteristics such as amplitude, frequency, or wave shape.

steep wavefront A rapid rise in voltage of a given signal, indicating the presence of high frequency odd harmonics of a fundamental wave frequency.

step up (or down) The process of increasing (or decreasing) the voltage of an electrical signal, as in a step-up (or step-down) transformer.

straight-line capacitance A capacitance employing a variable capacitor with plates so shaped that capacitance varies directly with the angle of rotation.

stray capacitance An unintended—and usually undesired—capacitance between wires and components in a circuit or system.

stray current A current through a path other than the intended one.

stress The force per unit of cross-sectional area on a given object or structure

subassembly A functional unit of a system.

subcarrier (SC) A carrier applied as modulation on another carrier, or on an intermediate subcarrier.

subharmonic A frequency equal to the fundamental frequency of a given signal divided by a whole number.

submodule A small circuit board or device that mounts on a larger module or device.

subrefraction A refraction for which the refractivity gradient is greater than standard.

subsystem A functional unit of a system.

superheterodyne receiver A radio receiver in which all signals are first converted to a common frequency for which the intermediate stages of the receiver have been optimized, both for tuning and filtering. Signals are converted by mixing them with the output of a local oscillator whose output is varied in accordance with the frequency of the received signals so as to maintain the desired *intermediate frequency*.

suppressor grid The fifth grid of a pentode electron tube, which provides screening between plate and screen grid.

surface leakage A leakage current from line to ground over the face of an insulator supporting an open wire route.

surface refractivity The refractive index, calculated from observations of pressure, temperature, and humidity at the surface of the earth.

surge A rapid rise in current or voltage, usually followed by a fall back to the normal value.

susceptance The reciprocal of reactance, and the imaginary component of admittance, expressed in siemens.

sweep The process of varying the frequency of a signal over a specified bandwidth.

sweep generator A test oscillator, the frequency of which is constantly varied over a specified bandwidth.

switching The process of making and breaking (connecting and disconnecting) two or more electrical circuits.

synchronization The process of adjusting the corresponding significant instants of signals—for example, the zero-crossings—to make them synchronous. The term *synchronization* is often abbreviated as *sync*.

synchronize The process of causing two systems to operate at the same speed.

synchronous In step or in phase, as applied to two or more devices; a system in which all events occur in a predetermined timed sequence.

synchronous detection A demodulation process in which the original signal is recovered by multiplying the modulated signal by the output of a synchronous oscillator locked to the carrier.

synchronous system A system in which the transmitter and receiver are operating in a fixed time relationship.

system standards The minimum required electrical performance characteristics of a specific collection of hardware and/or software.

systems analysis An analysis of a given activity to determine precisely what must be accomplished and how it is to be done.

T

tetrode A four element electron tube consisting of a cathode, control grid, screen grid, and plate.

thyratron A gas-filled electron tube in which plate current flows when the grid voltage reaches a predetermined level. At that point, the grid has no further control over the current, which continues to flow until it is interrupted or reversed.

tolerance The permissible variation from a standard.

torque A moment of force acting on a body and tending to produce rotation about an axis.

total harmonic distortion (THD) The ratio of the sum of the amplitudes of all signals harmonically related to the fundamental versus the amplitude of the fundamental signal. THD is expressed in percent.

trace The pattern on an oscilloscope screen when displaying a signal.

tracking The locking of tuned stages in a radio receiver so that all stages are changed appropriately as the receiver tuning is changed.

trade-off The process of weighing conflicting requirements and reaching a compromise decision in the design of a component or a subsystem.

transceiver Any circuit or device that receives and transmits signals.

transconductance The mutual conductance of an electron tube expressed as the change in plate current divided by the change in control grid voltage that produced it.

transducer A device that converts energy from one form to another.

transfer characteristics The intrinsic parameters of a system, subsystem, or unit of equipment which, when applied to the input of the system, subsystem, or unit of equipment, will fully describe its output.

transformer A device consisting of two or more windings wrapped around a single core or linked by a common magnetic circuit.

transformer ratio The ratio of the number of turns in the secondary winding of a transformer to the number of turns in the primary winding, also known as the *turns ratio*.

transient A sudden variance of current or voltage from a steady-state value. A transient normally results from changes in load or effects related to switching action.

transient disturbance A voltage pulse of high energy and short duration impressed upon the ac waveform. The overvoltage pulse can be one to 100 times the normal ac potential (or more) and can last up to 15 ms. Rise times measure in the nanosecond range.

transient response The time response of a system under test to a stated input stimulus.

transition A sequence of actions that occurs when a process changes from one state to another in response to an input.

transmission The transfer of electrical power, signals, or an intelligence from one location to another by wire, fiber optic, or radio means.

transmission facility A transmission medium and all the associated equipment required to transmit information.

transmission loss The ratio, in decibels, of the power of a signal at a point along a transmission path to the power of the same signal at a more distant point along the same path. This value is often used as a measure of the quality of the transmission medium for conveying signals. Changes in power level are normally expressed in decibels by calculating ten times the logarithm (base 10) of the ratio of the two powers.

transmission mode One of the field patterns in a waveguide in a plane transverse to the direction of propagation.

transmission system The set of equipment that provides single or multichannel communications facilities capable of carrying audio, video, or data signals.

transmitter The device or circuit that launches a signal into a passive medium, such as the atmosphere.

transparency The property of a communications system that enables it to carry a signal without altering or otherwise affecting the electrical characteristics of the signal.

tray The metal cabinet that holds circuit boards.

tremendously high frequency (THF) The frequency band from 300 GHz to 3000 GHz.

triangular wave An oscillation, the values of which rise and fall linearly, and immediately change upon reaching their peak maximum and minimum. A graphical representation of a triangular wave resembles a triangle.

trim The process of making fine adjustments to a circuit or a circuit element.

trimmer A small mechanically-adjustable component connected in parallel or series with a major component so that the net value of the two can be finely adjusted for tuning purposes.

triode A three-element electron tube, consisting of a cathode, control grid, and plate.

triple beat A third-order beat whose three beating carriers all have different frequencies, but are spaced at equal frequency separations.

troposphere The layer of the earth's atmosphere, between the surface and the stratosphere, in which about 80 percent of the total mass of atmospheric air is concentrated and in which temperature normally decreases with altitude.

trouble A failure or fault affecting the service provided by a system.

troubleshoot The process of investigating, localizing and (if possible) correcting a fault.

tube (1—electron) An evacuated or gas-filled tube enclosed in a glass or metal case in which the electrodes are maintained at different voltages, giving rise to a controlled flow of electrons from the cathode to the anode. **(2—cathode ray, CRT)** An electron beam tube used for the display of changing electrical phenomena, generally similar to a television picture tube. **(3—cold-cathode)** An electron tube whose cathode emits electrons without the need of a heating filament. **(4—gas)** A gas-filled electron tube in which the gas plays an essential role in operation of the device. **(5—mercury-vapor)** A tube filled with mercury vapor at low pressure, used as a rectifying device. **(6—metal)** An electron tube enclosed in a metal case. **(7—traveling wave, TWT)** A wide band microwave amplifier in which a stream of electrons interacts with a guided electromagnetic wave moving substantially in synchronism with the electron stream, resulting in a net transfer of energy from the electron stream to the wave. **(8—velocity-modulated)** An electron tube in which the velocity of the electron stream is continually changing, as in a klystron.

tune The process of adjusting the frequency of a device or circuit, such as for resonance or for maximum response to an input signal.

tuned trap A series resonant network bridged across a circuit that eliminates ("traps") the frequency of the resonant network.

tuner The radio frequency and intermediate frequency parts of a radio receiver that produce a low level output signal.

tuning The process of adjusting a given frequency; in particular, to adjust for resonance or for maximum response to a particular incoming signal.

turns ratio In a transformer, the ratio of the number of turns on the secondary to the number of turns on the primary.

tweaking The process of adjusting an electronic circuit to optimize its performance.

twin-line A feeder cable with two parallel, insulated conductors.

two-phase A source of alternating current circuit with two sinusoidal voltages that are 90° apart.

U

ultra high frequency (UHF) The frequency range from 300 MHz to 3000 MHz.

ultraviolet radiation Electromagnetic radiation in a frequency range between visible light and high-frequency X-rays.

unattended A device or system designed to operate without a human attendant.

unattended operation A system that permits a station to receive and transmit messages without the presence of an attendant or operator.

unavailability A measure of the degree to which a system, subsystem, or piece of equipment is not operable and not in a committable state at the start of a mission, when the mission is called for at a random point in time.

unbalanced circuit A two-wire circuit with legs that differ from one another in resistance, capacity to earth or to other conductors, leakage, or inductance.

unbalanced line A transmission line in which the magnitudes of the voltages on the two conductors are not equal with respect to ground. A coaxial cable is an example of an unbalanced line.

unbalanced modulator A modulator whose output includes the carrier signal.

unbalanced output An output with one leg at ground potential.

unbalanced wire circuit A circuit whose two sides are inherently electrically unlike.

uncertainty An expression of the magnitude of a possible deviation of a measured value from the true value. Frequently, it is possible to distinguish two components: the *systematic uncertainty* and the *random uncertainty*. The random uncertainty is expressed by the standard deviation or by a multiple of the standard deviation. The systematic uncertainty is generally estimated on the basis of the parameter characteristics.

undamped wave A signal with constant amplitude.

underbunching A condition in a traveling wave tube wherein the tube is not operating at its optimum bunching rate.

Underwriters Laboratories, Inc. A laboratory established by the National Board of Fire Underwriters which tests equipment, materials, and systems that may affect insurance risks, with special attention to fire dangers and other hazards to life.

ungrounded A circuit or line not connected to ground.

unicoupler A device used to couple a balanced circuit to an unbalanced circuit.

unidirectional A signal or current flowing in one direction only.

uniform transmission line A transmission line with electrical characteristics that are identical, per unit length, over its entire length.

unit An assembly of equipment and associated wiring that together forms a complete system or independent subsystem.

unity coupling In a theoretically perfect transformer, complete electromagnetic coupling between the primary and secondary windings with no loss of power.

unity gain An amplifier or active circuit in which the output amplitude is the same as the input amplitude.

unity power factor A power factor of 1.0, which means that the load is—in effect—a pure resistance, with ac voltage and current completely in phase.

unterminated A device or system that is not terminated.

up-converter A frequency translation device in which the frequency of the output signal is greater than that of the input signal. Such devices are commonly found in microwave radio and satellite systems.

uplink A transmission system for sending radio signals from the ground to a satellite or aircraft.

upstream A device or system placed ahead of other devices or systems in a signal path.

useful life The period during which a low, constant failure rate can be expected for a given device or system. The *useful life* is the portion of a product life cycle between break-in and wear out.

user A person, organization, or group that employs the services of a system for the transfer of information or other purposes.

V

VA An abbreviation for *volt-amperes*, volts times amperes.

vacuum relay A relay whose contacts are enclosed in an evacuated space, usually to provide reliable long-term operation.

vacuum switch A switch whose contacts are enclosed in an evacuated container so that spark formation is discouraged.

vacuum tube An electron tube. The most common vacuum tubes include the diode, triode, tetrode, and pentode.

validity check A test designed to ensure that the quality of transmission is maintained over a given system.

varactor A semiconductor that behaves like a capacitor under the influence of an external control voltage.

varactor diode A semiconductor device whose capacitance is a function of the applied voltage. A varactor diode, also called a *variable reactance diode* or simply a *varactor*, is often used to tune the operating frequency of a radio circuit.

variable frequency oscillator (VFO) An oscillator whose frequency can be set to any required value in a given range of frequencies.

variable impedance A capacitor, inductor, or resistor that is adjustable in value.

variable-gain amplifier An amplifier whose gain can be controlled by an external signal source.

variable-reluctance A transducer in which the input (usually a mechanical movement) varies the magnetic reluctance of a device.

variation monitor A device used for sensing a deviation in voltage, current, or frequency, which is capable of providing an alarm and/or initiating transfer to another power source when programmed limits of voltage, frequency, current, or time are exceeded.

varicap A diode used as a variable capacitor.

VCXO (voltage controlled crystal oscillator) A device whose output frequency is determined by an input control voltage.

vector A quantity having both magnitude and direction.

vector diagram A diagram using vectors to indicate the relationship between voltage and current in a circuit.

vector sum The sum of two vectors which, when they are at right angles to each other, equal the length of the hypotenuse of the right triangle so formed. In the general case, the vector sum of the two vectors equals the diagonal of the parallelogram formed on the two vectors.

velocity of light The speed of propagation of electromagnetic waves in a vacuum, equal to 299,792,458 m/s, or approximately 186,000 mi/s. For rough calculations, the figure of 300,000 km/s is used.

velocity of propagation The velocity of signal transmission. In free space, electromagnetic waves travel at the speed of light. In a cable, the velocity is substantially lower.

vernier A device that enables precision reading of a measuring set or gauge, or the setting of a dial with precision.

very low frequency (VLF) A radio frequency in the band 3 kHz to 30 kHz.

vestigial sideband A form of transmission in which one sideband is significantly attenuated. The carrier and the other sideband are transmitted without attenuation.

vibration testing A testing procedure whereby subsystems are mounted on a test base that vibrates, thereby revealing any faults resulting from badly soldered joints or other poor mechanical design features.

volt The standard unit of electromotive force, equal to the potential difference between two points on a conductor that is carrying a constant current of one ampere when the power dissipated between the two points is equal to one watt. One *volt* is equivalent to the potential difference across a resistance of one ohm when one ampere is flowing through it. The volt is named for the Italian physicist Alessandro Volta (1745-1827).

volt-ampere (VA) The apparent power in an ac circuit (volts times amperes).

volt-ohm-milliammeter (VOM) A general purpose multirange test meter used to measure voltage, resistance, and current.

voltage The potential difference between two points.

voltage drop A decrease in electrical potential resulting from current flow through a resistance.

voltage gradient The continuous drop in electrical potential, per unit length, along a uniform conductor or thickness of a uniform dielectric.

voltage level The ratio of the voltage at a given point to the voltage at an arbitrary reference point.

voltage reference circuit A stable voltage reference source.

voltage regulation The deviation from a nominal voltage, expressed as a percentage of the nominal voltage.

voltage regulator A circuit used for controlling and maintaining a voltage at a constant level.

voltage stabilizer A device that produces a constant or substantially constant output voltage despite variations in input voltage or output load current.

voltage to ground The voltage between any given portion of a piece of equipment and the ground potential.

voltmeter An instrument used to measure differences in electrical potential.

vox A voice-operated relay circuit that permits the equivalent of push-to-talk operation of a transmitter by the operator.

VSAT (very small aperture terminal) A satellite Ku-band earth station intended for fixed or portable use. The antenna diameter of a VSAT is on the order of 1.5 m or less.

W

watt The unit of power equal to the work done at one joule per second, or the rate of work measured as a current of one ampere under an electric potential of one volt. Designated by the symbol *W*, the watt is named after the Scottish inventor James Watt (1736-1819).

watt meter A meter indicating in watts the rate of consumption of electrical energy.

watt-hour The work performed by one watt over a one hour period.

wave A disturbance that is a function of time or space, or both, and is propagated in a medium or through space.

wave number The reciprocal of wavelength; the number of wave lengths per unit distance in the direction of propagation of a wave.

waveband A band of wavelengths defined for some given purpose.

waveform The characteristic shape of a periodic wave, determined by the frequencies present and their amplitudes and relative phases.

wavefront A continuous surface that is a locus of points having the same phase at a given instant. A *wavefront* is a surface at right angles to rays that proceed from the wave source. The surface passes through those parts of the wave that are in the same phase and travel in the same direction. For parallel rays the wavefront is a plane; for rays that radiate from a point, the wavefront is spherical.

waveguide Generally, a rectangular or circular pipe that constrains the propagation of an acoustic or electromagnetic wave along a path between two locations. The dimensions of a waveguide determine the frequencies for optimum transmission.

wavelength For a sinusoidal wave, the distance between points of corresponding phase of two consecutive cycles.

weber The unit of magnetic flux equal to the flux that, when linked to a circuit of one turn, produces an electromotive force of one volt as the flux is reduced at a uniform rate to zero in one second. The *weber* is named for the German physicist Wilhelm Eduard Weber (1804-1891).

weighted The condition when a correction factor is applied to a measurement.

weighting The adjustment of a measured value to account for conditions that would otherwise be different or appropriate during a measurement.

weighting network A circuit, used with a test instrument, that has a specified amplitude-versus-frequency characteristic.

wideband The passing or processing of a wide range of frequencies. The meaning varies with the context.

Wien bridge An ac bridge used to measure capacitance or inductance.

winding A coil of wire used to form an inductor.

wire A single metallic conductor, usually solid-drawn and circular in cross section.

working range The permitted range of values of an analog signal over which transmitting or other processing equipment can operate.

working voltage The rated voltage that may safely be applied continuously to a given circuit or device.

X

x-band A microwave frequency band from 5.2 GHz to 10.9 GHz.

x-cut A method of cutting a quartz plate for an oscillator, with the x-axis of the crystal perpendicular to the faces of the plate.

X ray An electromagnetic radiation of approximately 100 nm to 0.1 nm, capable of penetrating nonmetallic materials.

Y

y-cut A method of cutting a quartz plate for an oscillator, with the y-axis of the crystal perpendicular to the faces of the plate.

yield strength The magnitude of mechanical stress at which a material will begin to deform. Beyond the *yield strength* point, extension is no longer proportional to stress and rupture is possible.

yoke A material that interconnects magnetic cores. *Yoke* can also refer to the deflection windings of a CRT.

yttrium-iron garnet (YIG) A crystalline material used in microwave devices.

Index of Figures

Power Vacuum Tube Applications

Figure 1.1 Patent drawing of a multistage depressed-collector klystron using six collector elements (numbered 16–21). 7

Figure 1.2 Vacuum tube operating parameters based on power and frequency. 13

Figure 1.3 Operating frequency bands for shortwave broadcasting. 14

Figure 1.4 Composite baseband stereo FM signal. A full left-only or right-only signal will modulate the main (L+R) channel to a maximum of 45 percent. The stereophonic subchannel is composed of upper sideband (USB) and lower sideband (LSB) components. 16

Figure 1.5 Principal elements of a satellite communications link. 19

Figure 1.6 The power levels in transmission of a video signal via satellite. 20

Figure 1.7 Simplified block diagram of a pulsed radar system. 21

Figure 1.8 Direction-finding error resulting from beacon reflections. 25

Figure 1.9 The concept of two-way distance ranging. 26

Figure 1.10 The concept of differential distance ranging (hyperbolic). 27

Figure 1.11 The concept of one-way distance ranging. 27

Figure 1.12 Basic schematic of a 20 kW induction heater circuit. 30

Figure 1.13 The electromagnetic spectrum. 31

Modulation Systems and Characteristics

Figure 2.1 Characteristics of a series resonant circuit as a function of frequency for a constant applied voltage and different circuit Qs: (a) magnitude, (b) phase angle. 40

Figure 2.2 Characteristics of a parallel resonant circuit as a function of frequency for different circuit Qs: (a) magnitude, (b) phase angle. 41

Figure 2.3 Cylindrical resonator incorporating a coupling loop: (a) orientation of loop with respect to cavity, (b) equivalent coupled circuit. 44

Figure 2.4 Cross section of a quartz crystal taken in the plane perpendicular to the optical axis: (a) Y-cut plate, (b) X-cut plate. 44

Figure 2.5 The effects of temperature on two types of AT-cut crystals. 45

Figure 2.6 Equivalent network of a piezoelectric resonator. 46

Figure 2.7 Frequency vs. temperature characteristics for a typical temperature-compensated crystal oscillator. 48

Figure 2.8 Plate efficiency as a function of conduction angle for an amplifier with a tuned load. 51

Figure 2.9 A plot of signal power vs. frequency for white noise and frequency limited noise. 54

Figure 2.10 Frequency-domain representation of an amplitude-modulated signal at 100 percent modulation. E_c = carrier power, F_c = frequency of the carrier, and F_m = frequency of the modulating signal. 56

Figure 2.11 Time-domain representation of an amplitude-modulated signal. Modulation at 100 percent is defined as the point at which the peak of the waveform reaches twice the carrier level, and the minimum point of the waveform is zero. 57

Figure 2.12 Simplified diagram of a high-level amplitude-modulated amplifier. 59

Figure 2.13 Idealized amplitude characteristics of the FCC standard waveform for monochrome and color TV transmission. 60

Figure 2.14 Types of suppressed carrier amplitude modulation: (*a*) the modulating signal, (*b*) double-sideband AM signal, (*c*) double-sideband suppressed carrier AM, (*d*) single-sideband suppressed carrier AM. 61

Figure 2.15 Simplified QAM modulator. 62

Figure 2.16 Simplified QAM demodulator. 63

Figure 2.17 Plot of Bessel functions of the first kind as a function of modulation index. 65

Figure 2.18 RF spectrum of a frequency-modulated signal with a modulation index of 5 and other operating parameters as shown. 67

Figure 2.19 RF spectrum of a frequency-modulated signal with a modulation index of 15 and operating parameters as shown. 67

Figure 2.20 Preemphasis curves for time constants of 50, 75, and 100 μs. 69

Figure 2.21 Block diagram of an FM exciter. 70

Figure 2.22 Simplified reactance modulator. 71

Figure 2.23 Equivalent circuit of the reactance modulator. 72

Figure 2.24 Vector diagram of a reactance modulator producing FM. Note: $V_g = V_{c2}$. 72

Figure 2.25 Voltage-controlled direct-FM modulator. 72

Figure 2.26 Mechanical layout of a common type of $\frac{1}{4}$-wave PA cavity for FM service. 73

Figure 2.27 Pulse amplitude modulation waveforms: (*a*) modulating signal; (*b*) square-topped sampling, bipolar pulse train; (*c*) topped sampling, bipolar pulse train; (*d*) square-topped sampling, unipolar pulse train; (*e*) top sampling, unipolar pulse train. 76

Figure 2.28 Pulse time modulation waveforms: (*a*) modulating signal and sample-and-hold (S/H) waveforms, (*b*) sawtooth waveform added to S/H, (*c*) leading-edge PTM, (*d*) trailing-edge PTM. 77

Figure 2.29 Pulse frequency modulation. 78

Figure 2.30 The quantization process. 78

Figure 2.31 Delta modulation waveforms: (*a*) modulating signal, (*b*) quantized modulating signal, (*c*) pulse train, (*d*) resulting delta modulation waveform. 80

Figure 2.32 Binary FSK waveform. 81

Figure 2.33 Binary PSK waveform. 81

Figure 2.34 Various baseband modulation formats: (*a*) non-return-to zero, (*b*) unipolar-return-to-zero, (*c*) differential encoded (NRZ-mark), (*d*) split phase. 83

Vacuum Tube Principles

Figure 3.1 Variation of electron emission as a function of absolute temperature for a thoriated-tungsten emitter. 91

Figure 3.2 Common types of heater and cathode structures. 91

Figure 3.3 Ratio of secondary emission current to primary current as a function of primary electron velocity. 92

Figure 3.4 Vacuum diode: (*a*) directly heated cathode, (*b*) indirectly heated cathode. 93

Figure 3.5 Anode current as a function of anode voltage in a two-electrode tube for three cathode temperatures. 93

Figure 3.6 Mechanical configuration of a power triode. 95

Figure 3.7 Constant-current characteristics for a triode with a μ of 12. 97

Figure 3.8 Constant-current characteristics for a triode with a μ of 20. 98

Figure 3.9 Grounded-grid constant-current characteristics for a zero-bias triode with a μ of 200. 98

Figure 3.10 Internal configuration of a planar triode. 99

Figure 3.11 Tetrode plate current characteristics. Plate current is plotted as a function of plate voltage, with grid voltages as shown. 100

Figure 3.12 Radial beam power tetrode (4CX15000A). 101

Figure 3.13 Principal dimensions of 4CX15000A tetrode. 103

Figure 3.14 5CX1500B power pentode device. 105

Figure 3.15 Principal dimensions of 5CX1500B pentode. 108

Figure 3.16 Continuous wave output power capability of a gridded vacuum tube. 109

Figure 3.17 Performance of a class C amplifier as the operating frequency is increased beyond the design limits of the vacuum tube. 109

Figure 3.18 Transit-time effects in a class C amplifier: (*a*) control grid voltage, (*b*) electron position as a function of time (triode case), (*c*) electron position as a function of time (tetrode case), (*d*) plate current (triode case). 111

Figure 3.19 The relationship between anode dissipation and cooling method. 113

Figure 3.20 A typical transmitter PA stage cooling system. 113

Figure 3.21 Water-cooled anode with grooves for controlled water flow. 114

Figure 3.22 Typical vapor-phase cooling system. 115

Figure 3.23 Vapor-phase cooling system for a 4-tube transmitter using a common water supply. 116

Figure 3.24 Photo of a typical oxide cathode. 118

Figure 3.25 Spiral-type tungsten filament. 120

Figure 3.26 Bar-type tungsten filament. 120

Figure 3.27 Mesh tungsten filament. 121

Figure 3.28 Mesh grid structure. 122

Figure 3.29 Pyrolytic graphite grid: (*a*) before laser processing, (*b*) completed assembly. 124

Figure 3.30 The screen grid assembly of a typical tetrode PA tube. 125

Figure 3.31 The internal arrangement of the anode, screen, control grid, and cathode assemblies of a tetrode power tube. 125

Figure 3.32 Circuit for providing a low-impedance screen supply for a tetrode. 126

Figure 3.33 Series power supply configuration for swamping the screen circuit of a tetrode. 127

Figure 3.34 Typical curve of grid current as a function of control grid voltage for a high-power thoriated-tungsten filament tetrode. 128

Figure 3.35 Secondary emission characteristics of various metals under ordinary conditions. 128

Figure 3.36 A cutaway view of the anode structure of an RF power amplifier tube. 130

Figure 3.37 Thermal conductivity of common ceramics. 131

Figure 3.38 Cross section of the base of a tetrode socket showing the connection points. 134

Figure 3.39 The elements involved in the neutralization of a tetrode PA stage. 135

Figure 3.40 A graphical representation of the elements involved in the self-neutralization of a tetrode RF stage. 136

Figure 3.41 Push-pull grid neutralization. 137

Figure 3.42 Symmetrical stage neutralization for a grounded-grid circuit. 138

Figure 3.43 Push-pull grid neutralization in a single-ended tetrode stage. 138

Figure 3.44 Single-ended grid neutralization for a tetrode. 139

Figure 3.45 The previous figure redrawn to show the elements involved in neutralization. 139

Figure 3.46 Single-ended plate neutralization. 140

Figure 3.47 Single-ended plate neutralization showing the capacitance bridge present. 140

Figure 3.48 A grounded-screen PA stage neutralized through the use of stray inductance between the screen connector and the cavity deck: (*a*) schematic diagram, (*b*) mechanical layout of cavity. 142

Figure 3.49 Circuit of grounded-grid amplifier having grid impedance and neutralized for reactive currents. 143

Figure 3.50 Neutralization by cross-connected capacitors of a symmetrical cathode-excited amplifier wth compensation of lead inductance. 144

Figure 3.51 Graphical representation of the elements of a tetrode when self-neutralized. 145

Figure 3.52 Components of the output voltage of a tetrode when neutralized by adding series screen-lead capacitance. 146

Figure 3.53 Components of the output voltage of a tetrode when neutralized by adding external grid-to-plate capacitance. 146

Figure 3.54 Components of the output voltage of a tetrode neutralized by adding inductance common to the screen and cathode return. 147

Designing Vacuum Tube Circuits

Figure 4.1 Plate efficiency as a function of conduction angle for an amplifier with a tuned load. 152

Figure 4.2 Basic grounded-cathode RF amplifier circuit. 153

Figure 4.3 Typical amplifier circuit using a zero-bias triode. Grid current is measured in the return lead from ground to the filament. 154

Figure 4.4 Variation of plate voltage as a function of grid voltage. 155

Figure 4.5 Relationship between grid voltage and plate voltage plotted on a constant-current curve. 155

Figure 4.6 Instantaneous values of plate and grid current as a function of time. 156

Figure 4.7 Schematic of a tetrode RF power amplifier showing tube element bypass capacitors. 162

Figure 4.8 Schematic of a triode RF power amplifier showing tube element bypass capacitors: (*a*) control grid at dc ground potential, (*b*) control grid bypassed to ground. 163

Figure 4.9 Interelectrode capacitances supporting VHF parasitic oscillation in a high-frequency RF amplfier. 164

Figure 4.10 Placement of parasitic suppressors to eliminate VHF parasitic oscillations in a high-frequency RF amplifier: (*a*) resistor-coil combination, (*b*) parasitic choke. 165

Figure 4.11 A section of honeycomb shielding material used in an RF amplifier. 168

Figure 4.12 A selection of common fingerstock. 169

Figure 4.13 Shorted ¼-wavelength line: (*a*) physical circuit, (*b*) electrical equivalent. 172

Figure 4.14 Shorted line less than ¼ wavelength: (*a*) physical circuit, (*b*) electrical equivalent. 172

Figure 4.15 A simplified representation of the grid input circuit of a PA tube. 174

Figure 4.16 Three common methods for RF bypassing of the cathode of a tetrode PA tube: (*a*) grounding the cathode below 30 MHz, (*b*) grounding the cathode above 30 MHz, (*c*) grounding the cathode via a ½-wave transmission line. 175

Figure 4.17 RF circulating current path between the plate and screen in a tetrode PA tube. 176

Figure 4.18 The physical layout of a common type of ¼-wavelength PA cavity designed for VHF operation. 177

Figure 4.19 The RF circulating current paths for the ¼-wavelength cavity shown in Figure 4.18. 178

Figure 4.20 Graph of the RF current, RF voltage, and RF impedance for a ¼-wavelength shorted transmission line. At the feed point, RF current is zero, RF voltage is maximum, and RF impedance is infinite. 178

Figure 4.21 Graph of the RF current, RF voltage, and RF impedance produced by the physically foreshortened coaxial transmission line cavity. 179

Figure 4.22 Basic loading mechanism for a VHF ¼-wave cavity. 180

Figure 4.23 The ½-wavelength PA cavity in its basic form. 181

Figure 4.24 The distribution of RF current, voltage, and impedance along the inner conductor of a ½-wavelength cavity. 182

Figure 4.25 The configuration of a practical ½-wavelength PA cavity. 183

Figure 4.26 Coarse-tuning mechanisms for a ½-wave cavity. 184

Figure 4.27 Using the cavity's movable section to adjust for resonance: (*a*) the rotary section at maximum height, (*b*) the rotary section at minimum height. 185

Figure 4.28 The basic design of a folded ½-wavelength cavity. 186

Figure 4.29 Double-tuned overcoupled broadband cavity amplifier. 188

Figure 4.30 The use of inductive coupling in a ¼-wavelength PA stage. 189

Figure 4.31 Top view of a cavity showing the inductive coupling loop rotated away from the axis of the magnetic field of the cavity. 189

Figure 4.32 Cutaway view of the cavity showing the coupling loop rotated away from the axis of the magnetic field. 190

Figure 4.33 A ¼-wavelength cavity with capacitive coupling to the output load. 191

Figure 4.34 The equivalent circuit of a ¼-wavelength cavity amplifier with capacitive coupling. 192

Figure 4.35 The equivalent circuit of a ¼-wavelength cavity amplifier showing how series capacitive coupling appears electrically as a parallel circuit to the PA tube. 192

Figure 4.36 Typical VHF cavity amplifier. 194

Figure 4.37 Electrical equivalent of the cavity amplifier shown in Figure 4.36. 195

Figure 4.38 VHF cavity amplifiers: (*a*, above) cross-sectional view of a broadband design, (*b*, below) cavity designed for television service, (*c*, next page) FM broadcast cavity amplifier. 196

Figure 4.39 Conventional capacitor input filter full-wave bridge. 198

Figure 4.40 A portion of a high-voltage series-connected rectifier stack. 199

Figure 4.41 A 3-phase delta-connected high-voltage rectifier. 199

Figure 4.42 High-voltage rectifier assembly for a 3-phase delta-connected circuit. 200

Figure 4.43 Using buildout resistances to force current sharing in a parallel rectifier assembly. 201

Figure 4.44 Inverse-parallel thyristor ac power control: (*a*) circuit diagram, (*b*) voltage and current waveforms for full and reduced thyristor control angles. The waveforms apply for a purely resistive load. 202

Figure 4.45 Waveforms in an ac circuit using thyristor power control. 203

Figure 4.46 Voltage and current waveforms for inverse-parallel thyristor power control with an inductive load: (*a*) full conduction, (*b*) small-angle phase reduction, (*c*) large-angle phase reduction. 204

Figure 4.47 Modified full-thyristor 3-phase ac control of an inductive delta load. 205

Figure 4.48 A 6-phase boost rectifier circuit. 206

Figure 4.49 Thyristor-controlled high-voltage servo power supply. 207

Figure 4.50 Phase-controlled power supply with primary regulation. 208

Figure 4.51 Simplified block diagram of the gating circuit for a phase-control system using back-to-back SCRs. 209

Figure 4.52 Basic 3-phase rectifier circuits: (*a*) half-wave wye, (*b*) full-wave bridge, (*c*) 6-phase star, (*d*) 3-phase double-wye. 211

Figure 4.53 Voltage and current waveshapes for an inductive-input filter driven by a 3-phase supply. 214

Figure 4.54 Characteristics of a capacitive input filter circuit: (*a*) schematic diagram, (*b*) voltage waveshape across the input capacitor, (*c*) waveshape of current flowing through the diodes. 217

Figure 4.55 Plate voltage sampling circuits: (*a*) single-ended, (*b*) differential. 218

Figure 4.56 Differential amplifier circuits: (*a*) basic circuit, (*b*) instrumentation amplifier. 219

Figure 4.57 Transmitter high-voltage power supply showing remote plate current and voltage sensing circuits. 220

Applying Vacuum Tube Devices

Figure 5.1 The classic plate-modulated PA circuit for AM applications. 224

Figure 5.2 Basic methods of screen and plate modulation: (*a*) plate modulation, (*b*) plate and screen modulation, (*c*) plate modulation using separate plate/screen supplies. 225

Figure 5.3 Grid modulation of a class C RF amplifier. 226

Figure 5.4 Cathode modulation of a class C RF amplifier. 227

Figure 5.5 Classic vacuum tube class C amplifier: (*a*) representative waveforms, (*b*) schematic diagram. Conditions are as follows: sinusoidal grid drive, 120° anode current, resonant anode load. 228

Figure 5.6 Class B push-pull modulator stage for a high-level AM amplifier. 229

Figure 5.7 The pulse duration modulation (PDM) method of pulse width modulation. 230

Figure 5.8 The principal waveforms of the PDM system. 231

Figure 5.9 Block diagram of a PDM waveform generator. 232

Figure 5.10 Spectrum at the output of a nonlinear device with an input of two equal-amplitude sine waves of $f_1 = 2.001$ MHz and $f_2 = 2.003$ MHz. 235

Figure 5.11 Ideal grid-plate transfer curve for class AB operation. 236

Figure 5.12 Simplified schematic diagram of the Chireix outphasing modulated amplifier: (a) output circuit, (b) drive signal generation circuit. 238

Figure 5.13 Doherty high-efficiency AM amplifier: (a) operating theory, (b) schematic diagram of a typical system. 239

Figure 5.14 Screen-modulated Doherty-type circuit. 241

Figure 5.15 Practical implementation of a screen-modulated Doherty-type 1 MW power amplifier (Continental Electronics). 242

Figure 5.16 Terman-Woodyard high-efficiency modulated amplifier. 243

Figure 5.17 Dome high-efficiency modulated amplifier. 244

Figure 5.18 Basic block diagram of a TV transmitter. The three major subassemblies are the exciter, IPA, and PA. The power supply provides operating voltages to all sections, and high voltage to the PA stage. 247

Figure 5.19 Schematic diagram of a 60 kW klystron-based TV transmitter. 249

Figure 5.20 Schematic diagram of the modulator section of a 60 kW TV transmitter. 250

Figure 5.21 Cathode-driven triode power amplifier. 251

Figure 5.22 Grid-driven grounded-screen tetrode power amplifier. 254

Figure 5.23 Input matching circuits: (a) inductive input matching, (b) capacitive input matching. 256

Figure 5.24 Interstage RF coupling circuits: (a) variable capacitance matching, (b) variable tank matching, (c) variable pi matching, (d) fixed broadband impedance-transformation network. 257

Figure 5.25 Two common approaches to automatic frequency control of a magnetron oscillator. 259

Figure 5.26 Digital synchronizer system for radar applications. 260

Figure 5.27 A line-type modulator for radar. 261

Figure 5.28 Active-switch modulator circuits: (a) direct-coupled system, (b) capacitor-coupled, (c) transformer-coupled, (d) capacitor- and transformer-coupled. 262

Figure 5.29 A magnetic modulator circuit. 262

Microwave Power Tubes

Figure 6.1 Microwave power tube type as a function of frequency and output power. 267

Figure 6.2 Types of linear-beam microwave tubes. 268

Figure 6.3 Schematic diagram of a linear-beam tube. 268

Figure 6.4 Types of crossed-field microwave tubes. 269

Figure 6.5 Magnetron electron path looking down into the cavity with the magnetic field applied. 269

Figure 6.6 Family tree of the distributed emission crossed-field amplifier (CFA). 270

Figure 6.7 Cross section of a 7289 planar triode. 272

Figure 6.8 Grounded-grid equivalent circuits: (a) low-frequency operation, (b) microwave-frequency operation. The cathode-heating and grid-bias circuits are not shown. 273

Figure 6.9 Cutaway view of the tetrode (*left*) and the Diacrode (*right*). Note that the RF current peaks above and below the Diacrode center while on the tetrode there is only one peak at the bottom. 274

Figure 6.10 The elements of the Diacrode, including the upper cavity. Double current, and consequently double power, is achieved with the device because of the current peaks at the top and bottom of the tube, as shown. 275

Figure 6.11 Schematic representation of a reflex klystron. 276

Figure 6.12 Position-time curves of electrons in the anode-repeller space, showing the tendency of the electrons to bunch around the electron passing through the anode at the time when the alternating gap voltage is zero and becoming negative. 277

Figure 6.13 Schematic cross section of a reflex klystron oscillator. 279

Figure 6.14 Cross section of a classic two-cavity klystron oscillator. 280

Figure 6.15 Principal elements of a klystron. 282

Figure 6.16 Klystron cavity: (*a*) physical arrangement, (*b*) equivalent circuit. 283

Figure 6.17 Bunching effect of a multicavity klystron. 284

Figure 6.18 Diode section of a klystron electron gun. 285

Figure 6.19 Beam forming in the diode section of a klystron electron gun. 285

Figure 6.20 Klystron beam-current variation as a function of emitter temperature. 286

Figure 6.21 Modulating-anode electrode in a multicavity klystron. 287

Figure 6.22 Beam-current variation as a function of modulating-anode voltage. 288

Figure 6.23 The effect of a magnetic field on the electron beam of a klystron. 288

Figure 6.24 Field pattern of a klystron electromagnet. 289

Figure 6.25 Field pattern and beam shape in a properly adjusted magnetic field. 289

Figure 6.26 Distortion of the field pattern and beam shape due to magnetic material in the magnetic field. 290

Figure 6.27 Klystron cavity: (*a*) physical element, (*b*) equivalent electric circuit. 290

Figure 6.28 Capacitance-tuned klystron cavity: (*a*) physical element, (*b*) equivalent circuit. 291

Figure 6.29 Schematic equivalent circuit of a four-cavity klystron. 292

Figure 6.30 Klystron loop coupling: (*a*) mechanical arrangement, (*b*) equivalent electric circuit. 292

Figure 6.31 Plot of the relative phase of the reference electrons as a function of axial distance in a high-efficiency klystron. 293

Figure 6.32 Plot of the normalized RF beam currents as a function of distance along the length of a high-efficiency klystron. 293

Figure 6.33 Typical gain, output power, and drive requirements for a klystron. 294

Figure 6.34 Klystron beam current as a function of annular control electrode voltage. 297

Figure 6.35 Block diagram of a switching mod-anode pulser. 297

Figure 6.36 Mechanical design of the multistage depressed collector assembly. Note the "V" shape of the 4-element system. 301

Figure 6.37 A partially assembled MSDC collector. 302

Figure 6.38 Beam current as a function of annular control electrode voltage for an MSDC klystron. 302

Figure 6.39 Collector electron trajectories for the carrier-only condition. Note that nearly all electrons travel to the last electrode (4), producing electrode current I_4. 303

Figure 6.40 Collector trajectories at 25 percent saturation. With the application of modulation, the electrons begin to sort themselves out. 303

Figure 6.41 Collector electron trajectories at 50 percent saturation, approximately the blanking level. The last three electrodes (2, 3, and 4) share electrons in a predictable manner, producing currents I_2, I_3, and I_4. 304

Figure 6.42 Collector trajectories at 90 percent saturation, the sync level. Note the significant increase in the number of electrons attracted to the first electrode, producing I_1. 304

Figure 6.43 The distribution of electrode current as a function of drive power. Note the significant drop in I_4 as drive power is increased. I_5 is the electrode at cathode potential. 305

Figure 6.44 The MSDC collector assembly with the protective shield partially removed. 306

Figure 6.45 Mechanical construction of an external-cavity MSDC tube: (a) device, (b) device in cavity assembly. 306

Figure 6.46 Parallel configuration of the MSDC power supply. Each supply section has an output voltage that is an integral multiple of 6.125 kV. 307

Figure 6.47 Series configuration of the MSDC power supply, which uses four identical 6.125 kV power supplies connected in series to achieve the needed voltages. 308

Figure 6.48 Device bandwidth as a function of frequency and drive power. Beam voltage is 24.5 kV, and beam current is 5.04 A for an output power of 64 kW. These traces represent the full-power test of the MSDC device. 309

Figure 6.49 Device bandwidth in the beam-pulsing mode as a function of frequency and drive power. Beam voltage is 24.5 kV, and beam current is 3.56 A with an output power of 34.8 kW. 309

Figure 6.50 Simplified block diagram of a 60 kW (TVT) MSDC transmitter. The aural klystron may utilize a conventional or MSDC tube at the discretion of the user. 311

Figure 6.51 Simplified schematic diagram of the power supply for an MSDC klystron transmitter. A single high-voltage transformer with multiple taps is used to provide the needed collector voltage potentials. 311

Figure 6.52 Simplified schematic diagram of the Klystrode tube. 313

Figure 6.53 A 60 kW Klystrode tube mounted in its support stand with the output cavity attached. 313

Figure 6.54 A close-up view of the double-tuned output cavity of a 60 kW Klystrode tube. 314

Figure 6.55 Overall structure of the IOT gun and output cavity. 315

Figure 6.56 Pyrolytic graphite grid: (a) shell, (b) grid blank, (c) grid cut. 316

Figure 6.57 IOT input cavity. 317

Figure 6.58 IOT output cavity. 318

Figure 6.59 Simplified block diagram of the crowbar circuit designed into the Klystrode SK-series (Comark) transmitters. The circuit is intended to protect the tube from potentially damaging fault currents. 319

Figure 6.60 Schematic overview of the MSDC IOT or *constant efficiency amplifier*. 320

Figure 6.61 Calculated efficiency of an MSDC IOT compared to an unmodified IOT. 322

Figure 6.62 Basic elements of a traveling wave tube. 323

Figure 6.63 Magnetic focusing for a TWT: (*a*) solenoid-type, (*b*) permanent-magnet-type, (*c*) periodic permanent-magnet structure. 324

Figure 6.64 Helix structures for a TWT: (*a*) ring-loop circuit, (*b*) ring-bar circuit. 325

Figure 6.65 Coupled-cavity interaction structures: (*a*) forward fundamental circuit or "cloverleaf," (*b*) single-slot space harmonic circuit. 326

Figure 6.66 Generic TWT electron gun structure. 327

Figure 6.67 Cross-sectional view of a TWT electron gun with a curved hot cathode. 328

Figure 6.68 Generic configuration of a traveling wave tube. 328

Figure 6.69 Cross-sectional view of various collector configurations for a TWT. 329

Figure 6.70 Power supply configuration for a multistage depressed collector TWT. 330

Figure 6.71 Linear injected-beam microwave tube. 332

Figure 6.72 Reentrant emitting-sole crossed-field amplifier tube. 333

Figure 6.73 Conventional magnetron structure. 334

Figure 6.74 Cavity magnetron oscillator: (*a*) cutaway view, (*b*) cross section view perpendicular to the axis of the cathode. 335

Figure 6.75 Cavity magnetron oscillator anode: (*a*) hole-and-slot type, (*b*) slot type, (*c*) vane type. 335

Figure 6.76 Structure of a coaxial magnetron. 336

Figure 6.77 Structure of a tunable coaxial magnetron. 336

Figure 6.78 Cross-sectional view of a linear magnetron. 337

Figure 6.79 Functional schematic of the M-type radial BWO. 338

Figure 6.80 Strap-fed devices: (*a*) strapped radial magnetron, (*b*) nonreentrant strapping of a BWO. 339

Figure 6.81 Basic structure of a helical beam tube. 341

Figure 6.82 Functional schematic diagram of the gyrotron: (*a*) single coiled helical beam gyrotron, (*b*) electron trajectory of double-coil helical beam gyrotron. 342

Figure 6.83 Basic structure of the gyroklystron amplifier. 343

Figure 6.84 Basic structure of the gyroklystron traveling wave tube amplifier. 343

Figure 6.85 Basic structure of the gyrotron backward oscillator. 344

Figure 6.86 Basic structure of the gyrotwystron amplifier. 344

Figure 6.87 Basic structure of the ubitron. 346

Figure 6.88 Functional schematic diagrams of modified klystrons: (*a*) laddertron, (*b*) twystron. 346

Figure 6.89 Protection and metering system for a klystron amplifier. 349

RF Interconnection and Switching

Figure 7.1 Skin effect on an isolated round conductor carrying a moderately high frequency signal. 356

Figure 7.2 Semiflexible coaxial cable: (*a*) a section of cable showing the basic construction, (*b*) cable with various terminations. 358

Figure 7.3 Magnitude and power factor of line impedance with increasing distance from the load for the case of a short-circuited receiver and a line with moderate attenuation: (*a*) voltage distribution, (*b*) impedance magnitude, (*c*) impedance phase. 361

Figure 7.4 Power rating data for a variety of coaxial transmission lines: (*a*) 50 Ω line, (*b*) 75 Ω line. 364

Figure 7.5 Effects of transmission line pressurization on peak power rating. Note that P' = the rating of the line at the increased pressure, and P = the rating of the line at atmospheric pressure. 366

Figure 7.6 Attenuation characteristics for a selection of coaxial cables: (a) 50 Ω line, (b) 75 Ω line. 367

Figure 7.7 Conversion chart showing the relationship between decibel loss and efficiency of a transmission line: (a) high-loss line, (b) low-loss line. 369

Figure 7.8 The variation of coaxial cable attenuation as a function of ambient temperature. 370

Figure 7.9 The effect of load VSWR on transmission line loss. 370

Figure 7.10 Field configuration of the dominant or $TE_{1,0}$ mode in a rectangular waveguide: (a) side view, (b) end view, (c) top view. 374

Figure 7.11 Field configuration of the dominant mode in circular waveguide. 375

Figure 7.12 Ridged waveguide: (a) single-ridged, (b) double-ridged. 376

Figure 7.13 The effects of parasitic energy in circular waveguide: (a) trapped cross-polarization energy, (b) delayed transmission of the trapped energy. 377

Figure 7.14 Physical construction of doubly truncated waveguide. 378

Figure 7.15 A probe configured to introduce a reflection in a waveguide that is adjustable in magnitude and phase. 379

Figure 7.16 Waveguide obstructions used to introduce compensating reflections: (a) inductive window, (b) capacitive window, (c) post (inductive) element. 380

Figure 7.17 Waveguide stub elements used to introduce compensating reflections: (a) series *tee* element, (b) shunt *tee* element. 381

Figure 7.18 Cavity resonator coupling: (a) coupling loop, (b) equivalent circuit. 383

Figure 7.19 Filter characteristics by type: (a) low-pass, (b) high-pass, (c) bandpass, (d) bandstop. 385

Figure 7.20 Filter characteristics by alignment, third-order, all-pole filters: (a) magnitude, (b) magnitude in decibels. 386

Figure 7.21 The effects of filter order on rolloff (Butterworth alignment). 387

Figure 7.22 Physical model of a 90° hybrid combiner. 388

Figure 7.23 Operating principles of a hybrid combiner. This circuit is used to add two identical signals at inputs A and B. 389

Figure 7.24 The effects of load VSWR on input VSWR and isolation: (a) respective curves, (b) coupler schematic. 391

Figure 7.25 The effects of power imbalance at the inputs of a hybrid coupler. 392

Figure 7.26 Phase sensitivity of a hybrid coupler. 392

Figure 7.27 Nonconstant-impedance branch diplexer. In this configuration, two banks of filters feed into a coaxial tee. 393

Figure 7.28 Band-stop (notch) constant-impedance diplexer module. This design incorporates two 3 dB hybrids and filters, with a terminating load on the isolated port. 394

Figure 7.29 Bandpass constant-impedance diplexer. 395

Figure 7.30 Schematic diagram of a six-module bandpass multiplexer. This configuration accommodates a split antenna design and incorporates patch panels for bypass purposes. 397

Figure 7.31 Reverse-coupled $\frac{1}{4}$-wave hybrid combiner. 400

Figure 7.32 Common RF switching configurations. 401

Figure 7.33 Hybrid switching configurations: (*a*) phase set so that the combined energy is delivered to port 4, (*b*) phase set so that the combined energy is delivered to port 3. 401

Figure 7.34 Additional switching and combining functions enabled by adding a second hybrid and another phase shifter to a hot switching combiner. 402

Figure 7.35 The dielectric vane switcher, which consists of a long dielectric sheet mounted within a section of rectangular waveguide. 404

Figure 7.36 Dielectric post waveguide phase shifter. 405

Figure 7.37 Variable-phase hybrid phase shifter. 406

Figure 7.38 Optical antenna feed systems: (*a*) lens, (*b*) reflector. 408

Figure 7.39 Series-feed networks: (*a*) end feed, (*b*) center feed, (*c*) separate optimization, (*d*) equal path length, (*e*) series phase shifters. 409

Figure 7.40 Types of parallel-feed networks: (*a*) matched corporate feed, (*b*) reactive corporate feed, (*c*) reactive stripline, (*d*) multiple reactive divider. 410

Figure 7.41 The Butler beam-forming network. 410

Figure 7.42 Basic concept of a Reggia-Spencer phase shifter. 411

Figure 7.43 Switched-line phase shifter using *pin* diodes. 412

Figure 7.44 Typical construction of a TR tube. 412

Figure 7.45 Balanced duplexer circuit using dual TR tubes and two short-slot hybrid junctions: (*a*) transmit mode, (*b*) receive mode. 413

Figure 7.46 Construction of a dissipative waveguide filter. 414

Figure 7.47 Basic characteristics of a circulator: (*a*) operational schematic, (*b*) distributed constant circulator, (*c*) lump constant circulator. 415

Figure 7.48 Output of a klystron operating into different loads through a high-power isolator: (*a*) resistive load, (*b*) an antenna system. 417

Figure 7.49 Use of a circulator as a diplexer in TV applications. 418

Figure 7.50 Using multiple circulators to form a multiplexer. 418

Cooling Considerations

Figure 8.1 Blackbody energy distribution characteristics. 425

Figure 8.2 Nukiyama heat-transfer curves: (*a*) logarithmic, (*b*) linear. 426

Figure 8.3 Cooling system design for a power grid tube. 429

Figure 8.4 The use of a chimney to improve cooling of a power grid tube. 429

Figure 8.5 A poorly designed cooling system in which circulating air in the output compartment reduces the effectiveness of the heat-removal system. 430

Figure 8.6 A manometer, used to measure air pressure. 430

Figure 8.7 Cooling airflow requirements for a power grid tube. 431

Figure 8.8 Combined correction factors for land-based tube applications. 434

Figure 8.9 Conversion of mass airflow rate to volumetric airflow rate. 435

Figure 8.10 Functional schematic of a water cooling system for a high-power amplifier or oscillator. 436

Figure 8.11 Typical configuration of a water purification loop. 437

Figure 8.12 The effect of temperature on the resistivity of ultrapure water. 438

Figure 8.13 Vapor-phase-cooled tubes removed from their companion boilers. 441

Figure 8.14 Vapor-phase cooling system for a power tube. 442

Figure 8.15 Cross section of a *holed anode* design for vapor-phase cooling. 444

Figure 8.16 A 4CV35,000A tetrode mounted in its boiler assembly. 444

Figure 8.17 Control box for a vapor-phase cooling system. 445

Figure 8.18 Cutaway view of a boiler and tube combination. 446

Figure 8.19 Cutaway view of a control box. 447

Figure 8.20 Typical four-tube vapor cooling system with a common water supply. 448

Figure 8.21 Cutaway view of a pressure equalizer fitting. 449

Figure 8.22 Typical four-tube vapor cooling system using "steam-out-the-bottom" boilers. 450

Figure 8.23 Vapor cooling system incorporating a reservoir of distilled water. 451

Figure 8.24 Cutaway view of a "steam-out-the-bottom" boiler. 452

Figure 8.25 Airflow system for an air-cooled power tube. 457

Figure 8.26 A manometer device used for measuring back pressure in the PA compartment of an RF generator. 458

Figure 8.27 A typical heating and cooling arrangement for a high-power transmitter installation. Ducting of PA exhaust air should be arranged so that it offers minimum resistance to airflow. 459

Figure 8.28 Case study in which excessive summertime heating was eliminated through the addition of a 1 hp exhaust blower to the building. 460

Figure 8.29 Case study in which excessive back pressure to the PA cavity was experienced during winter periods, when the rooftop damper was closed. The problem was eliminated by repositioning the damper as shown. 461

Figure 8.30 Case study in which air turbulence at the exhaust duct resulted in reduced airflow through the PA compartment. The problem was eliminated by adding a 4-ft extension to the output duct. 462

Figure 8.31 Closed site ventilation design, with a backup inlet/outlet system. 463

Figure 8.32 Open site ventilation design, using no air conditioner. 465

Figure 8.33 Transmitter exhaust-collection hood. 466

Figure 8.34 Hybrid site ventilation design, using an air conditioner in addition to filtered outside air. 467

Figure 8.35 Results of the "foaming test" for water purity: (*a*) pure sample, (*b*) sample that is acceptable on a short-term basis, (*c*) unacceptable sample. 470

Reliability Considerations

Figure 9.1 Example fault tree analysis diagram: (*a*) process steps, (*b*) fault tree symbols, (*c*) example diagram. 479

Figure 9.2 The statistical distribution of equipment or component failures vs. time for electronic systems and devices. 481

Figure 9.3 Distribution of component failures identified through burn-in testing: (*a*) steady-state high-temperature burn-in, (*b*) temperature cycling. 482

Figure 9.4 The failure history of a piece of avionics equipment vs. time. Note that 60 percent of the failures occurred within the first 200 hours of service. 483

Figure 9.5 The roller-coaster hazard rate curve for electronic systems. 484

Figure 9.6 The effects of environmental conditions on the roller-coaster hazard rate curve. 484

Figure 9.7 The effects of environmental stress screening on the reliability *bathtub* curve. 485

Figure 9.8 Venn diagram representation of the relationship between flaw precipitation and applied environmental stress. 487

Figure 9.9 Estimation of the probable time to failure from an abnormal solder joint. 488

Figure 9.10 Proper fingerstock scoring of filament and control grid contacts on a power tube. 493

Figure 9.11 Damage to the grid contact of a power tube caused by insufficient fingerstock pressure. 493

Figure 9.12 The effects of filament voltage management on a thoriated-tungsten filament power tube: (*a*) useful life as a function of filament voltage, (*b*) available power as a function of filament voltage. Note the dramatic increase in emission hours when filament voltage management is practiced. 500

Figure 9.13 The effect of klystron emitter temperature on beam current. 501

Figure 9.14 Arc protection circuit utilizing an energy-diverting element. 503

Figure 9.15 A new, unused 4CX15,000A tube. Contrast the appearance of this device with the tubes shown in the following figures. 506

Figure 9.16 Anode dissipation patterns on two 4CX15,000A tubes: (*a*) excessive heating, (*b*) normal wear. 507

Figure 9.17 Base heating patterns on two 4CX15,000A tubes: (*a*) excessive heating, (*b*) normal wear. 507

Figure 9.18 A 4CX5000A tube that appears to have suffered socketing problems. 508

Figure 9.19 The interior elements of a 4CX15,000A tube that had gone to air while the filament was lit. 508

Figure 9.20 A 4CX15,000A tube showing signs of external arcing. 509

Figure 9.21 Electrical connections for reconditioning the gun elements of a klystron. 515

Device Performance Criteria

Figure 10.1 Root-mean-square (rms) voltage measurements: (*a*) the relationship of rms and average values, (*b*) the rms measurement circuit. 525

Figure 10.2 Average voltage measurements: (*a*) illustration of average detection, (*b*) average measurement circuit. 526

Figure 10.3 Comparison of rms and average voltage characteristics. 527

Figure 10.4 Peak voltage measurements: (*a*) illustration of peak detection, (*b*) peak measurement circuit. 528

Figure 10.5 Illustration of the crest factor in voltage measurements. 529

Figure 10.6 The effects of instrument bandwidth on voltage measurements. 530

Figure 10.7 Example of the equivalence of voltage and power decibels. 531

Figure 10.8 Illustration of the measurement of a phase shift between two signals. 534

Figure 10.9 Illustration of total harmonic distortion (THD) measurement of an amplifier transfer characteristic. 535

Figure 10.10 Example of reading THD from a spectrum analyzer. 536

Figure 10.11 Common test setup to measure harmonic distortion with a spectrum analyzer. 537

Figure 10.12 Simplified block diagram of a harmonic distortion analyzer. 537

Figure 10.13 Typical test setup for measuring the harmonic and spurious output of a transmitter. The notch filter is used to remove the fundamental frequency to prevent overdriving the spectrum analyzer input and to aid in evaluation of distortion components. 538

Figure 10.14 Conversion graph for indicated distortion and true distortion. 538

Figure 10.15 Example of interference sources in distortion and noise measurements. 539

Figure 10.16 Transfer-function monitoring configuration using an oscilloscope and distortion analyzer. 539

Figure 10.17 Illustration of problems that occur when measuring harmonic distortion in band-limited systems. 540

Figure 10.18 Frequency-domain relationships of the desired signals and low-order IMD products. 541

Figure 10.19 Test setup for measuring the intermodulation distortion of an RF amplifier using a spectrum analyzer. 542

Figure 10.20 Typical two-tone IMD measurement, which evaluates the third-order products within the bandpass of the original tones. 543

Figure 10.21 Illustration of the addition of distortion components: (*a*) addition of transfer-function nonlinearities, (*b*) addition of distortion components. 543

Figure 10.22 Illustration of the cancellation of distortion components: (*a*) cancellation of transfer-characteristic nonlinearities, (*b*) cancellation of the distortion waveform. 544

Figure 10.23 Basic model of an intermodulation correction circuit. 545

Figure 10.24 Digital decoding process: (*a*) 16-QAM, (*b*) 8-VSB. 546

Figure 10.25 Digitally modulated RF signal spectrum before and after amplification. 547

Figure 10.26 RF output power variation as a function of dc beam input power for a klystron. 554

Figure 10.27 RF output power of a klystron as a function of RF drive power. 555

Figure 10.28 Output power variation of a klystron as a function of drive power under different tuning conditions. 556

Figure 10.29 Ideal grid-plate transfer curve for class AB operation. 557

Figure 10.30 Frequency spectrum of third-order IM with the interfering level equal to the carrier level. 558

Figure 10.31 Illustration of how the VSWR on a transmission line affects the attenuation performance of the line. Scale *A* represents the VSWR at the transmitter; scale *B* represents the corresponding reflected power in watts for an example system. 561

Figure 10.32 Illustration of how VSWR causes additional loss in a length of feedline. 562

Figure 10.33 The generation of synchronous AM in a bandwidth-limited FM system. Note that minimum synchronous AM occurs when the system operates in the center of its passband. 564

Figure 10.34 IPA plate-to-PA-grid matching network designed for wide bandwidth operation. 565

Figure 10.35 Synchronous AM content of an example FM system as a function of system bandwidth. 566

Figure 10.36 Total harmonic distortion (THD) content of an example demodulated FM signal as a function of transmission-channel bandwidth. 567

Figure 10.37 The spectrum of an ideal AM transmitter. THD measures 0 percent; there are no harmonic sidebands. 578

Figure 10.38 The spectrum of an ideal transmitter with the addition of IPM. 569

Figure 10.39 Carrier amplitude regulation (carrier level shift) defined. 570

Figure 10.40 The mechanics of intermod generation at a two-transmitter installation. 571

Safe Handling of Vacuum Tube Devices

Figure 11.1 Electrocardiogram of a human heartbeat: (*a*) healthy rhythm, (*b*) ventricular fibrillation. 577

Figure 11.2 Effects of electric current and time on the human body. Note the "let-go" range. 578

Figure 11.3 Basic design of a ground-fault interrupter (GFI). 579

Figure 11.4 Example of how high voltages can be generated in an RF load matching unit. 580

Figure 11.5 Basic first aid treatment for electric shock. 581

Figure 11.6 AC service entrance bonding requirements: (*a*) 120 V phase-to-neutral (240 V phase-to-phase), (*b*) 3-phase 208 V wye (120 V phase-to-neutral), (*c*) 3-phase 240 V (or 480 V) delta. Note that the main bonding jumper is required in only two of the designs. 585

Figure 11.7 The power density limits for nonionizing radiation exposure for humans. 588

Figure 11.8 ANSI/IEEE exposure guidelines for microwave frequencies. 590

Index of Tables

Table 1.1 Channel Designations for VHF- and UHF-TV Stations in the United States 17

Table 1.2 Typical Radar Applications 22

Table 1.3 Standard Radar Frequency Bands 22

Table 1.4 Radio Frequencies Used for Electronic Navigation 25

Table 1.5 Common-Carrier Microwave Frequencies Used in the United States 29

Table 1.6 Frequency Band Designations 32

Table 2.1 Typical Equivalent Circuit Parameter Values for Quartz Crystals 47

Table 2.2 Comparison of Amplitude Modulation Techniques 64

Table 3.1 Minimum Cooling Airflow Requirements for the 4CX15000A Power Tetrode at Sea Level 102

Table 3.2 General Characteristics of the 4CX15000A Power Tetrode 104

Table 3.3 General Characteristics of the 5CX1500B Power Pentode 106

Table 3.4 Minimum Cooling Airflow Requirements for the 5CX1500B Power Tetrode 107

Table 3.5 Characteristics of Common Thermonic Emitters 117

Table 4.1 Protection Guidelines for Tetrode and Pentode Devices 170

Table 4.2 Protection Guidelines for a Triode 171

Table 4.3 Operating Parameters of 3-Phase Rectifier Configurations 212

Table 5.1 Approximate Input Capacitance of Common Power Grid Tubes Used in Grounded-Grid and Grounded-Cathode Service 253

Table 6.1 Typical Operating Parameters for an Integral-Cavity Klystron for UHF-TV Service 295

Table 6.2 Typical Operating Parameters for an External-Cavity MSDC Klystron for UHF-TV service 310

Table 6.3 Typical Operating Parameters of the IOT in TV Service 318

Table 7.1 Representative Specifications for Various Types of Flexible Air-Dielectric Coaxial Cable 363

Table 7.2 Electrical Characteristics of Common RF Connectors 366

Table 7.3 Summary of Standard Filter Alignments 387

Table 7.4 Single 90° Hybrid System Operating Modes 390

Table 7.5 Operating Modes of the Dual 90° Hybrid/Single-Phase Shifter Combiner System 403

Table 7.6 Typical Performance of a Dielectric Vane Phase Shifter 405

Table 7.7 Typical Performance of a Variable-Phase Hybrid Phase Shifter 407

Table 8.1 Thermal Conductivity of Common Materials 423

Table 8.2 Relative Thermal Conductivity of Various Materials As a Percentage of the Thermal Conductivity of Copper 423

Table 8.3 Emissivity Characteristics of Common Materials at 80°F 425

Table 8.4 Variation of Electrical and Thermal Properties of Common Insulators As a Function of Temperature 428

Table 8.5 Correction Factors for Q and ΔP 433

Table 9.1 Comparison of Conventional Reliability Testing and Environmental Stress Screening 486

Table 10.1 Common Decibel Values and Conversion Ratios 532

Table 10.2 Three-Halves Current Values of Commonly Used Factors 550

Table 11.1 The Effects of Current on the Human Body 576

Table 11.2 Required Safety Practices for Engineers Working Around High-Voltage Equipment 579

Table 11.3 Basic First Aid Procedures for Burns (More detailed information can be obtained from any Red Cross office.) 583

Table 11.4 Sixteen Common OSHA Violations 584

Table 11.5 Commonly Used Names for PCB Insulating Material 592

Table 11.6 Definition of PCB Terms as Identified by the EPA 594

Cited References

"55 kW Integral Cavity Klystron Arc Shields," Technical Bulletin no. 4441, Varian Associates, Palo Alto, CA, April 1981.

Anderson, Leonard R., *Electric Machines and Transformers*, Reston Publishing Company, Reston, VA.

Andrew Corporation, "Broadcast Transmission Line Systems," Technical Bulletin 1063H, Orland Park, IL, 1982.

Andrew Corporation, "Circular Waveguide: System Planning, Installation and Tuning," Technical Bulletin 1061H, Orland Park, IL, 1980.

Badger, George, "The Klystrode: A New High-Efficiency UHF-TV Power Amplifier," *Proceedings of the NAB Engineering Conference*, National Association of Broadcasters, Washington, DC, 1986.

Battison, John, "Making History," *Broadcast Engineering*, Intertec Publishing, Overland Park, KS, June 1986.

Ben-Dov, O., and C. Plummer, "Doubly Truncated Waveguide," *Broadcast Engineering*, Intertec Publishing, Overland Park, KS, January 1989.

Benson, K. B., and J. C. Whitaker (eds.), *Television Engineering Handbook*, revised Ed., McGraw-Hill, New York, 1991.

Benson, K. B., and J. C. Whitaker, *Television and Audio Handbook for Technicians and Engineers*, McGraw-Hill, New York, 1989.

Besch, David F., "Thermal Properties," in *The Electronics Handbook*, Jerry C. Whitaker (ed.), CRC Press, Boca Raton, FL, pp. 127–134, 1996.

Brauer, D. C., and G. D. Brauer, "Reliability-Centered Maintenance," *Proceedings of the IEEE Reliability and Maintainability Symposium*, IEEE, New York, 1987.

Birdsall, C. K., *Plasma Physics via Computer Simulation*, Adam Hilger, 1991.

Block, R., "CPS Microengineers New Breed of Materials," *Ceramic Ind.*, pp. 51–53, April 1988.

Bordonaro, Dominic, "Reducing IPM in AM Transmitters," *Broadcast Engineering*, Intertec Publishing, Overland Park, KS, May 1988.

Brittain, James E., "Scanning the Past: Guglielmo Marconi," *Proceedings of the IEEE*, Vol. 80, no. 8, August 1992.

Buchanan, R. C., *Ceramic Materials for Electronics*, Marcel Dekker, New York, 1986.

Burke, William, "WLW: The Nation's Station," *Broadcast Engineering*, Intertec Publishing, Overland Park, KS, November 1967.

Cablewave Systems, "The Broadcaster's Guide to Transmission Line Systems," Technical Bulletin 21A, North Haven, CT, 1976.

Cablewave Systems, "Rigid Coaxial Transmission Lines," Cablewave Systems Catalog 700, North Haven, CT, 1989.

Cameron, D., and R. Walker, "Run-In Strategy for Electronic Assemblies," *Proceedings of the IEEE Reliability and Maintainability Symposium*, IEEE, New York, 1986.

Capitano, J., and J. Feinstein, "Environmental Stress Screening Demonstrates Its Value in the Field," *Proceedings of the IEEE Reliability and Maintainability Symposium*, IEEE, New York, 1986.

"Ceramic Products," Brush Wellman, Cleveland, OH.

Chaffee, E. L., *Theory of Thermonic Vacuum Tubes*, McGraw-Hill, New York, 1939.

Chireix, Henry, "High Power Outphasing Modulation," *Proceedings of the IRE*, Vol. 23, no. 11, pg. 1370, November 1935.

Clark, Richard J., "Electronic Packaging and Interconnect Technology Working Group Report (IDA/OSD R&M Study)," IDA Record Document D-39, Institute of Defense Analysis, August 1983.

Clayworth, G. T., H. P. Bohlen, and R. Heppinstall, "Some Exciting Adventures in the IOT Business," *NAB 1992 Broadcast Engineering Conference Proceedings*, National Association of Broadcasters, Washington, DC, pp. 200 - 208, 1992.

"Cleaning and Flushing Klystron Water and Vapor Cooling Systems," Application Engineering Bulletin no. AEB-32, Varian Associates, Palo Alto, CA, February 1982.

"Cleaning UHF-TV Klystron Ceramics," Technical Bulletin no. 4537, Varian Associates, Palo Alto, CA, May 1982.

Clerc, Guy, and William R. House, "The Case for Tubes," *Broadcast Engineering*, Intertec Publishing, Overland Park, KS, pp. 67 - 82, May 1991.

Code of Federal Regulations, 40, Part 761.

Coleman, J. T., *Microwave Devices*, Reston Publishing, Reston, VA, 1982.

Collins, G. B., *Radar System Engineering*, McGraw-Hill, New York, 1947.

"Combat Boron Nitride, Solids, Powders, Coatings," Carborundum Product Literature, form A-14, 011, September 1984.

"Conditioning of Large Radio-Frequency Power Tubes," Application Bulletin no. 21, Varian/Eimac, San Carlos, CA, 1984.

"Coors Ceramics—Materials for Tough Jobs," Coors Data Sheet K.P.G.-2500-2/87-6429.

Cote, R. E., and R. J. Bouchard, "Thick Film Technology," in *Electronic Ceramics*, L. M. Levinson (ed.), Marcel Dekker, New York, pp. 307–370, 1988.

Crutchfield, E. B. (ed.), *NAB Engineering Handbook*, 8th Ed., National Association of Broadcasters, Washington, DC, 1991.

"Current Intelligence Bulletin #45," National Institute for Occupational Safety and Health, Division of Standards Development and Technology Transfer, February 24, 1986.

DeComier, Bill, "Inside FM Multiplexer Systems," *Broadcast Engineering*, Intertec Publishing, Overland Park, KS, May 1988.

DesPlas, Edward, "Reliability in the Manufacturing Cycle," *Proceedings of the IEEE Reliability and Maintainability Symposium*, IEEE, New York, 1986.

Dettmer, E. S., and H. K. Charles, Jr., "AIN and SiC Substrate Properties and Processing Characteristics," *Advances in Ceramics*, Vol. 31, American Ceramic Society, Columbus, OH, 1989.

Dettmer, E. S., H. K. Charles, Jr., S. J. Mobley, and B. M. Romenesko, "Hybrid Design and Processing Using Aluminum Nitride Substrates," *ISHM '88 Proc.*, pp. 545–553, 1988.

Devaney, John, "Piece Parts ESS in Lieu of Destructive Physical Analysis," *Proceedings of the IEEE Reliability and Maintainability Symposium*, IEEE, New York, 1986.

Dick, Bradley, "Caring for High-Power Tubes," *Broadcast Engineering*, Intertec Publishing, Overland Park, KS, pp. 48–58, November 1991.

Dick, Bradley, "New Developments in RF Technology," *Broadcast Engineering*, Intertec Publishing, Overland Park, KS, May 1986.

Doherty, W. H., "A New High Efficiency Power Amplifier for Modulated Waves," *Proceedings of the IRE*, Vol. 24, no. 9, pg. 1163, September 1936.

Dolgosh, Al, and Ron Jakubowski, "Controlling Interference at Communications Sites," *Mobile Radio Technology*, Intertec Publishing, Overland Park, KS, February 1990.

Dorschug, Harold, "The Good Old Days of Radio," *Broadcast Engineering*, Intertec Publishing, Overland Park, KS, May 1971.

Douglass, Barry G., "Thermal Noise and Other Circuit Noise," in *The Electronics Handbook*, Jerry C. Whitaker (ed.), CRC Press, Boca Raton, FL, pp. 30–36, 1996.

Doyle, Edgar, Jr., "How Parts Fail," *IEEE Spectrum*, IEEE, New York, October 1981.

Dye, D. W., *Proc. Phys. Soc.*, Vol. 38, pp. 399–457, 1926.

Eastman, Austin V., *Fundamentals of Vacuum Tubes*, McGraw-Hill, 1941.

"Electrical Standards Reference Manual," U.S. Department of Labor, Washington, D.C.

"Extending Transmitter Tube Life," Application Bulletin no. 18, Varian/Eimac, San Carlos, CA, 1990.

"Fault Protection," Application Bulletin no. 17, Varian/Eimac, San Carlos, CA, 1987.

Ferris, Clifford D., "Electron Tube Fundamentals," in *The Electronics Handbook*, Jerry C. Whitaker (ed.), CRC Press, Boca Raton, FL, pp. 295–305, 1996.

Fink, D., and D. Christiansen (eds.), *Electronics Engineers' Handbook*, 3rd ed., McGraw-Hill, New York, 1989.

Fink, D., and D. Christiansen (eds.), *Electronics Engineer's Handbook*, 2nd ed., McGraw-Hill, New York, 1982.

Fisk, J. B., H. D. Hagstrum, and P. L. Hartman, "The Magnetron As a Generator of Centimeter Waves," *Bell System Tech J.*, Vol. 25, pg. 167, April 1946.

Floyd, J. R., "How to Tailor High-Alumina Ceramics for Electrical Applications," *Ceramic Ind.*, pp. 44–47, February 1969; pp. 46–49, March 1969.

"Foaming Test for Water Purity," Application Engineering Bulletin no. AEB-26, Varian Associates, Palo Alto, CA, September 1971.

Fortna, H., R. Zavada, and T. Warren, "An Integrated Analytic Approach for Reliability Improvement," *Proceedings of the IEEE Reliability and Maintainability Symposium*, IEEE, New York, 1990.

Frerking, M. E., *Crystal Oscillator Design and Temperature Compensation*, Van Nostrand Reinhold, New York, 1978.

Frey, R., G. Ratchford, and B. Wendling, "Vibration and Shock Testing for Computers," *Proceedings of the IEEE Reliability and Maintainability Symposium*, IEEE, New York, 1990.

Gilmore, A. S., *Microwave Tubes*, Artech House, Dedham, MA, pp. 196–200, 1986.

Ginzton, E. L., and A. E. Harrison, "Reflex Klystron Oscillators," *Proc. IRE*, Vol. 34, pg. 97, March 1946.

Goldstein, Irving, "Communications Satellites: A Revolution in International Broadcasting," *Broadcast Engineering*, Intertec Publishing, Overland Park, KS, May 1979.

Gray, Truman S., *Applied Electronics*, John Wiley & Sons, New York, 1954.

Griffin, P., "Analysis of the F/A-18 Hornet Flight Control Computer Field Mean Time Between Failure," *Proceedings of the IEEE Reliability and Maintainability Symposium*, IEEE, New York, 1985.

Hallikainen, Harold, "Transmitter Control Systems," in *NAB Engineering Handbook*, 9th Ed., Jerry C. Whitaker (ed.), National Association of Broadcasters, Washington, D.C., 1998.

Hammar, Willie, *Occupational Safety Management and Engineering*, Prentice Hall, New York.

Hammett, William F., "Meeting IEEE C95.1-1991 Requirements," *NAB 1993 Broadcast Engineering Conference Proceedings*, National Association of Broadcasters, Washington, D.C., pp. 471 - 476, April 1993.

Hansen, M. D., and R. L. Watts, "Software System Safety and Reliability," *Proceedings of the IEEE Reliability and Maintainability Symposium*, IEEE, New York, 1988.

Harnack, Kirk, "Airflow and Cooling in RF Facilities," *Broadcast Engineering*, Intertec Publishing, Overland Park, KS, pp. 33–38, November 1992.

Harper, Charles A., *Electronic Packaging and Interconnection Handbook*, McGraw-Hill, New York, NY, 1991.

Harrison, Cecil, "Passive Filters," in *The Electronics Handbook*, Jerry C. Whitaker (ed.), CRC Press, Boca Raton, FL, pp. 279–290, 1996.

Heising, R. A., "Modulation in Radio Telephony," *Proceedings of the IRE*, Vol. 9, no. 3, pg. 305, June 1921.

Heising, R. A., "Transmission System," U.S. Patent no. 1,655,543, January 1928.

Hermason, Sue E., Major, USAF. Letter dated December 2, 1988. From: Yates, W., and D. Shaller, "Reliability Engineering as Applied to Software," *Proceedings of the IEEE Reliability and Maintainability Symposium*, IEEE, New York, 1990.

Hershberger, David, and Robert Weirather, "Amplitude Bandwidth, Phase Bandwidth, Incidental AM, and Saturation Characteristics of Power Tube Cavity Amplifiers for FM," application note, Harris Corporation, Quincy, IL, 1982.

Heymans, Dennis, "Hot Switches and Combiners," *Broadcast Engineering*, Overland Park, KS, December 1987.

"High Power Transmitting Tubes for Broadcasting and Research," Philips Technical Publication, Eindhoven, the Netherlands, 1988.

Hobbs, Gregg K., "Development of Stress Screens," *Proceedings of the IEEE Reliability and Maintainability Symposium*, IEEE, New York, 1987.

Honey, J. F., "Performance of AM and SSB Communications," *Tele-Tech.*, September 1953.

Horn, R., and F. Hall, "Maintenance Centered Reliability," *Proceedings of the IEEE Reliability and Maintainability Symposium*, IEEE, New York, 1983.

Hulick, Timothy P., "60 kW Diacrode UHF TV Transmitter Design, Performance and Field Report," *Proceedings of the 1996 NAB Broadcast Engineering Conference*, National Association of Broadcasters, Washington, DC, pg. 442, 1996.

Hulick, Timothy P., "Using Tetrodes for High Power UHF," *Proceedings of the Society of Broadcast Engineers*, Vol. 4, SBE, Indianapolis, IN, pp. 52-57, 1989.

Hulick, Timothy P., "Very Simple Out-of-Band IMD Correctors for Adjacent Channel NTSC/DTV Transmitters," *Digital Television '98*, Intertec Publishing, Overland Park, KS, 1998.

Hutter, R. G. E., *Beam and Wave Electronics in Microwave Tubes*, Interscience, London, 1960.

IEEE Standard Dictionary of Electrical and Electronics Terms, Institute of Electrical and Electronics Engineers, Inc., New York, 1984.

"Integral Cavity Klystrons for UHF-TV Transmitters," Varian Associates, Palo Alto, CA.

Irland, Edwin A., "Assuring Quality and Reliability of Complex Electronic Systems: Hardware and Software," *Proceedings of the IEEE*, Vol. 76, no. 1, IEEE, New York, January 1988.

Ishii, T. K., *Microwave Engineering*, Harcourt Brace Jovanovich, San Diego, CA, 1989.

Ishiik, T. K., "Other Microwave Vacuum Devices," in *The Electronics Handbook*, Jerry C. Whitaker (ed.), CRC Press, Boca Raton, FL, pp. 444–457, 1996.

Ishii, T. K., "Traveling Wave Tubes," in *The Electronics Handbook*, Jerry C. Whitaker (ed.), CRC Press, Boca Raton, FL, pp. 428–443, 1996.

Iwase, N., and K. Shinozaki, "Aluminum Nitride Substrates Having High Thermal Conductivity," *Solid State Technol.*, pp. 135–137, October 1986.

Johnson, Walter, "Techniques to Extend the Service Life of High-Power Vacuum Tubes," *NAB Engineering Conference Proceedings*, National Association of Broadcasters, Washington, D.C., pp. 285–293, 1994.

Jordan, Edward C. (ed.), *Reference Data for Engineers: Radio, Electronics, Computer and Communications*, 7th Ed., Howard W. Sams Company, Indianapolis, IN, 1985.

Kenett, R., and M. Pollak, "A Semi-Parametric Approach to Testing for Reliability Growth, with Application to Software Systems," *IEEE Transactions on Reliability*, IEEE, New York, August 1986.

Kimmel, E., "Temperature Sensitive Indicators: How They Work, When to Use Them," *Heat Treating*, March 1983.

Kingery, W. D., H. K. Bowen, and D. R. Uhlmann, *Introduction to Ceramics*, John Wiley & Sons, New York, pp. 637, 1976.

Kinley, Harold, "Base Station Feedline Performance Testing," *Mobile Radio Technology*, Intertec Publishing, Overland Park, KS, April 1989.

Kohl, Walter, *Materials Technology for Electron Tubes*, Reinhold, New York.

Krohe, Gary L., "Using Circular Waveguide," *Broadcast Engineering*, Intertec Publishing, Overland Park, KS, May 1986.

Kubichek, Robert, "Amplitude Modulation," in *The Electronics Handbook*, Jerry C. Whitaker (ed.), CRC Press, Boca Raton, FL, pp. 1175–1187, 1996.

Laboratory Staff, *The Care and Feeding of Power Grid Tubes*, Varian Eimac, San Carlos, CA, 1984.

Lafferty, J. M., *Vacuum Arcs*, John Wiley & Sons, New York, 1980.

Liao, S. Y., *Microwave Electron Tube Devices*, Prentice-Hall, Englewood Cliffs, NJ, 1988.

Markley, Donald, "Complying with RF Emission Standards," *Broadcast Engineering*, Intertec Publishing, Overland Park, KS, May 1986.

Martin, T. L., Jr., *Electronic Circuits*, Prentice Hall, Englewood Cliffs, NJ, 1955.

Mattox, D. M., and H. D. Smith, "The Role of Manganese in the Metallization of High Alumina Ceramics," *J. Am. Ceram. Soc.*, Vol. 64, pp. 1363–1369, 1985.

Maynard, Egbert, "OUSDRE Working Group, VHSIC Technology Working Group Report (IDA/OSD R&M Study)," Document D-42, Institute of Defense Analysis, November 1983.

McCroskey, Donald, "Setting Standards for the Future," *Broadcast Engineering*, Intertec Publishing, Overland Park, KS, May 1989.

McCune, Earl, "Final Report: The Multi-Stage Depressed Collector Project," *Proceedings of the NAB Engineering Conference*, National Association of Broadcasters, Washington, DC, 1988.

McCune, E. W., "Fision Plasma Heating with High-Power Microwave and Millimeter Wave Tubes," *Journal of Microwave Power*, Vol. 20, pp. 131–136, April, 1985.

Meeldijk, Victor, "Why Do Components Fail?" *Electronic Servicing and Technology*, Intertec Publishing, Overland Park, KS, November 1986.

Mendenhall, Geoffrey N., "A Study of RF Intermodulation Between FM Broadcast Transmitters Sharing Filterplexed or Colocated Antenna Systems," Broadcast Electronics, Quincy, IL, 1983.

Mendenhall, Geoffrey N., "A Systems Approach to Improving Subcarrier Performance," *Proceedings of the 1986 NAB Engineering Conference*, National Association of Broadcasters, Washington, DC, 1986.

Mendenhall, Geoffrey N., "Fine Tuning FM Final Stages," *Broadcast Engineering*, Intertec Publishing, Overland Park, KS, May 1987.

Mendenhall, Geoffrey N., "Improving FM Modulation Performance by Tuning for Symmetrical Group Delay," Broadcast Electronics, Quincy IL, 1991.

Mendenhall, Geoffrey N., "Techniques for Measuring Synchronous AM Noise in FM Transmitters," Broadcast Electronics, Quincy IL, 1988.

Mendenhall, Geoffrey N., "Testing Television Transmission Systems for Multichannel Sound Compatibility," Broadcast Electronics, Quincy, IL, 1985.

Mendenhall, G. N., M. B. Shrestha, and E. J. Anthony, "FM Broadcast Transmitters," in *NAB Engineering Handbook*, 8th Ed., E. B. Crutchfield (ed.), National Association of Broadcasters, Washington, DC, pp. 429 - 474, 1992.

Mina and Parry, "Broadcasting with Megawatts of Power: The Modern Era of Efficient Powerful Transmitters in the Middle East," *IEEE Transactions on Broadcasting*, Vol. 35, no. 2, IEEE, Washington, DC, June 1989.

Mistler, R. E., D. J. Shanefield, and R. B. Runk, "Tape Casting of Ceramics," in G. Y. Onoda, Jr., and L. L. Hench (eds.), *Ceramic Processing Before Firing*, John Wiley & Sons, New York, pp. 411–448, 1978.

Muller, J. J., "Cathode Excited Linear Amplifiers," *Electrical Communications*, Vol. 23, 1946.

Nelson, Cindy, "RCA Demolishes Last Antenna Tower at Historic Radio Central," *Broadcast Engineering*, Intertec Publishing, Overland Park, KS, February 1978.

Nenoff, Lucas, "Effect of EMP Hardening on System R&M Parameters," *Proceedings of the 1986 Reliability and Maintainability Symposium*, IEEE, New York, 1986.

Neubauer, R. E., and W. C. Laird, "Impact of New Technology on Repair," *Proceedings of the IEEE Reliability and Maintainability Symposium*, IEEE, New York, 1987.

"Occupational Injuries and Illnesses in the United States by Industry," OSHA Bulletin 2278, U.S. Department of Labor, Washington, DC, 1985.

OSHA, "Electrical Hazard Fact Sheets," U.S. Department of Labor, Washington, DC, January 1987.

OSHA, "Handbook for Small Business," U.S. Department of Labor, Washington, DC.

Ostroff, Nat S., "A Unique Solution to the Design of an ATV Transmitter," *Proceedings of the 1996 NAB Broadcast Engineering Conference*, National Association of Broadcasters, Washington, DC, pg. 144, 1996.

Ostroff, N., A. Kiesel, A. Whiteside, and A. See, "Klystrode-Equipped UHF-TV Transmitters: Report on the Initial Full Service Station Installations," *Proceedings of the NAB Engineering Conference*, National Association of Broadcasters, Washington, DC, 1989.

Ostroff, N., A. Whiteside, A. See, and A. Kiesel, "A 120 kW Klystrode Transmitter for Full Broadcast Service," *Proceedings of the NAB Engineering Conference*, National Association of Broadcasters, Washington, DC, 1988.

Ostroff, N., A. Whiteside, and L. Howard, "An Integrated Exciter/Pulser System for Ultra High-Efficiency Klystron Operation," *Proceedings of the NAB Engineering Conference*, National Association of Broadcasters, Washington, DC, 1985.

Pappenfus, E. W., W. B. Bruene, and E. O. Schoenike, *Single Sideband Principles and Circuits*, McGraw-Hill, New York, 1964.

Paulson, Robert, "The House That Radio Built," *Broadcast Engineering*, Intertec Publishing, Overland Park, KS, April 1989.

Pearman, Richard, *Power Electronics*, Reston Publishing Company, Reston, VA, 1980.

Perelman, R., and T. Sullivan, "Selecting Flexible Coaxial Cable," *Broadcast Engineering*, Intertec Publishing, Overland Park, KS, May 1988.

Peterson, Benjamin B., "Electronic Navigation Systems," in *The Electronics Handbook*, Jerry C. Whitaker (ed.), CRC Press, Boca Raton, FL, pp. 1710–1733, 1996.

Pfrimmer, Jack, "Identifying and Managing PCBs in Broadcast Facilities," *1987 NAB Engineering Conference Proceedings*, National Association of Broadcasters, Washington, DC, 1987.

Pierce, J. R., "Reflex Oscillators," *Proc. IRE*, Vol. 33, pg. 112, February 1945.

Pierce, J. R., "Theory of the Beam-Type Traveling Wave Tube," *Proc. IRE*, Vol. 35, pg. 111, February 1947.

Pierce, J. R., and L. M. Field, "Traveling-Wave Tubes," *Proc. IRE*, Vol. 35, pg. 108, February 1947.

Pond, N. H., and C. G. Lob, "Fifty Years Ago Today or On Choosing a Microwave Tube," *Microwave Journal*, pp. 226 - 238, September 1988.

Powell, Richard, "Temperature Cycling vs. Steady-State Burn-In," *Circuits Manufacturing*, Benwill Publishing, September 1976.

"Power Grid Tubes for Radio Broadcasting," Thomson-CSF Publication #DTE-115, Thomson-CSF, Dover, NJ, 1986.

Powers, M. B., "Potential Beryllium Exposure While Processing Beryllia Ceramics for Electronics Applications," Brush Wellman, Cleveland, OH.

Priest, D. H., and M.B. Shrader, "The Klystrode—An Unusual Transmitting Tube with Potential for UHF-TV," *Proc. IEEE*, Vol. 70, No. 11, pp. 1318–1325, November 1982.

"Protecting Integral-Cavity Klystrons Against Water Leakage," Technical Bulletin no. 3834, Varian Associates, Palo Alto, CA, July 1978.

"Recommendations for Protection of 4KM Series Klystrons During Prime Power Interruptions," Technical Bulletin no. 3542, Varian Associates, Palo Alto, CA.

"Reconditioning of UHF-TV Klystron Electron Gun Elements," Technical Bulletin no. 4867, Varian Associates, Palo Alto, CA, July 1985.

"Reduced Heater Voltage Operation," Technical Bulletin no. 4863, Varian Associates, Palo Alto, CA, June 1985.

Reich, Herbert J., *Theory and Application of Electronic Tubes*, McGraw-Hill, New York, 1939.

Ridgwell, J. F., "A New Range of Beam-Modulated High-Power UHF Television Transmitters," *Communications & Broadcasting*, no. 28, Marconi Communications, Clemsford Essex, England, 1988.

Riggins, George, "The Real Story on WLW's Long History," *Radio World*, Falls Church, VA, March 8, 1989.

Robinson, D., and S. Sauve, "Analysis of Failed Parts on Naval Avionic Systems," Report no. D180-22840-1, Boeing Company, October 1977.

Roth, A., *Vacuum Technology*, 3rd Ed., Elsevier Science Publishers B. V., 1990.

"Safety Precautions," Publication no. 3386A, Varian Associates, Palo Alto, CA, March 1985.

Sawhill, H. T., A. L. Eustice, S. J. Horowitz, J. Gar-El, and A. R. Travis, "Low Temperature Co-Fireable Ceramics with Co-Fired Resistors," in *Proc. Int. Symp. on Microelectronics*, pp. 173–180, 1986.

Schow, Edison, "A Review of Television Systems and the Systems for Recording Television," *Sound and Video Contractor*, Intertec Publishing, Overland Park, KS, May 1989.

Schubin, Mark, "From Tiny Tubes to Giant Screens," *Video Review*, April 1989.

Schwartz, B., "Ceramic Packaging of Integrated Circuits," in *Electronic Ceramics*, L. M. Levinson (ed.), Marcel Dekker, New York, pp. 1–44, 1988.

Seymour, Ken, "Frequency Modulation," in *The Electronics Handbook*, Jerry C. Whitaker (ed.), CRC Press, Boca Raton, FL, pp. 1188–1200, 1996.

Shrader, Merrald B., "Klystrode Technology Update," *Proceedings of the NAB Engineering Conference*, National Association of Broadcasters, Washington, DC, 1988.

Shrestha, Mukunda B., "Design of Tube Power Amplifiers for Optimum FM Transmitter Performance," *IEEE Transactions on Broadcasting*, Vol. 36, no. 1, IEEE, New York, pp. 46 - 64, March 1990.

Sims, G. D., and I. M Stephenson, *Microwave Tubes and Semiconductor Devices*, Interscience, London, 1963.

Skolnik, M. I. (ed.), *Radar Handbook*, McGraw-Hill, New York, 1980.

Smith, Milford K., Jr., "RF Radiation Compliance," *Proceedings of the Broadcast Engineering Conference*, Society of Broadcast Engineers, Indianapolis, IN, 1989.

Smith, William B., "Integrated Product and Process Design to Achieve High Reliability in Both Early and Useful Life of the Product," *Proceedings of the IEEE Reliability and Maintainability Symposium*, IEEE, New York, 1987.

Spangenberg, Karl, *Vacuum Tubes*, McGraw-Hill, New York, 1947.

Spradlin, B. C., "Reliability Growth Measurement Applied to ESS," *Proceedings of the IEEE Reliability and Maintainability Symposium*, IEEE, New York, 1986.

Stenberg, James T., "Using Super Power Isolators in the Broadcast Plant," *Proceedings of the Broadcast Engineering Conference*, Society of Broadcast Engineers, Indianapolis, IN, 1988.

Stokes, John W., *Seventy Years of Radio Tubes and Valves*, Vestal Press, New York, 1982.

Strong, C. E., "The Inverted Amplifier," *Electrical Communications*, Vol. 19, no. 3, 1941.

Surette, Robert A., "Combiners and Combining Networks," in *The Electronics Handbook*, Jerry C. Whitaker (ed.), CRC Press, Boca Raton, FL, pp. 1368–1381, 1996.

Svet, Frank A., "Factors Affecting On-Air Reliability of Solid State Transmitters," *Proceedings of the SBE Broadcast Engineering Conference*, Society of Broadcast Engineers, Indianapolis, IN, October 1989.

Symons, Robert S., "The Constant Efficiency Amplifier," *Proceedings of the NAB Broadcast Engineering Conference*, National Association of Broadcasters, Washington, DC, pp. 523–530, 1997.

Symons, Robert S., "Tubes—Still Vital After All These Years," *IEEE Spectrum*, IEEE, New York, pp. 53–63, April 1998.

Symons, R., M. Boyle, J. Cipolla, H. Schult, and R. True, "The Constant Efficiency Amplifier—A Progress Report," *Proceedings of the NAB Broadcast Engineering Conference*, National Association of Broadcasters, Washington, DC, pp. 77–84, 1998.

Tate, Jeffrey P., and Patricia F. Mead, "Crystal Oscillators," in *The Electronics Handbook*, Jerry C. Whitaker (ed.), CRC Press, Boca Raton, FL, pp. 185–199, 1996.

Technical Staff, Military/Aerospace Products Division, *The Reliability Handbook*, National Semiconductor, Santa Clara, CA, 1987.

"Temperature Measurements with Eimac Power Tubes," Application Bulletin no. 20, Varian Associates, San Carlos, CA, January 1984.

Terman, Frederick E., *Radio Engineering*, 3rd Ed., McGraw-Hill, New York, 1947.

Terman, Frederick E., and J. R. Woodyard, "A High Efficiency Grid-Modulated Amplifier, *Proceedings of the IRE*, Vol. 26, no. 8, pg. 929, August 1938.

"Toxics Information Series," Office of Toxic Substances, July 1983.

Tustin, Wayne, "Recipe for Reliability: Shake and Bake," *IEEE Spectrum*, IEEE, New York, December 1986.

"UHF-TV Klystron Focusing Electromagnets," Technical Bulletin no. 3122, Varian Associates, Palo Alto, CA, November 1973.

Varian Associates, "An Early History," Varian, Palo Alto, CA.

Varian, R., and S. Varian, "A High-Frequency Oscillator and Amplifier," *J. Applied Phys.*, Vol. 10, pg. 321, May 1939.

Vaughan, T., and E. Pivit, "High Power Isolator for UHF Television," *Proceedings of the NAB Engineering Conference*, National Association of Broadcasters, Washington, DC, 1989.

"Water Purity Requirements in Liquid Cooling Systems," Application Bulletin no. 16, Varian Associates, San Carlos, CA, July 1977.

"Water Purity Requirements in Water and Vapor Cooling Systems," Application Engineering Bulletin no. AEB-31, Varian Associates, Palo Alto, CA, September 1991.

Webster, D. L., "Cathode Bunching," *J. Applied Physics*, Vol. 10, pg. 501, July 1939.

Weirather, R. R., G. L. Best, and R. Plonka, "TV Transmitters," in *NAB Engineering Handbook*, 8th Ed., E. B. Crutchfield (ed.), National Association of Broadcasters, Washington, DC, pp. 499 - 551, 1992.

Weldon, J. O., "Amplifiers," U. S. Patent no. 2,836,665, May 1958.

Whitaker, Jerry C., *AC Power Systems*, 2nd Ed., CRC Press, Boca Raton, FL, 1998.

Whitaker, Jerry C., and T. Blankenship, "Comparing Integral and External Cavity Kly-strons," *Broadcast Engineering*, Intertec Publishing, Overland Park, KS, November 1988.

Whitaker, Jerry C., G. DeSantis, and C. Paulson, *Interconnecting Electronic Systems*, CRC Press, Boca Raton, FL, 1993.

Whitaker, Jerry C., *Maintaining Electronic Systems*, CRC Press, Boca Raton, FL, 1992.

Whitaker, Jerry C., *Radio Frequency Transmission Systems: Design and Operation*, McGraw-Hill, New York, 1991.

Whitaker, Jerry C. (ed.), *NAB Engineering Handbook*, 9th Ed., National Association of Broadcasters, Washington, DC, 1998.

Wilson, M. F., and M. L. Woodruff, "Economic Benefits of Parts Quality Knowledge," *Proceedings of the IEEE Reliability and Maintainability Symposium*, IEEE, New York, 1985.

Wilson, Sam, "What Do You Know About Electronics," *Electronic Servicing and Technology*, Intertec Publishing, Overland Park, KS, September 1985.

Witte, Robert A., "Distortion Measurements Using a Spectrum Analyzer," *RF Design*, Cardiff Publishing, Denver, CO, pp. 75 - 84, September 1992.1. National Electrical Code, NFPA #70.

Wong, Kam L., "Demonstrating Reliability and Reliability Growth with Environmental Stress Screening Data," *Proceedings of the IEEE Reliability and Maintainability Symposium*, IEEE, New York, 1990.

Wong, K. L., I. Quart, L. Kallis, and A. H. Burkhard, "Culprits Causing Avionics Equipment Failures," *Proceedings of the IEEE Reliability and Maintainability Symposium*, IEEE, New York, 1987.

Woodard, George W., "AM Transmitters," in *NAB Engineering Handbook*, 8th Ed., E. B. Crutchfield (ed.), National Association of Broadcasters, Washington, DC, pp. 353 - 381, 1992.

Woodard, George W., "AM Transmitters," in *NAB Engineering Handbook*, 9th Ed., Jerry C. Whitaker (ed.), National Association of Broadcasters, Washington, DC, 1998.

Worm, Charles M., "The Real World: A Maintainer's View," *Proceedings of the IEEE Reliability and Maintainability Symposium*, IEEE, New York, 1987.

Wozniak, Joe, "Using Tetrode Power Amplifiers in High Power UHF-TV Transmitters," *NAB 1992 Broadcast Engineering Conference Proceedings*, National Association of Broadcasters, Washington, DC, pp. 209–211, 1992.

Ziemer, Rodger E., "Pulse Modulation," in *The Electronics Handbook*, Jerry C. Whitaker (ed.), CRC Press, Boca Raton, FL, pp. 1201–1212, 1996.

Subject Index

1/2-wave cavity, 181
1/4 matching, 52
1/4-wave cavity, 176
2-cavity klystron oscillator, 280
3-phase thyristor control, 207

A

accelerating anode, 326
active crowbar, 317
air dielectric, 359
air-cooled tubes, 112
airflow volume, 459
all-pole filters, 386
aluminum nitride ceramic, 132
aluminum oxide-based ceramic, 129
AM/PM conversion, 331
amplification factor, 87
amplifier balance, 551
amplifier bandwidth, 171
amplifier efficiency, 50
amplifier electrical considerations, 158
amplifier mechanical considerations, 158
amplifier tuning, 548
amplitron, 339
amplitude modulation, 11,55
annular beam control electrode pulsing, 296
annular cathode, 326
annular control electrode, 246
ANSI C95.1-1982, 589
ANSI/IEEE C95.1-1992, 589
antenna efficiency, 38
anti-intermodulation signa, 545
arc detector, 347
array coordinate, 406
Arthur C. Clarke, 9
attenuation band, 385

automatic damper, 466
average power rating (coax), 363
average-responding voltmeter, 526
axial ratio compensator, 381

B

back-heating, 110,550
backward wave, 340
backward wave amplifier, 340
backward wave oscillator, 337
bake-out process, 133
balanced duplexer, 411
balun transformer, 52
bandpass constant-impedance diplexer, 395
band-stop constant-impedance diplexer, 393
bandwidth modification, 38
bathtub curve, 480
beam feedback, 278
beam focusing, 287
beam perveance, 249
beam pulsing, 246, 295
beam-reconditioning, 300
beryllium oxide, 586
beryllium oxide-based ceramic, 131
Bessel function, 65
binary frequency-shift keying, 79
binary on-off keying, 79
binary phase-shift keying, 80
biphase coding, 82
black heat, 498
blackbody, 424
boiler, 440, 443
boiler design, 442
boron nitride ceramic, 132
branch diplexer, 391

broadband amplifier, 51
broadband impedance matching, 256
broadband matching, 52
broadband noise, 572
buncher, 280
burn-in failure, 474
Butler matrix, 407
bypassing, 160

C

calefaction, 426
capacitive coupling, 190
capacitive input filter, 216
Carcinotron, 338
carrier amplitude regulation, 569
carrier level shift, 569
carrier wave, 58
catastrophic failure, 352,427,505
cathode, 115
cathode modulation, 226
cathode-driven linear amplifier, 234
cathode-driven triode amplifier, 251
cavity amplifier, 73, 171, 173
cavity amplifier design, 194
cavity bandwidth, 193
cavity input circuit, 173
cavity loading, 193
cavity output circuit, 175
cavity resonator, 42 - 43, 382
cavity tuning, 179
C-band satellite channel, 19
ceramic element cleaning, 514
ceramics, 129
characteristic impedance, 562
charged particles, 88
Chireix amplifier, 237
chirp spread spectrum, 84
chlorinated hydrocarbons, 591
circuit Q, 39
circuit resistance, 39
circular guide, 375
circular waveguide, 372
circulating current, 193
circulating water system, 452
circulation rate, 439

circulator, 413
class A amplification, 49
class A amplifier, 152
class AB amplifier, 152
class B amplification, 50
class B modulator, 228
class C amplification, 50
class C amplifier, 152
class C plate modulation, 224
class D amplification, 50
class of operation, 49
cloverleaf circuit, 325
coax mechanical parameters, 371
coaxial cable attenuation, 362
coaxial cable connectors, 365
coaxial magnetron, 335
coaxial transmission line, 356
coherent radar system, 259
collector depression, 329
collector electron trajectory, 302
combiner, 383
common amplification, 245
common-mode failure, 490
compartment shielding, 168
condensation, 439
condenser, 447
conduction cooling, 422
conjugate match, 563
constant efficiency amplifier, 320
constant-current device, 100
constant-impedance diplexer, 391
control grid modulation, 224
controlling harmonic energy, 553
convection cooling, 424
convection cooling zone, 424
coolant line cleaning, 513
coolant purification, 437
coolant purity testing, 469
cooling airflow requirement, 431
cooling blower selection, 432
cooling efficiency, 438
cooling mechanisms, 427
cooling system, 350,468
cooling system flushing, 511
cooling system performance, 459

corporate feed, 407
coupled-cavity circuit, 325
C-QUAM, 14
crest factor, 528
critical inductance, 216
crossed-field amplifier, 269
crossed-field tube, 259, 266, 331
cross-polar energy, 381
cross-polar energy (waveguide), 376
cross-polar filter, 381
crosstalk, 563
crush strength (cable), 371
crystal temperature compensation, 47
cumulative failure rate, 487
cutoff frequency, 385
cutoff frequency (waveguide), 373
cyclotron resonance maser, 340

D

David Sarnoff, 4
decibel measurement, 531
deemphasis, 69
degradation failure, 474
degree of modulation, 56
deionized water, 440
delta modulation, 79
demultiplexing, 38
derating, 489
Diacrode, 274
dielectric, 359
dielectric constant, 362,378
dielectric heating, 31
dielectric post phase shifter, 404
differential amplifier, 218
differential distance ranging, 26
differential encoding, 82
digital data system, 75
diode, 92
diplexer, 384
direct modulation, 69
direct paralleling (of amplifiers), 52
direct-FM, 70
direction finding, 24
directional wattmeter, 530
distance-measuring equipment, 24

distortion components, 542
distributed amplifier, 258
distributed constant circulator, 415
ditungsten carbide, 119
Doherty modulated amplifier, 238
Dome high-efficiency amplifier, 242
dominant mode, 372,374
dominant resonance, 42
double-coil helical beam gyrotron, 341
double-sideband amplitude modulation, 55
doubly truncated waveguide, 376
drift tube, 6,278
dual-polarity waveguide, 374
dual-polarized transition, 374
duplexer, 409,416
dynatron oscillation, 166

E

E vector, 372
Early Bird, 9
Echo 1, 8
effective power, 525
effective radiated power, 15
electric shock, 575
electric vector, 372
electrical axes, 43
electrical gloves, 580
electrolysis, 440
electrolytic target, 453
electromechanical transducer, 46
electron beam focusing, 327
electron bunching, 275
electron gun reconditioning, 514
electron motion, 89
electron optics, 88
electron properties, 87
electron transit time, 99,107
Electronic Industries Association , 10
electrostatic lens, 287
emissions floor, 499
emissivity, 424
emitting-sole tube, 331
enabling event phenomenon, 490
encoder, 38

energy diverter, 503
Environmental Protection Agency, 593
environmental stress screening, 483
equalizer line, 446
equipment grounding, 584
error-correction, 75
ethylene glycol, 435
exciter, 70
exhaust duct, 466
external-cavity klystron, 248, 296

F

failure mode and effects analysis, 474, 477
fault protection, 502
fault tree analysis, 474,477
FC-75, 587
ferrite isolator, 413
field emission, 495
figure of merit, 294
filament bypass capacitor, 254
filament bypassing, 161
filament voltage, 497
filament voltage control, 350
filament voltage management, 499
film vaporization zone, 426
filter alignment, 385
filter type, 384
first aid treatment, 580
flange buildup, 362
flat topping, 159
flicker effect, 533
flushing water lines, 512
FM power amplifier, 250
FM radio, 15
foam-dielectric, 359
focusing electrode, 326
focusing electromagnet, 516
folded 1/2-wave cavity, 186
forced convection cooling, 424
forward fundamental circuit, 325
forward wave amplifier, 341
four-port hybrid combiner, 388
free electron laser, 345
Freon 116, 365

frequency hopping, 84
frequency modulation, 11,63
frequency response, 563
frequency synthesizer, 70
frequency translation, 38
frequency-agile magnetron, 335
frequency-division multiplexing, 29
frequency-domain multiplexing, 38
fundamental resonance, 42,382
fusible temperature indicators, 453
fusible temperature-indicating lacquer, 454
fusing, 209

G

gas noise, 533
geostationary orbit, 18
getter, 133, 499
Global Positioning System, 28
grid structure, 124
grid tubes for microwave applications, 270
grid-driven linearamplifier, 233
grounded-grid amplifier neutralization, 141
grounded-grid circuit, 251
ground-fault current interrupter, 577
group delay, 398, 534
group delay compensation, 399
Guglielmo Marconi, 2
gyroklystron amplifier, 342
gyrotron, 340 - 341
gyrotron backward oscillator, 343
gyrotron traveling wave tube amplifier, 342
gyrotwystron amplifier, 344

H

H vector, 372
half-power frequency, 385
hard limiting, 68
harmonic amplifying stage, 553
harmonic distortion, 536
harmonic energy, 166

hazard rate, 482
heat capacity, 422
heat transfer, 422
heater power supply, 348
height above average terrain, 17
Heinrich Hertz, 2
Heliarc welding, 132
helical beam tube, 340
helix circuit, 324
high frequency (HF) band, 32
high-level anode modulation, 58
high-voltage beam supply, 348
high-voltage tube conditioning, 495
hi-potting, 514
holed anode, 443
honeycomb shield, 167
hot switching combiner, 399
HVAC, 460
hybrid combiner, 387, 399
hybrid coupler, 52
hybrid splitting/combining, 52
hyperbolic line of position, 26

I

impedance matching, 255
impedance matching (waveguide), 377
incidental phase modulation, 567
induced grid noise, 533
induction heating, 30
inductive coupling, 188
inductive input filter, 213
inductive output tube, 7, 246, 310
inductive reactance, 39
injected-beam crossed-field tube, 331
in-line directional wattmeter, 559
insertion loss, 368
intake air supply, 456
integral-cavity klystron, 248, 296
Intelsat 1, 9
interaction space, 333
intercarrier phase distortion, 523
interference, 53
intermod, 570
intermod study, 570

intermodulation distortion, 234, 330, 540, 563
intermodulation distortion measurement, 541
intermodulation distortion products, 555
International Standards Organization, 10
interstage coupling, 256
inverted magnetron, 337
ion bombardment, 117
IOT electron gun, 312
IOT input cavity, 315
IOT output cavity, 316
isolation amplifier, 218

J

J. Ambrose Fleming, 2

K

K-grid, 123
Klystrode tube, 246, 310
klystron, 5,275
klystron cooling system, 510
klystron filament voltage management, 499
klystron interaction gap, 296
klystron power control, 516
klystron reliability, 510
klystron resonant circuit, 289
klystron tuning, 553
Ku-band satellite channel, 19

L

laddertron, 345
latent defect, 487
latent defects, 483
lead length, 159
Lee De Forest, 2
Leidenfrost point, 424
let-go range, 576
level measurement, 524
line stretcher, 404
linear amplification, 232, 237
linear amplifying devices, 233
linear magnetron, 337

linear-beam device, 275
linear-beam tube, 259, 266
linearity precorrection, 544
loaded Q, 173, 193
LORAN C navigation system, 26
low frequency (LF) band, 32
lower-sideband signal, 55
lump constant circulator, 416

M

magnet power supply, 349
magnetic coupling, 383
magnetic vector, 372
magnetron, 258, 266, 332
maintainability, 351
major orbit, 340
Manchester coding, 82
manometer, 457
mean time between failure, 351, 474
mean time to repair, 351, 474
measurement bandwidth, 529
mechanical axes, 43
mechanically tuned magnetron, 335
medium frequency (MF) band, 32
message corruption, 84
metal base shell, 167
meter accuracy, 529
microdust, 457
microwave heating, 332
microwave radio relay, 29
microwave tube life, 347, 350
Miller effect, 71
Miller-Larson effect, 498
minor orbit, 340
mixing loss, 558
mod-anode pulsing, 246
mode of failure, 474
mode optimizer, 381
modulated waveform, 38
modulating anode, 286
modulating-anode pulsing, 296
modulation dissymmetry, 548
modulation index, 63
modulation system, 38
molybdenum, 122

MSDC IOT, 321
MSDC power supply design, 307
M-tubes, 331
M-type radial BWO, 337
multicavity klystron, 281
multiphase rectification, 210
multiple-beam network, 407
multiplexer, 417
multiplexing, 38
multistage depressed collector, 246
multistage depressed collector klystron, 300
multistage depressed-collector, 7

N

National Environmental Policy Act, 588
navigation systems, 23
NAVSTAR GPS, 28
negative resistance, 166
neutralization, 134, 257
neutralization adjustment, 147
neutralization at VHF, 141
nickle plated tube, 509
noise, 53
noise measurement, 532
noncoherent radar system, 258
nonionizing radiation, 588
nonionizing radio frequency radiation, 587
nonlinear distortion, 535
nonresonant feedline, 562
non-return-to-zero, 82
NRZ-mark, 82
nuclear magnetic resonance (NMR), 8
nucleate boiling, 424
Nukiyama curve, 424

O

Occupational Safety and Health Administration, 582
Omega navigation system, 28
one-port tests, 524
one-way distance ranging, 26
operating efficiency, 49

operating environment, 456,489
optical feed, 406
orange heat, 498
order of a filter, 386
orthogonal energy (waveguide), 376
output coupling, 187
oven-controlled crystal oscillator, 45
oxide cathode, 117

P

PA stage tuning, 501, 548
parabolic antenna, 18
parallel resonant circuit, 40
parallel tube RF amplifier, 552
parasitic energy (waveguide), 376
parasitic oscillation, 162
part failure rate, 474
partition noise, 533
parts-count reliability method, 477
passive filter, 384
patent defect, 488
PCB, 591
PCB components, 593
peak emission (of a cathode), 118
peak of sync, 245
peak power rating (coax), 363
peak-equivalent sine, 528
peak-of-sync efficiency, 294
peak-response meter, 527
peniotron, 341
pentode, 102
penultimate cavity, 555
percentage of modulation, 56
performance parameters, 523
perveance, 286
phase control, 204
phase locked loop, 70
phase measurement, 533
phase modulation, 68
phase shifter, 400
phase stability (coax), 368
phase-changing fluid, 454
phase-compensated cable, 371
phased-array antenna, 406
phase-stabilized cable, 371

physical plant design, 462
Pierce gun, 327
pierced shields, 167
piezoelectric effect, 43
piezoelectric properties, 43
planar triode, 99, 271
plate assembly, 129
plate current conduction angle, 151
plate line, 183
plate modulation, 223
plating types, 509
platinotron, 339
PLL, 70
polyphase power supply, 210
power combining, 52
power factor, 360
power supply filter circuit, 213
power supply requirements, 159
power tube conditioning, 494
Precise Positioning Service (PPS), 28
preemphasis, 69
preventive maintenance, 456,490
primary electrons, 90
primary power interruption, 517
principal mode, 373
probability density, 79
propagation mode (waveguide), 372
protective panel, 583
pulse amplitude modulation, 75
pulse code modulation, 78
pulse duration modulation, 75, 230
pulse frequency modulation, 77
pulse length modulation, 75
pulse modulation, 74
pulse position modulation, 75
pulse time modulation, 75
pulse width modulation, 75, 230
pulsed magnetron, 332
purification loop, 437
pyrolytic graphite, 121
pyrolytic graphite grid, 123, 314

Q

quadrature amplitude modulation, 62
quadrature hybrid, 399

quadriphase-shift keying, 81
quality assurance, 475
quality audit, 476
quantization, 78
quantization error, 79
quantization noise, 79
quartz crystal, 43
quartz resonator, 46
quasi-peak detector, 528

R

radar, 20,258
radar system, 22
radial magnetron, 338
radiation cooling, 424
Radio Central, 4
Radio Corporation of America, 3
radio detection and ranging, 20
reactance modulator, 69,71
receiver intermod, 571
receiving tubes, 1
reciprocal phase shifter, 409
rectangular waveguide, 372
rectifiers, parallel connected, 200
rectifiers, series connected, 198
reference electron, 277,283
reflector electrode, 276
reflex klystron, 276
regeneration loop, 439
Reggia-Spencer phase shifter, 408
reject port, 52
relative measurement, 524
relative velocity of propagation, 357
reliability and maintainability, 478
reliability engineering, 473
reliability growth, 475
reliability prediction, 475 - 476
remote metering, 216
repeller electrode, 276
resistance noise, 533
resistivity, 438
resonant circuit, 360
resonant feedline, 562
resonant frequency, 39
resonant tank circuit, 73

return piping, 449
RF combiner, 383
RF interconnection components, 355
RF output circuit, 350
RF power amplifier, 153
RF spectrum, 31
RF system performance, 563
ridged waveguide, 375
rigid coaxial cable, 356
root-mean-square, 525

S

sampling principle, 74
sandwich capacitor, 255
satellite communication, 18
satellite downlink, 18
satellite uplink, 18
satellites, 8
saturated efficiency (of a klystron), 294
saturation, klystron, 554
scale formation, 440, 510
scattering matrix, 388
Schottky effect, 90
screen bypassing, 161
screen grid, 99
screen-modulated Doherty amplifier, 240
secondary coolant, 442
secondary electron emission, 87
secondary electrons, 90, 124
secondary emission, 90, 126
secondary emission noise, 533
secondary guides, 379
self-neutralizing frequency, 145
semiconductor fuses, 209
semiflexible transmission line, 357
sensible-heat load, 463
sequential sampling, 475
shielding, 167
shortwave broadcasting, 14
shot effect, 533
shot noise, 54
signal multiplexing, 38
signal processing, 38
signal-to-noise ratio, 66, 532, 563
silicon avalanche diode, 201

silicon carbide ceramic, 132
silicon rectifier, 197
silver plated tube, 508
single-sideband suppressed carrier AM, 60
single-slot space harmonic circuit, 325
skin effect, 355
soft failure, 351
space charge, 93
space current, 275
specific heat, 422
spectrum-efficient transmission, 74
spin-tuned magnetron, 335
split-phase coding, 82
spot-knocking, 495
spread spectrum, 82
stabilotron, 339
stagger tuning, 51
Standard Positioning Service (SPS), 28
standardization, 478
static pressure, 459
statistical quality control, 476
steam pressure interlock, 449
steam-out-the-bottom boiler, 451
steerable reflector, 407
stop band, 385
strap-fed device, 338
strapping, 338
stress-analysis technique, 477
subsidiary communications authorization, 15
sulfur hexafluoride, 365
superhigh frequency (SHF) band, 34
suppressor grid, 102
suppressor grid modulation, 225
surface leakage resistance, 439
switch panels, 583
sync pulsing, 295
synchronous AM, 564
system grounding, 584

T

tee section, 379
television broadcasting, 16
Telstar 1, 9

temperature measuring-techniques, 453
temperature-compensated crystal oscillator, 45
temperature-sensing methods, 453
tensile strength (cable), 371
Terman-Woodyard amplifier, 242
test to failure, 475
tetrode, 99
thermal capacity, 422
thermal cycling, 492
thermal noise, 54
thermal properties, 421
thermal-agitation noise, 533
thermonic emission, 87,89
thermosiphoning, 441
thoriated-tungsten cathode, 117
thoriated-tungsten filament, 118
three-halves power law, 549
thyratron, 260
thyristor control circuit, 210
thyristor power control, 205
thyristor servo systems, 201
time hopping, 84
time-domain multiplexing, 38
tornadotron, 341
total harmonic distortion, 523, 563
Toxic Substances Control Act, 593
transistor, 8
transistor effect, 8
transition region, 300
transmission line efficiency, 365
transmission line impedance, 360
transmitter cooling system, 457
transverse electromagnetic mode, 359
transverse-electric mode, 372
transverse-magnetic mode, 372
traveling wave tube, 322
triode, 94
tube connection points, 133
tube construction, 132
tube life expectancy, 504
tube performance observations, 505
tube protection systems, 348
tube shipping requirements, 509
tube socket, 133

tube-changing procedure, 492
tuned feedline, 562
tuned-anode amplifier, 226
tungsten-impregnated cathode, 117, 119
tungsten-inert gas welding, 132
tuning, screen considerations, 549
turnaround loss, 398, 558
turnover temperature, 45
TV translator, 16
TV transmitter, 244 - 245
two-cavity klystron, 278
two-cavity klystron amplifier, 281
two-cavity klystron oscillator, 280
two-way distance ranging, 24
TWT circuit wave, 323
TWT collector assembly, 327
TWT efficiency, 329
TWT electron gun, 326
TWT interaction circuit, 324
TWT interaction structure, 323
TWT pulsing, 325
twystron, 346

U

ubitron, 345
UHF tetrode, 272
UHF transmitter, 245
UHF-TV broadcasting, 5
ultrahigh frequency (UHF) band, 34
unipolar return-to-zero, 82
unloaded Q, 173
untuned feedline, 562
upper-sideband signal, 55

V

vacuum tube design, 110
vacuum tube life, 421
vacuum tube protection, 169
vacuum tube standards, 10
vacuum tubes, 1

vapor cooling system, 451
vapor-cooled tubes, 112
vapor-phase cooling, 114, 440
varactor, 47
variable-dielectric vane, 404
variable-phase hybrid, 405
Varian, 5
velocity modulation, 278
ventricular fibrillation, 576
very high frequency (VHF) band, 33
vestigial-sideband AM, 60
voltage controlled oscillator, 72
voltage standing wave ratio, 357, 560
VSWR, 362, 559

W

warm-up cycle, 492
water insulating lines, 445
water-cooled tubes, 112, 114, 434
waveguide, 372
waveguide efficiency, 374
waveguide filter, 379
waveguide installation, 380
waveguide tuning, 381
waveguide-beyond-cutoff attenuator, 167
wear-out, 504
white noise, 53
wideband cavity, 187
William Hansen, 6
William Shockley, 8
wire grid, 122
WLW, 4

X

X-cut crystal, 43
X-ray shielding, 591
X rays, 591

Z

zero-bias triode, 97